Studies in Systems, Decision and Control

Volume 205

The series “Studies in Systems, Decision and Control” (SSDC) covers both new developments and advances, as well as the state of the art, in the various areas of broadly perceived systems, decision making and control–quickly, up to date and with a high quality. The intent is to cover the theory, applications, and perspectives on the state of the art and future developments relevant to systems, decision making, control, complex processes and related areas, as embedded in the fields of engineering, computer science, physics, economics, social and life sciences, as well as the paradigms and methodologies behind them. The series contains monographs, textbooks, lecture notes and edited volumes in systems, decision making and control spanning the areas of Cyber-Physical Systems, Autonomous Systems, Sensor Networks, Control Systems, Energy Systems, Automotive Systems, Biological Systems, Vehicular Networking and Connected Vehicles, Aerospace Systems, Automation, Manufacturing, Smart Grids, Nonlinear Systems, Power Systems, Robotics, Social Systems, Economic Systems and other. Of particular value to both the contributors and the readership are the short publication timeframe and the world-wide distribution and exposure which enable both a wide and rapid dissemination of research output.
** Indexing: The books of this series are submitted to ISI, SCOPUS, DBLP, Ulrichs, MathSciNet, Current Mathematical Publications, Mathematical Reviews, Zentralblatt Math: MetaPress and Springerlink.

More information about this series at http://www.springer.com/series/13304

Hyo-Sung Ahn

Formation Control

Approaches for Distributed Agents

Springer

Hyo-Sung Ahn
School of Mechanical Engineering
Gwangju Institute of Science
and Technology
Buk-gu, Gwangju, Korea (Republic of)

ISSN 2198-4182 ISSN 2198-4190 (electronic)
Studies in Systems, Decision and Control
ISBN 978-3-030-15186-7 ISBN 978-3-030-15187-4 (eBook)
https://doi.org/10.1007/978-3-030-15187-4

Library of Congress Control Number: 2019933700

This Springer imprint is published by the registered company Springer Nature Switzerland AG
The registered company address is: Gewerbestrasse 11, 6330 Cham, Switzerland

To my parents (Hak-Sun Kim & Chang-Soo Ahn) and to my family (Min-Hui Kim & Seung-Hyun Ahn)

Foreword

The cooperative control of autonomous systems is a cornerstone of the emerging field of network science and engineering. Indeed, our vision of large-scale automation, from transportation networks to distributed power generation, at some level, requires cooperation among otherwise independently acting systems. As we learn to analyse, and more importantly *build*, such systems, it becomes essential that we develop a strong formal understanding and foundation for how these systems should function.

While cooperative control represents a broad field of study, there are a few canonical problems that serve simultaneously as prototypical examples of such systems, and also as necessary subroutines to achieve more sophisticated multi-agent system behavior. The *formation control* problem is one such example, and is the focus of this book. The basic task in formation control is for a team of autonomous systems (for example, unmanned vehicles) to achieve some predetermined spatial configuration—the *formation*. It is desirable that the team can solve this problem in a distributed manner. That is, each agent in the ensemble should decide its own control action based on only a subset of sensed and/or communicated information from the entire system.

This ambitious book presents, to my knowledge, the first and most complete treatise of this subject. The aim of this book is to systematically study the distributed formation control problem using the theory of gradient dynamical systems as the main analytical engine for approaching this topic. The author provides a rigorous background of all the mathematical tools required to solve the formation control problem, including dynamical systems theory and rigidity theory, a branch of discrete mathematics used to study the flexibility of rigid bodies connected by flexible links and joints. Using these tools, the author studies a wide range of formation control problems including those based on distance sensing, bearing sensing, unidirectional and bidirectional sensing, among others. An emphasis is also placed on the important implementation issue of relative sensing without knowledge of a common reference frame. The author details how an orientation alignment subroutine can be embedded in many of the gradient-based formation control strategies to address this point.

This book is well suited for the graduate student and researcher interested primarily in formation control problems. I envision this work as an excellent reference but may also form the basis for a graduate-level course. It is a wonderful and much-needed contribution to the controls community, and more broadly to the network sciences.

Haifa, Israel
January 2019

Daniel Zelazo
Technion—Israel Institute
of Technology

Preface

Formation control of distributed agents (in short, distributed formation control) has a unique feature, different from the traditional formation control problems. The traditional formation control problems attempt to control the positions of agents directly, which are usually modeled as nonlinear or high-order dynamical systems. Since the positions of agents are controlled, it is basically assumed that the position information is available to each agent. That is, each agent is assumed to be able to sense its position p_i with respect to a global coordinate frame. However, since the positions are expressed with respect to a common global coordinate frame, it is not a distributed approach from a sensing perspective. With the introduction of consensus algorithms, the interaction features between neighboring agents have been utilized for a formation control. The interactions are usually characterized by diffusive couplings such as $p_i - p_j$ where agents i and j are neighboring agents. The diffusive coupling can be considered as a displacement vector between two agents i and j. So, in consensus approaches, agents are assumed to be able to measure the displacements $p_i - p_j$. Let agent i be able to measure the displacement $p_{ji} \triangleq p_j - p_i$ and agent k can measure the displacement $p_{jk} \triangleq p_j - p_k$. But, we can see that the two displacements p_{ji} and p_{jk} are still expressed with respect to the same common coordinate frame. Thus, the consensus-based formation control requires a global coordinate frame as a base for defining the displacement vectors. However, if a global sensing information or interactions with a centralized coordinator are required, it may reduce the performance of formation systems or increase the implementation costs. Let us imagine a group of unmanned aerial vehicles that need to move or behave together in certain formation configurations. A leader in this group is moving autonomously or by a remote control. The other agents need to move with respect to the leader in specific relative positions and/or relative orientations. If the follower agents can sense their states with respect to other agents including the leader, then they do not need to have interactions with the central computer. Even though global position information is available to each agent, it should be transformed to relative information between agents since the formation configuration is defined with respect to the motion of the leader. Of course, we can assign the desired trajectories of agents in real-time or in an offline manner. In this case, for the real-time trajectory assignment, each agent needs to communicate with the central coordinator, and for the offline trajectory uploading, only the previously designed trajectories can be pursued by the group. Thus, the distributed formation control has more autonomy and can save the implementation costs since motions of agents are defined in a relative way by local sensing. This monograph is motivated by this observation and attempts to deliver the recent developments in the distributed formation control.

Historically, the distributed formation control began with gradient control laws that try to reduce a potential function that is a function of errors of distances between neighboring agents. In Part II, we introduce gradient control laws for global and local stabilizations for general undirected graph cases. These results are presented in Chaps. 3 and 4 respectively. When the interactions between neighboring agents are directed, without cycles, it is relatively clear to ensure a global convergence. Chapter 5 introduces global stabilizations for distributed formations with specific directed graphs called persistent formations. However, the gradient control laws introduced in Part II can only ensure local stabilizations for general cases. To remedy this weakness, in Part III, we introduce formation control via orientation alignment. Note that when the directions of axes of a local coordinate frame of an agent are not same to the directions of axes of a global reference coordinate frame, it is called that the orientation of an agent is not aligned to the global coordinate frame. Similarly, when the directions of axes of two local coordinate frames are not the same, it is said that the two local coordinate frames are not aligned. Thus, by performing orientation alignment as well as formation control simultaneously, the misaligned orientations of the agents can be aligned and a desired formation can be achieved. The orientation alignment implies that the formation control in distributed setups can be transformed into the consensus-based displacement control. Thus, global or quasi-global convergences (see Chap. 6 for the definition of quasi-global stability) can be guaranteed depending on initial angles and communication variables. In Chap. 6, quasi-global convergence is ensured while in Chap. 8, global convergence is ensured. Note that for the orientation alignments in distributed agent systems, it is required to have communications between neighboring agents for exchanging certain information. Thus, the orientation alignment-based formation control schemes introduced in Part III have advantages of (quasi-) global stabilization but has disadvantages of requiring communications between agents. Also, note that the quasi-global stabilization in Chap. 6 requires less information exchange and less amount of computations than the global stabilization schemes introduced in Chap. 8. In Chap. 7, a more advanced formation control scheme is presented, which estimates the positions of agents with only local relative displacements. Then, the estimated positions are directly used for the formation control of distributed agents. Thus, the formation control in distributed setups can be transformed into the position-based formation control. In Part IV, distributed formation control problems with bearing measurements are introduced. Also, in this part, for a global stabilization of agents to a desired formation configuration, the orientation alignment schemes are utilized. In Part V, recent developments in advanced topics are introduced. The advanced topics include formation control with a moving leader and formation control under a moving frame in Chap. 11 and global stabilization of any complete graphs under gradient control laws in Chap. 12. Since a leader is moving with a constant velocity, other agents need to estimate the velocity in their own local coordinate frames. It is also shown that when the agents are modeled by double integrators, a desired formation configuration can be achieved while making agents move together with a velocity matching consensus. Without orientation alignment, it is shown that formations modeled by complete graphs in any dimensional spaces can be stabilized to a desired formation. But, in this case, it is required to have all-to-all communications among agents. Depending upon applications, it may be necessary to expand and contract the size of formation only using local measurements. This topic is also introduced in Chap. 12. The objective of formation control is to achieve a desired configuration in Euclidean space; but when orientations of agents also need to be controlled relatively, the positions and orientations of a group of agents may be controlled simultaneously. This problem may be called *formation control of special Euclidean groups*, and will be partially addressed in Chaps. 8 and 9 seeking a synchronization of orientation angles of agents. Finally, it has been observed that the distributed formation control setups can be exactly used to localize agents' positions. This problem may be called *a duality between formation control and network localization*. This issue and some recent results are presented in Chap. 13. Fig. 1 shows the outline of the monograph.

The works of Chaps. 3, 4, 5, 11 and Sect. 12.1 can be classified as the results in gradient-based formation control laws. The works in Chaps. 6–8 can be classified as the results in orientation alignment-based formation control laws, and the works in Chaps. 9, 10 and Sect. 11.3 can be classified as the results in bearing-based formation control laws. The resizing and scaling problems in Sect. 12.2 are developed under gradient-based and bearing-based approaches, and the network localizations in Chap. 13 can be developed under all three classes.

In Part I, the preliminary notations and essential mathematics are included as background of this monograph. In Chap. 1, we include the basic concepts and definition for distributed formation control. It defines formation control problems motivated from collective behaviors in nature and in human society. Then, this chapter includes basic mathematical concepts and the properties of distributed agent systems. The distributed agent systems are characterized by distributed sensings, distributed communications, and distributed computation and control. With these concepts, in Chap. 1, we formulate the formation control problems of distributed multi-agent systems. Since the results in this monograph are highly based on concepts in graph theory, graph rigidity, persistence, consensus, and some control theory, these are included in Chap. 2 as mathematical background. These mathematical backgrounds are essential for understanding distributed formation control problems. If a reader does not have enough background, it is recommended to read

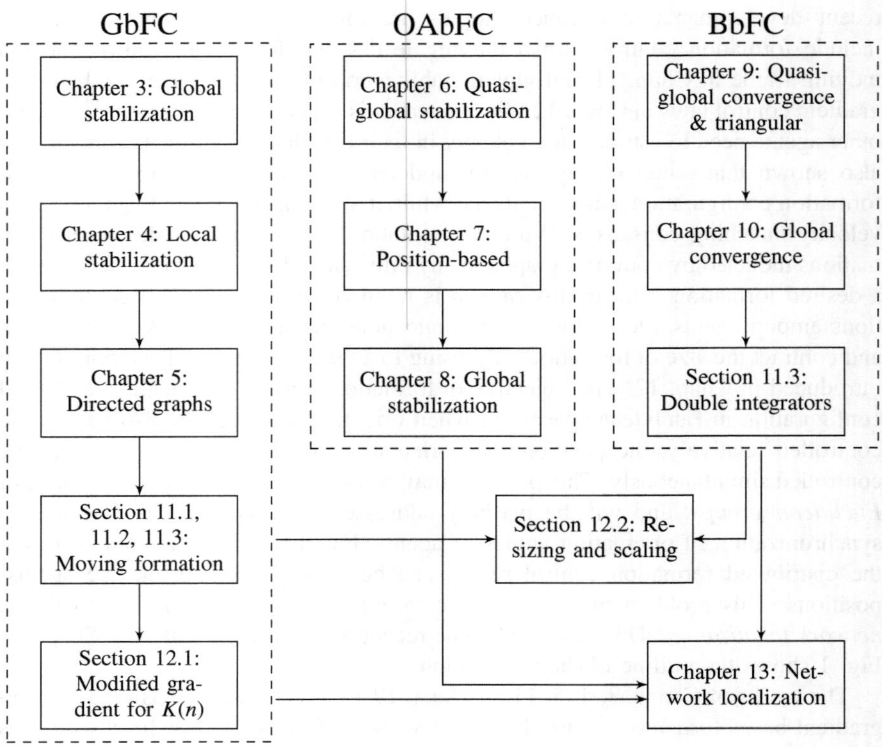

Fig. 1 Outline of monograph: GbFC—Gradient-based formation control laws. OAbFC—Orientation alignment-based formation control laws. BbFC—Bearing-based formation control laws

Chap. 2 first. But, if the author is familiar with rigidity theory, graph theory and nonlinear control, this chapter can be skipped. To increase the readability of monograph, some proofs which are lengthy or direct extensions of the previous proofs are placed at the Appendix.

Buk-gu, Gwangju, Korea (Republic of) Hyo-Sung Ahn

Acknowledgements

This monograph is based on my previous works, which were collaborated with my M.S. and Ph.D. students, and my colleague friends. I reused and reproduced the mathematical parts from my conference and journal papers with permissions from publishers including IEEE, Elsevier, Wiley, Springer, IET, IFAC, and ICROS. I have obtained the permissions for *reuse and reproduce* of the previous publications published by IEEE, Elsevier, Wiley, and Springer for this monograph through Copyright Clearance Center's RightsLink® service. I have obtained the permissions from the Institution of Engineering & Technolog (IET) and IFAC through email contacts and from the Institute of Control, Robotics and Systems (ICROS) of Korea by the *returned rights*. I clearly indicated and cited the appropriate conference or journal papers that were reused and reproduced for this monograph at the end of each chapter. This monograph is fully rewritten in a consistent way, with standard notations and symbols, and all the figures and simulation results are newly produced or regenerated. All the numerical simulations were performed by MATLAB®. The MATLAB source codes for numerical simulations and source files of figures can be downloaded from https://dcas.gist.ac.kr/new/book/. Some new results, background material, and examples have been added to this monograph. I would like to acknowledge that the works of this monograph have been supported from the National Research Foundation (NRF) of Korea under the grant NRF-2017R1A2B3007034.

I would like to appreciate my students for their works during their degree period at Gwangju Institute of Science and Technology (GIST). Without their works, I could not have completed this monograph. I appreciate my former students Hwan Hur, Kwang-Kyo Oh, Byeong-Yeon Kim, Young-Cheol Choi, Ji-Hwan Son, Sang-Chul Lee, Seung-Ju Lee, Young-Hun Lim, Myoung-Chul Park, Byung-Hun Lee, Sung-Mo Kang, Han-Eol Kim, Tae-Kyung Lee, Tong Duy Son, Sang-Hyuk Yun, Je-Young Jeong, Yun-Tae Kim, Seok-Young Han, Hae-Jun Kim, Viet Hoang Pham, Han-Young Park, Kyu-Hwan Kim, and Phuong Huu Hoang, and current students Gwi-Han Ko, Jae-Gyeong Lee, Minh Hoang Trinh, Koog-Hwan Oh, Yoo-Bin Bae, Seong-Ho Kwon, Quoc Van Tran, Young-Hun John, Jin-Hee Son,

Jae-Ryoung Byun, Hong-Jun Lee, Chuong Van Nguyen, Hong-Kyong Kim, Shin-Hyun Park, and Chan-Yeong Jeong. I also appreciate the colleagues Zhiyong Sun, Shiyu Zhao, Mengbin Ye, Daniel Zelazo, and Brian D. O. Anderson for their contributions on formation control theory presented in this monograph. I specially appreciate Prof. Brian D. O. Anderson for his encouragement, and insight, technical details and scholarly attitude throughout our collaborations. Special thanks also should go to Dr. Zhiyong Sun for his helpful and detailed technical comments in this monograph and go to the editor Oliver Jackson for his professional comments. This monograph is highly dependent on the works of my former students conducted at the distributed control and autonomous systems lab. (DCASL) of GIST and the collaborations with the colleagues.

Gwangju, Korea (Republic of)
January 2019

Hyo-Sung Ahn
Gwangju Institute of Science and Technology
email: hyosung@ieee.org

Contents

Part II Gradient Control Laws

Part III Orientation Alignment-based Approaches

Notations

x	A column vector $x = (x_1, \ldots, x_n)^T$
X	A matrix with its elements x_{ij}, denoted as $X = [x_{ij}]$
${}^g\Sigma$	Global coordinate frame
${}^i\Sigma$	Local coordinate frame attached to agent i
R_i^g	Coordinate transformation of special orthogonal group SO(3) from ${}^g\Sigma$ to ${}^i\Sigma$; it is simply written as R_i
θ_{ji}	Given two SO(2) agents, the relative orientation of agent j measured by agent i, i.e., $\theta_{ji} = \theta_j - \theta_i$
R_{ji}	Given two SO(3) agents, the relative orientation of agent j measured by agent i, i.e., $R_{ji} = R_j R_i^{-1}$
α_{ji}^k	Subtended angle between agents i and j measured at agent k
p_i	Position of agent i expressed w.r.t. ${}^g\Sigma$
p	Realization of n agents, i.e., $p = (p_1^T, p_2^T, \ldots, p_n^T)^T$
p_{ji}^i	Relative position of agent j measured in agent i's local coordinate frame; it is also written as $p_j^i = (p_j - p_i)^i$
z_{ij}	Edge vector defined as $z_{ij} = p_i - p_j$, where i is head and j is tail. Without notational confusion, we also use z_k to denote the k-th edge vector
g_{ji}	Bearing vector calculated as $g_{ji} = \frac{p_j - p_i}{\lVert p_j - p_i \rVert}$
$(i,j)^e$	An undirected edge connecting vertices i and j
$(i,j)^{\bar{e}}$	A directed edge with tail i and head j, i.e., $i \rightarrow j$, where j is an outgoing node of node i, and node i is an incoming node of node j
ed_k	The k-th edge, i.e., z_k
$\mathcal{V}$	Set of vertices; $\mathcal{V} \triangleq \{1, 2, \ldots, n\}$ and $\lvert\mathcal{V}\rvert = n$
$\mathcal{E}$	Set of undirected edges, i.e., $\mathcal{E} = \{\ldots, (i,j)^e, \ldots\}$ or $\mathcal{E} = \{\ldots, ed_k, \ldots\}$ and $\lvert\mathcal{E}\rvert = m$. The edge set can be decomposed as $\mathcal{E} = \mathcal{E}_+ \cup \mathcal{E}_-$ where $\mathcal{E}_+$ is the set of directed edges with arbitrary directions of all edges and $\mathcal{E}_+ \cap \mathcal{E}_- = \emptyset$
$\mathcal{E}^s$	Set of edges defining the sensing topology

$\mathcal{E}^a$	Set of edges defining the actuation topology
$\mathcal{E}^c$	Set of edges defining the communication topology
$\overrightarrow{\mathcal{E}}$	Set of directed edges
$\mathcal{G}$	A graph composed of the set of vertices, set of edges, and set of weights $\mathcal{G} = (\mathcal{V}, \mathcal{E}, \mathcal{W})$, where $\lvert\mathcal{V}\rvert = n$, $\lvert\mathcal{E}\rvert = m$, and $\lvert\mathcal{W}\rvert = m$
$\overrightarrow{\mathcal{G}}$	A directed graph composed of the set of vertices, set of directed edges, and set of weights $\overrightarrow{\mathcal{G}} = (\mathcal{V}, \overrightarrow{\mathcal{E}}, \mathcal{W})$, where $\lvert\mathcal{V}\rvert = n$, $\lvert\mathcal{E}\rvert = m$, and $\lvert\mathcal{W}\rvert = m$
$\mathbb{A}$	Adjacency matrix of graph $\mathcal{G}$
$\mathbb{H}$	Incidence matrix of a graph $\mathcal{G}$
$\mathbb{D}$	Degree matrix of graph $\mathcal{G}$
$\mathbb{L}$	Laplacian matrix of graph $\mathcal{G}$
$\mathbb{R}_{\mathcal{G}}$	Rigidity matrix of graph $\mathcal{G}$
$\mathbf{h}_{\mathcal{G}}(p)$	Edge function defined as $\mathbf{h}_{\mathcal{G}}(p) = (\ldots, \lVert z_{ij}\rVert^2, \ldots)^T$, $\forall (i,j)^e \in \mathcal{E}$
$\mathbf{h}_{\mathcal{G}_B}$	Bearing edge function defined as $\mathbf{h}_{\mathcal{G}_B} = (\ldots, g_{ji}^T, \ldots)^T$, $\forall (i,j)^e \in \mathcal{E}$; for a simplicity, it is also written as g
f_p	Framework composed of graph $\mathcal{G}$ and realization p, i.e., $f_p = (\mathcal{G}, p)$
p^*	Desired realization defined by control variables
$\mathcal{N}_i$	Set of neighboring agents of agent i
$\mathcal{N}_i^O$	Set of neighboring agents corresponding to out-degree edges (set of outgoing agents)
$\mathcal{N}_i^I$	Set of neighboring agents corresponding to in-degree edges (set of incoming agents)
$\mathbb{R}^d$	d-dimensional Euclidean space
$\mathbb{C}$	Set of complex numbers
$\mathbb{N}$	Set of natural numbers
d_{ij}	Distance of edge $(i,j)^e$, i.e., $d_{ij} = \lVert p_i - p_j\rVert = \lVert z_{ij}\rVert$
d_{ij}^*	Desired distance of edge $(i,j)^e$, i.e., $d_{ij}^* = \lVert p_i^* - p_j^*\rVert$
$\bar{d}_{ij}$	Squared distance of edge $(i,j)^e$, i.e., $\bar{d}_{ij} = \lVert p_i - p_j\rVert^2$
$\bar{d}_{ij}^*$	Squared desired distance of edge $(i,j)^e$, i.e., $\bar{d}_{ij}^* = \lVert p_i^* - p_j^*\rVert^2$
$\bar{e}_{ij}$	Error of the squared distance of edge $(i,j)^e$ defined as $\bar{e}_{ij} = d_{ij}^2 - (d_{ij}^*)^2$
e_{ij}	Distance error of edge $(i,j)^e$ defined as $e_{ij} = d_{ij} - d_{ij}^*$
$\mathcal{U}^c$	Complementary of the set $\mathcal{U}$
$\mathbf{0}_n$	A vector of length n with all elements being 0; or it is simply also written as 0 in appropriate dimension
$\mathbf{1}_n$	A vector of length n with all elements being 1
$\mathbb{I}_n$	$n \times n$ identity matrix
$\mathbf{0}_{n\times n}$	$n \times n$ zero matrix
$\text{diag}(x)$	A diagonal matrix with diagonal elements $x_1, \ldots, x_n$
$\text{blkdg}(X_i)$	A block diagonal matrix with diagonal element matrices $X_1, \ldots, X_n$

Part I
Background

Chapter 1
Preliminary Background

Abstract The formation control in distributed agent systems uses relative information for sensings and control including distances, angles, and relative displacements. Since the relative sensing measurements are used for controlling the behaviors of agents, it is essentially different from the traditional control methodologies and traditional decentralized control schemes that use states defined with respect to a global coordinate frame. This chapter provides the motivations for distributed formation control and basic mathematical concepts and definitions. To formulate the distributed formation control, it is first required to distinguish the global coordinate frame and distributed local coordinate frames. The distributed formation control laws are defined in distributed local coordinate frames. The distributed systems can be characterized by some attributes such as sensings, communications, computation, and control. This chapter explains the physical meaning of distributed sensings, distributed communications, distributed computation, and distributed control. Then, the key technical issues studied in the distributed formation control are outlined in this chapter.

1.1 Formations and Collective Behaviors

A formation can be considered as a triple composed of a group of agents, interactions between agents, and geographical positions of agents. The interactions between agents can be characterized by mutual or bilateral recognitions, or communications, or actuation behaviors. The recognitions can be obtained by sensing such as senses of vision, senses of hearing, or senses of smell. The communications can be realized via wired or wireless exchanges of information. The actuations can be determined by desired relative or absolute states of agents. The geographical positions (or in short, positions) are locations of agents in the group that may be defined also in a relative or an absolute sense. The positions may determine the interaction topology of agents, or the interaction topology may reversely determine the positions of agents. If interactions between agents may occur as a function of distances between agents' geographical positions, we may call such interactions as geographical interactions.

H.-S. Ahn, *Formation Control*, Studies in Systems, Decision and Control 205, https://doi.org/10.1007/978-3-030-15187-4_1

On the other hands, if the interactions between agents may take place as a topological property of the group, we call such interactions as topological interactions.

In nature, the topological interactions frequently occur in animals' collective behaviors, for example, in a group of starlings [4, 7]. In the topological interactions, an agent would be connected or interacted with a fixed number of agents regardless of metric distances, less depending on the positions of agents. However, for a decision-making in a group of animals, it is considered to have communications between agents of the group [8], although depending on the size of groups, either entire (all-to-all) or local communications are used. Thus, the geographical interactions should be also considered essential in nature. It is obvious that in nature or in artificial entities, the communication or sensing capability of a node toward other nodes will be reduced significantly if there are many other nodes in a short distance range. Consequently, both the topological interactions and geographical interactions may have to be considered to characterize the collective behaviors of nature or social systems.

The collective behaviors in nature can be defined as behaviors that generate meaningful patterns by actuating together. For examples, swarms of ants and bees, and synchronization of firefly and flogs can be considered collective behaviors. Formations are also one of the collective behaviors, which may be observed in a group of birds and a school of fishes. Also, although it was not examined much, a group of wildflowers shows a formation in geographical patterns. The most fascinating feature of collective behaviors is that they generate complicated and emergent actions and patterns via simple interactions. From the observations of various collective behaviors of nature (for example, as shown in Fig. 1.1), we may raise three main questions: Why do animals act or behave in a collective way? How do such behaviors take place? and are these behaviors applicable to engineering or technology?

The first question is more or less oriented to a scientific curiosity, while the second and the third questions are more related to engineering. There are some hypotheses for reasoning the collective behaviors in nature [3]. For examples, the V-shape formations of a group of flying birds may provide advantages of reducing aerodynamic drag forces, may enhance visibility, or may provide advantages for keeping the image of neighboring birds. It is hypothesized that the cluster flocks of small birds may offer a better protection against predators. It was also experimentally studied that the flock flying of birds could improve the homing performance significantly better than individual flights [11]. The second question can be further divided into mechanism aspect and control aspect. In mechanism aspect, there are three interesting views [3]. First, the members of collective behaviors organize specific patterns or actions in a self-organizing way without external command or interaction [8]. Second, the actuations occur via interaction-based group behaviors. Third, the interactions are conducted in a distributed way, which are not all-to-all, but via local interactions, i.e., topological way. From experimental studies of tracking of pigeons flying in flocks, it was observed that they fly under a hierarchical network such as directed leader-to-follower structure, only with local interactions [28]. From the control aspect, the collective behaviors show a robustness that keeps their patterns very safely in the presence of predators. Also, the group of agents can keep the patterns very

Fig. 1.1 Examples of collective behaviors in nature and animals (copied from copyright free site http://imagebase.net/): a flock of starlings (left-top), a school of fishes (right-top), flowers group (left-bottom), a colony of ants (right-bottom)

stable even with additions of new members or with loss of members, which is called scalability. Thus, it may be beneficial to imitate such collective features into artificial entities in the direction of increasing the robustness and scalability. Here, if we use the word *distributed* to denote such local mechanism and control aspects, the key features that need to be imitated are the properties of distributed sensing, distributed communications, and distributed decision in collective behaviors.

Analogous to nature, there are also various types of collective behaviors or formation systems in social or artificial systems (for examples, see Fig. 1.2). In humans, it may be important to coordinate each other to achieve a consensus or synchronization, or a common goal. For examples, in ensemble musical performance, each player of the group may be required to control the timing of tone onsets relative to those of other players. So, relative sensing and relative adjustments (actuations) would be essential for achieving a synchronized ensemble performance [40]. In soccer games, it was also asserted that the pattern of interactions between players is a key factor for characterizing the team performance [20]. It is argued that for a better performance, each member of a team needs to draw on other's ability and knowledge via interactions. It is interesting to see the arguments that high level of interactions (passing rate) would increase the performance, and centralized interaction patterns would decrease the team performance [20]. The same conclusion was also made for various team ball sports. In [32], it was also argued that a successful team would have collective

Fig. 1.2 Examples of collective behaviors in social and artificial entities (copied from copyright free site https://pixabay.com/): ensemble orchestra (left-top), soccer team (right-top), platooning vehicles (left-bottom), formation of flying LEDs (right-bottom)

behaviors with high probability of interacting with each other. Thus, for a better performance, it is recommended to increase the strength of couplings between key players. In vehicle's cruise control systems, a leader-to-follower topology is considered to improve roadway capacity, traffic safety, and fuel efficiency [13]. To realize a cooperative adaptive cruise control, it was required to have vehicle-to-vehicle communications, and to control the inter-vehicle distances with onboard vehicle sensors [13], which means that the vehicle control should be conducted via distributed sensings, distributed communications, and distributed actuations. In pedestrian dynamics of a group of persons, it was observed that a pedestrian attempts to keep a certain distance from other pedestrians and boarders [21]. Recently, with advances of unmanned aerial vehicles, a formation control of flying LEDs has attracted a great amount of attentions [31]. Although it is still ongoing research, it is preferable to realize such a large-scale system in a distributed way. In the current technology, each agent in the group of flying LEDs is localized by way of using a global reference sensor such as GPS. However, it is not desirable to use a global sensor since it requires a global reference frame, which could be a quite demanding requirement depending upon environment, and could be biased easily due to measurement errors. However, the most critical weakness of using global sensor arises when the trajectories of agents need to be changed. In such case, a centralized coordinator should contact all the agents individually to change the trajectories of individual agents. However, it is certain that if the agents have distributed sensing capabilities, they can control themselves on the base of local interactions. As another issue in centralized operation [31], agents need to be communicated with centralized operator to update their status, which would demand a lot of communication throughput. But, if agents can have agent-to-agent communication capability, they can organize their formation in a self-organizing way.

As addressed thus far, in collective behaviors in nature and social systems, the distributed (or relative) sensing and distributed actuations (adjustments) are key features for achieving a group task. Thus, formation control of artificial agents also needs to be formulated using distributed sensings, distributed communications, and distributed control, inspired by collective behaviors of nature and social systems. This monograph attempts to address these issues in mathematical setups and delivers the recent developments under the formulations of distributed formation control.

1.2 Concepts and Definitions

Throughout the monograph, we use notations of graph theory for mathematical formulations. A graph is composed of a set of vertices and a set of edges as $\mathcal{G} = (\mathcal{V}, \mathcal{E})$. A set of agents and the interactions among agents are conceptually represented by a graph $\mathcal{G} = (\mathcal{V}, \mathcal{E})$. When a directed edge is given as an element of the set of edges of the sensing graph $\vec{\mathcal{G}} = (\mathcal{V}, \vec{\mathcal{E}^s})$ as $(i, j)^{\bar{e}} \in \vec{\mathcal{E}^s}$, it means that agent i measures agent j. In this case, the edge vector is written as $z_{ji} = p_j - p_i$. Given a directed graph $\vec{\mathcal{G}} = (\mathcal{V}, \vec{\mathcal{E}})$ that may characterize the sensing topology, the communication topology, or the actuation topology, when two neighboring agents i and j are connected by an edge $(i, j)^{\bar{e}}$, it means that j is outgoing agent of agent i, and agent i is incoming agent of agent j. The edge $(i, j)^{\bar{e}}$ is the outgoing edge to agent i, while it is the incoming edge to agent j.

The states of an agent are represented by some magnitudes with respect to a reference coordinate frame. The positions and velocities of the origin of coordinate frames of agents, and directions of axes of coordinate frames of agents are considered main states of agents. The position of the origin of coordinate frame of an agent is simply called *position of agent*, and the direction of axes of coordinate frame of an agent is simply called *orientation of agent*. To compare the magnitude of states of agents, the positions and orientations need to be expressed with respect to a common reference coordinate frame. If they are represented with respect to different coordinate frames, they cannot be compared. In $\mathbb{R}^d$, the orientations are characterized as special orthogonal group SO(d), and the orientations as well as positions together are characterized by special Euclidean group SE(d). Since the agents have their own local coordinate frames attached to their origins, the orientations of local coordinate frames may not be aligned to the common coordinate frame. Figure 1.3 shows a global coordinate frame ${}^g\Sigma$, and the local coordinate frames ${}^i\Sigma$ and ${}^j\Sigma$ of agents i and j. The directions of axes of agents are not aligned to the directions of axes of

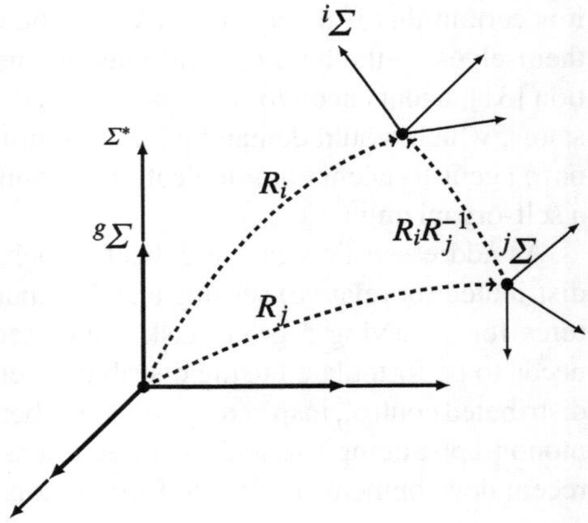

Fig. 1.3 Global reference coordinate frame ($^g\Sigma$) versus local reference coordinate frames ($^i\Sigma$ and $^j\Sigma$)

the global coordinate frame, which is also called that the orientations of agents are not aligned. The rotation matrix transforming the orientation representation from $^g\Sigma$ to $^i\Sigma$ is symbolically represented as R_i or R_i^g. The rotation matrix from $^j\Sigma$ to $^i\Sigma$, which is denoted as R_{ij}, is obtained as $R_{ij} \triangleq R_i^j = R_i R_j^{-1}$. Also, the inverse of the rotation matrix R_i is written as either R_i^{-1} or $R^i = R_g^i$. Here, $R^i = R_g^i$ represents a coordinate transformation from $^i\Sigma$ to $^g\Sigma$.

In Fig. 1.3, the symbol Σ^* represents a common reference direction, which is not necessarily to be aligned with $^g\Sigma$. If Σ^* is aligned to the direction of $^g\Sigma$, it is equivalent to the directions of the global reference frame. But it can be misaligned from $^g\Sigma$; then, we can obtain Σ^* from $^g\Sigma$ by the rotation matrix R_*. A coordinate frame whose directions of axes are aligned to Σ^* is called a common coordinate frame $^c\Sigma$. Note that $\Sigma^* \neq {}^c\Sigma$ because the origins of Σ^* and $^c\Sigma$ could be different. If agents have knowledge of $^c\Sigma$, it is called that the agents have an *aligned directional information*. If the directions of local coordinate frames are aligned to $^c\Sigma$ or Σ^*, i.e., $R_i = \mathbb{I}_d$ (in the case of $^g\Sigma = \Sigma^*$) or $R_i = R_*$, we say that the orientations of agents are physically aligned or just aligned to the global reference direction or to a common reference direction.

The estimation of the reference directions Σ^* is called *orientation estimation*, and physically aligning the directions of agents to Σ^* is called *orientation control*. For a notational convenience, both the orientation estimation and orientation control are called *orientation alignment*. Thus, by orientation alignment, we mean either orientation estimation, orientation control, or both.

Let agent i have a state or a signal x_i. It can be expressed with respect to $^g\Sigma$ or $^i\Sigma$ as x_i^g or x_i^i, respectively. The expression x_i^g is simply written as x_i. Let us denote the position value of agent i as p_i. Given positions of agents i and j, the displacement vector $z_{ij} = p_i - p_j$ is a vector expressed with respect to the global coordinate frame $^g\Sigma$. It can be also represented as $z_{ij}^i = p_i^i - p_j^i = -p_j^i$ in $^i\Sigma$ or $z_{ij}^j = p_i^j - p_j^j = p_i^j$ in $^j\Sigma$ due to $p_i^i = p_j^j = 0$. In the vector z_{ij}, the agent j is the tail and the agent i

is the head. Thus, from a sensing perspective, it means that the agent j measures the agent i, and the agent i is measured by the agent j. Similarly, the vector z_{ji} means that the agent i measures agent j, and it can be expressed as $z^i_{ji} = p^i_j$ and $z^j_{ji} = -p^j_i$. Thus, we can see that $z^i_{ij} = -p^i_j$ is the ith position measured by j, but expressed with respect to ${}^i\Sigma$. Similarly, we can see that $z^j_{ij} = p^j_i$ is the ith position measured by j, expressed with respect to ${}^j\Sigma$. It is noticeable that in general, although $p_j - p_i = z_{ji} = -z_{ij} = -(p_i - p_j)$, we have $p^j_i \neq -p^i_j$ due to

$$z^j_{ij} = p^j_i = R_j R_i^{-1} z^i_{ij} = R_j R_i^{-1}(-p^i_j)$$

From $z^i_{ji} = p^i_j$ and $z^j_{ji} = -p^j_i$, we also can see that $z^i_{ji} = -z^i_{ij}$ and $z^j_{ji} = -z^j_{ij}$. But, although the directions and reference frames of the vectors z^i_{ji}, z^i_{ij}, z^j_{ji}, and z^j_{ij} are different, the magnitudes of the vectors are same as $d_{ij} = \|p_i - p_j\| = \|z^i_{ji}\| = \|z^j_{ji}\| = \|z_{ij}\| = \|z_{ji}\|$. It is important to find a relationship between $z^i_{ji} = p^i_j$ and $z^i_{ij} = -p^i_j$ where the former p^i_j is the jth position with respect to i, expressed in ${}^i\Sigma$, and the latter $-p^i_j$ is the ith position with respect to j, expressed also in ${}^i\Sigma$. So, p^i_j and $-p^i_j$ can be expressed in ${}^g\Sigma$ as

$$(p^i_j)^g = R^i_g p^i_j$$
$$(-p^i_j)^g = R^i_g(-p^i_j)$$

From the observation $(p^i_j)^g = -(-p^i_j)^g$, we have $R^i_g p^i_j = -R^i_g(-p^i_j)$, which implies that $p^i_j = -(-p^i_j)$. Thus, we can see that the jth position measured in ${}^i\Sigma$ and expressed in ${}^i\Sigma$, and the ith position measured in ${}^j\Sigma$ but expressed in ${}^i\Sigma$ are the same vector with the opposite direction. Consequently, the negation of $-p^i_j$, i.e., $-(-p^i_j)$, can be expressed as $-(-p^i_j) = p^i_j$. Figure 1.4 depicts the relative sensings between neighboring agents i and j.

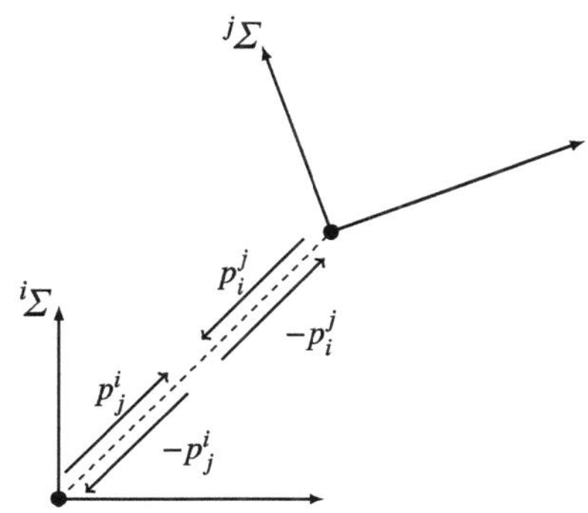

Fig. 1.4 The relative sensings, which are expressed in local coordinate frames, between neighboring agents

Let an agent i be updated under the single-integrator dynamics as

$$\dot{p}_i = u_i$$

The above equation means that the agent i is dynamically updated as per the input u_i. We need to distinguish the updates in ${}^g\Sigma$ and ${}^i\Sigma$. The equation $\dot{p}_i = u_i$ represents an update in ${}^g\Sigma$ because u_i is given in ${}^g\Sigma$. But, if the control input u_i is SE(d) invariant in d-dimensional space [39], by transforming the orientation to local frame as

$$R_i \dot{p}_i = \dot{p}_i^i = u_i^i = R_i u_i$$

we can have $\dot{p}_i^i = u_i^i$ which is the update in ${}^i\Sigma$. The signal u_i^i is the instantaneous input along the axes of the local coordinate frame, which is assumed available or assumed generated in the local coordinate frame. Thus, the instantaneous inputs u_i and u_i^i are physically the same vector; but just expressed in different coordinate frames. Let us suppose that agent i measures p_j^i in its own local coordinate frame. If the agent i provides the control input as $u_i^i = p_j^i$, then the agent i will approach toward the agent j along the vector z_{ji}. Otherwise, with $u_i^i = -p_j^i$, the agent will move along the opposite direction of agent j. With these observations, we can see that the agent i measures p_j^i and generates the control signal p_j^i with respect to agent j. Thus, in this case, agent j is a neighboring agent in both sensing and actuation topologies. This concept is formally defined as follows.

Definition 1.1 (*Sensing topology and actuation topology*) If an agent i measures the relative states of agent j, then the agent j is an outgoing neighboring agent of agent i in the sensing topology, and denoted as $j \in \mathcal{N}_i^O$, where $(i, j)^{\bar{e}} \in \mathcal{E}^s$. Similarly, if an agent i controls its motion with respect to agent j, then the agent j is an outgoing neighboring agent of agent i in the actuation topology, and denoted as $j \in \mathcal{N}_i^O$, where $(i, j)^{\bar{e}} \in \mathcal{E}^a$.

It is noticeable that the sensing topology and actuation topology do not need to be matched. For example, agent i measures p_j^i; but it controls itself toward other agent k ignoring j. Then, j is an outgoing agent in the sensing topology; but it is not a neighboring agent in the actuation topology. That is, in this case, $j \in \mathcal{N}_i^O$ in sensing topology; but $j \notin \mathcal{N}_i^O$ in actuation topology. Thus, we can have $(i, j)^{\bar{e}} \in \mathcal{E}^s$, but $(i, j)^{\bar{e}} \notin \mathcal{E}^a$. In Fig. 1.5, agent i senses the positions of agents j and k, but it controls only toward the agent k. This figure shows that the sensing topology is undirected, but the actuation topology is directed one. The set of directed edges shall be specifically denoted by $\overrightarrow{\mathcal{E}}$; but if there is no confusion, it can be also simply denoted as $\mathcal{E}$ even though it is a set of directed edges.

As mentioned, the measurement p_i^j is the position of agent i measured in the local coordinate frame of agent j; so it is available at agent j. Let agent j send the signal p_i^j to the agent i over a communication as depicted in Fig. 1.5. Also, let agent i send the signal p_k^i to the agent k over a communication. Then, agent i is an outgoing neighboring agent of agent j, and the agent k is an outgoing neighboring

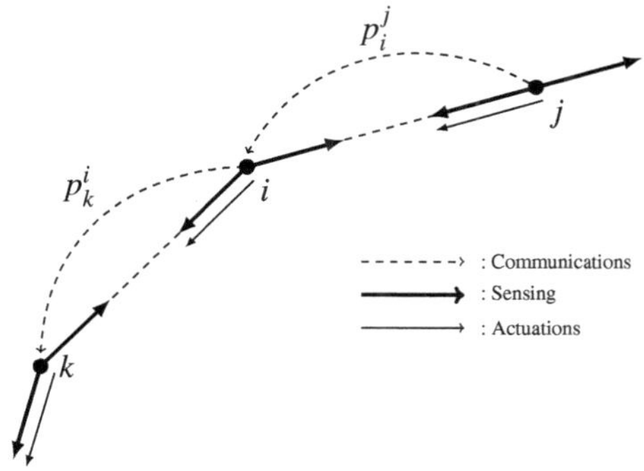

Fig. 1.5 The sensing, actuations, and communications interactions between neighboring agents

agent of agent i. So, we have $(j, i)^{\bar{e}} \in \mathcal{E}^c$ and $(i, k)^{\bar{e}} \in \mathcal{E}^c$ in terms of communication topology.

Definition 1.2 (*Communication topology*) If an agent i sends a piece of information to agent j, then j is an outgoing agent of agent i in communication topology and denoted as $(i, j)^{\bar{e}} \in \mathcal{E}^c$.

So, in general, it is assumed that $\mathcal{E}^c \neq \mathcal{E}^s$, $\mathcal{E}^a \neq \mathcal{E}^s$, and $\mathcal{E}^c \neq \mathcal{E}^a$. Given a set of agents $\mathcal{V} = \{1, 2, \ldots, n\}$, when they are related by a network topology $\mathcal{E}$, where $\mathcal{E} \subseteq \mathcal{E}^c \cup \mathcal{E}^a \cup \mathcal{E}^s$, it is together a graph $\mathcal{G} = (\mathcal{V}, \mathcal{E})$. Note that the network topology $\mathcal{E}$ is a combination of sensing topology, actuation topology, and communication topology, i.e., $\mathcal{E} = \mathcal{E}^s \cup \mathcal{E}^a \cup \mathcal{E}^c$. Thus, the topology characterizes the properties of sensing, communications, or actuations among agents. If the underlying topologies are assumed as $\mathcal{E}^s = \mathcal{E}^a = \mathcal{E}^c$, then we simply write the network topology as $\mathcal{E}$ without specifying the sensing, actuation, and communication topologies.

In a directed graph, if an edge $(i, j)^{\bar{e}} \in \overrightarrow{\mathcal{E}}$ exists, then the agent j is an outgoing agent of agent i; but the agent i is an incoming agent of agent j. Thus, we have $i \in \mathcal{N}_j^I$ and $j \in \mathcal{N}_i^O$. In terms of sensing, agents measure the sensing variables according to the sensing topology, and in terms of control, agents control the actuation variables according to the actuation topology. The actuation variables may be also called control variables. Also, according to the communication topology, agents exchange the communication variables. Thus, the sensing variables, actuation variables, and communication variables need to be distinguished according to the topologies. In Fig. 1.5, the agent i may sense $z_{ji}^i = p_j^i$ and $z_{ki}^i = p_k^i$, which are displacements. But agent i may control the distances such as $\|z_{ki}\| = \|p_k - p_i\|$. Thus, it is required to distinguish the network topologies according to sensing, control, and communications, and the considered variables also according to sensing, control, and communications. The relative displacement $z_{ji}^i = p_j^i$ includes two sensing variables, which are independent. As shown in Fig. 1.6, it can be decomposed as

$$z_{ji}^i = \|z_{ji}\| \angle \delta_{ji} = \|z_j^i\| \angle \delta_{ji}$$

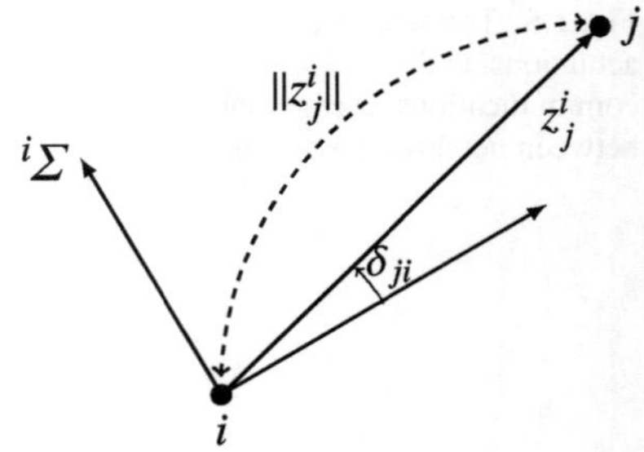

Fig. 1.6 The relative displacement vector composed of bearing angle δ_{ji} and distance $\|z^i_j\|$

where $\|z_{ji}\|$ is the distance between two agents and δ_{ji} is the bearing measurement. The bearing measurement is an angle of the direction of agent j measured with respect to ${}^i\Sigma$. Thus, the displacement z^i_{ji} includes distance and bearing angle variables.

Definition 1.3 Consider a set of n agents with sensing, actuation, and communication topologies as $\mathcal{G} = (\mathcal{V}, \mathcal{E}^s \cup \mathcal{E}^a \cup \mathcal{E}^c)$. If the group of agents is controlled in terms of distances $\|z_{ji}\|$ for all $(i, j)^e \in \mathcal{E}^a$, then it is called *distance-based* control whatever sensing and communication variables are. If the group is controlled by the bearing angles δ_{ji} for all $(i, j)^e \in \mathcal{E}^a$, then it is called *bearing-based* control whatever sensing and communication variables are. If the displacements between neighboring agents are controlled, then it is called *displacement-based* control.

It is important to remark that, in bearing-based formation control (see Chaps. 9 and 10) the orientations of agents are assumed to be aligned. In such case, the bearing angles can be simply considered as bearing vectors. Thus, it may be necessary to distinguish the bearing vectors and bearing angles; but in this monograph, the terminologies *bearing vectors* and *bearing angles* are considered to be interchangeable.

In the above definitions, the sensing and communication variables could be anything, i.e., distance, bearing, or displacement. But it is desirable to obtain or implement these variables in a distributed way. By *distributed*, we mean that only the neighboring agents could sense, actuate, and communicate relatively with each other. It may be required to emphasize that if agent i uses only the position information p_i to control its motion, without using information from other agents, then it may be called *decentralized*. By *decentralized*, the agent i uses only the information of its own state independent of other agents. But, in this decentralized concept, since the state p_i is defined with respect to a global coordinate frame, the sensing variables may include a piece of global information. For example, if agent i updates its state as $u_i(t) = p^i_j + p_i$, then it is not a distributed control law since it uses p_i defined in ${}^g\Sigma$. If all the sensing, actuation, and communication variables are obtained in a distributed way, it is a fully distributed control system. If some of variables are obtained in a global reference frame, it is a partially distributed control system. For example, let us suppose that agents control the distances between neighboring agents with sensing variables of relative displacements under the assumption that orientations of agents are aligned. Then it can be considered that the agents have a common global direction information. Thus, although the control variables are distributed one, it is not a fully distributed control system since the sensing variables include a piece of global information. In this sense, use of terminology of *decentralized* with a global information is not appropriate for defining a framework for distributed control systems.

Definition 1.4 (*Distributed versus centralized*) If the sensing, actuation, and communication variables are obtained with interactions between neighboring agents, then it is a fully distributed system. Otherwise, if some variables are obtained with a piece of global information, then it is a partially distributed system, or simply it is called a centralized system.

Given a set of n agents $1, \ldots, n$, the concatenation $p = (p_1, p_2, \ldots, p_n)$ is sometimes called a *point* or *realization* in a proper Euclidean space. More precisely, the realization p is the coordinate value assignment to the positions of agents. Thus, the realization is an algebraic concept. The vector $p = (p_1^T, p_2^T, \ldots, p_n^T)^T$ is a vector composed of position values. We use p to denote either a point (or a realization) or a position vector of the set of agents. Thus, without notational confusion, by p, we mean either the vector $p = (p_1^T, p_2^T, \ldots, p_n^T)^T \in \mathbb{R}^d$ or the realization $p = (p_1, p_2, \ldots, p_n)$. Note that in some case, we also use p to denote the matrix $p = [p_1, p_2, \ldots, p_n] \in \mathbb{R}^{d \times n}$. To properly deliver the meaning of *formation* of agents, it may be necessary to introduce the terminology *configuration*. The configuration is the shape of realized positions of agents up to trivial motions, which are common translations and rotations of the overall agents, or dilations of the size. Given a graph $\mathcal{G} = (\mathcal{V}, \mathcal{E})$, if the coordinate values are assigned to the agents, it is called *framework*, denoted as f_p. Thus, the framework is a pair of graph and realization, i.e., $f_p = (\mathcal{G}, p)$. It may be necessary to differentiate the concept of formation from the framework. Given a configuration (denoted as $\mathfrak{c}$), up to trivial motions, when the agents are conceptualized topologically by a graph concept, it is called formation, denoted as $f_{\mathcal{G}}$. Thus, the formation is a pair of configuration and graph, i.e., $f_{\mathcal{G}} = (\mathfrak{c}, \mathcal{G})$. In summary, the framework is a kind of algebraic concept and the formation is a kind of topological concept. By a formation, we may imagine a topological configuration; but to realize a desired formation in Euclidean space, we use the concept of framework. That is, by defining a desired framework $f_{p^*} = (\mathcal{G}, p^*)$, we solve formation control problems to achieve a desired configuration (i.e., $\mathfrak{c}^*$) in the sense of $p \to \mathfrak{c}^*$ in Euclidean space. It is worthwhile to mention that, in formation control problems, we may wish to consider positions as well as orientations in general dimensions. For examples, in 2-dimensional space or in 3-dimensional space, we can define a desired formation in $\mathbb{R}^2 \times \mathrm{SO}(2)$ or $\mathbb{R}^3 \times \mathrm{SO}(3)$. In Sects. 6.3, 7.1, and 8.3, a desired configuration along with desired synchronization in orientations is given in $\mathbb{R}^2 \times \mathrm{SO}(2)$. In Sects. 7.2 and 8.4, a desired configuration and orientation synchronization are given in $\mathbb{R}^3 \times \mathrm{SO}(3)$. Further, in Sect. 9.3, we focus on a synchronization of SO(3) groups for achieving a desired configuration in $\mathbb{R}^3$ under bearing setup. Thus, in these sections, the formation control problems are formulated in SE(2) and SE(3).

The following notations are used in this monograph. The *modulo* is defined as $i(\text{modulo}\, j) = i - jq$, where $i, j, i \geq j$ and $q = \text{integer}(i/j)$, which is the integer part of i/j, are natural numbers including 0. Given a square matrix X, the determinant is written as $det(X)$, the rank is written as $rank(X)$, the trace is written as $trace(X)$, and the null space is denoted as $null(X)$. The eigenvalues of X are denoted as $\lambda(X)$. When the eigenvalues are real, the minimum and maximum are written as λ_{min} and λ_{max}, respectively. The minimum and maximum singular values are denoted as $\sigma_{min}(X)$ and $\sigma_{max}(X)$. For a real matrix X, which is not necessarily symmetric, we call it a positive definite (or positive semi-definite) matrix if and only if $v^T Xv > 0$ (or $v^T Xv \geq 0$) for any nonzero vector v. Without notational confusion, for a symmetric or Hermitian matrix X, we also call it a positive definite (or positive semi-definite) matrix if and only if $v^T Xv > 0$ (or $v^T Xv \geq 0$) for any nonzero vector v. Note that if it is a symmetric or Hermitian matrix X, then it is positive definite (or positive semi-definite) if and only if $\lambda(X) > 0$ (or $\lambda(X) \geq 0$). The positive (semi)-definiteness or negative (semi)-definiteness is symbolically written as $X \succ 0$ ($X \succeq 0$) or $X \prec 0$ ($X \preceq 0$). The sign of a variable is denoted as sign$(\cdot)$ or sgn $\cdot$. Given a graph $\mathcal{G} = (\mathcal{V}, \mathcal{E}^s \cup \mathcal{E}^a \cup \mathcal{E}^c)$, the cardinality of sets are $|\mathcal{V}| = n$, $|\mathcal{E}^s| = m_s$, $|\mathcal{E}^a| = m_a$, and $|\mathcal{E}^c| = m_c$. Note that if an edge $(i, j)^e$ is an element for two edge sets, for example, $(i, j)^e \in \mathcal{E}^s$ and $(i, j)^e \in \mathcal{E}^c$, then it is considered a single edge, i.e., no multiple edges and no self-loop. With this notation, the cardinality of set of edges is m, i.e., $|\mathcal{E}^s \cup \mathcal{E}^a \cup \mathcal{E}^c| = |\mathcal{E}| = m$. For further concepts in graph theory, it is recommended to refer to Sect. 2.1.

1.3 Distributed Systems

The properties of distributed agents may be characterized in terms of sensings, communications, computation, and control. This section provides some concepts and definitions that characterize the essence of distributed systems.

1.3.1 Distributed Sensings

Let the position and orientation of an agent i be expressed as p_i and θ_i in ${}^g\Sigma$. To directly measure p_i and θ_i, the agent needs to have sensors that can measure the position and orientation with respect to a global coordinate frame. For the measurements with respect to global coordinate frame, the agents need to have some signals or global reference information provided by external sensors, which may be called *global sensings*. Contrary to the global sensings, the *distributed sensings* are defined from the interactions between neighboring agents. Let us consider Fig. 1.7. The coordinate frames of agents are not aligned. The agents can measure distances, bearing angles, and subtended angles in its own local coordinate frame. In this figure, for example, agent i measures the distances $d_{ij} = d_{ji}$ and $d_{ik} = d_{ki}$, and bearing angle δ_{ji}, and the agent k measures the subtended angle α^k_{ji}. The distance d_{ij} can be measured by local sensors such as time-of-flight-based wireless sensor network [22] or

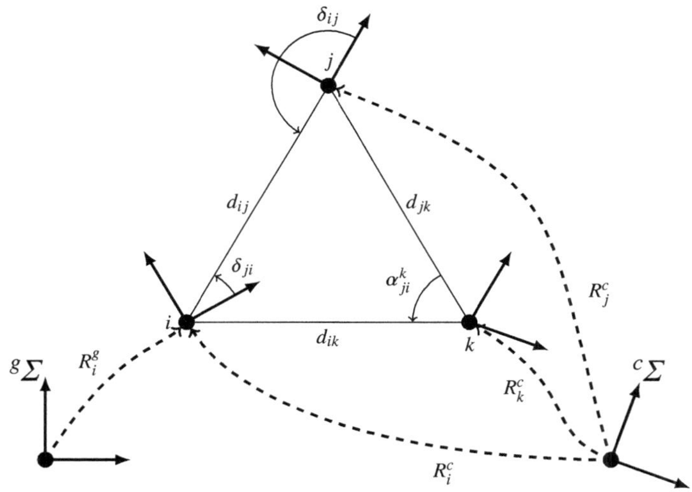

Fig. 1.7 The sensing variables in distributed agents. The local frames are not aligned (i.e., misaligned). The sensing variables are distances d_{ij}, d_{jk}, d_{ik}; bearing angles δ_{ij}, δ_{ji}; and subtended angles α^k_{ij}. The global reference frame is ${}^g\Sigma$ and a common reference frame is ${}^c\Sigma$

received-signal strength sensor [1], in GPS-denied environments assuming that the underlying sensing topology is undirected. The bearing angle δ_{ji} and the subtended angle α^k_{ji} may be measured by vision sensors mounted on mobile agents [12]. The pair of distance and bearing angle (d_{ij}, δ_{ji}) is the displacement measured on the local coordinate frame. Since d_{ij} and δ_{ji} are obtained in a local coordinate frame, the displacement can be considered obtainable in a local coordinate frame if d_{ij} and δ_{ji} are measured. The pair (d_{ij}, δ_{ji}) is simply written as $z^i_{ji} = p^i_j$, i.e., $z^i_{ji} = p^i_j \triangleq (d_{ij}, \delta_{ji})$. The bearing angle δ_{ji} is a directional information, so it can be considered as an angular information or as a directional vector. But the subtended angle α^k_{ji} is the distance angle between the vectors z_{jk} and z_{ik}. The three basic measurements d_{ji}, δ_{ji}, and α^i_{jk} are called distributed sensing variables. Combinations of these variables also can be called distributed sensing variables as far as a global sensing information is not directly utilized. Thus, the displacements $z^i_{ji} = p^i_j$ and the pair of distance and subtended angle (d_{ji}, α^i_{jk}) are also distributed sensing variables. In Part II and Part III of this monograph, we consider the displacement-based distributed sensings, while in Part IV we consider the bearing-based distributed sensings.

The local coordinate frames may be supposed to be expressed with respect to a fixed global reference frame ${}^g\Sigma$. As shown in Fig. 1.7, if the agent i needs to be expressed with respect to ${}^g\Sigma$, then the agent may have some sorts of global information via global sensings. For example, to have R^g_i, the agent i needs to have its orientation information expressed with respect to ${}^g\Sigma$. But let us consider another reference frame, which is called *common coordinate frame* ${}^c\Sigma$. The common coordinate frame is an arbitrary coordinate frame. If agents i, j, and k in Fig. 1.7 know their orientations with respect to ${}^c\Sigma$ as R_i^c, R_j^c, and R_k^c, respectively, then the relative orientations $R_i^c(R_j^c)^{-1}$, $R_i^c(R_k^c)^{-1}$, and $R_k^c(R_j^c)^{-1}$ are invariant for any common coordinate frame ${}^c\Sigma$. Thus, if we are interested in relative states, i.e., relative orientations, relative positions, or relative displacements, any arbitrary common coordinate frame can be used as a reference frame. It is important to remark that a common coordinate frame ${}^c\Sigma$ does not need to be fixed, and agents do not need to have any global sensings with respect to ${}^c\Sigma$. The common coordinate frame ${}^c\Sigma$ is used to formulate the distributed sensings or distributed control between neighboring agents. In this sense, the fixed global reference frame ${}^g\Sigma$ could be considered as a common reference frame. But the common reference frame does not need to be the global reference frame. We summarize it as follows:

- Global reference frame (${}^g\Sigma$): Fixed and used for referring the global information of agents
- A common reference frame (${}^c\Sigma$): Not necessarily fixed, and used for defining the relative information of neighboring agents

1.3.2 Distributed Communications

Agents of distributed systems may exchange some information with neighboring agents, which is called *distributed communications*, or with a centralized coordinator. If agents exchange a piece of information with a centralized coordinator, it is called *global communications*. The agents may need to have communications with each other to escape local minimum points or for a synchronization. If the agents could achieve a better performance by elaborating certain sensing variables, the agents may want to exchange the sensing variables. For example, let us suppose that the agents want to achieve a synchronization in their orientations. The synchronization can be achieved by way of consensus as

$$\dot{\theta}_i = \sum_{j \in \mathcal{N}_i} \theta_{ji}$$

where $\theta_{ji} = \theta_j - \theta_i$ and $(i, j)^e \in \mathcal{E}$, if the initial orientations of agents are pointing the same half circle, and the graph is connected (see Sect. 6.3). Note that the diffusive-coupling information of orientation angles cannot be obtained by local distributed sensings. However, if the bearing angles δ_{ji} and δ_{ij} are exchanged between agents i and j, it is possible to compute θ_{ji} as (see (6.4))

$$\theta_{ji} = [(\delta_{ji} - \delta_{ij} + 2\pi)\text{modulo}(2\pi)] - \pi$$

Thus, to compute θ_{ji} at agent i, it needs to have information of δ_{ij} that needs to be sent by agent j, and agent j needs to have information of δ_{ji} that also needs to be sent by agent i to agent j. These information exchanges motivate the distributed communications between neighboring agents.

Let the agents estimate their positions with respect to a common coordinate frame as $\hat{p}_i$ that may be estimated using some local measurements. Then, agents may exchange the estimated positions with neighboring agents and compare them as $\Delta\hat{p}_{ij} - \Delta p_{ij}$, where $\Delta\hat{p}_{ij} = \hat{p}_i - \hat{p}_j$ and $\Delta p_{ij} = p_i - p_j$ for a localization purpose. Here, the agent i needs to have information of $\hat{p}_j$ assuming that the relative displacements Δp_{ij} are measured in ${}^i\Sigma$. The information can be delivered to agent i from agent j via a local communication. Then, agents attempt to reduce $\Delta\hat{p}_{ij} - \Delta p_{ij}$ to zero, i.e., $\Delta\hat{p}_{ij} - \Delta p_{ij} \rightarrow 0$, which implies that the relative displacement difference in the estimation, i.e., $\Delta\hat{p}_{ij}$, and relative displacement difference in the actual positions, i.e., Δp_{ij}, become equivalent. Then, the estimated positions $\hat{p}_i$ of all agents will be satisfying $\hat{p}_i = p_i + p^c$, where p^c is a common offset of the estimated positions, which means that the positions are estimated up to a common offset. Thus, when we allow distributed communications between neighboring agents, we may achieve a network localization and may have a better performance in the control of agents. The distributed communications could be realized for examples, by using ad hoc networks [24]. Depending upon environments or purpose of the multi-agent systems, the communication also could be directed or undirected. For example, consider the communication in Fig. 1.5. In the attempt of reducing $\Delta\hat{p}_{ij} - \Delta p_{ij}$ by agent i, agent i needs information of $\hat{p}_j$. It does not need to send $\hat{p}_i$ to agent j; but it may need to send the information of $\hat{p}_i$ to another neighboring agent k. Then, the communication topology could be considered as a directed graph.

1.3.3 Distributed Computation and Control

The states of agents are propagated with control inputs as

$$\begin{aligned}\dot{p}_i &= u_i \\ \dot{\theta}_i &= \omega_i\end{aligned}$$

where u_i is the input for linear translation motions and ω_i is the input for angular rotational motions. Depending on control algorithms, the agent i computes the control inputs as functions of various variables that may include desired relative displacements and relative orientations, given to agent i. If the computations of u_i and ω_i are conducted in agent i, it is called *distributed computation*. Otherwise, if the computations are carried out in a centralized computer, it may be called *centralized computation*. It is obviously true that the agents need to have powerful onboard processors to process huge data such as images and database for a real-time learning. One possible solution for these onboard computations is to use embedded GPU processors [35]. Since the agents need to interact with neighboring agents, the computations would be expressed as

$$u_i = \sum_{j \in \mathcal{N}_i} u_i \left(p^i_{ji}, \theta_{ji}, p^{comm}_j, \theta^{comm}_j, p^{desired}_{ij}, \theta^{desired}_{ij} \right) \tag{1.1}$$

$$\omega_i = \sum_{j \in \mathcal{N}_i} \omega_i \left(p^i_{ji}, \theta_{ji}, p^{comm}_j, \theta^{comm}_j, p^{desired}_{ij}, \theta^{desired}_{ij} \right) \tag{1.2}$$

where the signals p^i_{ji} and θ_{ji} are measurements, the signals p^{comm}_j and θ^{comm}_j are two types of information delivered from agent j via communications, and the signals $p^{desired}_{ij}$ and $\theta^{desired}_{ij}$ are desired relative reference signals given to agent i.

Thus, as the agent i has a high network degree, the computational amount would increase proportionally. Hence, in terms of computational load, it would be better to have a network topology whose average degree is small; but in this case, the convergence time would increase as a trade-off. The control is nothing but an implementation of the computed control signals. But the agent i may implement the control efforts toward some specific neighboring agents. Actually, the set of neighboring agents that the control efforts are directed toward determines the actuation topology. The neighboring agents are determined by $\mathcal{N}_i$ in (1.1) and (1.2), whose state information are used for the computation. The neighboring agents in terms of actuation topology are decided by reference signals $p^{desired}_{ij}$ and $\theta^{desired}_{ij}$, because these reference signals determine the control directions. From this perspective, the network topologies conceptually defined in Definitions 1.1 and 1.2 can be more precisely defined by (1.1) and (1.2):

Definition 1.5 If information of agent j is used in the reference signals $p^{desired}_{ij}$ and $\theta^{desired}_{ij}$ of agent i, the agent j is a neighboring agent in terms of actuation topology. Similarly, if information of agent j is contained in the measurements of p^i_{ji} and θ_{ji} of agent i, then j is a neighboring agent in terms of sensing topology, and if information of agent j is contained in the communication information p^{comm}_j and θ^{comm}_j, then j is a neighboring agent in terms of communication topology.

Based on the above discussions, we can say that if the reference signals $p^{desired}_{ij}$ and $\theta^{desired}_{ij}$ include a piece of global information, then it is a centralized controller in terms of actuation topology. It is also remarkable that the neighboring agents j in the measurements, in the communication information, and in the desired reference signals could be different. Hence, the underlying network topologies for sensings, communications, and actuations could be different.

1.4 Formation Control of Multi-agent Systems

The formation control of multi-agent systems under distributed network topologies is, in short, called *distributed formation control*. A review on existing techniques for various formation control problems, including position-based approaches and displacement-based (consensus) approaches, can be found at [29]. In distributed formation control, a framework for determining a unique realization under local interactions was developed using a graph rigidity concept [17]. Since the edge constraints can be considered as local communication or local sensing, the framework given in [17] can be considered as a milestone of the distributed formation control. The more detailed control laws for dynamic agents were presented in [10] under a local sensing framework. Although the framework of [10] is very general in terms of sensing and local interactions, it focused on dynamic and kinematic perspectives rather than topological perspectives. In [18], the rigidity theory has been rigorously borrowed for defining the uniqueness of formations, with minimum number of local interactions. They employed the rigidity matrix concept, and edge split and vertex addition procedures for expanding the formation. Probably, one of the most intuitive analyses of distributed formation control problems can be found in [2] where a directed, undirected, and acyclic triangular formation was studied with agents governed by the first-order dynamics. Consider a triangular formation depicted in Fig. 1.8a where the measurements d_{12} and d_{21} are not matched, i.e., $d_{12} \neq d_{21}$. Then, in [2], they conclude that agents 1 and 2 will move continuously without stopping. So, they devoted their efforts to the analysis of the case of Fig. 1.8b that is acyclic triangular formation. In fact, even though there are some errors in the measurements, in the case of Fig. 1.8b, the follower agents will reach a steady state because an edge is not controlled mutually. For Fig. 1.8c, it was mentioned that the agents will not reach a stable equilibrium point when there is communication between agents. If agent 1 moves, then agent 2 will move. Then, agent 3 will successively move according to the movement of agent 2. Thus, agents will be moving forever when there are no communications with each other. But it was known that this issue could be resolved by gradient control laws later [6, 14–16]. It is also remarkable that the behavior of agents with mismatched distance measurements under a rigid formation graph was analyzed in [27, 36, 37], with a conclusion of convergence to an orbit. Therefore, in [2], with the above motivations, they designed control law using relative displacement measurements multiplied by the distance error $d_{ji}^* - d_{ji}$ as

$$\dot{p}_i = - \sum_{j \in \mathcal{N}^O} (d_{ji}^* - d_{ji})(p_j - p_i) \tag{1.3}$$

The stability of the formation dynamics then was analyzed via linearization. For acyclic general group of n agents and for nonholonomic dynamics, they reached the similar conclusion. Although the control law (1.3) is very intuitive, which can be derived as a gradient control law from a certain potential function, it is not a control law derived from a potential function that involves the squared distances.

Formation control of a group of agents can be formally defined as an interactive behavior to achieve a specified task. Let agents be updated as

$$\dot{x}_i = f_i(x_1, \ldots, x_n, u_i, \omega_i), \; i = 1, \ldots, n \tag{1.4}$$

$$z_i = h_i(x_1, \ldots, x_n) \tag{1.5}$$

Fig. 1.8 The triangular graph where **a** is undirected graph, **b** is acyclic directed graph, and **c** cyclic directed graph [2]

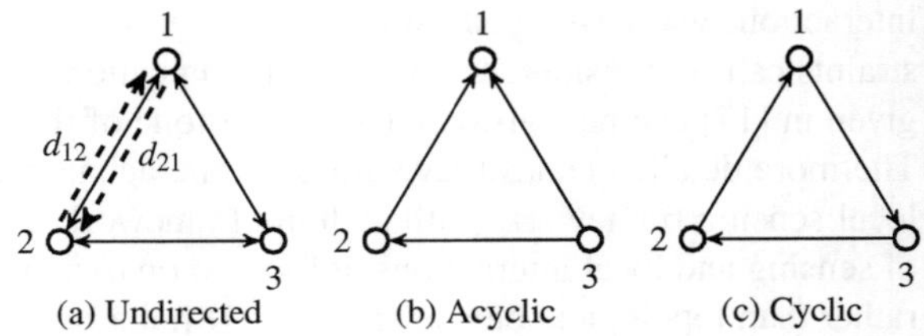

where $x_i = (p_i^T, \theta_i^T)^T$ and z_i is the reference output of interest. Then, the goal of formation systems is to achieve a collective task

$$f_{\mathcal{G}}(z_1, \ldots, z_n) = f_{\mathcal{G}}(z_1^*, \ldots, z_n^*) \tag{1.6}$$

where the function $f_{\mathcal{G}}$ defines a desired formation configuration, and the desired references of interest $z_i^*,\ i = 1, \ldots, n$ would be transformed to the desired relative reference signals. Consequently, the task of formation control is to design control laws for computing u_i and ω_i such that the reference outputs of interest would satisfy the constraint (1.6). Let the desired relative states $p_{ij}^{desired}$ and $\theta_{ij}^{desired}$ be given. Then, from these reference signals, we can obtain the desired references of interest $z_i^*,\ i = 1, \ldots, n$. For example, for two agents i and j, let us suppose that the desired distance is given as d_{ij}^*. Then, we can consider the reference outputs of interest as $z_i = p_i$ and $z_j = p_j$. Hence, any z_i^* and z_j^* satisfying $\|z_i^* - z_j^*\| = d_{ij}^*$ could be considered as the desired references of interest. The function $f_{\mathcal{G}}$ in (1.6) is a general formulation for defining the goal of formation control. Thus, we need to design the control laws u_i and ω_i such that the desired relative states $p_{ij}^{desired}$ and $\theta_{ij}^{desired}$ would be achieved if and (only if) the constraint (1.6) is satisfied. Note that it is definitely preferable to design a function f in (1.6) such that it is a necessary and sufficient condition for achieving the desired relative states $p_{ij}^{desired}$ and $\theta_{ij}^{desired}$. If it is only a sufficient condition, then even though the reference outputs of interest do not satisfy the constraint (1.6), there could be other constraint that may achieve the desired relative states $p_{ij}^{desired}$ and $\theta_{ij}^{desired}$.

To achieve the constraint (1.6), agents may use a piece of global information or rely on global communications. By *distributed agents*, we mean that agents only rely on distributed sensings, distributed communications, distributed computation, and/or distributed control, as described in Sect. 1.3. Thus, formation control of distributed agents can be formally defined as follows.

Definition 1.6 (*Formation control of distributed agents*) For a set of agents of (1.4) with the reference outputs of interest (1.5), the task of formation control of distributed agents is to design control laws u_i and $\omega_i, i = 1, \ldots, n$ such that the constraints (1.6) would be achieved, in distributed network topologies, i.e., by distributed sensings, distributed communications, distributed computation, and distributed control.

In the collective tasks (1.6), if the function $f_{\mathcal{G}}$ is for inter-agent distances d_{ij}^*, then the desired references of interest, $z_i^* \rightarrow p_i^*$, can be selected such that

$$d_{ij} \to d_{ij}^* = \|p_i^* - p_j^*\|, \ \forall (i, j)^e \in \mathcal{E} \tag{1.7}$$

Then, the formation control problem for the task (1.7) is called *distance-based formation control*. If the outputs of interest are bearing vectors such as

$$g_{ji} \to g_{ji}^* = \frac{p_j^* - p_i^*}{\|p_j^* - p_i^*\|}, \ \forall (i, j)^e \in \mathcal{E} \tag{1.8}$$

with the aligned orientations, it is called *bearing-based formation control*. Note that in the distance-based formation control (1.7), since the reference constraints are purely defined by local variables, i.e., inter-agent distances, it is a fully distributed control problem. However, in the bearing-based formation control (1.8), since the orientations of agents are assumed to be aligned, it contains a piece of global information. But, if we reformulate the constraint (1.8) in local frames as

$$g_{ji}^i \to g_{ji}^{i\,*} = \frac{p_j^{i\,*} - p_i^{i\,*}}{\|p_j^{i\,*} - p_i^{i\,*}\|}, \ \forall (i, j)^e \in \mathcal{E} \tag{1.9}$$

where the superscript i is used to denote the expression in ${}^i\Sigma$, it is a fully distributed control problem. The outputs of interest could be displacements as

$$z_{ji} \to z_{ji}^* = p_j^* - p_i^*, \ \forall (i, j)^e \in \mathcal{E} \tag{1.10}$$

or

$$z_{ji}^i \to z_{ji}^{i\,*} = p_j^{i\,*}, \ \forall (i, j)^e \in \mathcal{E} \tag{1.11}$$

The constraint (1.10) assumes that the orientations of agents are aligned, while the constraint (1.11) does not. Both (1.10) and (1.11) can be called *displacement-based formation control* although the former is a centralized control formulation and the latter is a distributed one. It is noticeable that the consensus approaches in formation control [33] can be classified as a displacement-based approach of the form (1.10). Also note that in the setup (1.11), if the orientations of agents are aligned, the problem becomes (1.10). The most widely studied one in formation control territory is given in the following setup:

$$p_i \to p_i^* \tag{1.12}$$

which controls the positions of agents directly. For the above setup (1.12), agents need to have global position information. Thus, it is a fully centralized approach, and it is called *position-based formation control*. In the position-based formation control, the major interests are collision avoidance or formation maintenance. The *subtended angle-based formation control* problems can be defined similarly. When the formation configuration can be uniquely determined by the constraints of subtended angles, the desired configuration can be achieved by $\alpha_{ji}^{k} \to \alpha_{ji}^{k\,*}$, where $\alpha_{ji}^{k\,*}$ are the desired subtended angles. Summarizing the above discussions, we can classify the formation control problems as *distance-based*, *bearing-based*, *displacement-based*, *position-based*, and *subtended angle-based* approaches. Note that these classifications are based on control variables rather than sensing variables. Of course, according to the sensing variables, we can also classify the formation control setups as *distance-based*, *bearing-based*, *displacement-based*, *position-based*, and *subtended angle-based* approaches. Thus, if we consider the sensing variables as inputs $\mathcal{I}^{fc}$ and the control variables $\mathcal{O}^{fc}$ as outputs, where $\mathcal{I}^{fc} = \mathcal{O}^{fc} \triangleq$ {distances, bearing angles, misaligned displacements, aligned displacements, positions, subtended angles}, the classification of formation control problems can be considered as a function:

$$\mathcal{FC} : \mathcal{I}^{fc} \mapsto \mathcal{O}^{fc} \tag{1.13}$$

As a dual problem of formation control in multi-agent systems, network localization under distributed network topology has been studied. The network localization problems can be formulated as

$$\dot{\hat{x}}_i = f_i(\hat{x}_1, \ldots, \hat{x}_n, \hat{u}_i, \hat{\omega}_i), \; i = 1, \ldots, n \tag{1.14}$$

$$\hat{z}_i = h_i(\hat{x}_1, \ldots, \hat{x}_n) \tag{1.15}$$

where $\hat{x}_i = (\hat{p}_i^T, \hat{\theta}_i^T)^T$ is the estimated position and/or orientation of agent i, $\hat{z}_i$ is the reference output of interest in network localization, and $\hat{u}_i$ and $\hat{\omega}_i$ are input signals to update the estimation laws. The goal of network localization is then to achieve a collective task

$$f_{\mathcal{G}}(\hat{z}_1, \ldots, \hat{z}_n) = f_{\mathcal{G}}(z_1, \ldots, z_n) \tag{1.16}$$

where the function $f_{\mathcal{G}}(z_1, \ldots, z_n)$ is defined from actual measurements $z_1, \ldots, z_n$ satisfying the constraint $z_i = h_i(x_1, \ldots, x_n)$. In network localization, we are typically interested in estimating the positions and/or orientations of agents up to a common offset. Thus, the output in network localization could be $\mathcal{O}^{nl} \triangleq$ {positions, orientations}. Thus, classification of network localization problems can be considered as a function:

$$\mathcal{NL} : \mathcal{I}^{nl} \mapsto \mathcal{O}^{nl} \tag{1.17}$$

where $\mathcal{I}^{nl} = \mathcal{I}^{fc}$.

Table 1.1 shows a classification of formation control and network localization. According to the control and sensing variables, we can classify the problems as position-, displacement-, distance-, bearing-, and subtended angle-based formation control. Also, according to the control (estimation) variables, we can consider local-

Table 1.1 Classification of formation control (O.A.: Orientation alignment. Disp.: Displacement. w/o: without.)

	Control variables	Sensing variables	Chapters
Position-based	Positions	Positions	
Disp.-based	Disp.	Disp. w/ O.A.	
	Disp.	Disp. w/o O.A.	Chapters 6, 8, 11
	Positions	Disp. w/o O.A.	Chapter 7
Distance-based	Distances	Disp. w/o O.A.	Chapters 3–5, 12
	Distances	Distances	
Bearing-based	Positions	Bearing vectors w/ O.A.	Chapter 10
	Bearing vectors	Bearing vectors w/o O.A.	Chapter 9
Subtended angle-based	Subtended angles	Subtended angles	
	Subtended angles	Subtended angles and Disp.	Section 2.2.4
Network localization	Positions	Disp. w/o O.A.	Sections 13.2, 13.3.1
	Positions	Bearing vectors w/o O.A.	Section 13.3.2
	Edges	Subtended angles	Section 13.4

ization problems as displacement-, distance-, bearing-, and subtended angle-based network localization.

In this monograph, distance-based, displacement-based, and bearing-based formation control problems will be covered. But remark that the displacement-based (as sensing variable) displacement control (as control variable) with aligned orientations is a consensus problem (see [34]), as already mentioned. This problem will not be treated since it has been widely reported already in a lot of literature and it cannot be categorized as a distributed formation control. This monograph provides recent developments in the displacement-based formation control under misaligned orientations. Note that this monograph does not consider pure distance-based formation control that uses pure distances as sensing variables. This problem has been partially studied in [5, 30] under formation control setups and in [9, 19] for target localization and target pursuit via an adaptive update method. But, recently, the distance-based formation control problem with distance measurements was solved using Lie Bracket approximations [38]. The subtended angle-based approaches are also not studied in this monograph; but when it is necessary, it will be mentioned throughout the monograph. It is remarkable that pure distances and subtended angles are scalar components; so, without using a sequence of samplings or perturbations, it looks not possible to realize a pure distance-based distance control, or pure subtended angle-based angle control into Cartesian coordinate frames. It appears that either sensing variables or control variables need to be vector components for realizing a distributed formation system. That is why we mainly focus on distance-based, displacement-based (under misaligned orientations), and bearing-based formation control problems in this monograph. It is also meaningful to mention that the control or sensing variables (displacement, distance, bearing, and angle) can be combined for a more general setup. That is, the control or sensing variables in some edges could be distances, but in other edges, they could be bearing, or subtended angles. Or, it could be some other combinations. These problems have been recently studied under the name of weak rigidity, and generalized weak rigidity [23, 25, 26] (also see Sect. 2.2.4). However, these approaches are a kind of extensions of basic classifications given in Table 1.1. So, we do not classify the weak rigidity approaches as a basic element in the classification of formation control.

Remark that the formation control laws presented in this monograph can be used for the control of multi-agent systems, including coordination of a group of mobile agents, formation flying of UAVs, platooning of a group of autonomous vehicles, and rendezvous of spacecrafts. Although these applications look quite promising, it is limited to distributed mobile multi-agent systems. There could be a further opportunity for applying the formation control algorithms to network systems including information fusion, analysis of social networks, complex network systems, and data encoding/decoding. However, in this monograph, these applications are not formulated in detail.

References

1. Ahn, H.-S., Yu, W.: Environmental-adaptive RSSI-based indoor localization. IEEE Trans. Autom. Sci. Eng. **6**(4), 626–633 (2009)
2. Baillieul, J., Suri, A.: Information patterns and hedging Brockett's theorem in controlling vehicle formations. In: Proceedings of the 42nd IEEE Conference on Decision and Control, vol. 1, pp. 556–563 (2003)
3. Bajec, I.L., Heppner, F.H.: Organized flight in birds. Anim. Behav. **78**(4), 777–789 (2009)
4. Ballerini, M., Cabibbo, N., Candelier, R., Cavagna, A., Cisbani, E., Giardina, I., Lecomte, V., Orlandi, A., Parisi, G., Procaccini, A., Viale, M., Zdravkovic, V.: Interaction ruling animal collective behavior depends on topological rather than metric distance: evidence from a field study. Proc. Natl. Acad. Sci. **105**(4), 1232–1237 (2008)
5. Cao, M., Yu, C., Anderson, B.D.O.: Formation control using range-only measurements. Automatica **47**(4), 776–781 (2011)
6. Cao, M., Yu, C., Morse, A.S., Anderson, B.D.O., Dasgupta, S.: Generalized controller for directed triangle formations. In: Proceedings of the IFAC 17th World Congress, pp. 6590–6595 (2008)
7. Cavagna, A., Cimarelli, A., Giardina, I., Parisi, G., Santagati, R., Stefanini, F., Tavarone, R.: From empirical data to inter-individual interactions: unveiling the rules of collective animal behavior. Math. Models Methods Appl. Sci. **20**(supp01), 1491–1510 (2010)
8. Conradt, L., Roper, T.J.: Consensus decision making in animals. TRENDS Ecol. Evol. **20**(8), 449–456 (2017)
9. Dandach, S.H., Fidan, B., Dasgupta, S., Anderson, B.D.O.: A continuous time linear adaptive source localization algorithm robust to persistent drift. Syst. Control Lett. **58**(1), 7–16 (2009)
10. Das, A.K., Fierro, R., Kumar, V., Ostrowski, J.P., Spletzer, J., Taylor, C.J.: A vision-based formation control framework. IEEE Trans. Robot. Autom. **18**(5), 813–825 (2002)

11. Dell'Ariccia, G., Dell'Omo, G., Wolfer, D.P., Lipp, H.-P.: Flock flying improves pigeons' homing: GPS track analysis of individual flyers versus small groups. Anim. Behav. **76**(4), 1165–1172 (2008)
12. Desouza, G.N., Kak, A.C.: Vision for mobile robot navigation: a survey. IEEE Trans. Pattern Anal. Mach. Intell. **24**(2), 237–267 (2002)
13. Dey, K.C., Yan, L., Wang, X., Wang, Y., Shen, H., Chowdhury, M., Yu, L., Qiu, C., Soundararaj, V.: A review of communication, driver characteristics, and controls aspects of cooperative adaptive cruise control (CACC). IEEE Trans. Intell. Transp. Syst. **17**(2), 491–509 (2016)
14. Dimarogonas, D.V., Johansson, K.H.: On the stability of distance-based formation control. In: Proceedings of the 47th IEEE Conference on Decision and Control, pp. 1200–1205 (2008)
15. Dimarogonas, D.V., Johansson, K.H.: Further results on the stability of distance-based multi-robot formations. In: Proceedings of the 2009 American Control Conference, pp. 2972–2977 (2009)
16. Dörfler, F., Francis, B.: Geometric analysis of the formation problem for autonomous robots. IEEE Trans. Autom. Control **55**(10), 2379–2384 (2010)
17. Eren, T., Belhumeur, P.N., Anderson, B.D.O., Morse, A.S.: A framework for maintaining formations based on rigidity. In: Proceedings of the 2002 IFAC World Congress, pp. 499–504 (2002)
18. Eren, T., Belhumeur, P.N., Morse, A.S.: Closing ranks in vehicle formations based on rigidity. In: Proceedings of the 41st IEEE Conference on Decision and Control, pp. 2959–2964 (2002)
19. Fidan, B., Dasgupta, S., Anderson, B.D.O.: Adaptive range-measurement-based target pursuit. Int. J. Adapt. Control Signal Process. **27**(1), 66–81 (2013)
20. Grund, T.U.: Network structure and team performance: the case of English Premier League soccer teams. Soc. Netw. **34**(4), 682–690 (2012)
21. Helbing, D., Molnár: Social force model for pedestrian dynamics. Phys. Rev. E **51**(5), 4282–4286 (1995)
22. Hur, H., Ahn, H.-S.: A circuit design for ranging measurement using chirp spread spectrum waveform. IEEE Sens. J. **10**(11), 1774–1778 (2010)
23. Jing, G., Zhang, G., Lee, H.W.J., Wang, L.: Weak rigidity theory and its application to multi-agent formation stabilization, pp. 1–26. arXiv:1804.02795 [cs.SY]
24. Khedher, D.B., Glitho, R.H., Dssouli, R.: Media handling aspects of multimedia conferencing in broadband wireless ad hoc networks. IEEE Netw. **20**(2), 42–49 (2006)
25. Kwon, S.-H., Ahn, H.-S.: Generalized weak rigidity: theory, and local and global convergence of formations, pp. 1–17. arXiv:1809.02562 [cs.SY]
26. Kwon, S.-H., Trinh, M.H., Oh, K.-H., Zhao, S., Ahn, H.-S.: Infinitesimal weak rigidity, formation control of three agents, and extension to 3-dimensional space, pp. 1–12. arXiv:1803.09545
27. Mou, S., Belabbas, M.-A., Morse, A.S., Sun, Z., Anderson, B.D.O.: Undirected rigid formations are problematic. IEEE Trans. Autom. Control **61**(10), 2821–2836 (2016)
28. Nagy, M., Ákos, Z., Biro, D., Vicsek, T.: Hierarchical group dynamics in pigeon flocks. Nature **464**(April), 890–893 (2010)
29. Oh, K.-K., Park, M.-C., Ahn, H.-S.: A survey of multi-agent formation control. Automatica **53**(3), 424–440 (2015)
30. Park, M.-C., Ahn, H.-S.: Displacement estimation by range measurements and application to formation control. In: 2015 IEEE International Symposium on Intelligent Control, pp. 888–893 (2015)
31. Park, M.C., Kang, S.-M., Lee, J.-G., Ko, G.-H., Oh, K.-H., Ahn, H.-S., Choi, Y.-C., Son, J.-H.: Realization of distributed formation flying using a group of autonomous quadcopters and application to visual performance show. In: Proceedings of the IEEE Transportation Electrification Conference and Expo, Asia-Pacific (ITEC), pp. 877–882 (2016)
32. Passos, P., Davids, K., Araújo, D., Pazc, N., Minguéns, J., Mendes, J.: Networks as a novel tool for studying team ball sports as complex social systems. J. Sci. Med. Sport **14**(2), 170–176 (2011)
33. Ren, W.: Consensus strategies for cooperative control of vehicle formations. IET Control Theory Appl. **1**(2), 505–512 (2007)

34. Ren, W., Beard, R.W., Atkins, E.M.: Information consensus in multivehicle cooperative control. IEEE Control Syst. Mag. **27**(2), 71–82 (2007)
35. Satria, M.T., Gurumani, S., Zheng, W., Tee, K.P., Koh, A., Yu, P., Rupnow, K., Chen, D.: Real-time system-level implementation of a telepresence robot using an embedded GPU platform. In: Proceedings of the Design, Automation & Test in Europe Conference & Exhibition (DATE), pp. 1445–1448 (2016)
36. Sun, Z.: Cooperative Coordination and Formation Control for Multi-agent Systems. Ph.D. thesis, Australian National University (2017)
37. Sun, Z., Mou, S., Anderson, B.D.O., Morse, A.S.: Rigid motions of 3-D undirected formations with mismatch between desired distances. IEEE Trans. Autom. Control **62**(8), 4151–4158 (2017)
38. Suttner, R., Sun, Z.: Formation shape control based on distance measurements using Lie Bracket approximations. SIAM J. Control Optim. 1–29 (2018). arXiv:1803.01606
39. Vasile, C.-I., Schwager, M., Belta, C.: Translational and rotational invariance in networked dynamical systems. IEEE Trans. Control Netw. Syst. **5**(3), 822–832 (2018)
40. Wing, A.M., Endo, S., Bradbury, A., Vorberg, D.: Optimal feedback correction in string quartet synchronization. J. Roy. Soc. Interface **11**(93), 1–11 (2014)

Chapter 2
Mathematical Background

Abstract The theory of formation control of distributed agents is developed on the base of mathematical concepts from graph theory. Specifically, algebraic graph theory and graph rigidity theory are two main mathematical backgrounds. But the graph theory acts as a basic topological setup; so to add a control flavor into these theories, we also need to have some background from nonlinear control theory along with consensus dynamics. This chapter provides essential mathematical background that can be used for the developments of distributed formation control theory. The detailed topics included in this chapter are basic concepts in graph theory, rigidity theory (distance rigidity, persistence, bearing rigidity, and weak rigidity), key results in consensus, and basics of nonlinear control theory.

2.1 Basics of Graph Theory

The basics of graph theory given in this chapter are adopted from [1, 6, 9, 11, 13]. A graph is a triple $\mathcal{G} = (\mathcal{V}, \mathcal{E}, \mathcal{W})$ where $\mathcal{V}$ is the set of vertices, $\mathcal{E}$ is the set of edges, and $\mathcal{W}$ is the set of weights. The vertices i and j, $i, j \in \mathcal{V}$, are adjacent or neighbors if they are the ends of a common edge $ed = (i, j)^e \in \mathcal{E}$. The set of neighbor vertices of i is denoted as $\mathcal{N}_i$. Two distinct edges $(i, j)^e$ and $(j, k)^e$ are adjacent if they have a common end vertex j. If there is no edge from the vertex i to the same vertex i, it is called no loop, and if, for a pair of neighboring vertices i and j, there is no second edge from i to j, it is called no multiple edges. If a graph does not have multiple edge nor any loop, it is called a simple graph. This monograph considers only simple graphs. Let the cardinality of $\mathcal{V}$ and $\mathcal{E}$ be $|\mathcal{V}| = n$ and $|\mathcal{E}| = m$, respectively. Without notational confusion, the vertices are, in order or in arbitrary permutations, denoted as $v_1, v_2, \ldots, v_n$ or $1, 2, \ldots, n$, and the edges are denoted as $ed_1, ed_2, \ldots, ed_m$, or $1, 2, \ldots, m$. The path graph is a pair of vertex set and edge set as $\mathcal{V} = \{1, 2, \ldots, n_p\}$ and $\mathcal{E} = \{(1, 2)^e, (2, 3)^e, (3, 4)^e, \ldots, (n_p - 1, n_p)^e\}$, and the cycle graph is a combination of a path graph and an additional edge $(n_p, 1)^e$. A complete graph is a graph that has a connecting edge from any vertex to any other vertex. For a vertex i, the degree of i is $d_{\mathcal{G}}(i) = |\mathcal{N}_i|$. A graph $\mathcal{G}$ is called k-regular if $d_{\mathcal{G}}(i) = k$ for all $i \in \mathcal{V}$. A graph is called regular if it is k-regular, for a positive

H.-S. Ahn, *Formation Control*, Studies in Systems, Decision and Control 205, https://doi.org/10.1007/978-3-030-15187-4_2

integer k. When we consider the topological perspective of graphs, we can ignore the weights. For two graphs $\mathcal{G}' = (\mathcal{V}', \mathcal{E}')$ and $\mathcal{G} = (\mathcal{V}, \mathcal{E})$, the graph $\mathcal{G}'$ is a subgraph of $\mathcal{G}$ when $\mathcal{V}' \subseteq \mathcal{V}$ and $\mathcal{E}' \subseteq \mathcal{E}$. Then, for non-empty subset $\mathcal{V}'$, the subgraph $\mathcal{G}'$ is called an induced graph by the set of vertices $\mathcal{V}'$. Or, for non-empty subset $\mathcal{E}'$, the subgraph $\mathcal{G}'$ is called an induced graph by the set of edges $\mathcal{E}'$.

If an edge has a direction as $i \to j$, where i is the tail and j is the head, when connecting two vertices i and j, it is called a directed edge and is denoted as $(i, j)^{\bar{e}}$. If a part or all of edges have directions, then it is called a directed graph and denoted as $\vec{\mathcal{G}} = (\mathcal{V}, \vec{\mathcal{E}})$. For a vertex i, in-degree $d_{\mathcal{G}}^{I}(i)$ and out-degree $d_{\mathcal{G}}^{O}(i)$ are defined as

$$d_{\mathcal{G}}^{I}(i) = |\{j : (j, i)^{\bar{e}}, \forall (j, i)^{\bar{e}} \in \vec{\mathcal{E}}\}|$$
$$d_{\mathcal{G}}^{O}(i) = |\{j : (i, j)^{\bar{e}}, \forall (i, j)^{\bar{e}} \in \vec{\mathcal{E}}\}|$$

For an edge $(k, i)^{\bar{e}}$, the vertex k is called incoming vertex to the vertex i, and for an edge $(i, j)^{\bar{e}}$ the vertex j is an outgoing vertex of the vertex i. For the vertex i, the set of all incoming vertices is incoming neighborhood, denoted as $\mathcal{N}_i^I$, and the set of all outgoing vertices is outgoing neighborhood, denoted as $\mathcal{N}_i^O$. A walk is a sequence of vertices and edges as $(v_1, ed_1, \ldots, ed_k, v_{k+1})$ where $ed_i = (v_i, v_{i+1})^e$. The number k is the length of the walk. A trail in $\mathcal{G}$ is a walk with all of its edges being distinct, and a path is a walk with all of vertices being distinct. In the graph $\mathcal{G}$, when the initial vertex is v_1 and final vertex is v_k, we call such a walk $v_1 v_k$-walk. A walk or trail is closed if the initial vertex and final vertex are the same. A cycle is a closed walk with distinct vertices except for the initial and final vertices. For every pair of vertices, (v_i, v_j), of a graph, if it has $v_i v_j$-path, then the graph is connected.

Consider two graphs $\mathcal{G}$ and $\mathcal{G}'$. A homomorphism is a mapping

$$f : \mathcal{G} \to \mathcal{G}'$$

with the pair $f = (f_v, f_e)$, where $f_v : \mathcal{V} \to \mathcal{V}'$ and $f_e : \mathcal{E} \to \mathcal{E}'$ satisfying

$$ed_k = (v_i, v_j)^e \Rightarrow ed_k' = f_e(ed_k) = (f_v(v_i), f_v(v_j))^e$$

If both mappings f_v and f_e are bijective, then the mapping f is called an isomorphism. If two graphs are isomorphic, then it is denoted as $\mathcal{G} \cong \mathcal{G}'$. Then, we call two graphs are equivalent in terms of graph topology. An isomorphism is a deformation of graph from $\mathcal{G}$ to $\mathcal{G}'$. An isomorphism from $\mathcal{G}$ to itself is called an automorphism. An automorphism is nothing but a permutation of labels, without deforming the graph. A tree is a connected graph that has no cycle as a subgraph. Given a graph $\mathcal{G}$, a subtree is called spanning tree of $\mathcal{G}$ if it is a tree having all vertices of $\mathcal{G}$.

A digraph $\vec{\mathcal{G}} = (\mathcal{V}, \vec{\mathcal{E}})$ is weakly connected if its underlying graph $\mathcal{G}$ is connected. It is called strongly connected if for every pairs of distinct vertices $v_i, v_j \in \mathcal{V}$, there is a directed walk from v_i to v_j. For a digraph $\vec{\mathcal{G}}$, if there are directed paths from a node v_i to all other nodes, then the digraph is called having a directed rooted tree. Here, the root node is v_i. The directed rooted tree is called arborescence.

For an edge $(i, j)^e \in \mathcal{E}$, let the graph have a weighting $w_{ij} \in \mathcal{W}$. Then, the weighted adjacency matrix of $\mathcal{G}$ is defined as

$$\mathbb{A} \triangleq [w_{ij}]$$

If the weights are 1 or 0, then we simply write the weight as $a_{ij} \in \mathcal{W}$ where $a_{ij} = 1$ or 0, and the weighted adjacency matrix is simply called adjacency matrix $\mathbb{A} = [a_{ij}]$. The two matrices $\mathbb{X}$ and $\mathbb{Y}$ are orthogonally equivalent if there is a permutation matrix $\mathbb{P}$ such that $\mathbb{X} = \mathbb{P}^{-1}\mathbb{Y}\mathbb{P}$ holds. The following statements are equivalent [1]:

- The two graphs $\mathcal{G}$ and $\mathcal{G}'$ are isomorphic.
- The corresponding adjacency matrices $\mathbb{A}_{\mathcal{G}}$ and $\mathbb{A}_{\mathcal{G}'}$ are orthogonally equivalent.

A matrix X is primitive if there is a positive integer k such that every elements of A^k are positive. For an undirected graph with vertex set $\mathcal{V} = \{v_1, \ldots, v_n\}$ and edge set $\mathcal{E} = \{ed_1, \ldots, ed_m\}$, the incidence matrix $\mathbb{H} = [h_{ik}]$ is defined as

$$h_{ik} = \begin{cases} h_{ik} = 1 \text{ if } v_k \text{ is an end vertex of the edge } ed_i \\ h_{ik} = 0 \text{ otherwise} \end{cases}$$

In the case of directed graph with the k-th directed edge $ed_k = (i, j)^{\bar{e}}$, we have $h_{ki} = -1$ and $h_{kj} = +1$. For an undirected edge set $\mathcal{E}$, let us decompose it as $\mathcal{E} = \mathcal{E}_+ \cup \mathcal{E}_-$ where $\mathcal{E}_+$ and $\mathcal{E}_-$ are subsets of directed edges of $\mathcal{E}$ satisfying the condition that if $(i, j)^{\bar{e}} \in \mathcal{E}_+$ holds, then $(j, i)^{\bar{e}} \in \mathcal{E}_-$ holds. Then, according to the edge sets $\mathcal{E}_+$ and $\mathcal{E}_-$, we can obtain the incidence matrices as $\mathbb{H}_+$ and $\mathbb{H}_-$, respectively.

Lemma 2.1 ([1]) *The two graphs $\mathcal{G}$ and $\mathcal{G}'$ that have the same sizes of vertices and edges are isomorphic if and only if there are permutation matrices $\mathbb{P} \in \mathbb{R}^{n \times n}$ and $\mathbb{Q} \in \mathbb{R}^{m \times m}$ such that*

$$\mathbb{H}_{\mathcal{G}'}^T = \mathbb{P}^T \mathbb{H}_{\mathcal{G}}^T \mathbb{Q}$$

Lemma 2.2 ([1]) *For any submatrix $\mathbb{H}_+' \in \mathbb{R}^{(n-1) \times n}$ of the arbitrarily directed incidence matrix $\mathbb{H}_+$ of an undirected graph $\mathcal{G}$, the followings are equivalent:*

- *$rank(\mathbb{H}_+') = n - 1$.*
- *The $n - 1$ edges corresponding to the rows of $\mathbb{H}_+'$ form a spanning tree of $\mathcal{G}$.*

An edge cut $\mathcal{S}$ is a set of edges $\mathcal{S} \in \mathcal{E}_{\mathcal{G}}$ satisfying that the graph $\mathcal{G}' = (\mathcal{V}, \mathcal{E} \setminus \mathcal{S})$ is disconnected, but $\mathcal{G}'' = (\mathcal{V}, \mathcal{E} \setminus \mathcal{S}')$ is connected for any subset $\mathcal{S}' \subset \mathcal{S}$. The smallest cardinality of an edge cut is called edge connectivity, and denoted as $\kappa'(\mathcal{G})$. If $k \leq \kappa'(\mathcal{G})$, the graph $\mathcal{G}$ is called k-edge connected. For example, if at least three edges need to be removed to have a disconnected graph, it is 3-edge connected. The minimum number of vertices of $\mathcal{G}$, whose removal disconnects the graph $\mathcal{G}$ is called the vertex connectivity, and denoted as $\kappa(\mathcal{G})$. If $k \leq \kappa(\mathcal{G})$, the graph $\mathcal{G}$ is called k-vertex connected.

Let the degree matrix of a graph $\mathcal{V}$ be denoted as $\mathbb{D} = \text{diag}(d_{\mathcal{G}}(i))$. The Laplacian matrix is defined as

$$\mathbb{L} = \mathbb{D} - \mathbb{A} = \mathbb{H}^T \mathbb{H}$$

which means that the Laplacian matrix is positive semi-definite. Let $\mu_1 \leq \mu_2 \leq \cdots \leq \mu_n$ be the eigenvalues of the Laplacian matrix $\mathbb{L}$. Then, if the graph is connected, we have $\mu_1 = 0$ with a corresponding eigenvector $(1, 1, \ldots, 1)^T$ and $\mu_2 > 0$. Furthermore, if the graph is regular of degree k, then $\mu_i = k - \lambda_i$ where λ_i are eigenvalues of the adjacency matrix $\mathbb{A}$ of the graph $\mathcal{G}$ in decreasing order. Consequently, from the above relationship, we can have $\text{rank}(\mathbb{L}) = n - 1$, and $\text{null}(\mathbb{L}) = \text{null}(\mathbb{H}) = \text{span}\{\mathbf{1}_n\}$.

A matrix A is called reducible if there exists a permutation matrix $\mathbb{P}$ such that

$$\mathbb{P} A \mathbb{P}^T = \left[\begin{array}{c|c} A_1 & \mathbf{0} \\ \hline A_{21} & A_2 \end{array}\right]$$

Given a graph $\mathcal{G}$, let us suppose that the adjacency matrix $\mathbb{A}$ is reducible. Then, the set of vertices can be partitioned as $\mathcal{V} = \mathcal{V}_1 \cup \mathcal{V}_2$, where $\mathcal{V}_1 \neq \emptyset$ and $\mathcal{V}_2 \neq \emptyset$. Thus, all edges between $\mathcal{V}_1$ and $\mathcal{V}_2$ are edge cuts of $\mathcal{G}$. Define the directed cut as an edge cut with a set of directed edges whose tails are in one component composed of $\mathcal{V}_1$, and heads are in another component composed of $\mathcal{V}_2$. Then, in the case of directed graphs, we can see that the adjacency matrix $\mathbb{A}$ is reducible if and only if $\overrightarrow{\mathcal{G}}$ has a directed cut. Otherwise, it is irreducible. The irreducibility in directed graphs implies that the graph is strongly connected. Consequently, a digraph is strongly connected if and only if it has no directed cut.

Lemma 2.3 ([6]) *Given a graph $\mathcal{G}$, let $f(v_i)$ be a value at vertex v_i, and let $f = (f(v_1), \ldots, f(v_n))^T$. Then, the function $g = \mathbb{L} f$, where $\mathbb{L}$ is the Laplacian matrix of the graph $\mathcal{G}$, is obtained as*

$$g_i = (\mathbb{L} f)_i = \sum_{j \in \mathcal{N}_i} \mathbb{L}_{ij} f_j = \sum_{j \in \mathcal{N}_i} a_{ij}(f_i - f_j)$$

The sum of eigenvalues of Laplacian matrix satisfies the following equality:

$$\sum_{i=1}^{n} \mu_i(\mathbb{L}) = \sum_{i \in \mathcal{V}} d_{\mathcal{G}}(i) = 2|\mathcal{E}_{\mathcal{G}}|$$

where $\mu_i(\mathbb{L})$ are the eigenvalues of the Laplacian and $|\mathcal{E}_{\mathcal{G}}|$ can be replaced by the sum of w_{ij} when the graph is a weighted graph.

2.2 Rigidity Theory

For a set of n-agents, $1, \ldots, n$, let the position values of agents be denoted as $p_i \in \mathbb{R}^d$ in Euclidean space. Then, the concatenation $p = (p_1, \ldots, p_n)$ is called a realization of agents. A framework is a pair of graph $\mathcal{G} = (\mathcal{V}, \mathcal{E})$ and a realization p of agents. Thus, the framework is a triple $f_p = (\mathcal{V}, \mathcal{E}, p)$. The rigidity theory studies conditions for a unique realization of framework when the interagent constraints between neighboring agents are specified by some scalar or vector magnitude. The magnitudes could be *distances* in undirected edges (it is called graph rigidity, or distance rigidity), *distances* in directed edges (persistence), *bearing* measurements (bearing rigidity), or *subtended* angles (when it includes subtended angles, it is called weak rigidity), or combinations of these components. The rigidity with all these components (distances, bearing angles, and subtended angles) may be called generalized rigidity.

2.2.1 *Graph Rigidity (Distance Rigidity)*

The graph rigidity in this section is summarized from [2, 15, 18]. From a framework f_p, let us calculate distances between a pair of points as $d_{ij} = \|p_i - p_j\|, i, j \in \mathcal{V}$. Let us select another framework as $f_q = (\mathcal{V}, \mathcal{E}, q)$ where q is a realization different from p in the same dimensional Euclidean space. Then, if $\|p_i - p_j\| = \|q_i - q_j\|$, $\forall i, j \in \mathcal{V}$, the two frameworks f_p and f_q are called congruent. When it is limited to the set of edges, if $\|p_i - p_j\| = \|q_i - q_j\|, \forall (i, j)^e \in \mathcal{E}$, the two frameworks are called equivalent to each other. Clearly for two congruent frameworks, there must be more distance constraints than for equivalent frameworks. The graph theoretical conditions that two different frameworks become congruent imply a (global) rigidity of a graph. A rigid framework cannot be "continuously" deformed in the same dimensional space. From a framework f_p, the distance constraints $\|p_i - p_j\| = const$ for edges make quadratic equations as $(p_i - p_j)^T (p_i - p_j) = const^2$, $(i, j)^e \in \mathcal{E}$. So, there are m quadratic constraints. The set of solutions for these m quadratic constraints defines an algebraic solution set $\mathcal{A}_{\mathcal{E}}$ for points p in a dn-space (n agents in d-dimensional space). It is then clear that a realization p is a solution of $\mathcal{A}_{\mathcal{E}}$, i.e., $p \in \mathcal{A}_{\mathcal{E}}$. For the same vertex set, we replace the edge set $\mathcal{E}$ by the edges of $K(n)$ graph. Then, we can define another algebraic solution set $\mathcal{A}_{K(n)}$. Then, obviously, we have $\mathcal{A}_{K(n)} \subset \mathcal{A}_{\mathcal{E}}$ because the algebraic set $\mathcal{A}_{K(n)}$ has more constraints than the set $\mathcal{A}_{\mathcal{E}}$. Actually, the algebraic solution set $\mathcal{A}_{\mathcal{E}}$ determines equivalent frameworks and the set $\mathcal{A}_{K(n)}$ determines congruent frameworks. The framework $f_p = (\mathcal{V}, \mathcal{E}, p)$ is rigid if $\mathcal{A}_{\mathcal{E}}$ and $\mathcal{A}_{K(n)}$ are equal in a neighborhood of p, and globally rigid if they are equal in an entire domain.

The algebraic set $\mathcal{A}_{K(n)}$ has $d(d+1)/2$ independent motions, which is sometimes called dimension, for moving a point $p \in \mathcal{A}_{K(n)}$ to another point $q \in \mathcal{A}_{K(n)}$, in $\mathbb{R}^d$. Thus, a framework $f_p = (\mathcal{V}, \mathcal{E}, p)$ is rigid if the corresponding algebraic set $\mathcal{A}_{\mathcal{E}}$ has dimension $d(d+1)/2$ in a neighborhood of p. Since each distance constraint has one independent motion, for a framework, we can have dn dimensions at maximum. Thus, a framework with $dn - d(d+1)/2$ properly distributed edges will be rigid.

A framework $f_p = (\mathcal{V}, \mathcal{E}, p)$ is generic if all frameworks of the points in a neighborhood of p are rigid or not rigid. So, for the two realizations p and q in a neighborhood, if a framework f_p is generic, then both the frameworks f_p and f_q will be rigid or nonrigid. If a framework f_p with the realization p is generic, the realization p is called generic realization. With this concept, we can see that a graph $\mathcal{G} = (\mathcal{V}, \mathcal{E})$ is rigid if the frameworks corresponding to generic realizations of p are rigid. Thus, we can equalize the rigidity and generic rigidity as follows:

Definition 2.1 A framework $f_p = (\mathcal{V}, \mathcal{E}, p)$ in $\mathbb{R}^d$ is generically rigid if the graph is rigid in $\mathbb{R}^d$.

The above definition is nothing but a statement that we consider only the generic realizations to characterize the rigidity of a graph. Or, the rigidity concept is a topological concept, and the generic rigidity is a geographical (algebraic) concept, but these two concepts are merged when we consider only the generic realizations. Figure 2.1 shows examples of rigid graphs. Figure 2.1a is rigid since it cannot be deformed, but it is not generic since it becomes flexible in a small neighborhood of the realization as shown in Fig. 2.1b. The formation of Fig. 2.1b is not rigid, but it is generic since all the frameworks in a very small neighborhood of the current realization are flexible. The formation of Fig. 2.1c is rigid since it cannot be deformed, and it is generic since the rigidity property is preserved in a small neighborhood of the realization p.

In fact, the rigidity and generic property are both graph algebraic conditions. By algebraic condition, we mean that the property should be checked taking account of the realization of agents. But the combination of rigidity and generic property, which is called *generic rigidity*, is a topological concept. By topological condition, we mean that the rigidity can be examined without relying upon the realization of agents. Hence, from the above definition, we can see that the graph algebraic concept is simply replaced by the graph topological condition. This will save us tremendous amount of efforts in defining a unique realization of graphs since we only need to consider the graph topology. The generic rigidity in $\mathbb{R}^2$ can be characterized by Laman lemma:

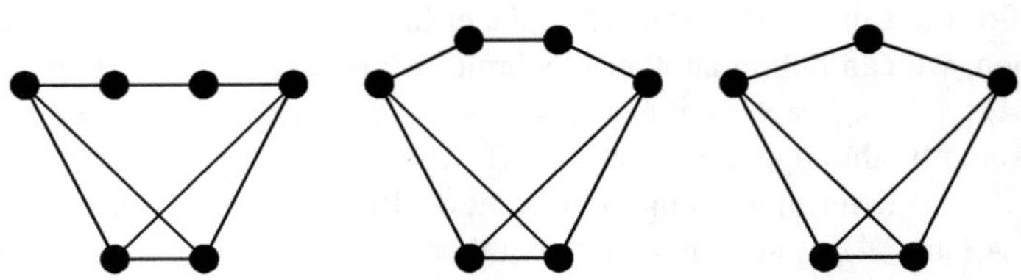

(a) Rigid, but not generic (b) Not rigid, but generic (c) Rigid and generic

Fig. 2.1 Rigid versus not rigid, and generic versus not generic

Lemma 2.4 ([28]) *A graph $\mathcal{G} = (\mathcal{V}, \mathcal{E})$ is (generically) rigid in $\mathbb{R}^2$ if and only if there is an induced graph $\mathcal{G} = (\mathcal{V}, \mathcal{E}')$ where $\mathcal{E}' \subseteq \mathcal{E}$ satisfying*

- $|\mathcal{E}'| = 2|\mathcal{V}| - 3$.
- $|\mathcal{E}''| \leq 2|\mathcal{V}''| - 3$, *where $\mathcal{E}'' \subseteq \mathcal{E}'$ and $\mathcal{V}''$ is the set of vertices induced by the edge set $\mathcal{E}''$.*

The Lemma 2.4 shows that the characteristics of rigidity can be evaluated by topological conditions rather than algebraic evaluation. Without notational confusion, when we discuss a rigidity, it means a generic rigidity. Therefore, we are interested in the topological concept rather than the algebraic concept. That is, since it is rather not so feasible to determine the rigidity property under a topological evaluation, we use the concept of the generic realization of a graph for an algebraical evaluation, which can be considered as equivalent to a topological evaluation.

From $(p_i - p_j)^T(p_i - p_j) = const^2$, taking a derivative along a time, we can have

$$(u_i - u_j)^T(p_i - p_j) = 0, \ \forall (i, j)^e \in \mathcal{E} \tag{2.1}$$

where $u_i = (dp_i)/(dt)$. The set of infinitesimal motions of the framework $f_p = (\mathcal{V}, \mathcal{E}, p)$ is a subspace of dn-vector space. That is, in $\mathbb{R}^d$, each node can have d motions at maximum. Thus, we have a total of $n \times d$ motions for the framework. But, since the distances act as constraints for the motions, the number of infinitesimal motions will be reduced when the nodes are connected more. For a given framework f_p, let the set of all possible infinitesimal motions be denoted as $\mathcal{M}_\mathcal{E}$ and, after adding edges to the graph so that it becomes a complete graph, let the set of all possible infinitesimal motions be denoted as $\mathcal{M}_{K(n)}$. Then, the set $\mathcal{M}_\mathcal{E}$ corresponds to the set $\mathcal{A}_\mathcal{E}$ and the set $\mathcal{M}_{K(n)}$ corresponds to the set $\mathcal{A}_{K(n)}$ such as

$$\mathcal{A}_\mathcal{E} \rightarrow \mathcal{M}_\mathcal{E}$$
$$\mathcal{A}_{K(n)} \rightarrow \mathcal{M}_{K(n)}$$

We can consider that $\mathcal{M}_\mathcal{E}$ is the tangent space of $\mathcal{A}_\mathcal{E}$ and $\mathcal{M}_{K(n)}$ is the tangent space of $\mathcal{A}_{K(n)}$. It is clear that $\mathcal{M}_{K(n)} \subseteq \mathcal{M}_\mathcal{E}$.

Definition 2.2 A framework $f_p = (\mathcal{V}, \mathcal{E}, p)$ is infinitesimally rigid if $\mathcal{M}_{K(n)} = \mathcal{M}_\mathcal{E}$. Otherwise, the framework is called infinitesimally flexible, or simply flex.

Thus, from the above definition, we can see that the (generic) rigidity is a concept defined for the spaces $\mathcal{A}_\mathcal{E}$ and $\mathcal{A}_{K(n)}$, while the infinitesimal rigidity is a concept defined for $\mathcal{M}_\mathcal{E}$ and $\mathcal{M}_{K(n)}$, which are tangent spaces of $\mathcal{A}_\mathcal{E}$ and $\mathcal{A}_{K(n)}$. The rigidity is a concept defined for the set of points, while the infinitesimal rigidity is a concept defined for the tangent motions. Thus, the infinitesimal rigidity implies the (generic) rigidity, but the converse is not true. But, when the set of realized points is generic, the concepts of rigidity, generic rigidity, and infinitesimal rigidity become merged for all frameworks. Let p be a realization of agents in $\mathbb{R}^d$, and let $\mathcal{P}$ denote the set of all realizations.

Definition 2.3 The set $\mathcal{P}$ is in general position and a realization $p \in \mathcal{P}$ is a general realization in $\mathbb{R}^d$ if for any subset $\mathcal{P}' \subseteq \mathcal{P}$ composed of q elements, where $q < d+1$, the affine space spanned by $\mathcal{P}'$ has dimension $q-1$.

From the squared distance functions $\bar{d}_{ij} = \|p_i - p_j\|^2$, we define the edge function as $\mathbf{h}_{\mathcal{G}}(p) = (\underbrace{\ldots, \bar{d}_{ij}, \ldots}_{=|\mathcal{E}|})^T$. Then, we take a derivative of $\mathbf{h}_{\mathcal{G}}(p)$ as

$$\frac{\partial \mathbf{h}_{\mathcal{G}}(p)}{\partial p} = 2\mathbb{R}(p)$$

where $\mathbb{R}(p) \in \mathbb{R}^{|\mathcal{E}| \times dn}$ is called rigidity matrix. Using the above relationship, we can obtain

$$\frac{\partial \mathbf{h}_{\mathcal{G}}(p)}{\partial t} = \frac{\partial \mathbf{h}_{\mathcal{G}}(p)}{\partial p}\frac{\partial p}{\partial t} = 2\mathbb{R}(p)\dot{p} = 0$$

Thus, the null space of the rigidity matrix is a non-empty set. The trivial motions of the framework, i.e., translations and rotations, would satisfy the above relationship. Now, denoting $\mathbf{h}_{K(n)}(p)$ as the edge function of the complete graph that is generated from $\mathcal{G}(p)$ by adding edges to all the pairs of non-neighboring agents, and using the null space of rigidity matrix, we can also define the rigidity as follows:

Definition 2.4 A framework $f_p = (\mathcal{V}, \mathcal{E}, p)$ is infinitesimally rigid in $\mathbb{R}^d$ if the tangent space of $\mathbf{h}_{\mathcal{G}}^{-1}(\mathbf{h}_{\mathcal{G}}(p))$ is equal to the tangent space of $\mathbf{h}_{K(n)}^{-1}(\mathbf{h}_{K(n)}(p))$, which includes only the trivial motions [32].

With the rigidity matrix, we can obtain a condition for rigidity for generic realizations.

Lemma 2.5 ([18]) *The framework $f_p = (\mathcal{V}, \mathcal{E}, p)$ is infinitesimally rigid in $\mathbb{R}^d$ if and only if*

$$rank(\mathbb{R}(p)) = \begin{cases} dn - d(d+1)/2 \text{ if } n \geq d \\ n(n-1)/2 \text{ otherwise} \end{cases}$$

The above lemma means that for a given framework, we can check the rigidity by evaluating the rank of the rigidity matrix. Thus, if we assume that all the realizations of a graph are generic, then by checking the rank of the rigidity matrix of a particular framework, we can confirm that the graph is rigid from a topological perspective. A framework is called minimally infinitesimally rigid in $\mathbb{R}^d$ if the framework is infinitesimally rigid and no single edge can be removed without losing the rigidity property. Thus, in the sense of keeping a rigidity in minimally infinitesimally rigid graphs, it is 1-edge connected. That is, if an edge is removed, the rigidity will be lost. If a framework is minimally infinitesimally rigid, then row vectors of rigidity matrix, which correspond to the edges of the graph, are independent. Thus, the rigidity matrix of minimally infinitesimally rigid frameworks has a full row rank. Under the

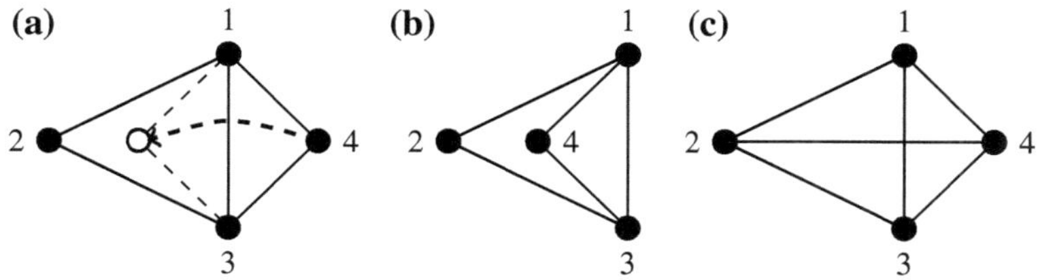

Fig. 2.2 Rigid graphs: **a** and **b** are equivalent but not congruent in the entire domain. For example, the vertex 4 can be located in two different points while satisfying the distance constrains with respect to vertices 1 and 2. Globally rigid graph: graph **a** becomes a globally rigid graph **c** by having additional distance constraint between vertices 2 and 4

assumption that there exists at least one edge that can be removed, if any edge can be removed without losing the rigidity, the graph is called redundantly rigid. Actually, a realization of infinitesimally rigid graphs is not unique in a coordinate frame. As shown in Fig. 2.2, for two realizations of a rigid graph, we can see that d_{24} in Fig. 2.2a is not equal to d_{24} in Fig. 2.2b, even though the underlying topology of the graph is rigid. That is, the two realizations are equivalent and rigid, but they are not congruent in the entire domain. It is remarkable that in a small neighborhood of p in the sense of $\|p' - p\| \leq \epsilon$, if the graph is rigid, then the realization p', which is equivalent to p, is also congruent to p. Thus, the congruence concept can be used in either a local neighborhood or in the entire domain.

Now in the entire domain, we can see that for having congruent realizations for a framework, we need to have a stronger concept than rigidity. To treat this concept, the concept of global rigidity has been defined and investigated. It is, by definition, said that if a graph is globally rigid, then there could be a unique realization up to translations and rotations in $\mathbb{R}^d$. More precisely, if the two equivalent frameworks f_p and f_q are also congruent to each other in $\mathbb{R}^d$, the framework f_p is globally rigid. For example, see Fig. 2.2c. For generic realizations in 2-dimensional space, we can have the following result.

Theorem 2.1 ([22]) *A graph $\mathcal{G}$ is globally rigid in $\mathbb{R}^2$ if and only if the graph is 3-vertex connected and redundantly rigid.*

From the above discussions, we can summarize the rigidity as

- Rigidity $\Leftrightarrow$ Locally congruent.
- Global rigidity $\Leftrightarrow$ Congruent in entire domain.

The above theorem provides a necessary and sufficient condition for a global rigidity in $\mathbb{R}^2$. But, in higher dimensional space, it is not clear to find an exact condition although a necessary condition has been reported as follows.

Theorem 2.2 ([5]) *Let us suppose that a graph in $\mathbb{R}^d$ is generically globally rigid. Then, it is $d + 1$ vertex connected and redundantly rigid.*

If any edge cannot be removed without losing the (global) rigidity property of the graph, it is called *(globally) minimally rigid* graph. The minimally rigid or globally minimally rigid graph can be produced by Henneberg extensions. In Henneberg extensions, there are two operations, i.e., 0-extension and 1-extension. Given an initial graph $\mathcal{G} = (\mathcal{V}, \mathcal{E})$, the 0-extension is to add a vertex v with two edges $(v, u_1)^e$ and $(v, u_2)^e$, where u_1, u_2 are vertices of the graph $\mathcal{G}$, to the graph $\mathcal{G}$. So, the graph after an 0-extension is the graph $\mathcal{G}' = (\mathcal{V} \cup \{v\}, \mathcal{E} \cup \{(v, u_1)^e, (v, u_2)^e\})$. Given an initial graph $\mathcal{G} = (\mathcal{V}, \mathcal{E})$, the 1-extension is to add a vertex v with three new edges $(v, u_1)^e$, $(v, u_2)^e$, and $(v, u_3)^e$, and to remove an edge $(u_2, u_3)^e$ from the initial graph $\mathcal{G}$ under the condition of $u_1 \neq u_2$, and $u_1 \neq u_3$. Thus, after an 1-extension, we have the graph $\mathcal{G}' = (\mathcal{V} \cup \{v\}, \mathcal{E} \cup \{(v, u_1)^e, (v, u_2)^e, (v, u_3)^e\} \setminus \{(u_2, u_3)^e\})$.

Theorem 2.3 ([22]) *The newly generated graph $\mathcal{G}$ is minimally rigid if and only if it can be produced by a sequence of* 0*-extensions from $K(2)$ graph.*

Theorem 2.4 ([7]) *A graph $\mathcal{G} = (\mathcal{V}, \mathcal{E})$ is globally minimally rigid if and only if it can be built up from a $K(4)$ graph by a sequence of* 1*-extensions.*

2.2.2 Persistence

The concept of rigidity of Sect. 2.2.1 is developed for undirected graphs. It is also natural to ask a possibility of a unique realization of directed graphs in a given space. This concept has been investigated under the name of persistence [19]. To motivate the persistence concept, let us consider Fig. 2.3. In Fig. 2.3a, the constraint determined by agents 1 and 2 can be realized in Euclidean space. Since the agent 2 is responsible for the distance constraint, by reaching a point satisfying the distance constraint, it can be realized in any dimensional space. In Fig. 2.3b, we have one more agent that is responsible for two constraints, i.e., the distances between agents 1 and 3, and agents 2 and 3. Although the agent 3 can stay in the position 3 and 3′, the constraints still can be realized in any d-dimensional space, where $d \geq 2$. Such constraints may be called feasible constraints or realizable constraints for a given dimensional space. In literature, such constraints have been termed as *consistence*; it means that given a graph topology, the constraints can be realized in a given dimensional space. Notice that in Fig. 2.3b, the configuration determined by agents 1, 2, and 3 is unique. In Fig. 2.3c, agents 3 and 4 can stay at the positions 3 and 4, or 3′ and 4′, respectively. Thus, the constraints defined by the graph topology can be realized in any dimensional space; so, the constraints are consistent. The configuration of Fig. 2.3c is also unique since any framework satisfying the constraints is congruent to the framework of Fig. 2.3c. More precisely, ignoring the directions of edges, when the underlying graph topology is rigid, then the realization may be unique in a local sense. Thus, it is necessary to have a rigidity as its underlying graph topology.

The constraint consistence may be defined rigorously using the concept *fitting*. Let an edge $(i, j)^{\bar{e}}$, $j \in \mathcal{N}_i^O$ that satisfies the desired distance constraints be called active.

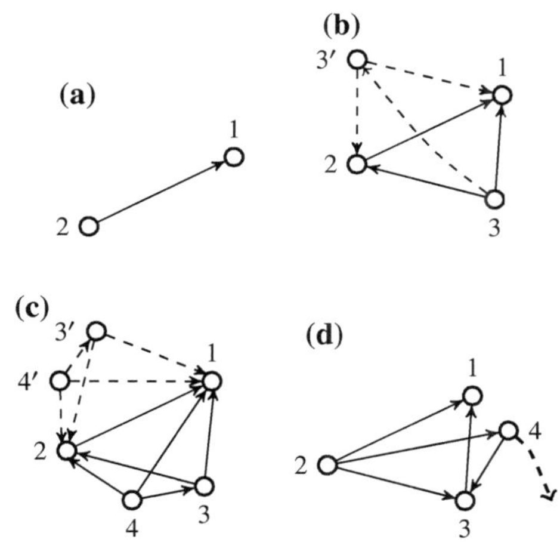

Fig. 2.3 Constraint consistent and rigidity in underlying topology: **a**, **b**, and **c** are constraint consistent, but **d** is not constraint consistent

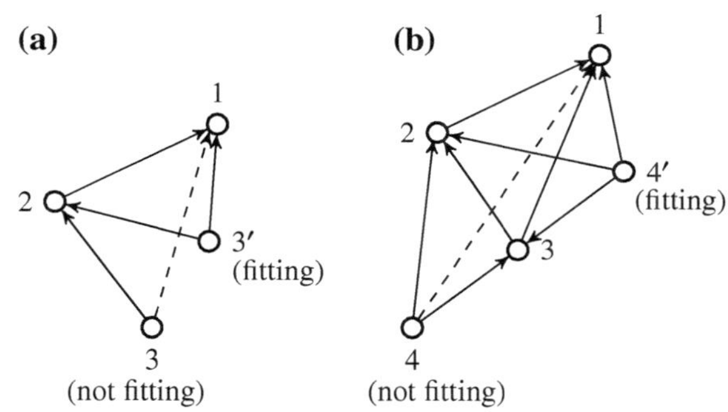

Fig. 2.4 Fitting in directed graphs. In **a**, node 3 has two desired constraints, and in **b**, node 4 has three desired constraints

Definition 2.5 ([19]) The position of agent i is called *fitting* to the constraints if there is no more additional active edge leaving agent i that can be achieved by modifying the position of agent i without loosing any existing active edge, while keeping the positions of other agents.

In Fig. 2.4a, suppose that the agent 3 has two desired constraints. At the position 3 (not fitting), it has one active edge constraint and at the position 3′ (fitting), it satisfies the two constraints. But, at position 3, it only satisfies the distance constraint between agents 2 and 3. So, the position 3 is not fitting. Let us consider agent 4 has three constraints as shown in Fig. 2.4b where the position 4′ satisfies the three constraints. But, at position 4, it only satisfies two distance constraints with respect to agents 2 and 3. Thus, the position 4 is not fitting while the position 4′ is fitting. The concept of *fitting* can be formally defined as [19]:

Definition 2.6 If there is no point $p_i^\dagger$ in the same dimension for agent i such that the following strict inclusion holds

$$\{(i,j)^{\bar{e}} \in \vec{\mathcal{E}} : \|p_i - p_j\| = d_{ij}^*,\ j \in \mathcal{N}_i^O\} \subset \{(i,j)^{\bar{e}} \in \vec{\mathcal{E}} : \|p_i^\dagger - p_j\| = d_{ij}^*,\ j \in \mathcal{N}_i^O\}$$

then the current position p_i of agent i is fitting.

The above definition means that the fitting position p_i of agent i attempts to satisfy as many as distance constraints, under the circumstance that other agents do not move. But, in the above definition, the set $\{(i,j)^{\bar{e}} \in \vec{\mathcal{E}} : \|p_i - p_j\| = d_{ij}^*,\ j \in \mathcal{N}_i^O\}$ is a subset of the set $\{(i,j)^{\bar{e}} \in \vec{\mathcal{E}} : \|p_i^\dagger - p_j\| = d_{ij}^*,\ j \in \mathcal{N}_i^O\}$. Thus, for example, if a position p_i^1 satisfies two distance constraints $d_{ij_1}^*$ and $d_{ij_2}^*$, with no other distance constraint, then the position p_i^1 is fitting. However, if another position p_i^2 satisfies one more distance $d_{ij_3}^*$ along with $d_{ij_1}^*$ and $d_{ij_2}^*$, then the position p_i^1 is not fitting; instead, the new position p_i^2 is fitting. But another new position p_i^3 satisfies only two distance constraints $d_{ij_1}^*$ and $d_{ij_3}^*$, then it is fitting. Consequently, we can see that the fitting position p_i of agent i tries to satisfy as many as distance constraints in a local sense rather than a global sense.

Similarly to the framework in undirected graphs, let us call the triplet $\vec{f}_p = (\mathcal{V}, \vec{\mathcal{E}}, p)$ a framework of digraphs. The fitting concept implies that the realized formation configuration cannot be deformed without changing the graph topology when all the agents are in fitting positions. The constraint consistence now can be defined using the fitting concept:

Definition 2.7 Given a digraph $\vec{\mathcal{G}} = (\mathcal{V}, \vec{\mathcal{E}})$, the graph with a realization p is constraint consistent if there exists $\epsilon > 0$ such that any framework $\vec{f}_{p'}$ with the realization p' fitting for the distance constraints $\bar{d}$ induced by the framework $\vec{f}_p = (\vec{\mathcal{G}}, p)$ and satisfying $\|p' - p\| < \epsilon$ is a realization of $\bar{d}$.

The above definition means that when a framework $\vec{f}_p$ is constraint consistent, each agent would satisfy as many as distance constraints, and any framework with fitting realization p' satisfying $\|p' - p\| < \epsilon$ would also have the same constraints. However, the constraint consistence is defined in a sequential way, i.e., an agent would satisfy the constraints as many as possible assuming that other agents are stationary. If agents move together, the number of constraints might be larger than the case of a constraint consistence. The persistence also can be defined using the fitting concept:

Definition 2.8 Given a digraph $\vec{\mathcal{G}} = (\mathcal{V}, \vec{\mathcal{E}})$, the graph with a realization p is persistent if there exists $\epsilon > 0$ such that any framework $\vec{f}_{p'}$ with the realization p' fitting for the distance constraints induced by the framework $\vec{f}_p = (\vec{\mathcal{G}}, p)$ and satisfying $\|p' - p\| < \epsilon$ is congruent to $\vec{f}_p$.

Based on the above definitions, we can see that the *persistence* is a combination of *constraint consistence* and *rigidity* of underlying topology. Let us consider Fig. 2.3d. The agent 4 only needs to satisfy a single distance constraint between agents 4 and 3. So, it can rotate around agent 3. But agent 2 needs to satisfy three constraints with regard to agents 1, 3, and 4. But, as agent 4 rotates, there could be a certain case where agent 2 cannot satisfy the three constraints. So, the underlying constraints for the given graph topology are not consistent. Thus, we can see that the persistence requires "consistence" in constraints and "rigidity" in underlying topology, which is summarized in the following lemma.

Lemma 2.6 ([19]) *A digraph is persistent if and only if it is rigid and constraint consistent.*

Similarly to the generic rigidity, the generic persistence can be defined and evaluated. A digraph is generically persistent if and only if it is generically rigid and generically constraint consistent. The generic rigidity can be checked by Laman lemma, Lemma 2.4. It seems difficult to find an exact condition for constraint consistent, but it is known that in $\mathbb{R}^2$, if all the nodes in a digraph have out-degree less than or equal to 2, then it is constraint consistent. Since the underlying topology is rigid, the sum of degrees of freedom of all nodes of a persistent graph is less than or equal to 3 in $\mathbb{R}^2$, i.e., translations and rotation. Since it is constraint consistent if all the nodes have out-degree less than or equal to 2, we can obtain a necessary and sufficient condition for the persistence in the following sense:

Theorem 2.5 ([19]) *A graph is (generically) persistent if and only if the underlying topology, which is obtained by removing outgoing edges from the nodes with out-degree larger than* 2 *until all the nodes have out-degree smaller than or equal to* 2, *is rigid.*

Definition 2.9 A graph is (generically) minimally persistent if we cannot remove any edge without breaking the persistence property.

From the above definition, we see that the number of edges in minimally persistent graphs is minimum. Also, we know that if all the nodes have out-degree less than or equal to 2, it is constraint consistent. Thus, to make the graph to be rigid, we also need to have a smallest number of edges as like the rigidity property of undirected graphs. So, we can make the following results:

Theorem 2.6 ([19]) *A graph with underlying topology being rigid is minimally persistent if and only if*

- *Three vertices have out-degree* 1 *and all other vertices have out-degree* 2.
- *Or, one vertex has no outgoing edge and another node has only an outgoing edge, and all other nodes have out-degree* 2.

Figure 2.5 shows an example of minimally persistent graph composed of seven vertices in $\mathbb{R}^2$. Figure 2.5 includes a cycle, i.e., $5 \rightarrow 6 \rightarrow 7 \rightarrow 5$. The conditions for the minimally persistent graph also have been scrutinized as [36]:

- (Leader-first-follower): The first agent (leader) has no outgoing edge. The second agent (first follower) has only one outgoing edge to the first agent. All other agents have two outgoing edges.

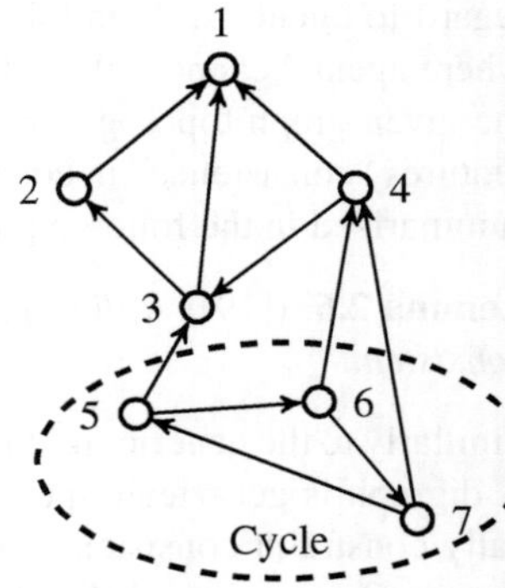

Fig. 2.5 An example of minimally persistent graphs. There is a cycle in nodes 5, 6, 7

- (Leader-remote-follower): The first agent (leader) has no outgoing edge. The second agent, known as the remote follower, has only one outgoing edge to an agent other than the first agent. All other agents have two outgoing edges.
- (Coleader): Three agents, known as coleaders, have one outgoing edges. All other agents have two outgoing edges.

As noted in [36], the leader-first-follower structure formation can be cyclic or acyclic (cycle-free), while the leader-remote-follower and coleader cases are cyclic. A minimally persistent formation with the acyclic leader-first-follower structure is called *acyclic minimally persistent formation.*

The cycle-free persistent formation can be obtained as follows:

Theorem 2.7 ([19]) *A cycle-free (acyclic) graph is persistent if and only if*

- *A leader node has no outgoing edge.*
- *A first follower has an outgoing edge to the leader.*
- *All other nodes have out-degree greater than or equal to* 2.

Unlike Theorem 2.6, in the definition of cycle-free persistent graph, the nodes are sequentially added. That is, in Theorem 2.7, the leader node is given, and the first follower is added to have an outgoing edge to the leader. Then, another node is added with two edges to the leader and the first follower. Then, another node is added to the graph composed of nodes 1, 2, and 3. For the remaining nodes, the same procedure is repeated. Thus, it is important to see that the condition for cycle-free (acyclic) persistent graphs is a kind of sequential concepts. When the out-degrees of agents except the leader and first follower are exactly two, under the cycle-free sequential setup as the Theorem 2.7, it is called *acyclic minimally persistent formation* in $\mathbb{R}^2$:

Theorem 2.8 ([19]) *An acyclic graph having more than one vertex is minimally persistent if and only if*

- *One vertex (leader) has an out-degree* 0.
- *Another vertex (follower) has an out-degree* 1 *and the corresponding edge is incident to the leader.*
- *Every other vertex has an out-degree* 2.

It is important to mention that the graph in Fig. 2.5 satisfies the conditions of Theorem 2.7, but the graph in Fig. 2.5 is not generated in a sequential way. Thus, there could exist cycles in Fig. 2.5 even with satisfying the conditions of Theorem 2.7. In $\mathbb{R}^3$, it is *acyclic minimally persistent formation* if the second follower has out-degree 2, and other agents have out-degree 3 under the same setup as the Theorem 2.7.

For higher dimensional space, we can have the following result.

Theorem 2.9 ([38]) *A framework with digraph* $\vec{\mathcal{G}} = (\mathcal{V}, \vec{\mathcal{E}})$ *is persistent in* $\mathbb{R}^d$, $d = 2, 3, \ldots$ *if and only if the underlying topology is rigid and the constraints are consistent. Also, the digraph* $\vec{\mathcal{G}}$ *is generically persistent in* $\mathbb{R}^d$, $d = 2, 3, \ldots$ *if and only if the underlying topology is generically rigid and the constraints are generically consistent.*

It is also known that a graph is generically constraint consistent if all the nodes have out-degree less than or equal to d in $\mathbb{R}^d$. Thus, with Theorem 2.9, we can evaluate the persistence of a graph by examining the rigidity property of underlying topology as well as the constraint consistent property by checking the degrees of the nodes. Starting from a minimally persistent graph, it was shown that there are Henneberg-style primitive operations for generating all minimally persistent graphs [20].

2.2.3 Bearing Rigidity

The basics of bearing rigidity are adopted from [10, 14, 40, 41]. Given two frameworks $f_p = (\mathcal{V}, \mathcal{E}, p)$ and $f_q = (\mathcal{V}, \mathcal{E}, q)$, where the realizations p and q are expressed with respect to a global coordinate frame, if the following constraint is satisfied

$$(p_i - p_j)^{\perp}(q_i - q_j) = 0, \ \forall (i, j)^e \in \mathcal{E} \tag{2.2}$$

where $(\cdot)^{\perp}$ is a rotated and transposed vector by $\pi/2$, then the two frameworks are called *parallel*. The *parallel* in (2.2) means that the edges of the framework f_p corresponding to the edges of f_q are parallel to each other. Since the edges of the framework under the aligned orientation setup are bearing vectors, the constraint (2.2) plays an essential role in the definition of the bearing rigidity. In the literature, that is the reason why the bearing rigidity has been also called *parallel rigidity*. Given a realization p for a graph $\mathcal{G}$, if it is translated or dilated (translations and dilations are trivial motions in bearing-based setup under the assumption that the orientations of agents have been aligned), then all the edges are moving in parallel. To define the bearing rigidity, taking a derivative for (2.2), we can have

$$(p_i - p_j)^{\perp}(u_i - u_j) = 0, \ \forall (i, j)^e \in \mathcal{E} \tag{2.3}$$

where $u_i = \dot{q}_i$, which is similar to (2.1). Then, (2.3) can be compactly written as

$$\mathbb{R}_B \dot{q} = 0, \ \forall (i, j)^e \in \mathcal{E} \tag{2.4}$$

where $\mathbb{R}_B$ is called bearing rigidity matrix [14].

Definition 2.10 ([10]) A framework f_p is bearing (parallel) rigid if all the parallel motions of f_p are trivial.

As mentioned already, there are two trivial translational motions along the two independent axes in a 2-dimensional Euclidean space and one trivial dilation motion for a bearing rigid graph. Since there are $2n$ motions for a graph composed of n agents, the maximum rank of $\mathbb{R}_B$ will be $2n - 3$.

Theorem 2.10 ([10]) *A framework $f_p = (\mathcal{V}, \mathcal{E}, p)$ is bearing rigid if rank$(\mathbb{R}_B) = 2n - 3$ in 2-dimensional space.*

As mentioned in [10], if we consider the formation topologically, the condition in the above theorem is also a necessary condition [41]. For the concepts in general d-dimensional space, with the notations (9.1) and (9.2), we define the bearing edge function as

$$\mathbf{h}_{\mathcal{G}_B} = \underbrace{(\ldots, g_{ji}^T, \ldots)^T}_{=|\mathcal{E}|} \in \mathbb{R}^{dm} \tag{2.5}$$

where $g_{ji} = \frac{p_j - p_i}{\|p_j - p_i\|}$. Then, the bearing rigidity matrix is also obtained as

$$\mathbb{R}_B = \frac{\partial \mathbf{h}_{\mathcal{G}_B}}{\partial p} = \frac{\partial \mathbf{h}_{\mathcal{G}_B}}{\partial z} \frac{\partial z}{\partial p} = \text{blkdg}\left(\frac{\mathbb{P}_{g_{ji}}}{\|z_{ji}\|} \right) (\mathbb{H} \otimes \mathbb{I}_d) \in \mathbb{R}^{dm \times dn} \tag{2.6}$$

The rigidity matrix in (2.6) has the same physical meaning as the rigidity matrix in (2.4), but the detailed outlooks may be different. In $\mathbb{R}^d$, the *infinitesimal bearing rigidity* is defined as follows.

Definition 2.11 ([41]) A framework $f_p = (\mathcal{V}, \mathcal{E}, p)$ is infinitesimally bearing rigid if all the infinitesimal motions are trivial (translations and dilations).

Note that infinitesimal bearing rigidity is an algebraical concept. Given a graph $\mathcal{G} = (\mathcal{V}, \mathcal{E})$, when the edge set is replaced by the edges of a complete graph $K(n)$, the bearing rigidity matrix $\mathbb{R}_B$ is denoted as $\mathbb{R}_{K,B}$. The bearing equivalence and bearing congruence are also similarly defined as in the equivalence and congruence in distance rigidity. The two frameworks f_p and f_q are said to be *bearing equivalent* if $\mathbb{P}_{p_i - p_j}(q_i - q_j) = 0, \ \forall (i, j)^e \in \mathcal{E}$ and called *bearing congruent* if $\mathbb{P}_{p_i - p_j}(q_i - q_j) = 0, \ \forall i, j \in \mathcal{V}$, where $\mathbb{P}$ is the orthogonal projection matrix (see (9.1)). If the bearing equivalence implies the bearing congruence, then it is called *globally bearing rigid*. Then, the bearing rigidity conditions and globally bearing rigidity conditions are given as follows.

Theorem 2.11 ([41]) *A framework $f_p = (\mathcal{V}, \mathcal{E}, p)$ is globally bearing rigid in $\mathbb{R}^d$ if and only if rank$(\mathbb{R}_{K,B}) =$ rank$(\mathbb{R}_B)$, and it is bearing rigid if and only if it is globally bearing rigid.*

The above theorem looks quite strong since the local bearing rigidity and global bearing rigidity are equivalent. But it is not difficult to see the reason why they are equivalent. For example, in $\mathbb{R}^2$, a position of a point will be unique if it has two bearing vector constraints. That is, when we add constraints to a point by bearing vectors, there is no ambiguity like the constraints under the distance setup. This is also true in higher dimensional space.

Note that the infinitesimal bearing rigidity implies the bearing rigidity and global bearing rigidity, but the converse is not true in general. Similarly to the distance rigidity, let a framework $f_p = (\mathcal{V}, \mathcal{E}, p)$ be generic if all frameworks of the points in a neighborhood of p are bearing rigid or not bearing rigid. This definition can be rephrased as follows.

Definition 2.12 A realization in a framework is called generic realization if and only if all the infinitesimal bearing motions of the framework are trivial.

Then, with the bearing rigidity matrix, we can make a condition for bearing rigidity for generic realizations.

Theorem 2.12 *The framework $f_p = (\mathcal{V}, \mathcal{E}, p)$ is infinitesimally bearing rigid in $\mathbb{R}^d$ if and only if rank$(\mathbb{R}_B) = dn - d - 1$.*

2.2.4 Weak Rigidity

The weak rigidity attempts to find conditions for unique realization of a graph when it has constraints of subtended angles, which was introduced in [31]. The main idea is to replace some of distance constraints by the corresponding-subtended angle constraints. For example, let us see Fig. 2.6. The triangular framework of Fig. 2.6a is constrained by three interagent distances; so it is rigid. Let us replace the distance constraints d_{23} by α_{23}^1, and d_{12} by α_{13}^2 as depicted in Fig. 2.6b. Although the triangle in Fig. 2.6b is constrained by one distance and two subtended angles, it is rigid.

Let a graph $\mathcal{G}$ have distance constraints and subtended angle constraints, which are determined by the set of edges $\mathcal{E}$ and set of subtended angles $\mathcal{A}$. Also, let a subtended

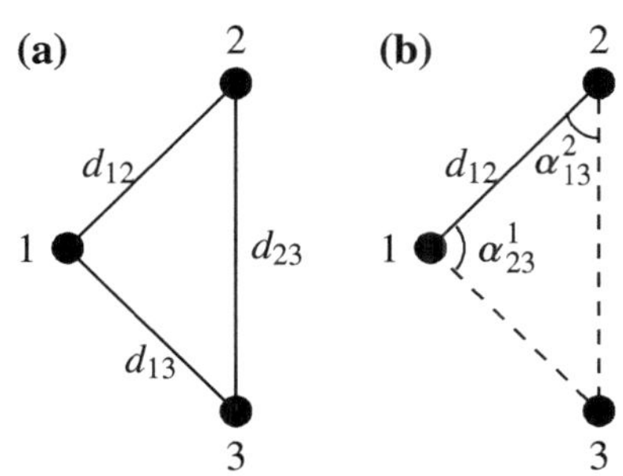

Fig. 2.6 Triangular formation: **a** is a distance rigid graph, and **b** is a weakly rigid graph

angle constraint be denoted as α^i_{jk}, where j and k are adjacent neighboring agents of agent i. The subtended angle constraints are topologically written as $(i, \{j, k\})$. For the definition of weak rigidity, we use the equivalence concept, with subtended angle constraints, as follows:

Definition 2.13 Consider two frameworks $f_p = (\mathcal{V}, \mathcal{E}, \mathcal{A}, p)$ and $f_q = (\mathcal{V}, \mathcal{E}, \mathcal{A}, q)$. The two frameworks are called *strongly equivalent* if

- $\|p_i - p_j\| = \|q_i - q_j\|, \forall (i, j)^e \in \mathcal{E}$.
- $\alpha(p)^i_{jk} = \alpha(q)^i_{jk}, \forall (i, \{j, k\}) \in \mathcal{A}$.

Now, with the above definition, we can generate a formal concept of weak rigidity.

Definition 2.14 A framework $f_p = (\mathcal{V}, \mathcal{E}, \mathcal{A}, p)$, with $|\mathcal{V}| = n, |\mathcal{E}| = m_e$, and $|\mathcal{A}| = m_a$, is weakly rigid if there exists a neighborhood $D_p \in \mathbb{R}^{2n}$ of p such that any framework $f_q = (\mathcal{V}, \mathcal{E}, \mathcal{A}, q)$, where $q \in D_p$, which is strongly equivalent to f_p is congruent to f_p. If the domain D_p is an entire domain as $D_p = \mathbb{R}^{2n}$, it is called globally weakly rigid.

To examine a relationship between weak rigidity and distance rigidity, let us focus on a graph $\mathcal{G} = (\mathcal{V}, \mathcal{E}, \mathcal{A})$, where the subtended angle constraints α^i_{jk} are given to the two adjacent edges $(i, j)^e$ and $(i, k)^e$. The graphs categorized in this form are called type-1 graphs.

Definition 2.15 Given an angle constraint $(i, \{j, k\})$, if there exist edges as $(i, j)^e$ and $(i, k)^e$, such graph is called type-1 graph. If there does not exist either edge $(i, j)^e$ or $(i, k)^e$, it is a more general graph. Such a graph, which is further generalized from the type-1-graph, is called type-2 graph.

Let us produce a graph without subtended angle constraints from the type-1 graph $\mathcal{G} = (\mathcal{V}, \mathcal{E}, \mathcal{A})$ such as given below.

Definition 2.16 Given a type-1 graph $\mathcal{G} = (\mathcal{V}, \mathcal{E}, \mathcal{A})$ with subtended angle constraints, an induced graph $\mathcal{G}^o = (\mathcal{V}^o, \mathcal{E}^o)$ is produced as follows:

- $\mathcal{V}^o = \mathcal{V}$.
- $\mathcal{E}^o = \{(i, j)^e | (i, j)^e \in \mathcal{E} \cup \{(i', j')^e \text{ if } (k, \{i', j'\}) \in \mathcal{A}\}\}$.

Theorem 2.13 *In $\mathbb{R}^2$, a framework $f_p = (\mathcal{V}, \mathcal{E}, \mathcal{A}, p)$ of a type-1 graph is weakly rigid if and only if $(\mathcal{G}^o, p)$ is rigid.*

Proof (Sufficient) Let $(\mathcal{G}^o, p)$ be rigid. Then, for two realizations p and q, we know that $(\mathcal{G}^o, p)$ and $(\mathcal{G}^o, q)$ are congruent if they are equivalent. For two frameworks $f_p = (\mathcal{V}, \mathcal{E}, \mathcal{A}, p)$ and $f_p = (\mathcal{V}, \mathcal{E}, \mathcal{A}, q)$, if they are strongly equivalent with properties of Definition 2.13, it is easy to see that $\|p_j - p_k\| = \|q_j - q_k\|$ from the cosine rule:

$$\begin{aligned}
\|p_{jk}\|^2 &= \|p_{ij}\|^2 + \|p_{ik}\|^2 - 2\|p_{ij}\|\|p_{ik}\| \cos \alpha^i_{jk} \\
&= \|q_{ij}\|^2 + \|q_{ik}\|^2 - 2\|q_{ij}\|\|q_{ik}\| \cos \alpha^i_{jk} \\
&= \|q_{jk}\|^2
\end{aligned}$$

where $p_{ij} = p_i - p_j$ and $q_{ij} = q_i - q_j$. Consequently, two frameworks $f_p = (\mathcal{V}, \mathcal{E}, \mathcal{A}, p)$ and $f_p = (\mathcal{V}, \mathcal{E}, \mathcal{A}, q)$ are congruent, which means that it is weakly rigid by the definition.

(Necessary) Let us suppose that a framework $f_p = (\mathcal{V}, \mathcal{E}, \mathcal{A}, p)$ is weakly rigid. Then, in D_p, any framework with a realization q which is strongly equivalent is congruent to f_p. Let us consider the induced graph $\mathcal{G}^o$. Then, due to the congruence of $\mathcal{G}$, we have $\|p_i - p_j\| = \|q_i - q_j\|, \forall (i, j)^e \in \mathcal{E}^o$, which is the equivalence condition in $\mathcal{G}^o$. Further, this equivalence property implies that $\alpha(p)^i_{jk} = \alpha(q)^i_{jk}$ for all $(i, \{j, k\}) \in \mathcal{A}$ due to the inverse cosine rule. Therefore, we can see that the two frameworks $(\mathcal{G}^o, p)$ and $(\mathcal{G}^o, q)$ are congruent in D_p, which completes the proof.

It is noticeable that the framework $(\mathcal{G}^o, p)$ of the type-1 graphs is rigid if it is infinitesimally rigid, which can be easily checked by the rigidity matrix of $(\mathcal{G}^o, p)$, assuming all the realizations are generic. It is important to remark that the generic property used here is the same as the generic property of distance rigidity in Sect. 2.2.1. In fact, when a graph has constraints both in distances and subtended angles, a new generic property may have to be defined for a rigorous mathematical concept. If it is restricted to generic realizations in the sense of distance rigidity concept, we can make the following result.

Theorem 2.14 *Let us consider a framework $f_p = (\mathcal{V}, \mathcal{E}, \mathcal{A}, p)$ of a type-1 graph in $\mathbb{R}^2$, where p is generic. The framework f_p is weakly rigid if and only if a rigidity matrix determined from the induced framework $(\mathcal{V}, \mathcal{E}^o, p)$ has rank $2n - 3$.*

Proof The proof is direct from the Theorem 2.13.

Now, based on the above theorem, and based on [41], we can provide the following remark:

Remark 2.1 It would be important to clarify the relationships between distance rigidity, persistence, bearing rigidity, and weak rigidity. In $\mathbb{R}^2$, we can see that if limited to generic realizations, then a framework is infinitesimally bearing rigid and weakly rigid if and only if it is infinitesimally distance rigid. Thus, for the type-1 graphs, it is infinitesimally bearing rigid if and only if it is infinitesimally weakly rigid.

Next, let us further generalize the above weak rigidity concepts to general graphs with general constraints. To this aim, we first define a generalized edge function in $\mathbb{R}^2$, taking account of the distance, bearing, and subtended angles constraints, as

$$\mathbf{h}_{\mathcal{G}_{dab}} = (\underbrace{\ldots, \|z_{ji}\|^2, \ldots}_{=m_d}\underbrace{\ldots, \alpha^k_{ji}, \ldots}_{=m_a}\underbrace{\ldots, g^T_{ji}, \ldots}_{=m_b})^T \in \mathbb{R}^{m_d+m_a+2m_b} \tag{2.7}$$

where $m_d + m_a + m_b = m$. Then, the generalized rigidity matrix is defined as

$$\mathbb{R}_{\mathcal{G}_{dab}} = \frac{\partial \mathbf{h}_{\mathcal{G}_{dba}}}{\partial p} \in \mathbb{R}^{(m_d+m_a+2m_b)\times 2n} \tag{2.8}$$

If the null space of $\mathbb{R}_{\mathcal{G}_{dab}}$ is equivalent to the set of infinitesimal trivial motions of the generalized graph, that is constrained by distance, bearing, and subtended angles constraints, it may be considered to be infinitesimally generalized rigid. Since it is out of the scope of this monograph to derive a fully generalized condition, let us narrow down the above idea to the distance and subtended angle constraints. Under this setup, we define a generalized edge function:

$$\mathbf{h}_{\mathcal{G}_{da}} = (\mathbf{h}_{\mathcal{G}_d}^T, \mathbf{h}_{\mathcal{G}_a}^T)^T = (\underbrace{\ldots, \|z_{ji}\|^2, \ldots}_{=m_d}\underbrace{\ldots, \alpha_{ji}^k, \ldots}_{=m_a})^T \in \mathbb{R}^m \tag{2.9}$$

where $m = m_d + m_a$. Then, we specifically define a generalized weak rigidity matrix as

$$\mathbb{R}_{\mathcal{G}_{da}} = \frac{\partial \mathbf{h}_{\mathcal{G}_{da}}}{\partial p} = \begin{bmatrix} \frac{\partial \mathbf{h}_{\mathcal{G}_d}}{\partial p} \\ \frac{\partial \mathbf{h}_{\mathcal{G}_a}}{\partial p} \end{bmatrix} \in \mathbb{R}^{m \times 2n} \tag{2.10}$$

For the definition of a generalized weak rigidity, we define the trivial motions of the framework $f_p = (\mathcal{V}, \mathcal{E}, \mathcal{A}, p)$ as follows.

Definition 2.17 The trivial motions of the framework $f_p = (\mathcal{V}, \mathcal{E}, \mathcal{A}, p)$ are translations, and rotations if $m_d \neq 0$. If $m_d = 0$, the dilations are also one of the trivial motions.

Let us formally define the generalized infinitesimal weak rigidity for a generalized graph where subtended angle constraints can be given to two neighboring nodes, without distance constraints. That is, for the angle constraint α_{ij}^k, there could be no distance constraints in the edges $(k, i)^e$ and $(k, j)^e$. Such generalized graphs are called the type-2 graphs as defined in Definition 2.15. Then, the infinitesimal weak rigidity for the type-2 graphs can be characterized by infinitesimal motions such as follows.

Definition 2.18 A framework $f_p = (\mathcal{V}, \mathcal{E}, \mathcal{A}, p)$ is infinitesimally weakly rigid if all the infinitesimal motions are trivial motions.

Given a type-2 graph $\mathcal{G}$, let us define a generalized induced graph $\mathcal{G}' = (\mathcal{V}', \mathcal{E}')$ as [27]:

- $\mathcal{V}' = \mathcal{V}$.
- $\mathcal{E}' = \{(i, j)^e | (i, j)^e \in \mathcal{E}\} \cup \{(i', j')^e \text{ if } (k, \{i', j'\}) \in \mathcal{A}\} \cup \{(i, k)^e \cup (j, k)^e \text{ if } \exists k$ such that $(k, \{i, j\}) \in \mathcal{A}\}$.
- $\mathcal{A}' = \mathcal{A}$.

It is clear to see that $m_d' = |\mathcal{E}'| \geq |\mathcal{E}| = m_d$. The displacement vectors z_{ij}' of the edges in $\mathcal{E}'$ are written as $z_{ij}' = z_k'$, where $k = 1, \ldots, m_d'$, and for simplicity, we order the vectors as $z_k = z_k'$ for $k = 1, \ldots, m_d$. Then, for the generalized induced graph, we can define an incidence matrix $\mathbb{H}_+'$ with arbitrary directions of the edges. Then, we can have the following relationship:

$$z' = (\mathbb{H}'_+ \otimes \mathbb{I}_2)p \tag{2.11}$$

which changes the generalized weak rigidity matrix of (2.10) as

$$\mathbb{R}_{\mathcal{G}_{da}} = \begin{bmatrix} \frac{\partial \mathbf{h}_{\mathcal{G}_d}}{\partial z'} \\ \frac{\partial \mathbf{h}_{\mathcal{G}_a}}{\partial z'} \end{bmatrix} \frac{\partial z'}{\partial p} = \begin{bmatrix} \frac{\partial \mathbf{h}_{\mathcal{G}_d}}{\partial z'} \\ \frac{\partial \mathbf{h}_{\mathcal{G}_a}}{\partial z'} \end{bmatrix} \mathbb{H}'_+ \otimes \mathbb{I}_2 \tag{2.12}$$

In (2.9) and (2.10), the subtended angles are directly considered as elements of the generalized edge function. But the angles α^k_{ji} are obtained as

$$\alpha^k_{ji} = \cos^{-1}\left(\frac{\|p_{kj}\|^2 + \|p_{ki}\|^2 - \|p_{ji}\|^2}{2\|p_{kj}\|\|p_{ki}\|}\right) \tag{2.13}$$

So, it is quite intractable to use the above angles as the constraints in the edge function. Thus, for a convenience of derivation, we use $c^k_{ji} \triangleq \cos\alpha^k_{ji}$ as the elements of the edge function. So, we rewrite (2.9) as

$$\mathbf{h}_{\mathcal{G}_{da}} = (\mathbf{h}^T_{\mathcal{G}_d}, \mathbf{h}^T_{\mathcal{G}_a})^T = (\underbrace{\ldots, \|z_{ji}\|^2, \ldots}_{=m_d}\underbrace{\ldots, c^k_{ji}, \ldots}_{=m_a})^T \in \mathbb{R}^m \tag{2.14}$$

With the above definition and concept, we now make the following main lemma for generalized infinitesimally weak rigidity [26]:

Lemma 2.7 *Given a framework $f_p = (\mathcal{V}, \mathcal{E}, \mathcal{A}, p)$ of a type-2 graph, let us compute the rigidity matrix $\mathbb{R}_{\mathcal{G}_{da}}$ as* (2.10) *with the edge function* (2.14). *Then,*

- *if $\mathcal{E} \neq \emptyset$, span$\{\mathbf{1}_n \otimes \mathbb{I}_2, (\mathbb{I}_n \otimes \mathbb{J}_2)p\} \subseteq$ null$(\mathbb{R}_{\mathcal{G}_{da}})$ and rank$(\mathbb{R}_{\mathcal{G}_{da}}) \leq 2n - 3$.*
- *otherwise, i.e., $\mathcal{E} = \emptyset$, span$\{\mathbf{1}_n \otimes \mathbb{I}_2, (\mathbb{I}_n \otimes \mathbb{J}_2)p, p\} \subseteq$ null$(\mathbb{R}_{\mathcal{G}_{da}})$ and rank$(\mathbb{R}_{\mathcal{G}_{da}})$ $\leq 2n - 4$.*

where $\mathbb{J}_2 = \begin{bmatrix} 0 & -1 \\ 1 & 0 \end{bmatrix}$ and $\mathbb{I}_n \otimes \mathbb{J}_2$ is the perpendicular rotation operator, which rotates the realization p by $\pi/2$ around the origin.

Proof It is easy to see that the column vectors of $\mathbf{1}_n \otimes \mathbb{I}_2 \in \mathbb{R}^{2n\times 2}$, the vector $(\mathbb{I}_n \otimes \mathbb{J}_2)p \in \mathbb{R}^{2n}$, and the vector $p \in \mathbb{R}^{2n}$ are linearly independent if p is not collocated. Thus, these vectors can be considered as bases of the span. From (2.12), since $\mathbb{H}'_+$ is the incidence matrix, it is clear that the vector $\mathbf{1}_n \otimes \mathbb{I}_2$ is in the null space of $\mathbb{R}_{\mathcal{G}_{da}}$. Also, using $(\mathbb{H}'_+ \otimes \mathbb{I}_2)(\mathbb{I}_n \otimes \mathbb{J}_2) = (\mathbb{H}'_+ \otimes \mathbb{J}_2) = (\mathbb{I}_{m'_d}\mathbb{H}'_+ \otimes \mathbb{J}_2\mathbb{I}_2) = (\mathbb{I}_{m'_d} \otimes \mathbb{J}_2)(\mathbb{H}'_+ \otimes \mathbb{I}_2)$, we can write (2.12) as

$$
\begin{aligned}
\mathbb{R}_{\mathcal{G}_{da}}(\mathbb{I}_n \otimes \mathbb{J}_2)p &= \begin{bmatrix} \frac{\partial \mathbf{h}_{\mathcal{G}_d}}{\partial z'} \\ \frac{\partial \mathbf{h}_{\mathcal{G}_a}}{\partial z'} \end{bmatrix} (\mathbb{H}'_+ \otimes \mathbb{I}_2)(\mathbb{I}_n \otimes \mathbb{J}_2)p \\
&= \begin{bmatrix} \frac{\partial \mathbf{h}_{\mathcal{G}_d}}{\partial z'} \\ \frac{\partial \mathbf{h}_{\mathcal{G}_a}}{\partial z'} \end{bmatrix} (\mathbb{I}_{m'_d} \otimes \mathbb{J}_2)(\mathbb{H}'_+ \otimes \mathbb{I}_2)p \\
&= \begin{bmatrix} \frac{\partial \mathbf{h}_{\mathcal{G}_d}}{\partial z'} \\ \frac{\partial \mathbf{h}_{\mathcal{G}_a}}{\partial z'} \end{bmatrix} (\mathbb{I}_{m'_d} \otimes \mathbb{J}_2)z' \\
&= \begin{bmatrix} \frac{\partial \mathbf{h}_{\mathcal{G}_d}}{\partial z'} \\ \frac{\partial \mathbf{h}_{\mathcal{G}_a}}{\partial z'} \end{bmatrix} \begin{bmatrix} \mathbb{J}_2 z'_1 \\ \vdots \\ \mathbb{J}_2 z'_{m'_d} \end{bmatrix}
\end{aligned} \tag{2.15}
$$

where $\frac{\partial \mathbf{h}_{\mathcal{G}_d}}{\partial z'} \in \mathbb{R}^{m_d \times m'_d}$ and $\frac{\partial \mathbf{h}_{\mathcal{G}_a}}{\partial z'} \in \mathbb{R}^{m_a \times m'_d}$. Let us consider $\frac{\partial \mathbf{h}_{\mathcal{G}_a}}{\partial z'}$ first. The elements of $\mathbf{h}_{\mathcal{G}_a}$ are c^k_{ji}; for a convenience, the edge vectors z'_{ik}, z'_{jk}, and z'_{ij} defining c^k_{ji} are denoted as z'_a, z'_b, and z'_c, respectively, where $a, b, c \in \{1, \ldots, m'_d\}$. Also, let the $h-$th cosine function c^k_{ji} be denoted as c_h, where $h \in \{1, \ldots, m_a\}$. We can consider that the h-th cosine function c_h is a function of z'_a, z'_b, and z'_c. Then, we can compute $\frac{\partial c_h}{\partial z'} \in \mathbb{R}^{1 \times m'_d}$ that is a row vector with all elements being zero except the $a-$, $b-$, and $c-$th elements. It is noted that

$$
\begin{aligned}
\left[\frac{\partial \mathbf{h}_{\mathcal{G}_a}}{\partial z'} \right] \begin{bmatrix} \mathbb{J}_2 z'_1 \\ \vdots \\ \mathbb{J}_2 z'_{m'_d} \end{bmatrix} &= \frac{\partial c_h}{\partial z'_a}\mathbb{J}_2 z'_a + \frac{\partial c_h}{\partial z'_b}\mathbb{J}_2 z'_b + \frac{\partial c_h}{\partial z'_c}\mathbb{J}_2 z'_c \\
&= \frac{\partial c_h}{\partial \|z'_a\|}\frac{\partial \|z'_a\|}{\partial z'_a}\mathbb{J}_2 z'_a + \frac{\partial c_h}{\partial \|z'_b\|}\frac{\partial \|z'_b\|}{\partial z'_b}\mathbb{J}_2 z'_b + \frac{\partial c_h}{\partial \|z'_c\|}\frac{\partial \|z'_c\|}{\partial z'_c}\mathbb{J}_2 z'_c \\
&= \frac{\partial c_h}{\partial \|z'_a\|}\frac{z'^T_a}{\|z'_a\|}\mathbb{J}_2 z'_a + \frac{\partial c_h}{\partial \|z'_b\|}\frac{z'^T_b}{\|z'_b\|}\mathbb{J}_2 z'_b + \frac{\partial c_h}{\partial \|z'_c\|}\frac{z'^T_c}{\|z'_c\|}\mathbb{J}_2 z'_c = 0
\end{aligned} \tag{2.16}
$$

Consequently, $(\mathbb{I}_n \otimes \mathbb{J}_2)p$ is in the null space of the generalized weak rigidity matrix $\mathbb{R}_{\mathcal{G}_{da}}$. Furthermore, due to $\frac{\partial \|z_i\|}{\partial z_i}\mathbb{J}_2 z_i = \frac{z_i^T}{\|z_i\|}\mathbb{J}_2 z_i = 0$, we also have

$$
\left[\frac{\partial \mathbf{h}_{\mathcal{G}_d}}{\partial z'} \right] \begin{bmatrix} \mathbb{J}_2 z'_1 \\ \vdots \\ \mathbb{J}_2 z'_{m'_d} \end{bmatrix} = 0 \tag{2.17}
$$

Therefore, if $\mathcal{E} \neq \emptyset$, we can see that $\text{span}\{\mathbf{1}_n \otimes \mathbb{I}_2, (\mathbb{I}_n \otimes \mathbb{J}_2)p\} \subseteq \text{null}(\mathbb{R}_{\mathcal{G}_{da}})$, which implies $\text{rank}(\mathbb{R}_{\mathcal{G}_{da}}) \leq 2n - 3$.

Next, let us consider the case of $\mathcal{E} = \emptyset$. We can compute the nonzero elements of $\frac{\partial c_h}{\partial z'_i}$, where $i = \{a, b, c\}$, as

$$\frac{\partial c_h}{\partial z'_a} z'_a = \frac{\|z'_a\|^2 - \|z'_b\|^2 + \|z'_c\|^2}{2\|z'_a\|\|z'_b\|} \tag{2.18}$$

$$\frac{\partial c_h}{\partial z'_b} z'_b = \frac{\|z'_b\|^2 - \|z'_a\|^2 + \|z'_c\|^2}{2\|z'_b\|\|z'_a\|} \tag{2.19}$$

$$\frac{\partial c_h}{\partial z'_c} z'_c = -\frac{\|z'_c\|^2}{\|z'_b\|\|z'_a\|} \tag{2.20}$$

In this case, the rigidity matrix becomes $\mathbb{R}_{\mathcal{G}_{da}} = \mathbb{R}_{\mathcal{G}_a} = \frac{\partial \mathbf{h}_{\mathcal{G}_a}}{\partial z'}(\mathbb{H}'_+ \otimes \mathbb{I}_2)$. Thus, summing up the right-hand sides of (2.18)–(2.20), we have

$$\begin{aligned}
\mathbb{R}_{\mathcal{G}_a} p &= \frac{\partial \mathbf{h}_{\mathcal{G}_a}}{\partial z'}(\mathbb{H}'_+ \otimes \mathbb{I}_2) p \\
&= \frac{\partial \mathbf{h}_{\mathcal{G}_a}}{\partial z'} z' \\
&= \frac{\partial c_h}{\partial z'_a} z'_a + \frac{\partial c_h}{\partial z'_b} z'_b + \frac{\partial c_h}{\partial z'_c} z'_c \\
&= 0
\end{aligned} \tag{2.21}$$

Thus, the point p is in the null space of $\mathbb{R}_{\mathcal{G}_a}$, which completes the poof.

With the above lemma, we now make the main results:

Theorem 2.15 *Consider a framework $f_p = (\mathcal{V}, \mathcal{E}, \mathcal{A}, p)$ of a type-2 graph, with $\mathcal{E} \neq \emptyset$, in $\mathbb{R}^2$. It is generalized infinitesimally weakly rigid if and only if rank$(\mathbb{R}_{\mathcal{G}_{da}}) = 2n - 3$. As a special case, if $\mathcal{E} = \emptyset$, the framework is generalized infinitesimally weakly rigid if and only if rank$(\mathbb{R}_{\mathcal{G}_{da}}) = 2n - 4$.*

Proof When $\mathcal{E} \neq \emptyset$, the basis vectors $\mathbf{1}_n \otimes \mathbb{I}_2$ and $(\mathbb{I}_n \otimes \mathbb{J}_2)p$ correspond to the translations and rotations. Thus, by Definitions 2.17 and 2.18, the proof is clear. Also, when $\mathcal{E} = \emptyset$, since the basis vector p corresponds to the dilations, the proof is also obvious.

It is remarkable that there are some further results on formation stabilizations for a generalized triangular system in [26] and for generalized infinitesimally weakly rigid graphs in [23].

2.3 Consensus

Given a set of agents $\dot{x}_i = u_i$, $x_i \in \mathbb{R}^d$, let us define a consensus set as

$$\mathcal{E}_{dn} \triangleq \{[\xi_1^T, \ldots, \xi_n^T]^T \in \mathbb{R}^{dn} : \xi_i = \xi_j, \forall i, j \in \{1, \ldots, n\}\}$$

Let agents be updated by local interactions as

$$u_i(t) = \sum_{j \in \mathcal{N}_i} w_{ij}(t)(x_j - x_i)$$

where the weights w_{ij} are time-variant. Then, the overall agents can be updated as

$$\dot{x} = -(\mathbb{L}(t) \otimes \mathbb{I}_d)x(t) \tag{2.22}$$

where $\mathbb{L}(t)$ is the time-variant Laplacian matrix. The time-variant digraph $\mathcal{G}(t)$ is called uniformly connected if, for any $t \geq t_0$, there exist a finite time T such that the graph has a directed rooted tree, i.e., arborescence, at the time interval $[t, T]$.

Theorem 2.16 ([29]) *Given a directed graph with Laplacian dynamics* (2.22)*, if the underlying graph is uniformly connected, then the agents converge to the set* $\mathcal{E}_{dn}$.

If the graph is directed, then the above theorem does not ensure a convergence to the average of initial values. For fixed digraphs, the following necessary and sufficient condition is developed:

Theorem 2.17 ([33]) *Given a fixed directed graph with Laplacian dynamics* (2.22)*, i.e.,* $w_{ij}(t) = w_{ij}$*, if the underlying graph is strongly connected and each node is balanced, then the agents converge to the average of the initial values.*

2.4 Basics of Control Theory

2.4.1 Essential Concepts for Nonlinear Control

In nonlinear control, to analyze a convergence of a dynamic system to a point, the set of equilibrium points of the system needs to be compact. The compactness of a set also plays a key role to ensure a convergence of formation control systems to a point. The following concepts are mainly summarized from [21, 25, 30, 34].

A metric space is a pair of objects $(X, d(x, y))$, where X is a set and d is a metric, and $x, y \in X$, satisfying basic axioms (positive, strictly positive, symmetry, and triangle inequality). Given a function $f(x)$ defined in a metric space X, which could be $\mathbb{R}$, the property of *continuously differentiable* of the function is essential for a stability analysis.

Definition 2.19 (*Continuity*) Let us assume that there exist $f(x)$ and $f(x_1)$, where $x, x_1 \in \mathbb{R}$. Then, the function $f(x) : \mathbb{R} \to \mathbb{R}$ is continuous on $x \in [0, \infty)$ if, for any $x_1 \geq 0$ and for any $\epsilon > 0$, there is a real number $\delta(\epsilon, x_1) > 0$ such that, for all $x \geq 0$, $|f(x) - f(x_1)| < \epsilon$, whenever $|x - x_1| < \delta$. It is called uniformly continuous if for any $\epsilon > 0$, there is a real number $\delta(\epsilon) > 0$ such that, for all $x \geq 0$, $|f(x) - f(x_1)| < \epsilon$, whenever $|x - x_1| < \delta$.

Definition 2.20 (*Differentiable*) A function $f(x) : D \to \mathbb{R}$, where $D \subset \mathbb{R}$, is differentiable at $x \in D$ if the limit $f'(x) \triangleq \lim_{z \to x} \frac{f(z)-f(x)}{z-x}$, which is also called *derivative*, exists. Further, if the function $f(x) : D \to \mathbb{R}$ is differentiable at every point $x \in D$, it is called differentiable on D.

Note that the *differentiable* implies *continuity*, but the converse is not true, in general.

Definition 2.21 (*Continuously differentiable*) A function $f(x) : D \to \mathbb{R}$, where $D \subset \mathbb{R}$, is continuously differentiable on D if it is differentiable and the derivative $f' : D \to \mathbb{R}$ is continuous.

The concept of convergence also needs to be defined precisely.

Definition 2.22 (*Convergence*) A sequence of real numbers $\{x_n\} \triangleq \{x_1, x_2, \ldots, \}$ is convergent if for any $\epsilon > 0$, there are a real number x_0 and an integer $N(\epsilon)$ such that $|x_n - x_0| < \epsilon$, wherever $n \geq N(\epsilon)$.

Definition 2.23 (*Convergence in metric space*) A sequence of points in a metric space (X, d) is called convergent if there is a point x_0 in (X, d) and if, for any real $\epsilon > 0$, there is an integer $N(\epsilon)$ such that $d(x_n, x_0) < \epsilon$, wherever $n \geq N(\epsilon)$.

Definition 2.24 (*Cauchy sequence*) A sequence $\{x_n\}$ in a metric space (X, d) is said to be a Cauchy sequence if, for each $\epsilon > 0$, there exists an integer N such that $d(x_{m_1}, x_{m_2}) \leq \epsilon$ for any choice of $m_1, m_2 \geq N$.

The *convergence* means that there exists a point in the metric space to which the sequence approaches. But, in the concept of Cauchy sequence, as the index of sequence increases, if the distance between two elements in the sequence is close enough, it is called *Cauchy*. Thus, a convergent sequence is a Cauchy, but a Cauchy in X does not imply a *convergent* in X. By merging these concepts, we can have a clear concept for a metric space as follows.

Definition 2.25 (*Complete*) A metric space (X, d) is said to be *complete* if each Cauchy sequence in (X, d) is a convergent sequence.

With the concept of *complete*, we can now narrow a metric space as follows.

Definition 2.26 (*Banach space*) A normed metric space (X, d) is called Banach space if it is complete.

Thus, a Banach space is a space having a norm as its metric, with a complete property. If a Banach space has a set of inner products, it is called Hilbert space. An equilibrium of dynamic systems is a set that does not have any hole. A space without any hole may be characterized by a concept of *compactness*. To further narrow a complete space to a compact space, we also need to have the following concepts:

Definition 2.27 (*Sequentially compact*) A metric space (X, d) is called sequentially compact if every sequence in (X, d) contains a convergent subsequence.

The above concept of *sequentially compact* means not only *complete* but something more, which is characterized by *totally boundedness*:

Definition 2.28 (*Totally bounded*) A subset A_ϵ of A is called ϵ-net of A if (1) A_ϵ is finite and (2) for each $x \in A$, there is a $y \in A_\epsilon$ such that $d(x, y) \leq \epsilon$. A set A is called *totally bounded* if for each $\epsilon > 0$, A contains an ϵ-net.

Theorem 2.18 *A metric space (X, d) is sequentially compact if and only if it is totally bounded and complete.*

Note that a compactness in metric space is equivalent to the concept of sequentially compactness. If we limit the space into real space, a compactness of a set can be characterized quite straightforwardly. But, for this, we need to be careful in the concepts of closedness and boundedness.

Definition 2.29 (*A point of adherence*) Let A be a subset of a metric space (X, d). A point x in (X, d) is called *a point of adherence* of A if each open set of (X, d) containing x also contains a point y in A.

The above definition means that each local neighborhood of x contains a point y of A. Let $\bar{A}$ denote the set of all points of adherence of A, which is called the *closure* of A.

Theorem 2.19 *The following statements are equivalent:*

- *A set A in a metric space (X, d) is closed.*
- *A set A is equal to its closure, i.e., $A = \bar{A}$.*
- *Every convergent sequence $\{x_n\}$ with $\{x_n\} \subset A$ has its limit in A.*

Definition 2.30 (*Bounded*) A subset $E \in \mathbb{R}$ is *bounded* above if there exists $\beta \in \mathbb{R}$ such that $x \leq \beta$ for every $x \in E$.

Theorem 2.20 *A set $A \in \mathbb{R}$ is compact if and only if it is closed and bounded. Also, a set $A \in \mathbb{R}^n$ (or $\mathbb{C}^n$) in a general dimensional space is compact if and only if it is closed and bounded.*

Theorem 2.21 *Let $f : X \to Y$ be a continuous function, where (X, d) and (Y, d) are metric spaces. If (X, d) is compact, then the range $f(x)$ is a compact set in (Y, d).*

2.4.2 LaSalle's Invariance Principle and Barbalat's Lemma

Let us consider the following autonomous systems:

$$\dot{x}(t) = f(x) \tag{2.23}$$

where $t \geq t_0$, where t_0 denotes the initial time. In the stability analysis of an equilibrium point of the above system, the concept of invariance set plays a key role.

Definition 2.31 ([35]) A set $\mathcal{U}$ is an invariant set for the dynamic systems (2.23) if every trajectory starting from a point in $\mathcal{U}$ stays in $\mathcal{U}$ for all $t \geq t_o$.

With the above concept, the LaSalle's invariance principle is stated as follows.

Theorem 2.22 ([24]) *Let $\Omega \subset D$ be a compact set that is invariant for the dynamic systems* (2.23). *Let $V : D \to \mathbb{R}$ be a continuously differentiable function such that $\dot{V}(x) \leq 0$ in Ω. Further, let $\mathcal{E} \triangleq \{x | \dot{V}(x) = 0,\ x \in \Omega\}$. If M is the largest invariant set in $\mathcal{E}$, then every solution starting in Ω approaches M as $t \to \infty$.*

Next, let us consider nonautonomous systems of the following form:

$$\dot{x}(t) = f(x, t) \tag{2.24}$$

The counterpart of LaSalle's invariance principle in nonautonomous systems can be produced by the Barbalat's lemma.

Lemma 2.8 ([24]) *Let $\phi(t) : \mathbb{R} \to \mathbb{R}$ be a uniformly continuous function on* $[0, \infty)$. *Suppose that* $\lim_{t\to\infty} \int_0^t \phi(\tau)d\tau$ *exists and is finite. Then, $\phi(t) \to 0$ as $t \to \infty$.*

With the help of Barbalat's lemma, the following theorem can be obtained:

Theorem 2.23 ([24]) *Let $D \subset \mathbb{R}^n$, and let D contain $x = 0$. Suppose that the function $f(x, t)$ in* (2.24) *is piecewise continuous in t and locally Lipschitz in x, uniformly in t. Further suppose that the function $f(t, 0)$ is uniformly bounded for all $t \geq 0$. Let us assume that there exists a continuously differentiable function $V : D \times [0, \infty) \to \mathbb{R}$ such that*

$$W_1(x) \leq V(t, x) \leq W_2(x)$$
$$\dot{V}(x, t) = \tfrac{\partial V}{\partial t} + \tfrac{\partial V}{\partial x} f(x, t) \leq -W_3(x)$$

where $W_1(x)$ and $W_2(x)$ are continuous positive definite functions and $W_3(x)$ is a continuous positive semi-definite function on D. Then, the trajectory of (2.24) *is bounded and satisfies $W_3(x) \to 0$ as $t \to \infty$ for all initial states $x(t_0)$ starting within the set $\mathcal{X}_0 \triangleq \{x \in B_r | W_2(x) \leq \rho\}$, where $B_r \in D$ is a ball with radius r and $\rho < \min_{\|x\|=r} W_1(x)$. Furthermore, if all the assumptions hold globally and $W_1(x)$ is radially unbounded, then $W_3(x) \to 0$ for all initial states.*

2.4.3 Input-to-State Stability

The following results are borrowed from [3, 4, 24].

Assumption 2.4.1 Let M be an n-dimensional C^2 (twice differentiable) and orientable Riemannian manifold without boundary, and let $f : M \times D \to T_x M$ be a C^1 (differentiable)-Lipschitz function and D be a closed subset of $\mathbb{R}^m$.

Let us consider the following nonlinear systems:

$$\dot{x}(t) = f(x, d(t)) \tag{2.25}$$

where $x \in M$, and $x = x_0$ at $t = t_0$ and $d, d \in D$, is a bounded disturbance. Without disturbance, the system (2.25) becomes a nominal system:

$$\dot{x}(t) = f(x(t), 0) \triangleq f_0(x) \tag{2.26}$$

Assumption 2.4.2 There exists a nonnegative C^1 function $V : M \to \mathbb{R}$ along the trajectory of (2.26) such that

$$\dot{V} = \frac{dV}{dx}\frac{dx}{dt} = \frac{dV}{dx}\dot{x}(t) < 0, \quad \forall x \in M : f_0(x) \neq 0 \tag{2.27}$$

Assumption 2.4.3 Let any equilibrium x^u which is not asymptotically stable, be isolated. Further, the Jacobian of $f_0(x)$ at x^u, i.e., $\frac{\partial f_0(x)}{\partial x}|_{x=x^u}$ has, at least, one eigenvalue with strictly positive real part.

Let the solution of (2.25) be denoted as $x(t, t_0, d(t))$. Then, we can define the almost globally input-to-state stability for the system (2.25).

Definition 2.32 ([4, Definition 1]) The system (2.25) is said to be almost globally input-to-state stable (ISS) with respect to a compact subset $M^o \subset M$ if M^o is locally asymptotically stable for $d = 0$ and there exists $\tilde{\gamma} \in \mathcal{K}$, which is the class $\mathcal{K}$ function [24], such that for each locally essentially bounded and measurable perturbation $d(t)$, there exists a zero volume set $\tilde{\mathcal{B}}_d \subset M$ such that, for all $x \in M \setminus \tilde{\mathcal{B}}_d$, the following inequality holds

$$\lim_{t\to+\infty} \sup \sigma(x(t, t_0, d(t)), M^o) \leq \tilde{\gamma}(\|d\|_\infty)$$

where $\sigma(x, M^o)$ denotes the Riemannian distance between x and $y \in M^o$.

Lemma 2.9 ([4, Proposition 3]) *Suppose that there exists a C^1 function $W : M \to \mathbb{R}_{\geq 0}$ along the trajectory of* (2.25) *such that*

$$\dot{W} = \frac{dW}{dx}\frac{dx}{dt} = \frac{dW}{dx} f(x, d(t)) \leq -\alpha(W(x)) + c + \delta(\|d\|), \tag{2.28}$$

for all $x \in M$ and all $d \in D$, where $\alpha > 0$, $c > 0$, and $\delta(\|d\|)$ is the class $\mathcal{K}$ function. Then, the states of system (2.25) *are ultimately bounded.*

Lemma 2.10 ([4, Proposition 2]) *For the systems* (2.25)*, let us suppose that Assumptions 2.4.1–2.4.3 are satisfied. Also, we suppose that the set of asymptotically stable equilibria of* (2.26)*, denoted by $\mathcal{X}_{eq}$, is finite. Then, if the ultimate boundedness property holds as per Lemma 2.9, then the original system* (2.25) *is almost globally ISS with respect to the set $\mathcal{X}_{eq}$.*

With the above lemmas, we can now have convergence properties for the cascade systems:

$$\dot{x}(t) = f(x, y) \tag{2.29}$$
$$\dot{y}(t) = g(y) \tag{2.30}$$

with appropriate dimensions in x and y.

Lemma 2.11 ([3, Theorem 2]) *Consider the cascade systems* (2.29)–(2.30) *with state* $z = (x^T, y^T)^T$, $x \in M$ *and* $y \in N$, *where the domains* M *and* N *satisfy the Assumption 2.4.1. Further assume that* f *and* g *satisfy* $f(x_{eq}, y_{eq}) = 0$ *and* $g(y_{eq}) = 0$, *for some points* $x_{eq} \in M$ *and* $y_{eq} \in N$. *Let the subsystem* (2.29) *be almost globally ISS with respect to the equilibrium* x_{eq} *with the input* y *and the subsystem* (2.30) *be almost globally asymptotically stable at* y_{eq}. *Then, the cascade system* (2.29)–(2.30) *is almost globally asymptotically stable at* $z_{eq} = (x_{eq}^T, y_{eq}^T)^T$.

Let us consider nonautonomous and cascade systems

$$\dot{x}(t) = f(t, x, y(t)) \tag{2.31}$$
$$\dot{y}(t) = g(t, y(t)) \tag{2.32}$$

where $f : [0, \infty) \times \mathbb{R}^n \times \mathbb{R}^m \rightarrow \mathbb{R}^n$ and $g : [0, \infty) \times \mathbb{R}^m \rightarrow \mathbb{R}^m$, and the nominal systems of (2.31) when $y(t) = 0$ as

$$\dot{x}(t) = f_0(t, x) \tag{2.33}$$

Lemma 2.12 ([24, Lemma 4.6]) *Let the vector field function* $f(t, x, y(t))$ *of the system* (2.31) *be continuously differentiable and globally Lipschitz in* (x, y), *uniformly in* t. *If the nominal system* (2.33) *is globally exponentially stable at* x_{eq}, *then* (2.31) *is input-to-state stable with the input* $y(t)$.

Lemma 2.13 ([24, Lemma 4.7]) *Let us suppose that the system* (2.31) *is input-to-state stable as per Lemma 2.12, and the origin of the system* (2.32) *is globally uniformly asymptotically stable. Then, the origin of the cascade system* (2.31)–(2.32) *is globally uniformly asymptotically stable.*

Figure 2.7 depicts the overall idea of the input-to-state stability of autonomous systems. The nonlinear dynamics $\dot{x} = f(x, 0)$ needs to be globally asymptotically stable, and $\dot{x} = f(x, d(t))$ needs to be ultimately bounded, both of which implies that $\dot{x} = f(x, d(t))$ is almost globally input-to-state stable. Then, replacing $d(t)$ by y, if $\dot{y} = g(y)$ is almost globally asymptotically stable, then we can say that $\dot{x} = f(x, y)$ is almost globally asymptotically stable. Figure 2.8 depicts the overall idea of the input-to-state stability of nonautonomous systems. In this case, we require a little stronger property in $\dot{x} = f(t, x, y)$, which should be globally exponentially stable when $y = 0$. Then, the overall cascade system will be globally uniformly asymptotically stable.

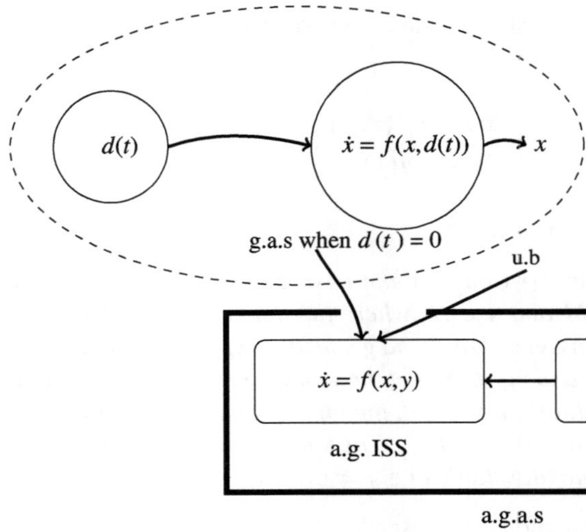

Fig. 2.7 The main idea of almost global input-to-state stability (ISS) and almost global asymptotic stability (a.g.a.s) of autonomous cascade systems (g.a.s: globally asymptotically stable. u.b.: ultimately bounded. a.g. ISS: almost globally input-to-state stable. a.g.a.s.: almost globally asymptotically stable)

Fig. 2.8 The main idea of input-to-state stability of nonautonomous cascade systems (g.e.s: globally exponentially stable. g.u.a.s.: globally uniformly asymptotically stable)

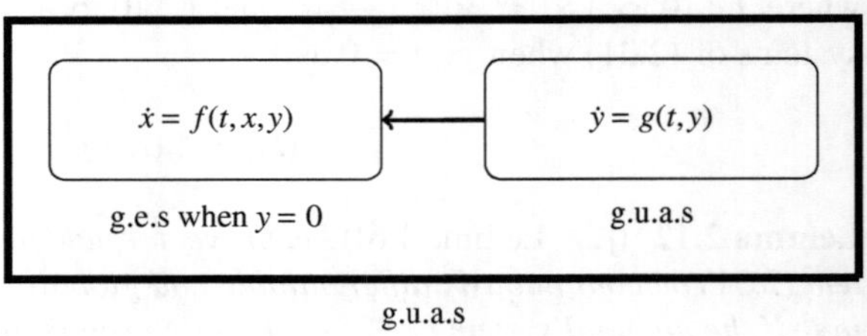

2.4.4 *Finite-Time Stability*

The contents of this section are adopted from [8, 12, 16, 17]. As a background of the finite-time stability analysis, consider the following differential inclusion:

$$\dot{x} \in F(x) \tag{2.34}$$

where $F(x)$ is a Filippov set value map $F(x) : D \to \mathbb{B}(\mathbb{R}^n)$, and $D \subseteq \mathbb{R}^n$ and $\mathbb{B}(\mathbb{R}^n)$ denotes a collection of subsets of $\mathbb{R}^n$ [16]. Next, consider the differential equation:

$$\dot{x} = f(x(t)) \tag{2.35}$$

where $f : \mathbb{R}^n \to \mathbb{R}^n$ is measurable and locally essentially bounded function. Denoting a ball with center at x, with radius δ, as $B_\delta(x)$, the solution $x(t) : [t_0, t_f) \to \mathbb{R}^n$

of (2.35) is called Filippov solution if $x(t)$ satisfies the differential inclusion (2.34) almost all $t \in [t_0, t_f)$, where $F(x)$ is defined as

$$F(x) = \mathcal{K}[f](x) \triangleq \bigcap_{\delta>0} \bigcap_{\mu(\chi)=0} \text{co}\{f(B_\delta(x)) \setminus \chi\} \tag{2.36}$$

where $\mu(\cdot)$ is the Lebesgue measure in $\mathbb{R}^n$, co denotes the convex closure, $\bigcap_{\mu(\chi)=0}$ denotes the intersection over all sets χ of Lebesgue measure zero [16]. When the domain of existence of the Filippov solution is maximal, it is a maximal solution. A set M is weakly invariant (strongly invariant) for (2.35) if for each $x_0 \in M$, M contains a maximal solution (all maximal solutions) of (2.35) [12].

Given a function $V(p) : \mathbb{R}^n \to \mathbb{R}$, denoting $\text{Critical}(V(p)) \triangleq \{p \in \mathbb{R}^n : \nabla_p V(p) = 0\}$ and $\underline{\lambda} = \min\{\lambda | \lambda > 0,\ \lambda(A)\}$ for a positive semi-definite matrix A, for the following gradient update

$$\dot{p} = -\frac{\nabla_p V(p)}{\|\nabla_p V(p)\|} \tag{2.37}$$

$$\dot{p} = -\text{sign}(\nabla_p V(p)) \tag{2.38}$$

the convergence properties were analyzed in [12] as follows.

Lemma 2.14 *Let a function $V(p) : \mathbb{R}^n \to \mathbb{R}$ be a second-order differentiable function. Let $x_0 \in S \subset \mathbb{R}^n$, where S is compact and strongly invariant for* (2.37) *(respectively, for* (2.38)*). Let us suppose that there exists a neighborhood $U_{Critical(V(p))}$ of Critical$(V(p)) \cap S$, in which either (i) for all $p \in U_{Critical(V(p))}$, the Hessian, $H_{V(p)}$, of $V(p)$ is positive semi-definite, or (ii) for all $p \in U_{Critical(V(p))} \setminus (Critical(V(p)) \cap S)$, $H_{V(p)}$ is positive semi-definite, the multiplicity of* 0 *eigenvalue is constant, and $\nabla_p V(p)$ is orthogonal to the eigenspace of $H_{V(p)}$ corresponding to* 0*. Then,*

- *Each solution of* (2.37) *(or* (2.38)*) converges to a critical point of $V(p)$ in a finite time.*
- *Moreover, if $U_{Critical(V(p))} = S$, the convergence time of the solution starting from p_0 is upper bounded as $1/\lambda_0 \|\nabla_p V(p_0)\|$ (or $1/\lambda_0 \|\nabla_p V(p_0)\|_1$ for* (2.38)*) where $\|\cdot\|_1$ is* 1*-norm and $\underline{\lambda}^* = \min_{p \in S} \underline{\lambda}(H_{V(p)})$.*

Lemma 2.15 ([8]) *Consider $\dot{p} = f(p(t))$. Let us suppose that there exists a positive definite and continuous function $V(p) : D \to \mathbb{R}$ such that*

$$\dot{V} + kV^\alpha \leq 0,\ \ p \in D^o \setminus \{0\} \tag{2.39}$$

where D^o is an open neighborhood as $D^o \subseteq D$, $k > 0$, and $\alpha \in (0, 1)$, and 0 *is an equilibrium point. Then, the equilibrium point is finite-time stable with the settling time $T \leq V(p(t_0))^{1-\alpha}/k(1-\alpha)$. Moreover, if $D = \mathbb{R}^n$, it is globally finite-time stable.*

2.5 Notes

The mathematical background of this chapter is quite common; so many reference sources might have been missed. But some results are newly added. For examples, the results of Sect. 2.2.4 are reused from [26, 27, 31]. Thus, the following copyright and permission notices are acknowledged explicitly.

- The statement of Theorem 2.13 with the proof is © [2017] IEEE. Reprinted, with permission, from [31].
- The statements of Lemma 2.7 and Theorem 2.15 with the proofs are © [2018] IEEE. Reprinted, with permission, from [27].

The infinitesimal rigidity of formation is decided by rigidity matrix $\mathbb{R}$, and the connectivity of a graph is determined by Laplacian matrix $\mathbb{L}$. There are some analogy between these two matrices. Based on the rigidity matrix $\mathbb{R}$, let us define a symmetric rigidity matrix $\mathbb{M} = \mathbb{R}^T \mathbb{R}$ [39], which has the same null space as the rigidity matrix $\mathbb{R}$. In 2-dimensional space, the matrix $\mathbb{M}$ has three zero eigenvalues with three corresponding eigenvectors. The three eigenvectors represent the three independent trivial motions of the realized formation in 2-dimensional space, i.e., translations along the x- and y-axes and a rotation around the virtual z-axis. The Laplacian matrix has a simple zero eigenvalue if and only if the network is connected. As the second eigenvalue of $\mathbb{L}$ is used to determine the connectivity of the graph, the fourth eigenvalue of $\mathbb{M}$ is also used to determine the rigidity of realized formation. It is well known that the bigger the second eigenvalue of $\mathbb{L}$ is, the faster the convergence speed of the graph to a consensus is. Similarly, the bigger the fourth eigenvalue of $\mathbb{M}$ is, the larger the degree of rigidity becomes. Consequently, by enlarging the fourth eigenvalue of $\mathbb{M}$, the realized formation would become more rigid [37]. There are several interesting open research works in this direction. For example, given a rigid framework, or rigid architecture, we may be able to analyze the degree of rigidity of the structure by analyzing the fourth eigenvalue of the rigid matrix $\mathbb{M}$. The physical interpretations of eigenvalues and eigenvectors of $\mathbb{M}$ would be also another interesting topic, related with the concepts of equilibrium force and resolvable force [37].

References

1. Agnarsson, G., Greenlaw, R.: Graph Theory: Modeling, Applications, and Algorithms. Pearson Education International, Upper Saddle River (2007)
2. Anderson, B.D.O., Yu, C., Fidan, B., Hendrickx, J.M.: Rigid graph control architectures for autonomous formations. IEEE Control Syst. Mag. **28**(6), 48–63 (2008)
3. Angeli, D.: An almost global notion of input to state stability. IEEE Trans. Autom. Control **49**(6), 866–874 (2004)
4. Angeli, D., Praly, L.: Stability robustness in the presence of exponentially unstable isolated equilibria. IEEE Trans. Autom. Control **56**(7), 1582–1592 (2011)
5. Aspnes, J., Eren, T., Goldenberg, D.K., Morse, A.S., Whiteley, W., Yang, Y.R., Anderson, B.D.O., Belhumeur, P.N.: A theory of network localization. IEEE Trans. Mob. Comput. **5**(12), 1663–1678 (2006)

6. Beineke, L.W., Wilson, R.J.: Topics in Algebraic Graph Theory. Cambridge University Press, Cambridge (2007)
7. Berg, A.R., Jordán, T.: A proof of Connelly's conjecture on 3-connected circuits of the rigidity matroid. J. Comb. Theory Ser. B **88**, 77–97 (2003)
8. Bhat, S.P., Bernstein, D.S.: Finite-time stability of continuous autonomous systems. SIAM J. Control Optim. **38**(3), 751–766 (2000)
9. Biggs, N.: Algebraic Graph Theory. Cambridge University Press, Cambridge (1974)
10. Bishop, A.N., Shames, I., Anderson, B.D.O.: Stabilization of rigid formations with direction-only constraints. In: Proceedings of the 50th IEEE Conference on Decision and Control and European Control Conference, pp. 746–752 (2011)
11. Chartrand, G., Zhang, P.: Introduction to Graph Theory. McGraw-Hill, New York City (2005)
12. Cortes, J.: Finite-time convergent gradient flows with applications to network consensus. Automatica **42**(11), 1993–2000 (2006)
13. Cvetković, D., Rowlinson, P., Simić, S.: Eigenspaces of Graphs. Cambridge University Press, Cambridge (1997)
14. Eren, T., Whiteley, W., Morse, A.S., Belhumeur, P.N., Anderson, B.D.O.: Sensor and network topologies of formations with direction, bearing, and angle information between agents. In: Proceedings of the 42nd IEEE Conference on Decision and Control, pp. 3064–3069 (2003)
15. Graver, J., Servatius, B., Servatius, H.: Combinatorial Rigidity. American Mathematical Society, Providence (1993)
16. Haddad, W.M., Sadikhov, T.: Dissipative differential inclusions, set-valued energy storage and supply rate maps, and stability of discontinuous feedback systems. Nonlinear Analysis: Hybrid Systems, pp. 83–108 (2013)
17. Haimo, V.T.: Finite time controllers. SIAM J. Control Optim. **24**(4), 760–770 (1986)
18. Hendrickson, B.: Conditions for unique graph realizations. SIAM J. Comput. **21**(1), 65–84 (1992)
19. Hendrickx, J.M., Anderson, B.D.O., Delvenne, J.-C., Blondel, V.D.: Directed graphs for the analysis of rigidity and persistence in autonomous agent systems. Int. J. Robust Nonlinear Control **17**(10–11), 960–981 (2007)
20. Hendrickx, J.M., Fidan, B., Yu, C., Anderson, B.D.O., Blondel, V.D.: Formation reorganization by primitive operations on directed graphs. IEEE Trans. Autom. Control **53**(4), 968–979 (2008)
21. Heuser, H.G.: Functional Analysis. Wiley, New York (1982)
22. Jackson, B., Jordán, T.: Connected rigidity matroids and unique realizations of graphs. J. Comb. Theory Ser. B **94**(1), 1–29 (2005)
23. Jing, G., Zhang, G., Lee, H.W.J., Wang, L.: Weak rigidity theory and its application to multi-agent formation stabilization, pp. 1–26. arXiv:1804.02795 [cs.SY]
24. Khalil, H.K.: Nonlinear Systems. Prentice Hall, Upper Saddle River (2002)
25. Kreyszig, E.: Introductory Functional Analysis with Applications. Wiley, New York (1978)
26. Kwon, S.-H., Trinh, M.H., Oh, K.-H., Zhao, S., Ahn, H.-S.: Infinitesimal weak rigidity, formation control of three agents, and extension to 3-dimensional space, pp. 1–12. arXiv:1803.09545
27. Kwon, S.-H., Trinh, M.H., Oh, K.-H., Zhao, S., Ahn, H.-S.: Infinitesimal weak rigidity and stability analysis on three-agent formations. In: Proceedings of the SICE Annual Conference, pp. 266–271 (2018)
28. Laman, G.: On graphs and the rigidity of plane skeletal structuress. J. Eng. Math. **4**(4), 331–340 (1970)
29. Moreau, L.: Stability of multiagent systems with time-dependent communication links. IEEE Trans. Autom. Control **50**(2), 169–182 (2005)
30. Naylor, A.W., Sell, G.R.: Linear Operator Theory in Engineering and Science. Springer, New York (1982)
31. Park, M.-C., Kim, H.-K., Ahn, H.-S.: Rigidity of distance-based formations with additional subtended-angle constraints. In: Proceedings of the International Conference on Control, Automation and Systems, pp. 111–116 (2017)
32. Roth, B.: Rigid and flexible frameworks. Am. Math. Mon. **88**(1), 6–21 (1981)

33. Saber, R.O., Murray, R.M.: Consensus problems in networks of agents with switching topology and time-delays. IEEE Trans. Autom. Control **49**(9), 1520–1533 (2004)
34. Schröder, B.S.W.: Mathematical Analysis - A Concise Introduction. Wiley, New York (2008)
35. Slotine, J.-J.E., Li, W.: Applied Nonlinear Control. Prentice-Hall International Inc., Upper Saddle River (2008)
36. Summers, T.H., Yu, C., Dasgupta, S., Anderson, B.D.O.: Control of minimally persistent leader-remote-follower and coleader formations in the plane. IEEE Trans. Autom. Control **56**(12), 2778–2792 (2011)
37. Trinh, M.-H., Park, M.-C., Sun, Z., Anderson, B.D.O., Pham, V.H., Ahn, H.-S.: Further analysis on graph rigidity. In: Proceedings of the 55th IEEE Conference on Decision and Control, pp. 922–927 (2016)
38. Yu, C., Hendrickx, J.M., Fidan, B., Anderson, B.D.O., Blondel, V.D.: Three and higher dimensional autonomous formations: rigidity, persistence and structural persistence. Automatica **43**(3), 387–402 (2007)
39. Zelazo, D., Franchi, A., Bülthoff, H.H., Giordano, P.R.: Decentralized rigidity maintenance control with range measurements for multi-robot systems. Int. J. Robot. Res. **34**(1), 105–128 (2015)
40. Zelazo, D., Franchi, A., Giordano, P.R.: Rigidity theory in SE(2) for unscaled relative position estimation using only bearing measurements. In: Proceedings of the European Control Conference, pp. 2703–2708 (2014)
41. Zhao, S., Zelazo, D.: Bearing rigidity and almost global bearing-only formation stabilization. IEEE Trans. Autom. Control **61**(5), 1255–1268 (2016)

Part II
Gradient Control Laws

Chapter 3
Global Stabilization

Abstract The distributed formation control can be studied from various approaches. In distance-based formation control, which uses relative displacements under misaligned orientations as sensing variables and distances as control variables, the most intuitive approach is the *gradient-based approach*. The gradient-based formation control laws use a potential function for generating local controllers for distributed agents. If the potential function is a function of distance errors that can be sensed in each agent, then the agent can implement a control law, which attempts to reduce the potential function, via a local coordinate frame. Thus, if the underlying topology ensures a unique configuration when the desired distances are satisfied, the desired formation can be considered as achieved. Since a gradient of the potential function has to lead a distributed formation controller, it is important to select an appropriate potential function. There are two solutions in gradient control laws. The first one is to stabilize the formation globally from any initial condition for some specific graphs. The second one is to stabilize the formation locally when extending to general n-agents, under a general rigidity topology. It is not possible to stabilize the formation to a desired one from any initial condition under gradient control laws. That is why the existing works consider specific formations for a global convergence. This chapter considers three-agent cases in 2-dimensional space, four-agent cases in 3-dimensional space, and polygon graphs for a global convergence. Although the control laws of this chapter use relative displacements as sensing variables and distances as control variables, there is no communication between neighboring agents.

3.1 Gradient Control Laws

The gradient-based formation control laws are well described in the works conducted in [12]. Along with the works in [12], there have been several creative works developed in [4–6, 9, 14]. To ensure a unique realization based on a characteristic of network topology, the concept of graph rigidity has been widely employed [15]. It is recommendable to refer to Sect. 2.2.1 for more background of graph rigidity. However, before these works, there are also some early works that foresee fundamental

H.-S. Ahn, *Formation Control*, Studies in Systems,
Decision and Control 205, https://doi.org/10.1007/978-3-030-15187-4_3

algebraic characteristics and control perspectives related to the graph rigidity [1, 3, 10, 20].

As aforementioned, in gradient control approaches, each agent attempts to reduce its own errors that are defined as a function of the distances between its neighboring agents. Thus, the underlying control topology of the networks is determined from graph rigidity for a uniqueness of realized configurations. Given the squared desired distance of edges $\bar{d}_{ij}^*$ and edge function $\mathbf{h}_{\mathcal{G}}(p)$, which can be also expressed by $\mathbf{h}_{\mathcal{G}}(z)$ as a function of z, the potential function is selected as

$$\phi(z) = \frac{1}{8}\|\mathbf{h}_{\mathcal{G}}(z) - \bar{d}^*\|^2 \tag{3.1}$$

where $z = (\ldots, z_k, \ldots)^T$ and $\bar{d}^* = (\ldots, \bar{d}_{ij}^*, \ldots)^T$. The set of desired distances can be given from a desired configuration. Then, taking the gradient of $\phi(z)$ as $\nabla\phi(z)$, the control law for each agent is designed as [12]

$$\begin{aligned} u_i &= -\nabla_{p_i}\phi(z) \\ &= -\left[\frac{\partial\phi(z)}{\partial p_i}\right]^T \\ &= -\sum_{j\in\mathcal{N}_i}\frac{1}{2}(\|z_{ij}\|^2 - \bar{d}_{ij}^*)z_{ij} = \sum_{j\in\mathcal{N}_i}\frac{1}{2}(\|z_{ij}\|^2 - \bar{d}_{ij}^*)z_{ji} \end{aligned} \tag{3.2}$$

Note that in the above control law, the component $\|z_{ij}\|^2 - \bar{d}_{ij}^*$ is a scalar; so it can be obtained by relative sensing in local coordinate frame. The vector z_{ij} is expressed in a global coordinate frame ${}^g\Sigma$. However, taking a coordinate transformation to the both sides of (3.2) from ${}^g\Sigma$ to the local coordinate frame ${}^i\Sigma$, it can be changed as

$$\begin{aligned} R_i^g u_i \triangleq u_i^i &= -\sum_{j\in\mathcal{N}_i}\frac{1}{2}(\|z_{ij}\|^2 - \bar{d}_{ji}^*)R_i^g z_{ij} \\ &= \sum_{j\in\mathcal{N}_i}\frac{1}{2}(\|z_{ij}\|^2 - \bar{d}_{ij}^*)z_{ji}^i \end{aligned} \tag{3.3}$$

Using ranging measurement sensors, the agent i can obtain the scalar components $\|z_{ij}\|^2 - \bar{d}_{ij}^*$ as well as the vector components z_{ij}^i with respect to its own local coordinate frame. Summing these scalars and vector components also in its local coordinate frame, it decomposes the control input u_i^i along the local x-axis and y-axis directions, respectively. Hence, the gradient control law (3.2) can be implemented in the local coordinate frame. It was analyzed that the center of the formation does not change under the gradient control law (3.2); thus, we can have a constraint of $(\mathbf{1}_n^T \otimes \mathbb{I}_2)p = \text{const}$. Then, using this property, we can reduce the size of the state vector from n to $n-1$. For example, consider three agents with positions p_1, p_2, and p_3. Then, we can have $p_1 + p_2 + p_3 = \text{const}$. Thus, $p_3 = \text{const} - p_1 - p_2$; hence, we can remove p_3 since it is dependent on p_1 and p_2. Since the reduced state vector is bounded, we could conduct a stability analysis under the concept of compactness [12] (see Sect. 2.4.1 for the concept of compactness).

For the stability analysis, it is necessary to determine equilibrium set that can be obtained as the solution of $\sum_{j\in\mathcal{N}_i}\frac{1}{2}(\|z_{ij}\|^2 - \bar{d}_{ij}^*\|)z_{ji} = 0$ from (3.2). If $\|z_{ij}\|^2 - \bar{d}_{ij}^* = 0$ for all $i \in \mathcal{V}$, then the equilibrium set can be considered as a desired one since the desired distances for all edges have been achieved. However, as discussed in [12], although there exists an edge $(i, j)^e$ such that $\|z_{ij}\|^2 \neq \bar{d}_{ij}^*$ holds, the right-hand side of (3.2) could be zero. This circumstance can happen when the control input forces are summed zero although each of them may not be zero. Such an undesired circumstance is called *undesired equilibrium* contrary to the *desired equilibrium*, which is the case of $\|z_{ij}\|^2 = \bar{d}_{ij}^*$ for all edges. They have used a center manifold theory for proving the local asymptotic stability after having a linearization [12]. It was shown that it is not easy to ensure a global convergence. That is why they have focused on a simple triangular formation in order to analyze a global convergence.

Focusing on triangular formations, there have been several different approaches to prove the global convergence of formations [1, 4–6, 9, 14]. In [1], the authors used a simple and intuitive control law for ensuring a global convergence of a cyclic triangular formation. They designed a movement rule for agent i toward the leader agent j with the speed of $s_i = -(d_{ij}^* - d_{ij})$. That is, the agent i is controlled to move toward its leader neighbor node with control input $u_i = -(d_{ij}^* - d_{ij})$; so it can be expressed as

$$u_i^i = (\|z_{ij}\| - d_{ij}^*)z_{ji}^i = -(\|z_{ij}\| - d_{ij}^*)z_{ij}^i \tag{3.4}$$

It is not a gradient control law of the form (3.3); but it is quite similar to the gradient control law. Under the condition that no two agents are collocated initially, it was proved that the desired triangular formation can be globally achieved under the control law (3.4).

In [4, 5, 9], they also considered the directed triangular formation in which each agent measures the relative positions of neighboring agents in its own coordinate frame under the gradient control form. Since each agent has one in-degree and one out-degree, it can define its potential function as $\phi_i(p) = \frac{1}{4}(\|z_i\|^2 - \bar{d}_i^*)^2$. Then, taking the gradient as $[\nabla_{p_i}\phi_i(p)]^T$, each agent updates its control input as

$$\dot{p}_i = u_i = \sum_{j\in\mathcal{N}_i^O}(\|z_{ij}\|^2 - \bar{d}_{ij}^*)z_{ji} = -\sum_{j\in\mathcal{N}_i^O}(\|z_{ij}\|^2 - \bar{d}_{ij}^*)z_{ij} \tag{3.5}$$

where $\mathcal{N}_i^O$ denotes the set of the nodes corresponding to the out-degree edges. The control law (3.5) is the directed version of the gradient law (3.2). Although the authors in [4, 5, 9] use the dynamics of line space to avoid non-compactness of the desired set of the system (3.5) and reach the same conclusion, the analyses in [4, 5, 9] are different. In [4], after defining a correct equilibrium set and incorrect equilibrium sets, they conducted stability analysis for these sets. They reached the same conclusions also using a generalized function of errors in [5]. In [9], they also defined the target formation in the link space as

$$\mathcal{E}_z = \{z \in \mathrm{Im}(\mathbb{H} \otimes \mathbb{I}_2) : \|z_k\| = d_k^*, k = 1, 2, 3\} \tag{3.6}$$

where $\mathbb{H}$ is the incidence matrix of the directed triangular graph, d_k^* is the desired length of kth edge, and z_k is the kth edge vector, which relates the link space z and state vector p as $z = (\mathbb{H} \otimes \mathbb{I}_2)p$. Using the transformation $z = (\mathbb{H} \otimes \mathbb{I}_2)p$, they generated the dynamics of link space as

$$\dot{z}_i = z_{i+1(\text{modulo3})}(\|z_{i+1(\text{modulo3})}\|^2 - \bar{d}^*_{i+1(\text{modulo3})}) - z_i(\|z_i\|^2 - \bar{d}_i^*),\ i = 1, 2, 3 \tag{3.7}$$

where $i + 1(\text{modulo3})$ means that it is 1 when $i + 1 = 4$. Then, using a manifold instability theory, they have shown that the system (3.7) is globally asymptotically stable to the desired target formation if the initial formation is not in degenerate spaces (i.e., a point or a line). In the next section, we introduce a generalized gradient-based global convergence law formulated under inter-agent dynamics. This result is extracted from [13, 14].

3.2 Global Convergence of Three-Agent Formations

Consider a set of n-agents in 2-dimensional space

$$\dot{p}_i = u_i, i = 1, \ldots, n \tag{3.8}$$

where p_i and u_i are position and control input of agent i, respectively, expressed with respect to a global coordinate frame ${}^g\Sigma$. The sensing topology of the system is described by an undirected graph $\mathcal{G} = (\mathcal{V}, \mathcal{E}^s)$, which implies that when $(i, j)^e \in \mathcal{E}^s$, agent i can sense the relative position of agent j in its own local coordinate frame ${}^i\Sigma$, and vice versa. Thus, if agent j is a neighbor of agent i in the sensing topology, then the following measurement is assumed available to agent i:

$$p^i_{ji} \triangleq (p_j - p_i)^i = p^i_j \tag{3.9}$$

where the superscript i is used to denote that the measurement is expressed with respect to the ith local coordinate frame. The measurement p^i_j includes the absolute distance between agents i and j and the direction of agent j measured from the ith local coordinate frame ${}^i\Sigma$. In distance-based distributed formation control, the absolute distances between neighboring agents are the control variables. Thus, the desired formation is defined as

$$E_{p^*} \triangleq \{p \in \mathbb{R}^{2n} : \|p_i - p_j\| = \|p_i^* - p_j^*\|, i, j \in \mathcal{V}\} \tag{3.10}$$

where the desired position vector $p^* = (p_1^{*T}, \ldots, p_n^{*T})^T \in \mathbb{R}^{2n}$ is a virtual vector used for defining desired distances $\|p_i^* - p_j^*\|$. In formation control, the formation needs to be defined as a set that satisfies the desired values of control variables. The desired formation with realization p^*, i.e., $(\mathcal{G}, p^*)$, is assumed to be infinitesimally rigid in order to formulate the problem topologically. If the topology of the underlying graph satisfies the congruence property, we can achieve the desired formation defined in (3.10) by making the lengths of edges be equal to the desired ones. Thus, the task of agent i is to control the motion of itself such that

$$\|p_i - p_j\| = \|p_i^* - p_j^*\|, \forall (i, j)^e \in \mathcal{E}^a \tag{3.11}$$

where the control variables are the lengths of the edges of $\mathcal{E}^a$, which is the set of edges characterizing the actuation topology. Since the agent i needs to generate the control input u_i using p_{ji}^i such that the desired equilibrium set E_{p^*} would be stable, the input u_i could be designed as the measured output feedback controller, i.e., $u_i = u_i(\ldots, p_{ji}^i, \ldots)$, $j \in \mathcal{N}_i$. The direct solution for this is the gradient control law outlined in the previous section. Another approach, which can be called indirect approach, is to use the inter-agent distances $\bar{d}_{ij} \triangleq \|p_i - p_j\|^2$ as virtual states. Then, we introduce the virtual control input u_{ij} as shown in the dynamics $\dot{\bar{d}}_{ij}(t) = u_{ij}(t)$. It is noticeable that the control signal u_i is the control input given to the agent i. But, the virtual signal $u_{ij}(t)$ is given to the edge connecting agents i and j. Since the edge $(i, j)^e$ does not have any control mechanism, it is called virtual input. The main idea of controlling $\bar{d}_{ij}$ is to assign the tasks of agents to the edge $(i, j)^e$. Then, after designing the control task in terms of edge dynamics, we again reassign the control task of edges to the agents by separating the control tasks of edges to the agents.

Let us suppose that we could control the inter-agent distances as

$$\begin{aligned}
\dot{\bar{d}}_{ij} &= u_{ij} \\
&= \frac{d}{dt}(\|p_i - p_j\|^2) \\
&= \frac{\partial(\|p_i - p_j\|^2)}{\partial p_i}\frac{\partial p_i}{\partial t} + \frac{\partial(\|p_i - p_j\|^2)}{\partial p_j}\frac{\partial p_j}{\partial t} \\
&= 2(p_i - p_j)^T(u_i - u_j)
\end{aligned} \tag{3.12}$$

So, given desired squared inter-agent distances as $\bar{d}_{ij}^* \triangleq \|p_i^* - p_j^*\|^2$, which are given from the desired realization p^*, we need to design a feedback controller as $2(p_i - p_j)^T(u_i - u_j) = -k_{ij}\bar{e}_{ij}$, where k_{ij} is a positive constant given to the edge $(i, j)^e$ and $\bar{e}_{ij} = \bar{d}_{ij} - \bar{d}_{ij}^*$. Note that in the case $\dot{\bar{d}}_{ij} = -k_{ij}\bar{e}_{ij}$, it holds that $\bar{e}_{ij} \to 0$ as $t \to \infty$. Thus, if we could control the inter-agent distances as shown in (3.12), it appears that the desired distance could be achieved rapidly. However, there are two technical issues in this idea. The first problem is the realizability of the set of inter-agent distances. For example, in the three-agent case of triangular formation, it should be true that

$$d_{ij} \leq d_{ik} + d_{kj},\ i, j, k \in \{1, 2, 3\} \tag{3.13}$$

which is called the triangular inequality. Thus, when the inter-agent distances are controlled by the control input u_{ij}, the above inequality (3.13) should be satisfied for all $t \geq t_0$. The second problem is to divide the control inputs u_{ij} into u_i and u_j, since u_{ij} is the virtual control input applied to the edge $(i, j)^e$. Since, for the triangular case, u_{12}, u_{13}, and u_{23} are coupled, it is not trivial to divide the virtual inputs to u_1, u_2, and u_3.

Continuing from (3.12), we can assign the control input u_{ij} as

$$u_{ij} = 2(p_i - p_j)^T(u_i - u_j) = -k_{ij}\bar{e}_{ij} \tag{3.14}$$

which leads to

$$\bar{d}_{ij}(t) = e^{-k_{ij}t}\bar{d}_{ij}(t_0) + (1 - e^{-k_{ij}t})\bar{d}_{ij}^* \tag{3.15}$$

From the above solutions, we see that $\bar{d}_{ij}(t)$ evolves along t monotonically from $\bar{d}_{ij}(t_0)$ to $\bar{d}_{ij}^*$. Here, we assume that the initial formation with distances $\bar{d}_{ij}(t_0)$ and the final target formation with the desired distances $\bar{d}_{ij}^*$ are both realizable in 2-dimensional space. Thus, the triangular inequality (3.13) is satisfied for $t = t_0$ and $t = t_f$, which implies that if the distances of edges are updated by (3.14), then the initial and final formations are realized in 2-dimensional space. However, note that even though the initial and desired formation configurations are realizable, it is not clear whether the realization $p(t)$, $\forall t > t_0$ is realizable in $\mathbb{R}^2$ under the dynamics of (3.15). In the following Lemma 3.2, we will investigate this property. Before generating this lemma, we need to reformulate the problem in a concise way. From $(p_i - p_j)^T u_i - (p_i - p_j)^T u_j = -\frac{1}{2}k_{ij}\bar{e}_{ij}$, let u_i and u_j be assigned as

$$(p_i - p_j)^T u_i = -\frac{1}{4}k_{ij}\bar{e}_{ij} \tag{3.16}$$

$$(p_j - p_i)^T u_j = -\frac{1}{4}k_{ij}\bar{e}_{ij} \tag{3.17}$$

The above assignment is to divide the control task of the edge $(i, j)^e$ to the agents i and j evenly. It may be divided as $(p_i - p_j)^T u_i = -\frac{\kappa}{2}k_{ij}\bar{e}_{ij}$ and $(p_j - p_i)^T u_j = -\frac{1-\kappa}{2}k_{ij}\bar{e}_{ij}$, where $0 < \kappa < 1$; in such case, the convergence also can be similarly proved. Then, with (3.16) and (3.17), u_{ij} will satisfy the relationship (3.14). Since the formation we study in this chapter is a triangle, each agent has two constraints. For agent i, it has constraints with agents j and k. Thus, for agent i, the two constraints can be combined as a single form:

$$\underbrace{\begin{bmatrix} (p_i - p_j)^T \\ (p_i - p_k)^T \end{bmatrix}}_{\triangleq A_i} u_i = -\frac{k_{ij}}{4} \underbrace{\begin{bmatrix} \bar{e}_{ij} \\ \bar{e}_{ik} \end{bmatrix}}_{\triangleq b_i} \tag{3.18}$$

where $p_i = (x_i, y_i)^T$, $p_j = (x_j, y_j)^T$, and $p_k = (x_k, y_k)^T$. For agents j and k, similar constraints could be formulated. The matrix A_i is a 2×2 matrix; it is clear that the matrix A_i is non-singular if and only if p_i, p_j, and p_k are noncollinear. Thus, under the noncollinearity assumption, u_i can be computed as

$$\begin{aligned} u_i &= -\frac{k_{ij}}{4} A_i^{-1} b_i \\ &= -\frac{k_{ij}}{4\det(A_i)} \begin{bmatrix} y_i - y_k & -(y_i - y_j) \\ -(x_i - x_k) & x_i - x_j \end{bmatrix} b_i \end{aligned} \tag{3.19}$$

where $\det(A_i) = (x_i - x_j)(y_i - y_k) - (y_i - y_j)(x_i - x_k)$. The following lemma is developed for using in Lemma 3.2.

Lemma 3.1 *Let $x^2 = \alpha x_0^2 + (1 - \alpha)x_d^2$ and $y^2 = \alpha y_0^2 + (1 - \alpha) y_d^2$ where $x > 0$, $y > 0$, and $0 < \alpha < 1$. Then, we have*

$$\alpha x_0 y_0 + (1 - \alpha) x_d y_d \leq xy \tag{3.20}$$

Proof Taking square to the both sides of (3.20), we have $\alpha^2(x_0y_0)^2 + 2\alpha(1 - \alpha)x_0y_0x_dy_d + (1-\alpha)^2(x_dy_d)^2 \leq (xy)^2$. Also, multiplying x^2 and y^2, we have $(xy)^2 = \alpha^2 x_0^2 y_0^2 + \alpha(1-\alpha)x_0^2y_d^2 + \alpha(1-\alpha)x_d^2y_0^2 + (1-\alpha)^2x_d^2y_d^2$. Since $\alpha(1-\alpha) x_0^2y_d^2 + \alpha(1-\alpha)x_d^2y_0^2 \geq 2\alpha(1-\alpha)x_0y_0x_dy_d$, we can have $\alpha x_0 y_0 + (1-\alpha)x_dy_d \leq xy$.

Lemma 3.2 *For the triangular formation with agents i, j, and k, the control law (3.19) preserves the noncollinearity of $p(t)$ for all $t \geq t_0$ if the initial framework $(\mathcal{G}, p(t_0))$ and the desired framework $(\mathcal{G}, p^*)$ are infinitesimally rigid.*

Proof For a simplicity of notation, let the initial time be zero, i.e., $t_0 = 0$. Due to the infinitesimally rigidity conditions of initial and desired frameworks, there exist positive constants ϵ_{ik}^0 and ϵ_{ik}^* such as $d_{ik}(0) = d_{ij}(0) + d_{jk}(0) - \epsilon_{ik}^0$ and $d_{ik}^* = d_{ij}^* + d_{jk}^* - \epsilon_{ik}^*$. Thus, with (3.15), $\bar{d}_{ik}(t)$ can be written as

$$\begin{aligned} \bar{d}_{ik}(t) &= e^{-k_{ik}t}\left(d_{ij}(0) + d_{jk}(0) - \epsilon_{ik}^0\right)^2 + (1 - e^{-k_{ik}t})\left(d_{ij}^* + d_{jk}^* - \epsilon_{ik}^*\right)^2 \\ &= \bar{d}_{ij}(t) + \bar{d}_{jk}(t) + 2e^{-k_{ik}t}d_{ij}(0)d_{jk}(0) + 2(1 - e^{-k_{ik}t})d_{ij}^*d_{jk}^* \\ &\quad - 2e^{-k_{ik}t}(d_{ij}(0) + d_{jk}(0))\epsilon_{ik}^0 - 2(1 - e^{-k_{ik}t})(d_{ij}^* + d_{jk}^*)\epsilon_{ik}^* \\ &\quad + e^{-k_{ik}t}(\epsilon_{ik}^0)^2 + (1 - e^{-k_{ik}t})(\epsilon_{ik}^*)^2 \\ &= \bar{d}_{ij}(t) + \bar{d}_{jk}(t) + 2e^{-k_{ik}t}d_{ij}(0)d_{jk}(0) + 2(1 - e^{-k_{ik}t})d_{ij}^*d_{jk}^* \\ &\quad + e^{-k_{ik}t}\epsilon_{ik}^0\left[\epsilon_{ik}^0 - 2(d_{ij}(0) + d_{jk}(0))\right] \end{aligned}$$

$$+ (1 - e^{-k_{ik}t})\epsilon_{ik}^* \left[\epsilon_{ik}^* - 2(d_{ij}^* + d_{jk}^*)\right] \tag{3.21}$$

Since $d_{ij}(0) + d_{jk}(0) > \epsilon_{ik}^0$ and $d_{ij}^* + d_{jk}^* > \epsilon_{ik}^*$, we have

$$e^{-k_{ik}t}\epsilon_{ik}^0 \left[\epsilon_{ik}^0 - 2(d_{ij}(0) + d_{jk}(0))\right] < 0 \tag{3.22}$$

and

$$(1 - e^{-k_{ik}t})\epsilon_{ik}^* \left[\epsilon_{ik}^* - 2(d_{ij}^* + d_{jk}^*)\right] < 0 \tag{3.23}$$

Thus, using the algebraic property of Lemma 3.1, we have

$$2e^{-k_{ik}t}d_{ij}(0)d_{jk}(0) + 2(1 - e^{-k_{ik}t})d_{ij}^* d_{jk}^* \leq 2d_{ij}(t)d_{jk}(t) \tag{3.24}$$

Hence, by (3.22), (3.23), and (3.24), we can have

$$\begin{aligned} \bar{d}_{ik}(t) &< \bar{d}_{ij}(t) + \bar{d}_{jk}(t) + 2d_{ij}(t)d_{jk}(t) \\ &= \left(d_{ij}(t) + d_{jk}(t)\right)^2 \end{aligned} \tag{3.25}$$

Thus, we have $d_{ik}(t) < d_{ij}(t) + d_{jk}(t)$, which is the desired triangular inequality. The triangular inequality is equivalent to the nonollinearity in three-agent case which completes the proof.

Using Lemma 3.2, the following result can be generated [14].

Theorem 3.1 *If the initial framework $(\mathcal{G}, p(t_0))$ and the desired framework $(\mathcal{G}, p^*)$ are infinitesimally rigid, then under the formation control law (3.19), the formation system globally exponentially converges to a realization in E_{p^*}, which is the desired configuration of (3.10).*

Proof Since the initial and desired formation realizations are infinitesimally rigid, by Lemma 3.2, the matrix A_i in (3.19) is non-singular for all $t \geq t_0$. Thus, u_i, $i = 1, 2, 3$ are uniquely determined, which implies e_{ij} goes to zero exponentially fast. Furthermore, since the matrix b_i converges to zero, u_i becomes zero. Thus, the realization to a point is achieved.

The control input $u_i = -\frac{k_{ij}}{4}A_i^{-1}b_i$ given in (3.19) may diverge when the triangular formation is close to a collinear configuration. The determinant of A_i is equal to the (signed) area of the region determined by the vectors $p_i - p_j$ and $p_i - p_k$.[1] Thus,

[1]Let a matrix A have row vectors as v_1, v_2 in the 2-D case, and as v_1, v_2, v_3 in the 3-D case. Let the area determined by the paralellogram by the vectors v_1, v_2 in 2-D be ξ_2, and the volume determined by the skewed cube by the vectors v_1, v_2, v_3 in 3-D be ξ_3. Then, the determinant of A, i.e.,det(A), is equal to the area or volume of A as $\det(A) = \xi_2$ in 2-D, and $\det(A) = \xi_3$ in 3-D. Thus, the area decided by the triangular defined by the vertex points of the two vectors v_1, v_2 is equal to $\frac{1}{2}\det(A) = \frac{1}{2}\xi_2$, and the tetrahedral defined by the vertex points of the three vectors v_1, v_2, v_3 in 3-D is equal to $\frac{1}{6}\det(A) = \frac{1}{6}\xi_3$.

if the formation is close to a collinear configuration, the determinant would be close to zero. To overcome such a drawback, the control law can be modified as follows:

$$\begin{aligned} u_i &= -\frac{k_{ij}}{4}|\det(A_i)|A_i^{-1}b_i \\ &= -\frac{k_{ij}}{4}\begin{bmatrix} y_i - y_k & -(y_i - y_j) \\ -(x_i - x_k) & x_i - x_j \end{bmatrix}\begin{bmatrix} \bar{e}_{ij} \\ \bar{e}_{ik} \end{bmatrix} \end{aligned} \tag{3.26}$$

which cancels the denominator term in the computation of A_i^{-1}. Thus, the control law (3.26) is similar to the traditional gradient control law (3.2) or (3.3). But, the law (3.26) is more twisted than the traditional gradient law.

Theorem 3.2 *Under the assumptions that the initial framework* $(\mathcal{G}, p(t_0))$ *and the desired framework* $(\mathcal{G}, p^*)$ *are infinitesimally rigid, the formation control law (3.26) ensures the agents to globally exponentially converge to a realization in* E_{p^*}.

Proof The determinant $|\det(A_i(t))| = |\det(A_j(t))|$ for all $i, j \in \mathcal{V}$ is the area of the triangular. From $\dot{\bar{d}}_{ij}(t) = -k_{ij}|\det(A_i(t))|(\bar{d}_{ij}(t) - \bar{d}_{ij}^*)$, we can have

$$\bar{d}_{ij}(t) = e^{-\int_{t_0}^{t} k_{ij}|\det(A_i(\tau))|d\tau}\bar{d}_{ij}(t_0) + (1 - e^{-\int_0^t k_{ij}|\det(A_i(\tau))|d\tau})\bar{d}_{ij}^* \tag{3.27}$$

Thus, from Lemma 3.2, the modified controller (3.26) also preserves the noncollinearity of $p(t)$ for all $t \geq t_0$. Following the same procedure as the proof of Theorem 3.1, we can see that the agents converge to a desired formation in E_{p^*}, which is an equilibrium configuration.

From the control law (3.26), the center of the triangular is updated as

$$\frac{1}{3}(\dot{p}_1 + \dot{p}_2 + \dot{p}_3) = -\frac{k_{ij}}{4}\begin{bmatrix} (y_1 - y_2)\bar{e}_{12} + (y_3 - y_1)\bar{e}_{13} + (y_2 - y_3)\bar{e}_{23} \\ (x_2 - x_1)\bar{e}_{12} + (x_1 - x_3)\bar{e}_{13} + (x_3 - x_2)\bar{e}_{23} \end{bmatrix} \tag{3.28}$$

In general, since $(y_1 - y_2)\bar{e}_{12} + (y_3 - y_1)\bar{e}_{13} + (y_2 - y_3)\bar{e}_{23} \neq 0$ and $(x_2 - x_1)\bar{e}_{12} + (x_1 - x_3)\bar{e}_{13} + (x_3 - x_2)\bar{e}_{23} \neq 0$ unless $\bar{e}_{ij} = 0, \forall (i, j) \in \mathcal{E}^e$, the center of triangular formation is not stationary. Also, the area of the triangular is not time-invariant due to $\frac{d(\det(A_i))}{dt} \neq 0$ before a convergence. The key idea of triangular formation using the inter-agent dynamics, introduced in this section, is to reassign the control forces computed for edges to the neighboring agents evenly. In certain cases, the control forces computed for the edges may be assigned to one of neighboring agents fully, or they may be assigned unevenly to the agents.

Example 3.1 (*Triangular network with directed and undirected edges*) Consider a triangular network depicted in Fig. 3.1. The control forces computed for the edges $(i, j)^e$ and $(j, k)^e$ are evenly reassigned to the neighboring agents, while the control force computed for the edge $(i, k)^e$ is fully assigned to the agent i. Let us design a distributed control law for agent k. For agents i and j, the control inputs are computed as

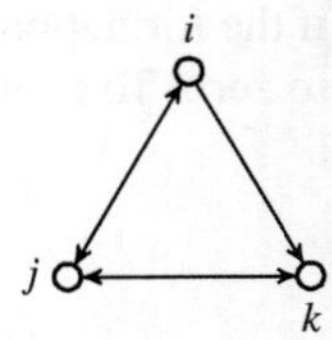

Fig. 3.1 The triangular graph with a combination of directed and undirected sensings

$$\begin{bmatrix} (p_i - p_j)^T \\ (p_i - p_k)^T \end{bmatrix} u_i = -\frac{k_o}{4} \begin{bmatrix} \bar{e}_{ij} \\ 2\bar{e}_{ik} \end{bmatrix} \text{ and } \begin{bmatrix} (p_j - p_i)^T \\ (p_j - p_k)^T \end{bmatrix} u_j = -\frac{k_o}{4} \begin{bmatrix} \bar{e}_{ji} \\ \bar{e}_{jk} \end{bmatrix}$$

However, for k, there is only one constraint as $(p_k - p_j)^T u_k = -\frac{k_o}{4}\bar{e}_{kj}$. With the above control inputs, due to Lemma 3.2, the collinearity is still preserved and u_i and u_j are uniquely determined, which implies that the triangular configuration is always realizable in 2-dimensional Euclidean space. To satisfy the single constraint for the agent k, we may be able to design the following distributed control law:

$$u_k^k = -\frac{k_o \bar{e}_{kj}}{4\|p_j^k\|^2} p_j^k \tag{3.29}$$

where u_k^k is the control input of agent k in its own local coordinate frame and $p_j^k = (p_k - p_j)^k$. The control law (3.29) for agent k means that the agent k controls its motion just along the edge $(k, j)^e$ until the desired inter-agent distance is achieved.

3.3 Global Convergence of Polygon Formations

In the previous section, we focus on three agents under undirected graphs. When extending the number of agents to more than four, the analysis of inter-agent distance-based approach would be quite challenging, because we need to ensure the realizability of inter-agent distances for all $t \geq t_0$ in d-dimensional Euclidean space. This topic is highly related with the realizability of Euclidean distance matrices (EDM) [8]. Actually, the lower the dimension d is, the more there are constraints. There is no complete solution for the graphs composed of n agents under the setups of traditional gradient control laws. In this section, we consider more than four agents under a specific setup. It is supposed that there is only one cycle in the network topology. Also, to make the problem simple, we consider directed graphs. Specifically, we consider cyclic directed formations of general n-agents in a 2-dimensional space. This result is summarized from [16, 17]. The directed graph is denoted as $\vec{\mathcal{G}} = (\mathcal{V}, \vec{\mathcal{E}})$. Each edge has a direction, all in clockwise or in counter-clockwise directions, and we suppose that the directions are aligned as $(i, j)^{\bar{e}} \in \vec{\mathcal{E}}$, $j = i + 1(\text{modulo } n)$. Figure 3.2 shows a directed cyclic formation composed of six agents.

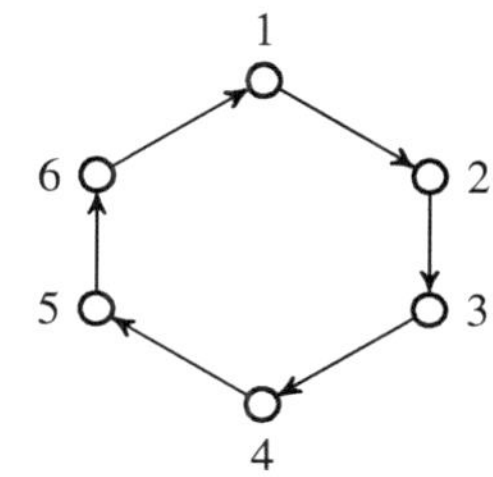

Fig. 3.2 A directed polygon formation

The relative displacement with direction is denoted by $z_{(i+1)i} = p_{i+1} - p_i$. Since $n+1$modulo n is 1, we have

$$\sum_{i=1}^{n} z_{(i+1)i} = \sum_{i=1}^{n} (p_{i+1} - p_i) = 0 \tag{3.30}$$

Let the desired squared inter-agent distance be denoted as $\bar{d}^*_{i(i+1)}$ for the edge $(i, i+1)^{\bar{e}}$. The concatenation of the desired inter-agent distances is a vector $\bar{d}^* = (\ldots, \bar{d}^*_{i(i+1)}, \ldots)^T$. Since the desired distances are assigned only to the pair of neighboring nodes, the overall underlying graph of the desired configuration is not rigid. Thus, the control task is to achieve an equivalent framework of a given desired framework. But, if the desired inter-agent distances are given arbitrary, the framework may not be realizable in 2-dimensional space. To be realizable, the generalized inequality can be given, which is the necessary and sufficient condition for a realization of a polygon formation in 2-dimensional space [11] as

$$2\max\{d^*_{1(2)}, d^*_{2(3)}, \ldots, d^*_{n-1(n)}\} \leq \sum_{i=1}^{n} d^*_{i(i+1)} \tag{3.31}$$

As the physical meaning of the above inequality, a summation of edge lengths from a point to another point should be greater than a length of the maximum edge length. Thus, to make a cyclic polygon formation, it is required to satisfy the inequality (3.31). The following example shows this interpretation.

Example 3.2 Let us consider a polygon with three desired edge lengths d^*_{12}, d^*_{23}, and d^*_{31}. Let $d^*_{12} = 1$, $d^*_{23} = 1$, and $d^*_{31} = 3$. Then, from $d^*_{12} = 1$ and $d^*_{23} = 1$, the maximum distance between agent 1 and agent 3 cannot be greater than 2. Thus, $d^*_{31} = 3$ cannot be realized. From (3.31), we also have $2\max\{d^*_{1(2)}, d^*_{2(3)}, \ldots, d^*_{n-1(n)}\} = 6$ and $\sum_{i=1}^{n} d^*_{i(i+1)} = 5$, which do not satisfy the inequality (3.31).

Let us denote the inter-agent squared distance error as $\bar{e}_{i(i+1)} = \|z_{(i+1)i}\|^2 - \bar{d}^*_{i(i+1)}$. The concatenation of the errors is a vector $\bar{e} = (\ldots, \bar{e}_{i(i+1)}, \ldots)^T \in \mathbb{R}^m$. To design a controller for each agent, we use the following potential function $\phi : \mathbb{R}^{2n} \mapsto \mathbb{R}$

$$\phi(\bar{e}) = \frac{1}{4}\sum_{i=1}^{n} \bar{e}_{i(i+1)}^2 \tag{3.32}$$

By taking the gradient of the above potential function, we can obtain a gradient control law as

$$u = -\left[\frac{\partial \phi(\bar{e})}{\partial p}\right]^T \tag{3.33}$$

which can be expressed for agent i as

$$\begin{aligned} u_i &= -\left[\frac{\partial \phi(\bar{e})}{\partial p_i}\right]^T \\ &= (\|z_{(i+1)i}\|^2 - \bar{d}^*_{i(i+1)})(p_{i+1} - p_i) - (\|z_{(i)i-1}\|^2 - \bar{d}^*_{i-1(i)})(p_i - p_{i-1}) \\ &= \bar{e}_{i(i+1)} z_{(i+1)i} - \bar{e}_{i-1(i)} z_{(i)i-1} \end{aligned} \tag{3.34}$$

Note that the above controller for agent i is of directional. To implement the second term $\bar{e}_{i-1(i)} z_{(i)i-1}$, the agent i needs to sense the relative displacement with respect to agent $i-1$. Since the sensing topology is also a directed cycle graph, the agent i cannot get $z_{(i)i-1}$. Thus, without including the second term, we use the following law:

$$u_i = \bar{e}_{i(i+1)} z_{(i+1)i} \tag{3.35}$$

It appears that $\sum_{i=1}^n \dot{p}_i = \sum_{i=1}^n u_i = \sum_{i=1}^n \bar{e}_{i(i+1)} z_{(i+1)i} \neq 0$ in general case. Thus, the center of formation may vary continuously. For the convergence analysis, let us define the following two sets:

$$\mathcal{C} \triangleq \{p : \text{rank}[z_{21}, z_{32}, \ldots, z_{1n}] < 2\} \tag{3.36}$$

$$\mathcal{O} \triangleq \{p : z_{(i+1)i} = 0, \bar{e}_{j(j+1)} = 0, \forall i \in \mathcal{I}_z, \forall j \in \mathcal{I}_e\} \tag{3.37}$$

where $\mathcal{I}_z$ and $\mathcal{I}_e$ partition the set $\{1, 2, \ldots, n\}$, and they are non-empty and disjoint. The set $\mathcal{C}$ is equivalent to a degenerate case of formation (see Sect. 12.1.1). That is, the set $\mathcal{C}$ means that the configuration of agents is a line or a point, and the set $\mathcal{O}$ means that some of agents are of collocation, while others are in the desired inter-agent distances with respect to the neighboring agents.

Example 3.3 Consider four agents p_1, p_2, p_3, p_4. Let their initial positions be $p_1 = (1, 1)^T, p_2 = (2, 2)^T, p_3 = (4, 4)^T$, and $p_4 = (7, 7)^T$, which is of collinear. Then, we have $\text{rank}[z_{21}, z_{32}, z_{43}, z_{14}] = \text{rank}[(1, 1)^T, (2, 2)^T, (3, 3)^T, (-6, -6)^T] = 1$. Let the positions be $p_1 = (1, 1)^T, p_2 = (1, 1)^T, p_3 = (1, 1)^T$, and $p_4 = (1, 1)^T$, which is of collocation. Then, we have $\text{rank}[z_{21}, z_{32}, z_{43}, z_{14}] = \text{rank}[(0, 0)^T, (0, 0)^T, (0, 0)^T, (0, 0)^T] = 0$. Lastly, let $p_1 = (1, 1)^T, p_2 = (2, 2)^T, p_3 = (4, 4)^T$, and $p_4 = (1, 2)^T$.

Then, we have $\text{rank}[z_{21}, z_{32}, z_{43}, z_{14}] = \text{rank}[(1,1)^T, (2,2)^T, (-3,-2)^T, (0,-1)^T] = 2$.

Taking the derivative of the potential function $\phi(\bar{e})$, we have

$$
\begin{aligned}
\dot{\phi} &= \frac{1}{2}\sum_{i=1}^{n} \bar{e}_{i(i+1)}\dot{\bar{e}}_{i(i+1)} \\
&= \frac{1}{2}\sum_{i=1}^{n} 2\bar{e}_{i(i+1)} z_{(i+1)i}^T(-\bar{e}_{i(i+1)} z_{(i+1)i} + \bar{e}_{i+1(i+2)} z_{(i+2)i+1}) \\
&= \frac{1}{2}\sum_{i=1}^{n} -\bar{e}_{i(i+1)} z_{(i+1)i}^T(\bar{e}_{i(i+1)} z_{(i+1)i}) + 2\bar{e}_{i(i+1)} z_{(i+1)i}^T(\bar{e}_{i+1(i+2)} z_{(i+2)i+1}) \\
&\quad - \bar{e}_{i+1(i+2)} z_{(i+2)i+1}^T(\bar{e}_{i+1(i+2)} z_{(i+2)i+1}) \\
&= -\frac{1}{2}\sum_{i=1}^{n} \|\bar{e}_{i(i+1)} z_{(i+1)i} - \bar{e}_{i+1(i+2)} z_{(i+2)i+1}\|^2
\end{aligned}
\tag{3.38}
$$

With the above result, it is shown that all the states are bounded. Furthermore, from (3.38), if $\bar{e}_{i(i+1)} z_{(i+1)i} \neq \bar{e}_{i+1(i+2)} z_{(i+2)i+1}$, we have $\dot{\phi} < 0$, which implies $\bar{e} \to 0$ as t goes on. But, from (3.38), since $\dot{\phi} \leq 0$, the error vectors converge to the cases of $\sum_{i=1}^{n} \|\bar{e}_{i(i+1)} z_{(i+1)i}^T - \bar{e}_{i+1(i+2)} z_{(i+2)i+1}\|^2 = 0$. In such cases, the errors may converge to zero; but in some cases, even with nonzero errors, it may be true that $\dot{\phi} = 0$.

Example 3.4 (*A weighted collinear case*) There are many possible cases for having $\bar{e}_{i(i+1)} z_{(i+1)i} - \bar{e}_{i+1(i+2)} z_{(i+2)i+1} = 0$ with nonzero error terms. For example, Fig. 3.3 depicts a case for this situation, where $\alpha_3 p_{i+2} + \alpha_2 p_i = -\alpha_1 p_{i+1}$ holds. From $\bar{e}_{i(i+1)} z_{(i+1)i} - \bar{e}_{i+1(i+2)} z_{(i+2)i+1} = 0$, we can have $(\bar{e}_{i(i+1)} + \bar{e}_{i+1(i+2)}) p_{i+1} - \bar{e}_{i(i+1)} p_i - \bar{e}_{i+1(i+2)} p_{i+2} = 0$. Thus, if $\alpha_1 = \bar{e}_{i(i+1)} + \bar{e}_{i+1(i+2)}$, $\alpha_2 = -\bar{e}_{i(i+1)}$, and $\alpha_3 = -\bar{e}_{i+1(i+2)}$, then although the errors are not equal to zero, the right-hand side of (3.38) could be zero.

To characterize the property of the convergence, we check all the possible cases when the right-hand side of (3.38) is equalized to zero. There are three cases as follows:

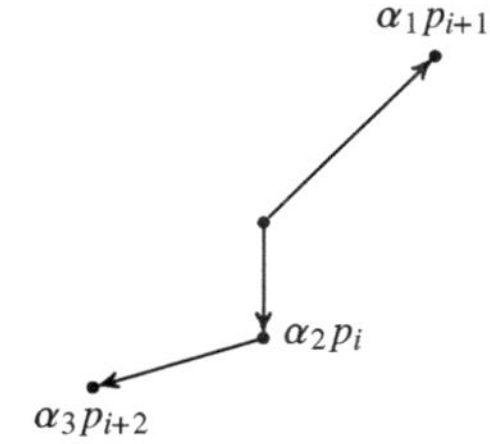

Fig. 3.3 A weighted collinear combination

- Case 1: $\bar{e} = 0$
- Case 2: Some elements of $\bar{e}$ are zero, while others are nonzero
- Case 3: $p \in \mathcal{C}$ and $\bar{e}_{i(i+1)}z_{(i+1)i} - \bar{e}_{i+1(i+2)}z_{(i+2)i+1} = 0$ for all i

The case 1 is the desired one, while the case 3 is the collinear case. The case 2 corresponds to the set $\mathcal{O}$ and the case 3 corresponds to the set $\mathcal{C}$. The stability of the above cases can be conducted using some repulsiveness properties of the undesired configurations [16, 17]. Let us first examine the repulsiveness of the set $\mathcal{O}$.

Lemma 3.3 *If the initial configuration is in $\mathcal{O}$, i.e., $p(t_0) \in \mathcal{O}$, then p stays there forever. Otherwise, if it is not in $\mathcal{O}$, i.e., $p(t_0) \notin \mathcal{O}$, $p(t)$ does not approach the set $\mathcal{O}$ for all $t \geq t_0$.*

Proof Without loss of generality, let us suppose that $z_{21} = 0$ and $e_{i(i+1)} = 0,\ i = 2, \ldots, n$. Then, $\dot{p}_i = u_i = 0$ for all $i \in \mathcal{V}$. Also, without loss of generality, suppose that $z_{21} = \Delta z_{21} \neq 0$ and $\|\Delta z_{21}\|$ is much smaller than d_{12}^*, where the symbol Δ is used to denote a very small value. Then, $u_1 = \bar{e}_{12}z_{21} = (\|\Delta z_{21}\|^2 - \bar{d}_{12}^*)\Delta z_{21}$. Since $\text{sgn}(\|\Delta z_{21}\|^2 - \bar{d}_{12}^*) = -1$, u_1 is controlled along the opposite direction of Δz_{21}, which implies that the agent 1 is forced to move away from the agent 2. Thus, it never converges to $\mathcal{O}$.

To check the attractiveness of $\mathcal{C}$, we need to develop several lemmas. To this aim, define det_{ij} as $\text{det}_{ij} \triangleq \text{det}[z_{(i+1)i}, z_{(j+1)j}]$ and define a vector q of length $\binom{n}{2}$ as

$$q \triangleq (\text{det}_{12}, \ldots, \text{det}_{1n}, \text{det}_{21}, \ldots, \text{det}_{2n}, \ldots, \text{det}_{n(n-1)})^T$$

From

$$\begin{aligned}
\frac{d\text{det}_{ij}}{dt} &= \text{det}\left[\frac{dz_{(i+1)i}}{dt}, z_{(j+1)j}\right] + \text{det}\left[z_{(i+1)i}, \frac{dz_{(j+1)j}}{dt}\right] \\
&= \text{det}[\dot{p}_{i+1} - \dot{p}_i, z_{(j+1)j}] + \text{det}[z_{(i+1)i}, \dot{p}_{j+1} - \dot{p}_j] \\
&= \text{det}[\bar{e}_{(i+1)(i+2)}z_{(i+2)(i+1)} - \bar{e}_{i(i+1)}z_{(i+1)i}, z_{(j+1)j}] \\
&\quad + \text{det}[z_{(i+1)i}, \bar{e}_{(j+1)(j+2)}z_{(j+2)(j+1)} - \bar{e}_{j(j+1)}z_{(j+1)j}] \\
&= \bar{e}_{(i+1)(i+2)}\text{det}[z_{(i+2)(i+1)}, z_{(j+1)j}] - \bar{e}_{i(i+1)}\text{det}[z_{(i+1)i}, z_{(j+1)j}] \\
&\quad + \bar{e}_{(j+1)(j+2)}\text{det}[z_{(i+1)i}, z_{(j+2)(j+1)}] - \bar{e}_{j(j+1)}\text{det}[z_{(i+1)i}, z_{(j+1)j}] \\
&= -(\bar{e}_{i(i+1)} + \bar{e}_{j(j+1)})\text{det}_{ij} + \bar{e}_{(i+1)(i+2)}\text{det}_{(i+1)j} + \bar{e}_{(j+1)(j+2)}\text{det}_{i(j+1)},
\end{aligned} \tag{3.39}$$

expanding all $\frac{d\text{det}_{ij}}{dt}$ in a vector form, we can have

$$\dot{q}(t) = \mathbb{M}(\bar{e})q(t) \tag{3.40}$$

where $\mathbb{M}$ is a matrix with elements that are functions of $\bar{e}_{i(i+1)}$ determined accordingly from (3.39).

Lemma 3.4 *It holds that $p(t) \in \mathcal{C}$ if and only if $q(t) = 0$.*

Proof If $p(t) \in \mathcal{C}$, then $\text{rank}[z_{(i+1)i}, z_{(j+1)j}] < 2$ for all i, j since $\text{rank}[z_{21}, z_{32}, \ldots, z_{1n}] < 2$. Thus, $\det_{ij} = 0$ for all i, j from the definition; hence, $q(t) = 0$. Next, let us suppose that $\det_{ij} = 0$ for all i, j, i.e., $q(t) = 0$. Then, z_i and z_j are linearly dependent for all i, j, which means that the vectors $z_{21}, z_{32}, \ldots, z_{1n}$ are linearly dependent. Thus, all the points are on the same line in the 2-dimensional space.

Lemma 3.5 *The vector $q(t) = 0$ for all t on finite horizon, i.e., $-\infty < t < \infty$, if there exists at least one instant $t = t'$ such that $q(t') = 0$, where $-\infty < t' < \infty$. Otherwise, $q(t) \neq 0$ for all t on $-\infty < t < \infty$.*

Proof Since $e(t)$ is bounded for all t, it is certain that $\|\mathbb{M}(\bar{e})\| \leq k$, where k is a positive constant. Taking δ' such that $\delta' < \min\{1, 1/k\}$ and assuming $q(t') = 0$, for any $t \in (t' - \delta', t' + \delta')$, where $-\infty < t' < \infty$, we have

$$\begin{aligned}
\|q(t)\| &= \|q(t) - q(t')\| \\
&= \left\| \int_{t'}^{t} \mathbb{M}(s_1) q(s_1) ds_1 \right\| \\
&\leq k \int_{t'}^{t} \|q(s_1)\| |ds_1| \\
&= k \int_{t'}^{t} \left\| \int_{t'}^{s_1} \mathbb{M}(s_2) q(s_2) ds_2 \right\| |ds_1| \\
&\leq k^2 \int_{t'}^{t} \int_{t'}^{s_1} \|q(s_2)\| |ds_2| \| |ds_1| \\
&\;\;\vdots \\
&\leq k^m \int_{t'}^{t} \int_{t'}^{s_1} \cdots \int_{t'}^{s_{m-1}} \|q(s_m)\| |ds_m| \cdots |ds_2| \| |ds_1| \\
&\leq k^m |t - t'| |s_1 - t'| \cdots |s_{m-1} - t'| L \\
&\leq (k\delta')^m L
\end{aligned} \tag{3.41}$$

where $L = \sup\{\|q(t)\| : t \in (t' - \delta', t' + \delta')\}$. Since $k\delta' < 1$, we have $(k\delta')^m \to 0$ as $m \to \infty$. Thus, if q is zero at t', then it is also zero for all t where $-\infty < t < \infty$.

Lemma 3.6 *If the initial configuration is not in $\mathcal{C}$, i.e., $p(t_0) \notin \mathcal{C}$, $p(t)$ does not stay in $\mathcal{C}$ for all t.*

Proof We know that the condition $p(t_0) \notin \mathcal{C}$ is equivalent to the condition $q(t_0) \neq 0$ by Lemma 3.4. Suppose that $p(t_0) \notin \mathcal{C}$, but $q(t) = 0$ at $t > t_0$. However, from Lemma 3.5, if $q(t') = 0$, then $q(t)$ has to be zero for all t. Thus, it is a contradiction. Thus, if $p(t_0) \notin \mathcal{C}$, then $p(t) \notin \mathcal{C}$ for all t where $t < \infty$.

Theorem 3.3 *If the initial configuration of the agents satisfies $p(t_0) \notin \mathcal{C} \cup \mathcal{O}$, then $e(t) \to 0$ as $t \to \infty$.*

Proof From Lemmas 3.3 and 3.6, if $p(t_0) \notin \mathcal{C} \cup \mathcal{O}$, then the configuration does not stay in the union set $\mathcal{C} \cup \mathcal{O}$. Thus, $\dot{\phi}(t) < 0$ except the point corresponding to $e = 0$, which implies that e converges to zero by Barbalat's lemma (see Lemma 2.8).

For undirected graphs, the analysis becomes more complicated. For example, if we use the gradient control law (3.34) with the potential function $\phi(\bar{e})$, we have

$$\begin{aligned}\dot{\phi} &= \frac{1}{2}\sum_{i=1}^{n} \bar{e}_{i(i+1)}\dot{\bar{e}}_{i(i+1)} \\ &= \frac{1}{2}\sum_{i=1}^{n} 2\bar{e}_{i(i+1)} z_{(i+1)i}^T (\bar{e}_{(i+1)(i+2)} z_{(i+2)(i+1)} - \bar{e}_{(i-1)i} z_{i(i-1)}) \\ &= 0 \end{aligned} \tag{3.42}$$

Thus, the summation of errors also does not change. Furthermore, due to $\sum_{i=1}^{n} \dot{p}_i = \sum_{i=1}^{n} u_i = 0$, the center of the formation does not change. Note that the polygon formations do not provide a unique configuration, i.e., although the desired inter-agent distances are achieved, the final formation configuration may be different from a desired one if there is any desired configuration.

3.4 Global Convergence of $K(3) + 1$ Edge Formations

In Sect. 3.2, we used inter-agent dynamics for designing an intuitive distributed control law and in Sect. 3.3, we considered a directed cyclic formation. Since it is still difficult to generalize the topology to general graphs under the gradient control law, we would like to further investigate some specific topologies that may be stabilized. It is natural to answer whether the gradient control law can stabilize a formation when there is an addition of a node with single edge to the triangular formation. Since the additional node has only one constraint, the overall graph will be flexible. The added node with a single edge may act as a perturbation to the graph. Such a graph can be called $K(3) + 1$ edge graph that is a $K(3)$ graph with a single edge addition. Intuitively, since the triangular formation is stabilizable by a gradient control law and a node can be readily added to the triangular, it looks that the $K(3) + 1$ edge graph could be stabilized. However, unlikely the expectation, the overall analysis is not trivial. The results of this section are reproduced from [19].

Figure 3.4 depicts the $K(3) + 1$ graph studied in this section. Since we consider four agents in 2-dimensional space, without loss of generality, let us consider neighboring sets of each agent as follows:

$$\mathcal{N}_1 = \{2, 3\},\ \mathcal{N}_2 = \{1, 3\},\ \mathcal{N}_3 = \{1, 2, 4\},\ \mathcal{N}_4 = \{3\} \tag{3.43}$$

As analyzed in Sect. 3.2, in the triangular formation, each agent has two connections. But, in (3.43), the agent 3 has three connections. Since the agent 3 has to control

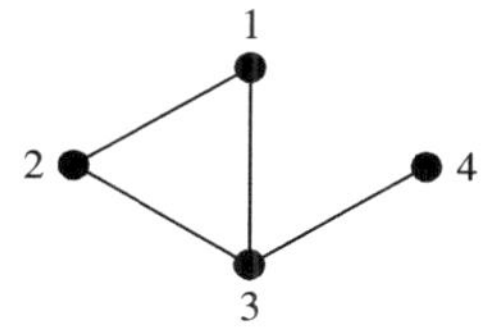

Fig. 3.4 Formation composed of a triangular shape and an additional node (i.e., $K(3)+1$ formation)

with respect to agents 1, 2, and 4, the dynamic characteristics of the $K(3)+1$ edge graph would be different from the $K(3)$ graph.

Similarly as (3.32), we use the following potential function:

$$\phi(\bar{e}) = \frac{1}{4} \sum_{(i,j)^e \in \mathcal{E}} \bar{e}_{ij}^2 \tag{3.44}$$

By taking the gradient, we can have the following control law:

$$u = -\left[\frac{\partial \phi(\bar{e})}{\partial p}\right]^T = -\mathbb{R}_{\mathcal{G}}^T \bar{e} \tag{3.45}$$

where $\mathbb{R}_{\mathcal{G}}$ is the rigidity matrix for the framework $(\mathcal{G}, p)$ and $\bar{e} = (\bar{e}_{12}, \bar{e}_{13}, \bar{e}_{23}, \bar{e}_{34})^T$.

For a use in the analysis of the $K(3)+1$ graph, let us first obtain the rigidity matrix of $K(4)$ graph, which is a complete graph of four agents. In the case of $K(4)$, by ordering edges as $\bar{e}_{12}, \bar{e}_{13}, \bar{e}_{14}, \bar{e}_{23}, \bar{e}_{24}, \bar{e}_{34}$, the rigidity matrix can be calculated as

$$\mathbb{R}_{\mathcal{G}_K} = \begin{bmatrix} p_1^T - p_2^T & p_2^T - p_1^T & 0 & 0 \\ p_1^T - p_3^T & 0 & p_3^T - p_1^T & 0 \\ p_1^T - p_4^T & 0 & 0 & p_4^T - p_1^T \\ 0 & p_2^T - p_3^T & p_3^T - p_2^T & 0 \\ 0 & p_2^T - p_4^T & 0 & p_4^T - p_2^T \\ 0 & 0 & p_3^T - p_4^T & p_4^T - p_3^T \end{bmatrix} \tag{3.46}$$

The control input on the right-hand side of (3.45) is the multiplication of rigidity matrix and error vector. As shown in (3.46), the rigidity matrix is a function of relative positions of agents and the error vector is a function of errors. If we can commutate the error terms and position terms in the multiplication, it may be helpful in performing an analysis. For this commutation, we define a matrix $\mathbb{E}_{\mathcal{G}}$, which is called *graph error matrix*, which in this case is given by

$$\mathbb{E}_{\mathcal{G}_K} = \begin{bmatrix} \bar{e}_{12} + \bar{e}_{13} + \bar{e}_{14} & -\bar{e}_{12} & -\bar{e}_{13} & -\bar{e}_{14} \\ -\bar{e}_{12} & \bar{e}_{12} + \bar{e}_{23} + \bar{e}_{24} & -\bar{e}_{23} & -\bar{e}_{24} \\ -\bar{e}_{13} & -\bar{e}_{23} & \bar{e}_{13} + \bar{e}_{23} + \bar{e}_{34} & -\bar{e}_{34} \\ -\bar{e}_{14} & -\bar{e}_{24} & -\bar{e}_{34} & \bar{e}_{14} + \bar{e}_{24} + \bar{e}_{34} \end{bmatrix} \tag{3.47}$$

Then, for the $K(4)$ graph, the control input (3.45) can be rewritten as

$$u = -\left[\frac{\partial\phi(\bar{e})}{\partial p}\right]^T = -\mathbb{R}_{\mathcal{G}_K}^T \bar{e} = -(\mathbb{E}_{\mathcal{G}_K} \otimes \mathbb{I}_2) p \tag{3.48}$$

It is remarkable that the graph error matrix has a similar form as a weighted Laplacian matrix. Thus, if $p = v \otimes \mathbf{1}_n$, where v is any vector in $\mathbb{R}^2$, then $u = 0$, which implies that if agents are collocated initially, then they will be collocated forever. It is also clear that the control input u can be implemented in a distributed way, by decomposing $-\mathbb{R}_{\mathcal{G}_K}^T \bar{e}$ in element-wise as

$$u_i = -\sum_{j \in \mathcal{N}_i} \bar{e}_{ij}(p_i - p_j) = \sum_{j \in \mathcal{N}_i} \bar{e}_{ij} z_{ji} \tag{3.49}$$

Let us consider the $K(3) + 1$ edge graph. For this graph, with the edges $\mathcal{E} = \{(1, 2)^e, (1, 3)^e, (2, 3)^e, (3, 4)^e\}$, we can change $\mathbb{R}_{\mathcal{G}_K}$ and $\mathbb{E}_{\mathcal{G}_K}$ matrices as

$$\mathbb{R}_{\mathcal{G}} = \begin{bmatrix} p_1^T - p_2^T & p_2^T - p_1^T & 0 & 0 \\ p_1^T - p_3^T & 0 & p_3^T - p_1^T & 0 \\ 0 & p_2^T - p_3^T & p_3^T - p_2^T & 0 \\ 0 & 0 & p_3^T - p_4^T & p_4^T - p_3^T \end{bmatrix} \tag{3.50}$$

$$\mathbb{E}_{\mathcal{G}} = \begin{bmatrix} \bar{e}_{12} + \bar{e}_{13} & -\bar{e}_{12} & -\bar{e}_{13} & 0 \\ -\bar{e}_{12} & \bar{e}_{12} + \bar{e}_{23} & -\bar{e}_{23} & 0 \\ -\bar{e}_{13} & -\bar{e}_{23} & \bar{e}_{13} + \bar{e}_{23} + \bar{e}_{34} & -\bar{e}_{34} \\ 0 & 0 & -\bar{e}_{34} & \bar{e}_{34} \end{bmatrix} \tag{3.51}$$

Then, the right-hand side of (3.45) can be expressed as

$$u = -\nabla_p \phi(\bar{e}) = -\left[\frac{\partial\phi(\bar{e})}{\partial p}\right]^T = -\mathbb{R}_{\mathcal{G}}^T \bar{e} = -(\mathbb{E}_{\mathcal{G}} \otimes \mathbb{I}_2) p \tag{3.52}$$

in which $u_i, i = 1, \ldots, 4$ are given in (3.49). For a clarity of analysis, we write down u_i explicitly as

$$u = \begin{bmatrix} -\bar{e}_{12} z_{12} - \bar{e}_{13} z_{13} \\ \bar{e}_{12} z_{12} - \bar{e}_{23} z_{23} \\ \bar{e}_{23} z_{23} + \bar{e}_{13} z_{13} - \bar{e}_{34} z_{34} \\ \bar{e}_{34} z_{34} \end{bmatrix} \tag{3.53}$$

The summation of input signals is zero, i.e., $\sum_{i=1}^4 u_i = 0$. Thus, the center of formation is fixed. From (3.45), the equilibrium set can be defined as

$$\mathcal{U}_{eq} \triangleq \{p : \mathbb{R}_{\mathcal{G}}^T \bar{e} = 0\} \tag{3.54}$$

The following provides stability properties of the equilibrium set $\mathcal{U}_{eq}$ [19]:

Theorem 3.4 *For the $K(3)+1$ edge graph described by (3.43), under gradient control law (3.49), the equilibrium set $\mathcal{U}_{eq}$ is globally asymptotically stable.*

Proof The derivative of $\phi(\bar{e})$ is given as

$$\dot{\phi}(\bar{e}) = -\bar{e}^T \mathbb{R}_{\mathcal{G}} \mathbb{R}_{\mathcal{G}}^T \bar{e} = -(\mathbb{R}_{\mathcal{G}}^T \bar{e})^T \mathbb{R}_{\mathcal{G}}^T \bar{e} \leq 0 \tag{3.55}$$

Thus, since all the states are bounded and continuously differentiable as per Definition 2.21, by Barbalat's lemma and by Theorem 2.23, the set $\mathcal{U}_{eq}$ is globally asymptotically stable.

The desired configuration is the case of $\bar{e} = 0$. But, with $\mathbb{R}_{\mathcal{G}}^T \bar{e} = 0$, the desired configuration, i.e., $\bar{e} = 0$ may not be ensured. That is, there may exist undesired equilibrium case (i.e., $\mathbb{R}_{\mathcal{G}}^T \bar{e} = 0$ but $\bar{e} \neq 0$). Thus, the equilibrium set $\mathcal{U}_{eq}$ can be divided as

$$\mathcal{U}_{eq}^C = \{p : \bar{e}_{ij} = 0, \ \forall (i,j)^e \in \mathcal{E}\} \tag{3.56}$$

$$\mathcal{U}_{eq}^I = \mathcal{U}_{eq} \setminus \mathcal{U}_{eq}^C = \{p : \mathbb{R}_{\mathcal{G}}^T \bar{e} = 0 \text{ and } \exists\, \bar{e}_{ij} \neq 0\} \tag{3.57}$$

where $\mathcal{U}_{eq}^C$ denotes the correct equilibrium set and $\mathcal{U}_{eq}^I$ denotes the incorrect equilibrium set. Let p^* and p^{*^c} denote a desired realization corresponding to the correct equilibrium set and an undesired realization corresponding to the incorrect equilibrium set, respectively.

To examine the stability of $\mathcal{U}_{eq}^I$, we check eigenvalues of Hessian of the potential function at the incorrect equilibrium points. The Jacobian of $\phi(\bar{e})$ with respect to p_j is given as

$$\begin{aligned} \frac{\partial \phi(\bar{e})}{\partial p_j^T} &= \frac{1}{2} \sum_{(i,j)^e \in \mathcal{E}} \left(\frac{\partial \phi(\bar{e})}{\partial \bar{e}_{ij}} \frac{\partial \bar{e}_{ij}}{\partial p_j} \right)^T \\ &= \sum_{(i,j)^e \in \mathcal{E}} \bar{e}_{ij} z_{ij}^T \frac{\partial z_{ij}}{\partial p_j} \end{aligned} \tag{3.58}$$

Taking one more derivative with respect to p_i, we obtain the Hessian of $\phi(\bar{e})$ given by

$$\begin{aligned} \frac{\partial^2 \phi(\bar{e})}{\partial p_i \partial p_j^T} &= \sum_{(i,j)^e \in \mathcal{E}} \left(\frac{\partial \bar{e}_{ij}}{\partial p_i} z_{ij}^T + \bar{e}_{ij} \frac{\partial z_{ij}^T}{\partial p_i} \right) \frac{\partial z_{ij}}{\partial p_j} \\ &= \sum_{(i,j)^e \in \mathcal{E}} \frac{\partial z_{ij}^T}{\partial p_i} \left(2 z_{ij} z_{ij}^T + \bar{e}_{ij} \otimes \mathbb{I}_2 \right) \frac{\partial z_{ij}}{\partial p_j} \end{aligned} \tag{3.59}$$

From the above relationship, the Hessian at point p is given as

$$H_\phi(p) = 2\mathbb{R}_\mathcal{G}(p)^T \mathbb{R}_\mathcal{G}(p) + \mathbb{E}_\mathcal{G}(p) \otimes \mathbb{I}_2 \tag{3.60}$$

If the Hessian $H_\phi(p^{*^c})$ is not positive semi-definite, then at p^{*^c}, the system can be said to be unstable. Note that the rigidity matrix $\mathbb{R}_\mathcal{G}$ is a 4×8 matrix. There exists a permutation matrix P that reorders the columns of the matrix $\mathbb{R}_\mathcal{G}$ such that the first four columns are x-axis components of the vectors p_i and the last four columns are y-axis components of the vectors. Thus, by multiplying the permutation matrix to $\mathbb{R}_\mathcal{G}$, we can have $\mathbb{R}_\mathcal{G} P = [\mathbb{R}^x_\mathcal{G} \mathbb{R}^y_\mathcal{G}]$, where $\mathbb{R}^x_\mathcal{G}$ is the matrix composed of the columns corresponding to the x-axis components and $\mathbb{R}^y_\mathcal{G}$ is the matrix composed of the columns corresponding to the y-axis components. Also, since $P^T \left[\mathbb{E}_\mathcal{G}(p) \otimes \mathbb{I}_2\right] P = \mathbb{I}_2 \otimes \mathbb{E}_\mathcal{G}(p)$ from a property of Kronecker product, we have

$$H = P^T H_\phi P = \left[\begin{array}{c|c} H_{11} & H_{12} \\ \hline H_{21} & H_{22} \end{array}\right] = \left[\begin{array}{c|c} 2\mathbb{R}^{x\,T}_\mathcal{G} \mathbb{R}^x_\mathcal{G} + \mathbb{E}_\mathcal{G} & 2\mathbb{R}^{x\,T}_\mathcal{G} \mathbb{R}^y_\mathcal{G} \\ \hline 2\mathbb{R}^{y\,T}_\mathcal{G} \mathbb{R}^x_\mathcal{G} & 2\mathbb{R}^{y\,T}_\mathcal{G} \mathbb{R}^y_\mathcal{G} + \mathbb{E}_\mathcal{G} \end{array}\right] \tag{3.61}$$

Example 3.5 Let the positions of the four agents be $p_1 = (0, 1)^T$, $p_2 = (-1, 0)^T$, $p_3 = (1, 0)^T$, $p_4 = (2, 1)^T$. Then, we have the rigidity matrix and permutation matrix as

$$\mathbb{R}_\mathcal{G} = \begin{bmatrix} (1,1) & (-1,-1) & (0,0) & (0,0) \\ (-1,1) & (0,0) & (1,-1) & (0,0) \\ (0,0) & (-2,0) & (2,0) & (0,0) \\ (0,0) & (0,0) & (-1,-1) & (1,1) \end{bmatrix} \text{ and } P = \begin{bmatrix} 1&0&0&0&0&0&0&0 \\ 0&0&0&0&1&0&0&0 \\ 0&1&0&0&0&0&0&0 \\ 0&0&0&0&0&1&0&0 \\ 0&0&1&0&0&0&0&0 \\ 0&0&0&0&0&0&1&0 \\ 0&0&0&1&0&0&0&0 \\ 0&0&0&0&0&0&0&1 \end{bmatrix}$$

from which we can compute

$$\mathbb{R}_\mathcal{G} P = \begin{bmatrix} 1 & -1 & 0 & 0 & 1 & -1 & 0 & 0 \\ -1 & 0 & 1 & 0 & 1 & 0 & -1 & 0 \\ 0 & -2 & 2 & 0 & 0 & 0 & 0 & 0 \\ 0 & 0 & -1 & 1 & 0 & 0 & -1 & 1 \end{bmatrix}$$

The first four column vectors correspond to the x-axis components while the last four column vectors correspond to the y-axis components of $\mathbb{R}_\mathcal{G}$.

Lemma 3.7 *If the sub-matrix $H_{22} = 2\mathbb{R}^{y\,T}_\mathcal{G} \mathbb{R}^y_\mathcal{G} + \mathbb{E}_\mathcal{G}$ of H given in (3.61) is not positive semi-definite, then the Hessian matrices H and H_ϕ are not positive semi-definite.*

Proof If the matrix H_{22} is not positive semi-definite, then there exists a nontrivial vector v such that $v^T H_{22} v < 0$. Then, using the nontrivial vector $u = (\mathbf{0}_4^T, v^T)^T$, we have $u^T H u = v^T H_{22} v < 0$. Thus, the matrices H and H_V are not positive semi-definite.

From (3.53), the incorrect equilibrium set can be further decomposed as

$$\mathcal{U}_{eq}^{I} = \mathcal{U}_{eq}^{I,0} \bigcap \left(\mathcal{U}_{eq}^{I,1} \bigcup \mathcal{U}_{eq}^{I,2} \bigcup \mathcal{U}_{eq}^{I,3}\right) \tag{3.62}$$

where

$$\begin{aligned}
\mathcal{U}_{eq}^{I,0} &\triangleq \{p : \bar{e}_{43} z_{43} = 0\} \\
\mathcal{U}_{eq}^{I,1} &\triangleq \{p : \bar{e}_{12} z_{12} = \bar{e}_{23} z_{23} = \bar{e}_{31} z_{31} \neq 0\} \\
\mathcal{U}_{eq}^{I,2} &\triangleq \{p : \bar{e}_{ij} = \bar{e}_{ik} = 0,\ p_j = p_k,\ i, j, k \in \{1, 2, 3\}\} \\
\mathcal{U}_{eq}^{I,3} &\triangleq \{p : z_{12} = z_{23} = z_{13} = 0\}
\end{aligned}$$

The set $\mathcal{U}_{eq}^{I,0}$ is the case of $\bar{e}_{43} = 0$ or z_{43}=0, where $\bar{e}_{43} = 0$ is the desired case in terms of the edge $(4, 3)^e$, but $z_{43} = 0$ is not the desired case since agents 3 and 4 are collocated. The set $\mathcal{U}_{eq}^{I,1}$ means that the vectors z_{12}, z_{23}, and z_{31} are on the same line in 2-dimensional space, which implies the points p_1, p_2, and p_3 are of collinear. The set $\mathcal{U}_{eq}^{I,2}$ contains points where two agents in the triangle are at the same position, and other agents stay at a point where the distance constraints from two others are satisfied. But, with the assumption of the desired distances being all different, it is not necessary to consider this set, since this set cannot happen. The last set $\mathcal{U}_{eq}^{I,3}$ means that all the agents are collocated. Notice that the sets $\mathcal{U}_{eq}^{I,1}$, $\mathcal{U}_{eq}^{I,2}$, and $\mathcal{U}_{eq}^{I,3}$ are different to each other; but, in these sets, agents $1, 2$, and 3 are on the same line. Thus, without loss of generality, let us suppose that agents $1, 2$, and 3 are on the x-axis in 2-dimensional space, which implies $y_1 = y_2 = y_3 = 0$ when denoting $p_i = (x_i, y_i)^T$. In this case, the sub-matrix H_{22} is expressed in a simple way as

$$H_{22} = \begin{bmatrix} \bar{e}_{12} + \bar{e}_{13} & -\bar{e}_{12} & -\bar{e}_{13} & 0 \\ -\bar{e}_{12} & \bar{e}_{12} + \bar{e}_{23} & -\bar{e}_{23} & 0 \\ -\bar{e}_{13} & -\bar{e}_{23} & \bar{e}_{13} + \bar{e}_{23} + \bar{e}_{34} + 2y_4^2 & -\bar{e}_{34} - 2y_4^2 \\ 0 & 0 & -\bar{e}_{34} - 2y_4^2 & \bar{e}_{34} + 2y_4^2 \end{bmatrix} \tag{3.63}$$

Theorem 3.5 *For the formation system characterized by (3.43), when agents are controlled by (3.49), any incorrect equilibrium point* $p^{*^c} \in \mathcal{U}_{eq}^{I}$ *is not stable.*

Proof The proof can be completed by showing that the matrix H_{22} at $p^{*^c} \in \mathcal{U}_{eq}^{I}$ is not positive semi-definite.

- Case of $\mathcal{U}_{eq}^{I,1}$: Without loss of generality, let us suppose that agent 2 is between agents 1 and 3. Then, from the condition $z_{12} = \bar{e}_{23} z_{23} = \bar{e}_{31} z_{31} \neq 0$, we have $\text{sign}(\bar{e}_{12}) = \text{sign}(\bar{e}_{23}) = -\text{sign}(\bar{e}_{31})$ and $\bar{e}_{ij} \neq 0$. Let us first suppose that $\bar{e}_{12} > 0$

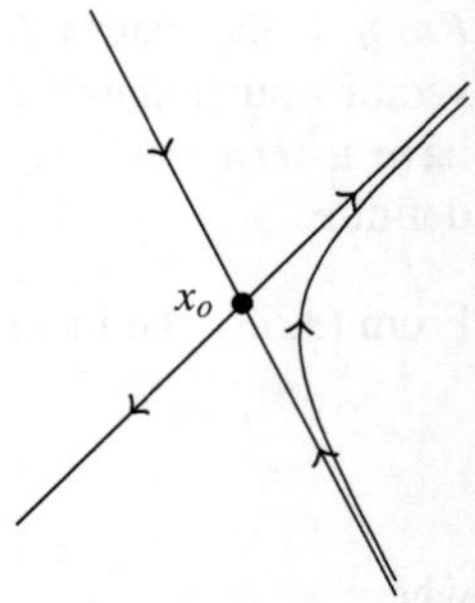

Fig. 3.5 Saddle points

and $\bar{e}_{23} > 0$, and $\bar{e}_{31} < 0$. But, this case cannot satisfy the triangular inequality of the desired formation. Thus, we have $\bar{e}_{12} < 0$ and $\bar{e}_{23} < 0$, and $\bar{e}_{31} > 0$. In this case, let us select a nontrivial vector $v = (0, 1, 0, 0)^T$. Then, we have $v^T H_{22} v = \bar{e}_{12} + \bar{e}_{23} < 0$. Thus, since there exist nontrivial vectors u and v such that $u^T H u = v^T H_{22} v < 0$, the equilibrium set $\mathcal{U}_{eq}^{I,1}$ is not positive semi-definite.

- Case of $\mathcal{U}_{eq}^{I,2}$: Without loss of generality, let us suppose that $\bar{d}_{13}^* = \bar{d}_{23}^*$, and $p_1^{*^c} = p_2^{*^c}$ and $\bar{e}_{13} = \bar{e}_{23} = 0$. In this case, since $\bar{e}_{12} = -\bar{d}_{12}^* < 0$, with $v = (1, 0, 0, 0)^T$, we have $v^T H_{22} v = \bar{e}_{12} + \bar{e}_{31} < 0$.
- Case of $\mathcal{U}_{eq}^{I,3}$: In this case, with $v = (1, 0, 0, 0)^T$, we have $v^T H_{22} v = \bar{e}_{12} + \bar{e}_{31} = -\bar{d}_{12}^* - \bar{d}_{13}^* < 0$.

From the above analysis, at any $p^{*^c} \in \mathcal{U}_{eq}^I$, the matrix H_{22} is not positive semi-definite. Consequently, the incorrect equilibrium set $\mathcal{U}_{eq}^I$ is not stable.

Since at a point $p^{*^c} \in \mathcal{U}_{eq}^I$, it is not stable, the Hessian matrix has some eigenvalues with negative real part, i.e., there exist negative eigenvalues $\lambda_i(H) < 0$; but it also has some zero eigenvalues or eigenvalues being positive. Thus, the point p^{*^c} could be a saddle point.[2] This implies that if the initial position is a point in $\mathcal{U}_{eq}^I$, then it will be staying there forever. Thus, we can make the following theorem.

Theorem 3.6 *For the formation system characterized by (3.43), when agents are controlled by (3.49), if the initial configuration is not in $\mathcal{U}_{eq}^I$, then the trajectory will converge to $\mathcal{U}_{eq}^C$.*

[2]The saddle points could be one of equilibrium points when the linearized dynamics has positive and negative eigenvalues. When a trajectory approaches toward the eigenvectors corresponding to the positive eigenvalues, it will converge to the origin asymptotically; but when it reaches to the eigenspace spanned by the eigenvectors corresponding to the negative eigenvalues, it will escape from the origin. Figure 3.5 depicts a saddle point, x_o.

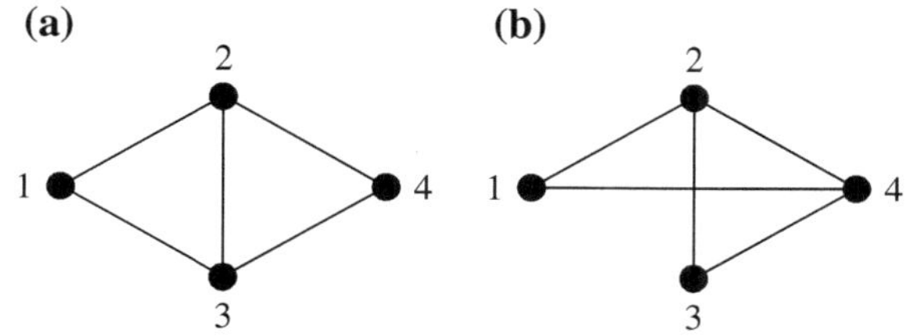

Fig. 3.6 Formation of $K(4)$ with an subtraction of an edge (i.e., $K(4) - 1$ formation): The graphs **a** and **b** are isomorphic

3.5 Global Convergence of $K(4) - 1$ Edge Formations

In Sect. 3.4, we examined the stability of $K(3) + 1$ edge formations under the gradient control law. In this section, we further attempt to generalize the topology of graphs by removing an edge from $K(4)$ graphs [19]. If we use the gradient control law with the stability analysis on the incorrect equilibrium points as like the approach of Sect. 3.3, we may be able to analyze $K(4) - 1$ edge graph that is a $K(4)$ graph with a single edge deletion, as depicted in Fig. 3.6. In Fig. 3.6, if we ignore the labels, the graphs in Fig. 3.6a, b are isomorphic. So, the analysis will be same in the two cases; we use the graph depicted in Fig. 3.6a for the analysis.

Since we consider four agents in 2-dimensional space, without loss of generality, let us consider the neighboring set of each agent as

$$\mathcal{N}_1 = \{2, 3\},\ \mathcal{N}_2 = \{1, 3, 4\},\ \mathcal{N}_3 = \{1, 2, 4\},\ \mathcal{N}_4 = \{2, 3\} \tag{3.64}$$

Using the potential function $\phi(\bar{e}) = \frac{1}{4}\sum_{(i,j)^e \in \mathcal{E}} \bar{e}_{ij}^2$, we can have the gradient control law as $u = -\left[\frac{\partial \phi(\bar{e})}{\partial p}\right]^T = -\mathbb{R}_{\mathcal{G}}^T \bar{e}$, where $\mathbb{R}_{\mathcal{G}}$ is the rigidity matrix for a given framework and $\bar{e} = (\bar{e}_{12}, \bar{e}_{13}, \bar{e}_{23}, \bar{e}_{24}, \bar{e}_{34})^T$. So, for $\bar{e} = (\bar{e}_{12}, \bar{e}_{13}, \bar{e}_{23}, \bar{e}_{24}, \bar{e}_{34})^T$, we can change $\mathbb{R}_{\mathcal{G}_K}$ and $\mathbb{E}_{\mathcal{G}_K}$ matrices given in (3.46) and (3.47) as

$$\mathbb{R}_{\mathcal{G}} = \begin{bmatrix} p_1^T - p_2^T & p_2^T - p_1^T & 0 & 0 \\ p_1^T - p_3^T & 0 & p_3^T - p_1^T & 0 \\ 0 & p_2^T - p_3^T & p_3^T - p_2^T & 0 \\ 0 & p_2^T - p_4^T & 0 & p_4^T - p_2^T \\ 0 & 0 & p_3^T - p_4^T & p_4^T - p_3^T \end{bmatrix} \tag{3.65}$$

$$\mathbb{E}_{\mathcal{G}} = \begin{bmatrix} \bar{e}_{12} + \bar{e}_{13} & -\bar{e}_{12} & -\bar{e}_{13} & 0 \\ -\bar{e}_{12} & \bar{e}_{12} + \bar{e}_{23} + \bar{e}_{24} & -\bar{e}_{23} & -\bar{e}_{24} \\ -\bar{e}_{13} & -\bar{e}_{23} & \bar{e}_{13} + \bar{e}_{23} + \bar{e}_{34} & -\bar{e}_{34} \\ 0 & -\bar{e}_{24} & -\bar{e}_{34} & \bar{e}_{24} + \bar{e}_{34} \end{bmatrix} \tag{3.66}$$

Then, with the gradient control law (3.52), we can calculate the control input explicitly as

$$u = \begin{bmatrix} -\bar{e}_{12}z_{12} - \bar{e}_{13}z_{13} \\ \bar{e}_{12}z_{12} - \bar{e}_{23}z_{23} - \bar{e}_{24}z_{24} \\ \bar{e}_{23}z_{23} + \bar{e}_{13}z_{13} - \bar{e}_{34}z_{34} \\ \bar{e}_{24}z_{24} + \bar{e}_{34}z_{34} \end{bmatrix} \tag{3.67}$$

With the above control inputs, the center of formation is fixed due to $\sum_{i=1}^{4} u_i = 0$. Also, similarly to the sets (3.56) and (3.57), the equilibrium set $\mathcal{U}_{eq} = \{p : \mathbb{R}_{\mathcal{G}}^T \bar{e} = 0\}$ can be divided into the correct equilibrium set $\mathcal{U}_{eq}^C$ and incorrect equilibrium set $\mathcal{U}_{eq}^I$. Then, at $p^{*^c} \in \mathcal{U}_{eq}^I$, we have $(\mathbb{E}_{\mathcal{G}}(p^{*^c}) \otimes \mathbb{I}_2)p^{*^c} = 0$ but with some of $\bar{e}_{ij}$ being nonzero. With the relationship $p^T(\mathbb{E}_{\mathcal{G}}(p^{*^c}) \otimes \mathbb{I}_2)p = \sum_{(i,j)^e \in \mathcal{E}} \bar{e}_{ij} \| p_i - p_j \|^2$, we can have

$$(p^{*^c})^T(\mathbb{E}_{\mathcal{G}}(p^{*^c}) \otimes \mathbb{I}_2)p^{*^c} = \sum_{(i,j)^e \in \mathcal{E}} \bar{e}_{ij} \| p_i^{*^c} - p_j^{*^c} \|^2 = 0 \tag{3.68}$$

and

$$(p^*)^T(\mathbb{E}_{\mathcal{G}}(p^{*^c}) \otimes \mathbb{I}_2)p^* = \sum_{(i,j)^e \in \mathcal{E}} \bar{e}_{ij} \| p_i^* - p_j^* \|^2 = \sum_{(i,j)^e \in \mathcal{E}} \bar{e}_{ij} \bar{d}_{ij}^* \tag{3.69}$$

Using the above two equalities, at p^{*^c}, we can obtain

$$\begin{aligned} (p^*)^T(\mathbb{E}_{\mathcal{G}}(p^{*^c}) \otimes \mathbb{I}_2)p^* &= (p^*)^T(\mathbb{E}_{\mathcal{G}}(p^{*^c}) \otimes \mathbb{I}_2)p^* - (p^{*^c})^T(\mathbb{E}_{\mathcal{G}}(p^{*^c}) \otimes \mathbb{I}_2)p^{*^c} \\ &= \sum_{(i,j)^e \in \mathcal{E}} \bar{e}_{ij} \bar{d}_{ij}^* - \sum_{(i,j)^e \in \mathcal{E}} \bar{e}_{ij} (d_{ij})^2 \\ &= - \sum_{(i,j)^e \in \mathcal{E}} \bar{e}_{ij}^2 < 0 \end{aligned} \tag{3.70}$$

Thus, the matrix $\mathbb{E}_{\mathcal{G}}(p^{*^c}) \otimes \mathbb{I}_2$ is not positive semi-definite, which is summarized in the following lemma.

Lemma 3.8 *For the formation topology characterized by (3.64), under the gradient control law, the graph error matrix $\mathbb{E}_{\mathcal{G}}$ is not positive semi-definite at any incorrect equilibrium point p^{*^c}.*

As like the previous section, we can obtain the Hessian matrix H_ϕ of the potential function $\phi(\bar{e})$. From (3.67), the incorrect equilibrium set could be determined by four constraints, with some $\bar{e}_{ij} \neq 0$, such as

(i) $\bar{e}_{12}z_{12} + \bar{e}_{13}z_{13} = 0$
(ii) $\bar{e}_{12}z_{12} - \bar{e}_{23}z_{23} - \bar{e}_{24}z_{24} = 0$
(iii) $\bar{e}_{23}z_{23} + \bar{e}_{13}z_{13} - \bar{e}_{34}z_{34} = 0$
(iv) $\bar{e}_{24}z_{24} + \bar{e}_{34}z_{34} = 0$

First, to satisfy (i) $\bar{e}_{12}z_{12} + \bar{e}_{13}z_{13} = 0$, (i-a) agents 1, 2, and 3 should be on the same line in 2-dimensional space, or (i-b) we should have $\bar{e}_{12} = 0$ and $z_{13} = 0$ or $z_{12} = 0$

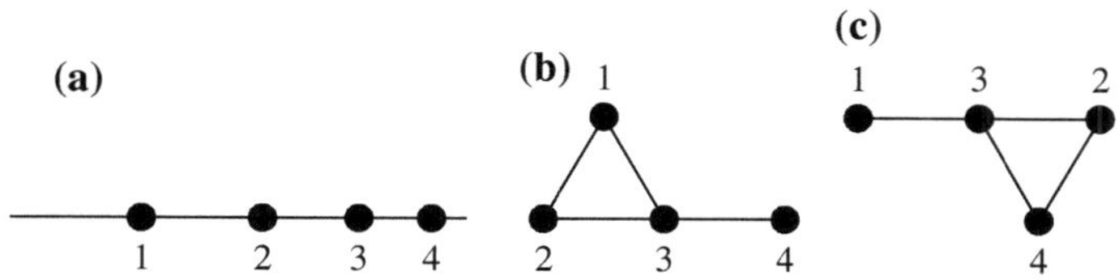

Fig. 3.7 The possible cases of collinear formations

and $\bar{e}_{13} = 0$, or (i-c) the errors should be zero as $\bar{e}_{12} = 0$ and $\bar{e}_{13} = 0$. The cases (i-a) and (i-b) represent the collinearity of agents 1, 2, and 3, while the case (i-c) implies the agent 1 is located at the desired distances from agents 2 and 3. Next, let us consider the case (iv). Likewise the case (i), to satisfy (iv) $\bar{e}_{24}z_{24} + \bar{e}_{34}z_{34} = 0$, agents 2, 3, and 4 should be collinear or the agent 4 should be located at the desired distances from agents 2 and 3. Thus, when combining the case (i) and the case (iv), besides the desired one, there are three possible cases as follows:

- Case (a): agents 1, 2, 3, and 4 are collinear,
- Case (b): agent 1 is located at the desired distances from agents 2 and 3, while agents 2, 3, and 4 are collinear, and
- Case (c): agent 4 is located at the desired distances from agents 2 and 3, while agents 1, 2, and 3 are collinear.

Figure 3.7 depicts these three cases. Note that the case (b) and the case (c) are automorphism. So, in the analysis, we consider only the case (c). In the case (c), since $\bar{e}_{24} = 0$ and $\bar{e}_{34} = 0$, the constraints (ii) and (iii) can be changed as $\bar{e}_{12}z_{12} - \bar{e}_{23}z_{23} = 0$ and $\bar{e}_{23}z_{23} + \bar{e}_{13}z_{13} = 0$, respectively, and the constraint (iv) is removed. So, for the case (c), we can have the following lemma.

Lemma 3.9 *When agents* 1, 2, *and* 3 *have the constraints* $\bar{e}_{12}z_{12} + \bar{e}_{13}z_{13} = \bar{e}_{12}z_{12} - \bar{e}_{23}z_{23} = \bar{e}_{23}z_{23} + \bar{e}_{13}z_{13} = 0$, *we can conclude that* $\bar{e}_{12} < 0$, $\bar{e}_{23} < 0$, $\bar{e}_{13} > 0$, *and* $\bar{e}_{23} + \bar{e}_{13} < 0$ *and* $\bar{e}_{12} + \bar{e}_{13} < 0$.

Proof See the appendix.

Theorem 3.7 *For the formation system (3.64), under the gradient control law (3.67), the undesired equilibrium set is unstable.*

Proof First, let us consider the case (a), where all agents are collinear. In this case, without loss of generality, we suppose that agents are on the x-axis. Then, since all the y values are zero, we can write $H_{22}(p^{*^c}) = 2\mathbb{R}_{\mathcal{G}}^{y\,T}(p^{*^c})\mathbb{R}_{\mathcal{G}}^{y}(p^{*^c}) + \mathbb{E}(p^{*^c}) = \mathbb{E}(p^{*^c})$. From (3.70), we can have that $\mathbb{E}(p^{*^c})$ is not positive semi-definite. Second, let us consider the case (c). Since agents 1, 2, and 3 are on the same line, without loss of generality, we also suppose that agents 1, 2 and 3 are on the x-axis, which implies that the y-axis values of agents 1, 2, and 3 are all zero. Then, we can have H_{22} matrix as

$$H_{22} = \begin{bmatrix} \bar{e}_{12} + \bar{e}_{13} & -\bar{e}_{12} & -\bar{e}_{13} & 0 \\ -\bar{e}_{12} & \bar{e}_{12} + \bar{e}_{23} + 2y_4^2 & -\bar{e}_{23} & -2y_4^2 \\ -\bar{e}_{13} & -\bar{e}_{23} & \bar{e}_{13} + \bar{e}_{23} + 2y_4^2 & -2y_4^2 \\ 0 & -2y_4^2 & -2y_4^2 & 2y_4^2 \end{bmatrix} \tag{3.71}$$

If we take a vector $v = (1, 0, 0, 0)^T$, then $v^T H_{22}(p^{*^c})v = \bar{e}_{12} + \bar{e}_{13} < 0$. Thus, $H_{22}(p^{*^c})$ is not positive semi-definite. Hence, with the above arguments, it is clear that the set $\mathcal{U}_{eq}^I$ is unstable.

With the above results, we can finally obtain the following result.

Theorem 3.8 *For the distributed systems characterized by (3.64), when agents are controlled by (3.67), if the initial configuration is not in $\mathcal{U}_{eq}^I$, then the trajectory will converge to $\mathcal{U}_{eq}^C$.*

Example 3.6 (*Undesired equilibrium point*) Let the four agents be located along a line with relative displacements as $z_{21} = (1, 0)^T$, $z_{32} = (1, 0)^T$, and $z_{43} = (1, 0)^T$. Then, we can assign their coordinate values on the x-axis simply as $p_1 = (1, 0)^T$, $p_2 = (2, 0)^T$, $p_3 = (3, 0)^T$, and $p_4 = (4, 0)^T$. Then, to characterize equilibrium points, from (3.67), ignoring the y-axis components, we can have four equations as

$$\begin{aligned} &\bar{e}_{12}(-1) + \bar{e}_{13}(-2) = 0 \\ &\bar{e}_{12}(-1) + \bar{e}_{23}(+1) + \bar{e}_{24}(+2) = 0 \\ &\bar{e}_{23}(-1) + \bar{e}_{13}(-2) + \bar{e}_{34}(+1) = 0 \\ &\bar{e}_{24}(-2) + \bar{e}_{34}(-1) = 0 \end{aligned}$$

which are reduced as $\bar{e}_{12} = -2\bar{e}_{13}$, $\bar{e}_{34} = -2\bar{e}_{24}$, and $2\bar{e}_{13} + \bar{e}_{23} + 2\bar{e}_{24} = 0$. Then, if there exist any nonzero errors satisfying these three equations, the configuration determined by the relationships $z_{21} = (1, 0)^T$, $z_{32} = (1, 0)^T$, and $z_{43} = (1, 0)^T$ will be considered as incorrect equilibrium states. For example, if $\bar{e}_{13} = 1$, $\bar{e}_{24} = 1$, $\bar{e}_{12} = -2$, $\bar{e}_{34} = -2$, and $\bar{e}_{23} = -4$, it will satisfy the above four constraints with $\bar{d}_{12}^* = 3$, $\bar{d}_{13}^* = 3$, $\bar{d}_{24}^* = 3$, $\bar{d}_{34}^* = 3$, and $\bar{d}_{23}^* = 5$, which are realizable desired distances in 2-dimensional Euclidean space.

3.6 Global Convergence in 3-Dimensional Space

In Sect. 3.2, $K(3)$ formation in 2-dimensional space was stabilized to a desired configuration under a gradient control law that was designed on the basis of inter-agent dynamics. As an extension of Sect. 3.2, this section considers $K(4)$ formation in 3-dimensional space. The approach of Sect. 3.2 cannot be used directly since the realizability of inter-agent distances (six edges) needs to be evaluated at all time instants. But, the case of $K(4)$ in 3-D is more complicated than the case of Sect. 3.2 since we need to evaluate the realizability of six inter-agent distances whenever the inter-agent dynamics are updated. Thus, in this section, we would like to develop a control law on the basis of the traditional gradient control law, with a little complicated stability analysis. Following the similar approach as in Sect. 3.5, we would analyze the stability of degenerate configurations. The result of this section is reproduced from [18].

Consider four agents $\mathcal{V} = \{1, 2, 3, 4\}$ in 3-dimensional space under a complete graph topology. The position of agent i is represented by $p_i = (x_i, y_i, z_i)^T$. It is assumed that the desired formation configuration has nonzero volume in $\mathbb{R}^3$. The rigidity matrix and graph error matrix are given in (3.46) and (3.47), respectively. The Hessian matrix $H_\phi(p)$ given in (3.60) can be simply changed as

$$H_\phi(p) = 2\mathbb{R}_\mathcal{G}(p)^T \mathbb{R}_\mathcal{G}(p) + \mathbb{E}_\mathcal{G}(p) \otimes \mathbb{I}_3 \tag{3.72}$$

where $\mathbb{E}_\mathcal{G}(p)$ is given in (3.47). From the gradient control law, the control inputs for the agents can be calculated as

$$u = \begin{bmatrix} -\bar{e}_{12}z_{12} - \bar{e}_{13}z_{13} - \bar{e}_{14}z_{14} \\ \bar{e}_{12}z_{12} - \bar{e}_{23}z_{23} - \bar{e}_{24}z_{24} \\ \bar{e}_{13}z_{13} + \bar{e}_{23}z_{23} - \bar{e}_{34}z_{34} \\ \bar{e}_{14}z_{14} + \bar{e}_{24}z_{24} + \bar{e}_{34}z_{34} \end{bmatrix} \tag{3.73}$$

Since the above law is a gradient control law for an undirected graph, the summation of control inputs is equal to zero. So, the center of formation is fixed. From an agent perspective, when agents are updated by the above control law, each agent attempts to reduce the distance errors with respect to three neighboring agents, in a combination way.

If the Hessian H_ϕ (3.72) has, at least, a negative eigenvalue at an incorrect equilibrium point, the incorrect equilibrium set is considered unstable. Using the column reordering transformation matrix P, the rigidity matrix can be decomposed as $\mathbb{R}_\mathcal{G} P = [\mathbb{R}_\mathcal{G}^x, \mathbb{R}_\mathcal{G}^y, \mathbb{R}_\mathcal{G}^z]$, where $\mathbb{R}_\mathcal{G}^x$, $\mathbb{R}_\mathcal{G}^y$, and $\mathbb{R}_\mathcal{G}^z$ are the matrices composed of the columns corresponding to the x-, y-, and z-axes, respectively. The transformed Hessian matrix H, which is computed as $H = P^T H_\phi(p) P$, is given as

$$H = P^T H_\phi(p) P = \left[\begin{array}{c|c|c} 2{\mathbb{R}_\mathcal{G}^x}^T \mathbb{R}_\mathcal{G}^x + \mathbb{E}_\mathcal{G} & 2{\mathbb{R}_\mathcal{G}^x}^T \mathbb{R}_\mathcal{G}^y & 2{\mathbb{R}_\mathcal{G}^x}^T \mathbb{R}_\mathcal{G}^z \\ \hline 2{\mathbb{R}_\mathcal{G}^y}^T \mathbb{R}_\mathcal{G}^x & 2{\mathbb{R}_\mathcal{G}^y}^T \mathbb{R}_\mathcal{G}^y + \mathbb{E}_\mathcal{G} & 2{\mathbb{R}_\mathcal{G}^y}^T \mathbb{R}_\mathcal{G}^z \\ \hline 2{\mathbb{R}_\mathcal{G}^z}^T \mathbb{R}_\mathcal{G}^x & 2{\mathbb{R}_\mathcal{G}^z}^T \mathbb{R}_\mathcal{G}^y & 2{\mathbb{R}_\mathcal{G}^z}^T \mathbb{R}_\mathcal{G}^z + \mathbb{E}_\mathcal{G} \end{array}\right] \tag{3.74}$$

Using the same process as (3.68)–(3.70), it can be shown that the matrix $\mathbb{E}_\mathcal{G}(p^{*^c}) \otimes \mathbb{I}_3$ is not positive semi-definite [18].

Theorem 3.9 *The formation system of $K(4)$ graph in 3-dimensional space is unstable at any incorrect equilibrium point p^{*^c}.*

Proof By equalizing the right-hand side of (3.73) to be zero, it can be easily shown that the incorrect equilibrium point occurs when agents are on the same plane. Thus, without loss of generality, let us suppose that the agents stay in the x-y plane, which means that $z_i = 0$, for $i = 1, 2, 3, 4$. Then, at any incorrect equilibrium point, the transformed Hessian matrix can be expressed as

$$H(p^{*^c}) = \left[\begin{array}{c|c|c} 2{\mathbb{R}_{\mathcal{G}}^x}^T \mathbb{R}_{\mathcal{G}}^x + \mathbb{E}_{\mathcal{G}} & 2{\mathbb{R}_{\mathcal{G}}^x}^T \mathbb{R}_{\mathcal{G}}^y & 0 \\ \hline 2{\mathbb{R}_{\mathcal{G}}^y}^T \mathbb{R}_{\mathcal{G}}^x & 2{\mathbb{R}_{\mathcal{G}}^y}^T \mathbb{R}_{\mathcal{G}}^y + \mathbb{E}_{\mathcal{G}} & 0 \\ \hline 0 & 0 & \mathbb{E}_{\mathcal{G}} \end{array}\right] \tag{3.75}$$

Since $\mathbb{E}_{\mathcal{G}}(p^{*^c})$ is not positive semi-definite, the matrix $H(p^{*^c})$ is also not positive semi-definite.

Theorem 3.10 *If the agents are on a plane initially, then it will stay over the plane forever when agents are controlled by the gradient control law.*

Proof First, based on Theorem 3.4, it can be shown that the equilibrium set $\mathcal{U}_{eq}$ given in (3.54) is globally asymptotically stable. Thus, the point $p(t)$ will converge to either $\mathcal{U}_{eq}^I$ or $\mathcal{U}_{eq}^C$, where $\mathcal{U}_{eq}^I$ or $\mathcal{U}_{eq}^C$ are given in (3.56) and (3.57), since $\mathcal{U}_{eq} = \mathcal{U}_{eq}^I \bigcup \mathcal{U}_{eq}^C$. Next, let $Z(p) = [z_{12}, z_{13}, z_{14}] \in \mathbb{R}^{3\times3}$ and $\Delta(p) = \det Z$. Then, the set of all realizations of coplanar four agents is defined by

$$\mathcal{C} = \{p = [p_1, p_2, p_3, p_4] \in \mathbb{R}^{3\times4} \mid \text{rank} Z(p) < 3\} \tag{3.76}$$

It is well known that the absolute value of $\Delta(p)$ is twice the area of the configuration of formation in $\mathbb{R}^2$. Thus, $\Delta(p) = 0$ for any $p \in \mathcal{C}$. By taking a time derivative of Δ, the following relationship is obtained:

$$\begin{aligned} \dot{\Delta} &= \det[\dot{z}_{12}, z_{13}, z_{14}] + \det[z_{12}, \dot{z}_{13}, z_{14}] + \det[z_{12}, z_{13}, \dot{z}_{14}] \\ &= -\text{trace}(\mathbb{E}_{\mathcal{G}})\Delta \end{aligned} \tag{3.77}$$

from which, $\Delta(p(t))$ is calculated as

$$\Delta(p(t)) = e^{-\int_{t_0}^{t} \text{trace}\mathbb{E}_{\mathcal{G}}(p(s))ds} \Delta(p(t_0)) \tag{3.78}$$

Hence, if $\Delta(p(t_0)) = 0$ at t_0, i.e., it is of coplanar initially, then $\Delta(p(t)) = 0$ for all $t \geq t_0$, which means $p(t) \in \mathcal{C}$.

Theorem 3.11 *If the agents are not on a plane initially and the desired configuration is given in 3-dimensional space, then the desired formation can be achieved by the gradient control law.*

Proof Similarly to Theorem 3.4, it is shown that the equilibrium set $\mathcal{U}_{eq}$ is globally asymptotically stable. Thus, $\mathbb{R}_{\mathcal{G}}^T \bar{e} = 0$, which means that the point $p(t)$ will converge to either $\mathcal{U}_{eq}^I$ or $\mathcal{U}_{eq}^C$. From Theorem 3.9, the incorrect equilibrium set $\mathcal{U}_{eq}^I$, which contains only the coplanar configurations, is unstable. But from Theorem 3.10, if the initial point is on a plane, then it will stay over there forever. By combining the above arguments, if the agents are not on a plane initially and the desired configuration is given in 3-dimensional space, the trajectory will converge to $\mathcal{U}_{eq}^{C}$.

It is remarkable that $\Delta(p(t_0)) = 0$ when $\text{rank}Z(p) = 0, 1, 2$. Thus, it would be of interest to answer for all cases when $\text{rank}Z(p) \leq 2$. To this aim, let us define the following sets:

$$\mathcal{C}_i = \{p = [p_1, p_2, p_3, p_4] \in \mathbb{R}^{3\times 4} | \text{rank}Z(p) = i\},\ i = 0, 1, 2 \tag{3.79}$$

Lemma 3.10 *Consider an incorrect equilibrium point p^{*^c} on a plane, i.e., $p^{*^c} \in \mathcal{U}_{eq}^{I} \bigcap \mathcal{C}_2$. Then, $\mathbb{E}_{\mathcal{G}}(p^{*^c})$ has rank 1 and it is negative semi-definite.*

Proof Since p^{*^c} is an incorrect equilibrium point on a plane, ignoring z-axis components, $(\mathbb{E}_{\mathcal{G}}(p^{*^c}) \otimes \mathbb{I}_3)p^{*^c} = 0$ can be rewritten as

$$\mathbb{E}_{\mathcal{G}}(p^{*^c})M = 0$$

where the matrix M is given as

$$\begin{bmatrix} 1 & x_1^{*^c} & y_1^{*^c} & 0 \\ 1 & x_2^{*^c} & y_2^{*^c} & 0 \\ 1 & x_3^{*^c} & y_3^{*^c} & 0 \\ 1 & x_4^{*^c} & y_4^{*^c} & 0 \end{bmatrix}$$

But, the points $(x_i^{*^c}, y_i^{*^c})$, $i = \{1, 2, 3, 4\}$ are not collocated. Thus, the matrix M has rank 3. But to satisfy $\mathbb{E}_{\mathcal{G}}(p^{*^c})M = 0$, the matrix $\mathbb{E}_{\mathcal{G}}(p^{*^c})$ must have rank less than or equal to 1. But, we know that since p^{*^c} is an incorrect equilibrium point, some distance errors would be nonzero. Thus, the rank cannot be zero, which implies that $\text{rank}\mathbb{E}_{\mathcal{G}}(p^{*^c}) = 1$. Also, from the fact that $\mathbb{E}_{\mathcal{G}}(p^{*^c})$ is not positive semi-definite, it is clear that $\mathbb{E}_{\mathcal{G}}(p^{*^c})$ has at least one negative eigenvalue. But since it is rank 1, all other eigenvalues are zero except one negative eigenvalue. Thus, it is negative semi-definite.

From the above lemma, it is clear that $\text{trace}\mathbb{E}_{\mathcal{G}}(p^{*^c}) < 0$ for $p^{*^c} \in \mathcal{U}_{eq}^{I} \bigcap \mathcal{C}_2$. Also since the trace of a matrix is the summation of diagonal terms, when $\text{rank}Z = 0$ ($\mathcal{C}_0$), i.e., all the agents are collocated, it is also true that $\text{trace}\mathbb{E}_{\mathcal{G}}(p^{*^c}) < 0$ due to $e_{ij} < 0$ for all $(i, j)^e \in \mathcal{E}$.

Lemma 3.11 *If the initial configuration is not in $\mathcal{C}$, i.e., $p(t_0) \notin \mathcal{C}$, then the trajectory of agents controlled by the gradient control law does not stay in $\mathcal{U}_{eq}^{I} \bigcap (\mathcal{C}_0 \bigcup \mathcal{C}_2)$.*

Proof Suppose that $p(t) \in \mathcal{U}_{eq}^{I} \bigcap (\mathcal{C}_0 \bigcup \mathcal{C}_2)$ at $t \geq t_f$. Then, $\Delta(p(t)) = 0$ must hold for $t \geq t_f$ by Lemma 3.10. From (3.78), at $t \geq t_f \geq t_0$, we have

$$\Delta(p(t)) = e^{-\int_{t_f}^{t} \text{trace}\mathbb{E}_{\mathcal{G}}(p(s))ds} e^{-\int_{t_0}^{t_f} \text{trace}\mathbb{E}_{\mathcal{G}}(p(s))ds} \Delta(p(t_0)) \tag{3.80}$$

As discussed above, at a point $p(t) \in \mathcal{U}_{eq}^I \bigcap (\mathcal{C}_0 \bigcup \mathcal{C}_2)$, since $\text{trace}\mathbb{E}_{\mathcal{G}} < 0$, it is true that $e^{-\int_{t_f}^{t} \text{trace}\mathbb{E}_{\mathcal{G}}(p(s))ds} \geq 1$ and $e^{-\int_{t_0}^{t_f} \text{trace}\mathbb{E}_{\mathcal{G}}(p(s))ds}$ is a nonzero constant. Thus, since agents are not in $\mathcal{C}$, it is also true that $\Delta(p(t_0)) \neq 0$. Thus, it must be true that $\Delta(p(t)) \neq 0$ at $t \geq t_f \geq t_0$. But, this is a contradiction to the claim that $\Delta(p(t)) = 0$ must hold for $t \geq t_f$.

Next, it is necessary to check whether the trajectory approaches the set $\mathcal{U}_{eq}^I \bigcap \mathcal{C}_1$ from some specific initial points. Without loss of generality, select a point $p^{*^c} \in \mathcal{U}_{eq}^I \bigcap \mathcal{C}_1$ such as $p_i^{*^c} = (x_i^{*^c}, 0, 0)^T$ on the x-axis. Then, the transformed Hessian matrix can be calculated as

$$H(p^{*^c}) = \left[\begin{array}{c|c|c} 2\mathbb{R}_{\mathcal{G}}^{x\ T}\mathbb{R}_{\mathcal{G}}^x + \mathbb{E}_{\mathcal{G}} & 0 & 0 \\ \hline 0 & \mathbb{E}_{\mathcal{G}} & 0 \\ \hline 0 & 0 & \mathbb{E}_{\mathcal{G}} \end{array}\right] \tag{3.81}$$

Then, the repulsiveness of $\mathcal{U}_{eq}^I \bigcap \mathcal{C}_1$ can be analyzed using eigenvalues of $H(p^{*^c})$ as follows [18]:

Lemma 3.12 *Let $p^{*^c} \in \mathcal{U}_{eq}^I \bigcap \mathcal{C}_1$. Select a point p in a neighborhood of p^{*^c} such that $p \notin \mathcal{C}$. Then, p cannot be attracted toward p^{*^c}.*

Proof Suppose that $\mathbb{E}_{\mathcal{G}}(p^{*^c})$ has a positive eigenvalue λ_+ with the corresponding eigenvector $s = (s_1, s_2, s_3, s_4)^T$ of unit length. Then, it is clear that λ_+ is a multiple eigenvalue of $H(p^{*^c})$, and thus there are two eigenvectors corresponding to the multiple eigenvalue λ_+ such as

$$h_y = (0, s_1, 0, 0, s_2, 0, 0, s_3, 0, 0, s_4, 0)^T$$
$$h_z = (0, 0, s_1, 0, 0, s_2, 0, 0, s_3, 0, 0, s_4)^T$$

Thus, by the positive eigenvalue λ_+, there are two subspaces along which a point p is attracted to the point p^{*^c}. Hence, the attracted point p can be expressed as $p = p^{*^c} + \epsilon_y h_y + \epsilon_z h_z$. Now, from $p_1 = (x_1^{*^c}, \epsilon_y s_1, \epsilon_z s_1)^T$, $p_2 = (x_2^{*^c}, \epsilon_y s_2, \epsilon_z s_2)^T$, $p_3 = (x_3^{*^c}, \epsilon_y s_3, \epsilon_z s_3)^T$, and $p_4 = (x_4^{*^c}, \epsilon_y s_4, \epsilon_z s_4)^T$, since the ratio of z-component over y-component is same as ϵ_z/ϵ_y, the four points p_1, p_2, p_3, and p_4 are on a plane. Thus, a point p should be on a plane in order to be attracted toward p^{*^c}. This contradicts the assumption that $p \notin \mathcal{C}$.

Theorem 3.12 *The region of attraction for the desired equilibrium set is $\mathbb{R}^{3\times 4} \setminus \mathcal{U}_{eq}^I$.*

Proof It is clear that the trajectory approaches either $\mathcal{U}_{eq}^C$ or $\mathcal{U}_{eq}^I$. By Lemmas 3.11 and 3.12, any trajectory starting from $\mathbb{R}^{3\times 4} \setminus \mathcal{U}_{eq}^I$ does not approach the incorrect equilibrium points. Also, it is clear that $\mathcal{C}$ is a positively invariant set. Therefore, it is clear that the region of attraction for $\mathcal{U}_{eq}^C$ is $\mathbb{R}^{3\times 4} \setminus \mathcal{U}_{eq}^I$.

The analysis presented in this section implies that the dimension of configuration of agents does not change, by a gradient control law. Although we have focused on four agents in the 3-D, it may be generalized to a general case. Related to this argument, it may be required to recommend a recent work [21].

3.7 Summary and Simulations

It was shown that the traditional gradient control laws presented in Sects. 3.2, 3.4, and 3.5 could ensure an almost global convergence of triangular formations, $K(3) + 1$ edge formations and $K(4) - 1$ edge formations in 2-D. It was also shown that the gradient control law can stabilize a $K(4)$ formation in 3-D, which is a tetrahedron, in Sect. 3.6. In these gradient control laws, the sensing variables are the relative displacements measured in a local coordinate frame, while the control variables are edge distances. To make the edge distances to be desired ones, the relative displacements have been measured and the distances are controlled. The key feature of the underlying topology for these control laws is a graph rigidity that ensures a unique configuration. Since both the neighboring agents need to control for an edge, the underlying graph is undirected one. Thus, the basic topologies for sensing and actuation for a unique configuration are undirected rigid graphs, i.e., $\mathcal{G} = (\mathcal{V}, \mathcal{E}^s)$ and $\mathcal{G} = (\mathcal{V}, \mathcal{E}^a)$ are rigid graphs, which is a topological perspective. Table 3.1 summarizes these points. In Sect. 3.3, we applied the gradient control law for directed edges, although the formation is limited to a polygon graph. The underlying topology for this case is a cycle, and the edges are directed ones. But, the sensing and control variables are still p_{ji}^i and $\|z_{ij}\|$.

Since a global convergence has been ensured for the considered graphs (i.e., triangular, $K(3) + 1$, $K(4) - 1$, and directed polygon, and tetrahedron in 3-D), when a specific graph, which has been converged to a desired configuration, is given, any isomorphic graphs will be converging to the desired one. That is, the formation control in this chapter has been achieved up to isomorphism. But, although a new labeling is given to the agents, the desired formation will be achieved. Thus, the formation control of this chapter is also achieved up to automorphism, as far as the desired inter-agent distances are given between neighboring agents. Of course, if the desired distances are given to a specific label, the formation control will not be up to automorphism.

Table 3.1 Variables and network properties of traditional gradient control laws

	Variables	Topology	Edge direction
Sensing	p_{ji}^i	Rigidity	Undirected
Control	$\|z_{ij}\|$	Rigidity	Undirected

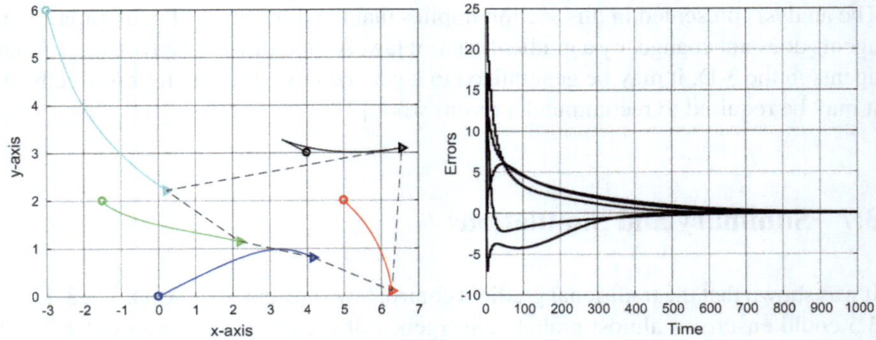

Fig. 3.8 Polygon formation: Agents converge to a desired configuration from any initial positions. Left—Trajectories. Right—Errors of distances

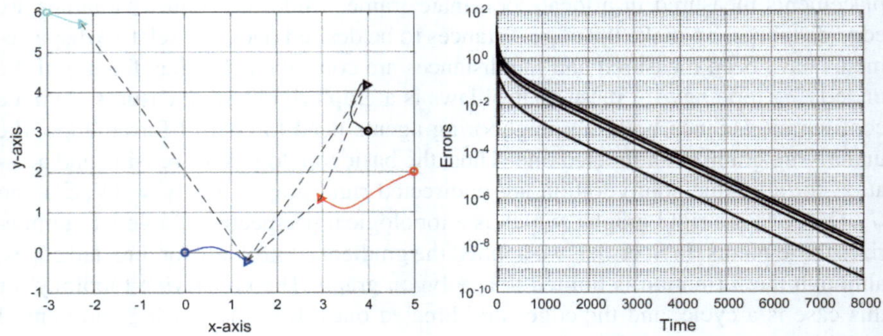

Fig. 3.9 $K(3)+1$ edge formation: Agents converge to a desired configuration from any initial positions. Left- Trajectories. Right- Errors of distances

Figures 3.8, 3.9, and 3.10 show the simulation results of gradient control laws. Figure 3.8 depicts the trajectories of polygon formation composed of five agents, Fig. 3.9 shows the trajectories of $K(3)+1$ formation composed of four agents, and Fig. 3.10 shows the trajectories of $K(4)-1$ formation composed of four agents. These figures depict the trajectories of agents and their final formation configurations. The initial positions of agents are marked by $\circ$ and the final positions are marked by $\triangleright$. From a comparison through simulations (see the right plots), we could observe that the $K(4)-1$ formation takes more time for the convergence than the $K(3)+1$ formation. That is, the $K(3)+1$ formation converges to the desired configuration more rapidly than the $K(4)-1$ formation.

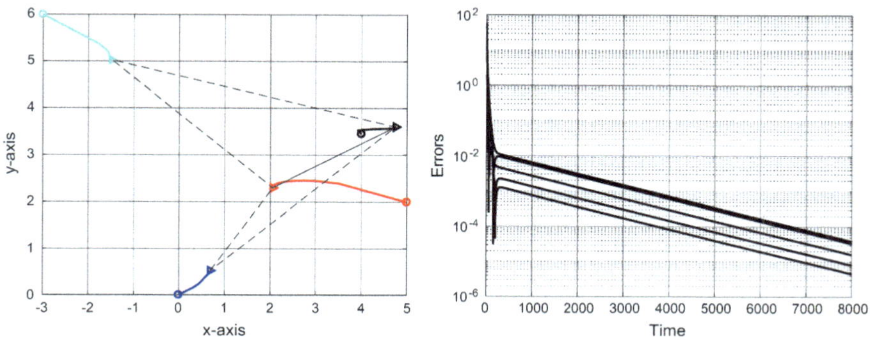

Fig. 3.10 $K(4) - 1$ edge formation: Agents converge to a desired configuration from any initial positions. Left—Trajectories. Right—Errors of distances

3.8 Notes

The results of Sects. 3.2, 3.3, and 3.6 are reused and reproduced from [14, 16–18]. The results of Sects. 3.4 and 3.5 are reused from [19]. Note that [19] also includes a tetrahedron with an extra node in 3-dimensional space, and analysis for generalized flex graphs which are composed of basic rigid groups and flex edges. Although it is difficult to achieve a unique formation with distance constraints, some other information such as area constraints can be combined to achieve a unique configuration [2]. Also, it was analyzed that triangulated Laman graphs may be globally stabilized by gradient-style control laws [7]. The following copyright and permission notices are acknowledged.

- The statements of Lemma 3.2, Theorems 3.1 and 3.2 with the proofs are © [2011] IEEE. Reprinted, with permission, from [13].
- The text of Sect. 3.3 is © [2012] IEEE. Reprinted, with permission, from [17].
- The text of Sect. 3.6 is © [2014] IEEE. Reprinted, with permission, from [18].
- The statements of Lemma 3.2, Theorems 3.1, and 3.2 with the proofs are © Reprinted from [14], with permission from Elsevier.
- The statements of Lemmas 3.3, 3.4, 3.5, 3.6, and Theorem 3.3 with the proofs are © Reproduced from [16] by permission of the Institution of Engineering & Technology.
- The statements of Lemma 3.7 and Theorem 3.5 with the proofs are © Reproduced from [19] by permission of John Wiley and Sons.

References

1. Anderson, B.D.O., Yu, C., Dasgupta, S., Morse, A.S.: Control of a three-coleader formation in the plane. Syst. Control Lett. **56**(9–10), 573–578 (2007)
2. Anderson, B.D.O., Sun, Z., Sugie, T., Azuma, S.-I., Sakurama, K.: Formation shape control with distance and area constraints. IFAC J. Syst. Control **1**, 2–12 (2017)
3. Baillieul, J., Suri, A.: Information patterns and hedging Brockett's theorem in controlling vehicle formations. In: Proceedings of the 42nd IEEE Conference on Decision and Control, vol. 1, pp. 556–563 (2003)
4. Cao, M., Morse, A.S., Yu, C., Anderson, B.D.O., Dasgupta, S.: Controlling a triangular formation of mobile autonomous agents. In: Proceedings of the 46th IEEE Conference on Decision and Control, pp. 3603–3608 (2007)
5. Cao, M., Yu, C., Morse, A.S., Anderson, B.D.O., Dasgupta, S.: Generalized controller for directed triangle formations. In: Proceedings of the IFAC 17th World Congress, pp. 6590–6595 (2008)
6. Cao, M., Morse, A.S., Yu, C., Anderson, B.D.O., Dasgupta, S.: Maintaining a directed, triangular formation of mobile autonomous agents. Commun. Inf. Syst. **11**(1), 1–16 (2011)
7. Chen, X., Belabbas, M.-A., Basar, T.: Global stabilization of triangulated formations. SIAM J. Control Optim. **55**(1), 172–199 (2017)
8. Dattorro, J.: Convex Optimization and Euclidean Distance Geometry. Meboo Publishing, USA (2017)
9. Dörfler, F., Francis, B.: Geometric analysis of the formation problem for autonomous robots. IEEE Trans. Autom. Control **55**(10), 2379–2384 (2010)
10. Eren, T., Belhumeur, P.N., Anderson, B.D.O., Morse, A.S.: A framework for maintaining formations based on rigidity. In: Proceedings of the 2002 IFAC World Congress, pp. 499–504 (2002)
11. Johnson, C.R.: The Euclidean distance completion problem: cycle completability. SIAM J. Matrix Anal. Appl. **16**(2), 646–654 (1995)
12. Krick, L., Broucke, M.E., Francis, B.A.: Stabilization of infinitesimally rigid formations of multi-robot networks. Int. J. Control **82**(3), 423–439 (2009)
13. Oh, K.-K., Ahn, H.-S.: Distance-based formation control using Euclidean distance dynamics matrix: three-agent cases. In: Proceedings of the American Control Conference, pp. 4810–4815 (2011)
14. Oh, K.-K., Ahn, H.-S.: Formation control of mobile agents based on inter-agent distance dynamics. Automatica **47**(10), 2306–2312 (2011)
15. Oh, K.-K., Park, M.-C., Ahn, H.-S.: A survey of multi-agent formation control. Automatica **53**(3), 424–440 (2015)
16. Park, M.-C., Ahn, H.-S.: Stabilisation of directed cycle formations and application to two-wheeled mobile robots. IET Control Theory Appl. **9**(9), 1338–1346 (2015)
17. Park, M.-C., Kim, B.-Y., Oh, K.-K., Ahn, H.-S.: Control of inter-agent distances in cyclic polygon formations. In: Proceedings of the 2012 IEEE Multi-conference on Systems and Control, pp. 951– 955 (2012)
18. Park, M.-C., Sun, Z., Anderson, B.D.O., Ahn, H.-S.: Stability analysis on four agent tetrahedral formations. In: Proceedings of the 53rd IEEE Conference on Decision and Control, pp. 631–636 (2014)
19. Pham, V.H., Trinh, M.H., Ahn, H.-S.: Formation control of rigid graphs with flex edges. Int. J. Robust Nonlinear Control **28**(6), 2543–2559 (2018)
20. Smith, S.L., Broucke, M.E., Francis, B.A.: Stabilizing a multi-agent system to an equilibrium polygon formation. In: Proceedings of the 17th International Symposium on Mathematical Theory of Networks and Systems, pp. 2415–2424 (2006)
21. Sun, Z., Yu, C.: Dimensional-invariance principles in coupled dynamical systems: a unified analysis and applications. IEEE Trans. Autom. Control 1–15 (2018). arXiv:1703.07955

Chapter 4
Local Stabilization

Abstract In Chap. 3, solutions for global stabilization of formation systems with several specific topologies have been presented. In this chapter, generalized formations with n-agents are studied. However, it is difficult to guarantee a global convergence for general n-agent formations without any communication. This chapter provides analysis for local convergence of general n agents under the traditional gradient control laws. Following Chap. 3, it is assumed that agents can sense locations of neighboring agents with respect to their own coordinate frames; but they cannot exchange information or cannot communicate with other neighboring agents. So, the control goal is to achieve the desired distances only based on the relative measurements. Since the control variables are distances, the formation control problems studied in this chapter are classified as distance-based control. By satisfying all the desired distances of rigid graphs, we can achieve a unique configuration in d-dimensional ($d = 2, 3$) space, or in general d-dimensional space, up to translations and rotations. This chapter presents a generalized gradient control law for n agents on the basis of Sect. 3.2.

4.1 Inter-agent Dynamics

We extend the formation control law (3.19), which is for three-agent case, to general n-agent systems. The results are extracted from [8, 9]. For agent i under undirected network topologies, the number of constraints is equal to the number of connections to neighboring agents, which is the degree of agent i. Since the graph is considered to be rigid in 2-dimensional space, the minimum degree of each agent is 2. From (3.19), the constraints for the control law u_i can be modified as

$$\underbrace{\begin{bmatrix} \vdots \\ (p_i - p_j)^T \\ \vdots \end{bmatrix}}_{\triangleq A_i} u_i = -\frac{k_{ij}}{4} \underbrace{\begin{bmatrix} \vdots \\ \bar{e}_{ij} \\ \vdots \end{bmatrix}}_{\triangleq b_i}, \ j \in \mathcal{N}_i, \ i \in \mathcal{V} \tag{4.1}$$

H.-S. Ahn, *Formation Control*, Studies in Systems,
Decision and Control 205, https://doi.org/10.1007/978-3-030-15187-4_4

where the control gains k_{ij} are fixed to a constant. If the agent i has more than three constraints, the matrix A_i is overdetermined; thus there may not exist solutions that satisfy (4.1). To minimize the difference between $\|A_i u_i\|$ and $\|\frac{k_{ij}}{4} b_i\|$, we find a solution u_i such that $A_i u_i$ becomes the projection of $\frac{k_{ij}}{4} b_i$ onto the column space of A_i, which is equivalent to the solution of the following least square problem:

$$u_i = \underset{u_i \in \mathbb{R}^2}{\operatorname{argmin}} \left\| A_i u_i + \frac{k_{ij}}{4} b_i \right\|^2, \; i \in \mathcal{V} \tag{4.2}$$

The solution to the above overdetermined equation is computed as

$$u_i = -\frac{k_{ij}}{4} (A_i^T A_i)^{-1} A_i^T b_i \tag{4.3}$$

when the matrix $A_i^T A_i$, which is expressed as $\sum_{j \in \mathcal{N}_i} (p_i - p_j)(p_i - p_j)^T$, is non-singular. By replacing the position vector p_i by $p_i = (x_i, y_i)^T$, the matrix $A_i^T A_i$ can be further changed as

$$A_i^T A_i = \sum_{j \in \mathcal{N}_i} \begin{bmatrix} (x_i - x_j)^2 & (x_i - x_j)(y_i - y_j) \\ (x_i - x_j)(y_i - y_j) & (y_i - y_j)^2 \end{bmatrix}, \; i \in \mathcal{V} \tag{4.4}$$

It is required to have an inverse of $A_i^T A_i$ to have a unique input u_i. The condition for this is ensured if the framework is infinitesimally rigid, which is summarized as follows [9].

Lemma 4.1 *For the general n-agent case, if the graph $\mathcal{G}$ with realization p, i.e., framework f_p, where $\mathcal{G} = (\mathcal{V}, \mathcal{E})$, is infinitesimally rigid, then $A_i^T A_i$ is positive definite for all $i \in \mathcal{V}$.*

Proof Since it is infinitesimally rigid, there is not a case where three agents are collinear. Thus, $\sum_{j \in \mathcal{N}_i} (x_i - x_j)^2 > 0$ and $\sum_{j \in \mathcal{N}_i} (y_i - y_j)^2 > 0$ for all i. The second leading principal minor of the matrix $A_i^T A_i$ is positive semi-definite by Cauchy–Schwarz inequality [14]:

$$\sum_{j \in \mathcal{N}_i} (x_i - x_j)^2 \sum_{j \in \mathcal{N}_i} (y_i - y_j)^2 - \left(\sum_{j \in \mathcal{N}_i} (x_i - x_j)(y_i - y_j) \right)^2 \geq 0 \tag{4.5}$$

Let the index of neighboring agents of agent i be denoted as $j_i^1, \ldots, j_i^{|\mathcal{N}_i|}$. Then, we know that the equality in (4.5) holds only when two vectors $(x_i - x_{j_i^1}, \ldots, x_i - x_{j_i^{|\mathcal{N}_i|}})^T$ and $(y_i - y_{j_i^1}, \ldots, y_i - y_{j_i^{|\mathcal{N}_i|}})^T$ are linearly dependent. This dependence occurs only when i and $\forall j,\ j \in \mathcal{N}_i$ are on the same line. But, the graph is infinitesimally rigid, and thus the equality from (4.5) can be eliminated by the assumption.

Thus, the matrix $A_i^T A_i$ is always positive definite, which implies there always exists the inverse of $A_i^T A_i$, which is also positive definite.

For the development of convergence analysis, we employ the following inequality, which is called Lojasiewicz's inequality [1].

Lemma 4.2 *Let f be a real analytic function on a neighborhood of z in $\mathbb{R}^n$. Then, there exist constants $c > 0$ and $\rho \in [0, 1)$ such that*

$$\|\nabla f(x)\| \geq c|f(x) - f(z)|^{\rho} \tag{4.6}$$

in some neighborhood of z.

In the inequality (4.6), the existence of $\nabla f(x)$ means that it is differentiable, and existence of $f(x) - f(z)$ in any small neighborhood means that it is continuous. Thus, if the function f is continuously differentiable, then the condition (4.6) always holds.

The desired configuration E_{p^*} defined in (3.10) is not compact since it can be moving to infinity (not bounded as per the Definition 2.30 and Theorem 2.20). But, if we define the desired formation configuration in terms of edge lengths, which is called link space, then the set is compact since it is closed and bounded (see Theorem 2.20). For a realization $p = (p_1^T, \ldots, p_n^T)^T$, we denote an edge vector as $z_{ji} = p_j - p_i$ for the edge $(i, j)^e$. Without loss of generality, let us denote the kth edge as ed_k, $ed_k \in \mathcal{E}$. For undirected graphs, we can consider that an edge has both directions. Now, we define the edge vector as $z = [\ldots, z_{ji}^T, \ldots]^T \in \mathbb{R}^m$, where $|\mathcal{E}| = m$, for all edges and $i < j$. Then, the desired formation configuration in the link space can be expressed as

$$E_{z^*} \triangleq \{z \in \mathbb{R}^m : \|z_{ji}\| = d_{ji}^*, (i, j)^e \in \mathcal{E}, i < j\} \tag{4.7}$$

The dynamics of the formation system can be now expressed in link space as

$$\dot{z} = (\mathbb{H}_+ \otimes \mathbb{I}_2)\dot{p} \tag{4.8}$$

where $\mathbb{H}_+$ is the incidence matrix corresponding to the edge set $\mathcal{E}_+$ of the realized graph and $\mathbb{I}_2$ is the 2×2 matrix. If the switching topology is a dynamical one, we need to consider

$$\dot{z} = (\mathbb{H}_+ \otimes \mathbb{I}_2)\dot{p} + (\dot{\mathbb{H}}_+ \otimes \mathbb{I}_2)p$$

under the condition that $\dot{\mathbb{H}}_+$ is well defined. If the topology is fixed, for example, for the graph of Fig. 4.4, we can set the incidence matrix, with arbitrary directions of edges, as

$$\mathbb{H}_+ = \begin{bmatrix} 1 & -1 & 0 & 0 & 0 \\ 0 & 1 & -1 & 0 & 0 \\ 0 & 0 & 1 & -1 & 0 \\ 0 & 0 & 0 & 1 & -1 \\ -1 & 0 & 0 & 0 & 1 \\ 1 & 0 & -1 & 0 & 0 \\ 1 & 0 & 0 & -1 & 0 \\ 0 & 1 & 0 & -1 & 0 \\ 0 & 0 & -1 & 0 & 1 \end{bmatrix}$$

In the above incidence matrix, according to the order of indices of edges and according to the directions of edge vectors, the row vectors can be permutated and the sign of 1 and -1 in each row can be switched. Thus, in an undirected graph, if we are interested in the inter-agent distances, the incidence matrix can be computed quite flexibly.

Let us define a potential function $V(z)$ in the link space as

$$V(z) = \frac{k_p}{4} \sum_{(i,j)^e \in \mathcal{E}} (\bar{e}_{ji})^2 \tag{4.9}$$

where the error $\bar{e}_{ij}$ is defined as $\bar{e}_{ji} \triangleq \|z_{ji}\|^2 - \bar{d}_{ji}^*$. Let us further define $\bar{e} = (\ldots, \bar{e}_{ji}, \ldots)^T$ and $D_{(A_i^T A_i)^{-1}} = \text{diag}[\ldots, (A_i^T A_i)^{-1}, \ldots]$, and denote the rigidity matrix in terms of edge vector as $\mathbb{R}_\mathcal{G}(z)$. Remark that the gradient of the potential function $V(z)$ can be represented by the rigidity matrix as $\nabla V(z) = k_p(\mathbb{R}_\mathcal{G}(z))^T \bar{e}$. Then, we can make the following result.

Lemma 4.3 *Suppose that the desired framework $f_{p^*} = (\mathcal{G}, p^*)$ is infinitesimally rigid. Then, there exists a level set $\Omega_c = \{z : V(z) \leq c\}$ such that $D_{(A_i^T A_i)^{-1}}$ is positive definite and $(\mathbb{R}_\mathcal{G}(z))^T \bar{e} \neq 0$ for any $z \in \Omega_c$ and $z \notin E_{z^*}$.*

Proof Since $(\mathcal{G}, p^*)$ is infinitesimally rigid, any p, which is close to p^*, is infinitesimally rigid from the definition of rigidity. Thus, by Lemma 4.1, the diagonal matrix $D_{(A_i^T A_i)^{-1}}$ is positive definite. Thus, it is clear that there exists $\rho_{max} > 0$ such that for any ρ satisfying $\rho_{max} \geq \rho > 0$, $D_{(A_i^T A_i)^{-1}}$ is positive definite for any $z \in \Omega_\rho$, where $\Omega_\rho = \{z : V(z) \leq \rho\}$. Next, the function $V(z) = \frac{k_p}{4} \sum_{(i,j)^e \in \mathcal{E}} (\bar{e}_{ij})^2$ is analytic in some neighborhood of $p^\dagger \in E_{p^*}$ since it is continuously differentiable. Thus, by Lojasiewicz lemma given in Lemma 4.2, there exists a neighborhood $\mathcal{U}_{p^\dagger}$ of $p^\dagger$ and constants $k_{p^\dagger}$ and $0 \leq \rho_{p^\dagger} < 1$ such that

$$\|\nabla V(z)\| = \|k_p(\mathbb{R}_\mathcal{G}(z))^T \bar{e}\| \geq k_{p^\dagger} \|V(p) - V(p^\dagger)\|^{\rho_{p^\dagger}} \tag{4.10}$$

for all $p \in \mathcal{U}_{p^\dagger}$. Due to the completeness of the set, we have

$$\|k_p(\mathbb{R}_\mathcal{G}(z))^T \bar{e}\| \geq k_{p^\dagger} \|V(p)\|^{\rho_{p^\dagger}} > 0 \tag{4.11}$$

for all $p \in \{p : p \in \mathcal{U}_{p^\dagger} \text{ and } p \notin E_{p^*}\}$ since $V(p^\dagger) = 0$. So, for any $\bar{z} \in E_{z^*}$, we can take a neighborhood $\mathcal{U}_{\bar{z}}$ of $\bar{z}$ so that

$$\|(\mathbb{R}_{\mathcal{G}}(z))^T \bar{e}\| > 0 \tag{4.12}$$

where $z \in \{z : z \in \mathcal{U}_{\bar{z}} \text{ and } z \notin E_{z^*}\}$.

Next, due to the rigidity of the graph, there could be a number of different (i.e., equivalence but not congruent) rigid configurations, which is finite. Let us denote the number of different configurations by N_z. Then, for any $k \in \{1, \ldots, N_z\}$, $\bar{z}_k \in E_{z^*}$ holds. Since the set E_{z^*} is compact as per Theorem 2.20, there exists a finite open cover $\mathcal{U}_{E_{z^*}} = \bigcup_{k=1}^{N_z} \mathcal{U}_{\bar{z}_k}$ such that (4.12) holds for all $z \in \mathcal{U}_{E_{z^*}}$. Thus, taking Ω_c and c such that $\Omega_c \subseteq \mathcal{U}_{E_{z^*}}$ and $c \leq \rho_{max}$, we can ensure that $D_{(A_i^T A_i)^{-1}}$ is positive definite for any $z \in \Omega_c$ and $(\mathbb{R}_{\mathcal{G}}(z))^T \bar{e} \neq 0$ for any $z \in \Omega_c^\wedge \triangleq \{z : z \in \Omega_c \text{ and } z \notin E_{z^*}\}$.

Now, with the above lemma, we can make the following theorem for local convergence of a general n-agent formation [9].

Theorem 4.1 *Suppose that the desired framework $(\mathcal{G}, p^*)$ is infinitesimally rigid. Then, the control law (4.3) ensures the local asymptotic stability of the desired set E_{p^*} and the control law can be implemented in a distributed way.*

Proof Let us use the potential function $V(z)$ of (4.9) as a Lyapunov candidate. The derivative can be obtained as

$$\dot{V}(z) = -k_p(\bar{e})^T \mathbb{R}_{\mathcal{G}}(z) D_{(A_i^T A_i)^{-1}} (\mathbb{R}_{\mathcal{G}}(z))^T \bar{e} \tag{4.13}$$

From Lemma 4.3, there exists a level set Ω_c such that $D_{(A_i^T A_i)^{-1}}$ is positive definite and $(\mathbb{R}_{\mathcal{G}}(z))^T \bar{e} \neq 0$ for any $z \in \Omega_c^\wedge = \{z : z \in \Omega_c \text{ and } z \notin E_{z^*}\}$. Thus, $\dot{V}(z)$ is negative definite in $\Omega_c^\wedge$, which implies a local asymptotic stability of E_{z^*}. Thus, it is direct to conclude that the set E_{p^*} is locally asymptotically stable. Next, in (4.3), with the assumption of $p_i^i = 0$ (i.e., agent i considers the origin of its local coordinate frame as zero), the term $A_i^T A_i = \sum_{j \in \mathcal{N}_i} (p_i - p_j)(p_i - p_j)^T$ can be changed as $(A_i^T A_i)^i = \sum_{j \in \mathcal{N}_i} (p_j^i)(p_j^i)^T$ in the local coordinate frame and the matrix A_i^T can be changed as $(A_i^T)^i = (\ldots, -p_j^i, \ldots)$ in the local coordinate frame. Thus, u_i^i is obtained as

$$u_i^i = \frac{k_{ij}}{4} \sum_{j \in \mathcal{N}_i} [p_j^i (p_j^i)^T]^{-1} (\ldots, p_j^i, \ldots) b_i \tag{4.14}$$

which is the implementation in the local coordinate frame.

Figure 4.1 depicts the main idea of the proof of Lemma 4.3 and Theorem 4.1. The positive definiteness of $D_{(A_i^T A_i)^{-1}}$ and $(\mathbb{R}_{\mathcal{G}}(z))^T \bar{e} \neq 0$ is ensured by Lemma 4.3 and the negative definiteness of $\dot{V}(z)$ is ensured in Theorem 4.1, which implies the local asymptotical stability of E_{p^*}.

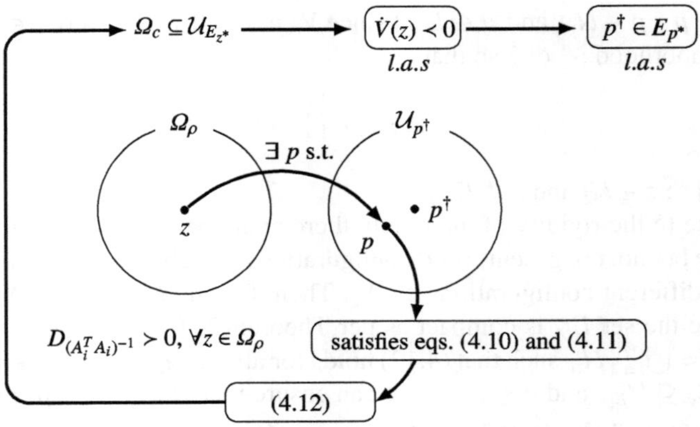

Fig. 4.1 Sketch of the proof of Lemma 4.3 and Theorem 4.1; the acronym *l.a.s* means locally asymptotically stable

The above theorem ensures a local asymptotic stability of the desired set in link space. So, it is not certain whether the final formation would be stationary or not. That is, we still need to show a convergence to a point p in E_{p^*}, which is $\dot{p} = 0$.

Theorem 4.2 *Suppose that the initial and desired frameworks $(\mathcal{G}, p^*)$ are infinitesimally rigid. Then, the control law (4.3) ensures a convergence to a point in E_{p^*}.*

Proof From the definition of the rigidity matrix, it can be shown that the control law (4.3) is represented as

$$u = -k_{gg} D_{(A_i^T A_i)^{-1}} (\mathbb{R}_{\mathcal{G}}(z))^T \bar{e} \tag{4.15}$$

where k_{gg} is a constant. From Lemma 4.3, $D_{(A_i^T A_i)^{-1}}$ is positive definite and $(\mathbb{R}_{\mathcal{G}}(z))^T \bar{e} \neq 0$. But from Theorem 4.1, we know that $\bar{e} \to 0$ asymptotically, which implies $u \to 0$ and $\dot{p} \to 0$.

The control law (4.15) is more general than the traditional gradient control law since the term $(\mathbb{R}_{\mathcal{G}}(z))^T \bar{e}$ is multiplied by a positive definite matrix. That is, even if the traditional gradient control law is multiplied by any positive definite matrix, it still ensures a local asymptotic convergence. In the next section, we introduce a gradient control law for local exponential convergence that was developed in [12].

4.2 Exponential Convergence

This section provides an exponential convergence of rigid graphs with a gradient-based modified control law. The main idea is to divide a graph with minimally rigid part and other remaining parts. The desired framework is given by a tuple $(\mathcal{G}, p^*)$ in $\mathbb{R}^2$, and supposed to be infinitesimally rigid. So, the number of edges of $\mathcal{G}$ is greater

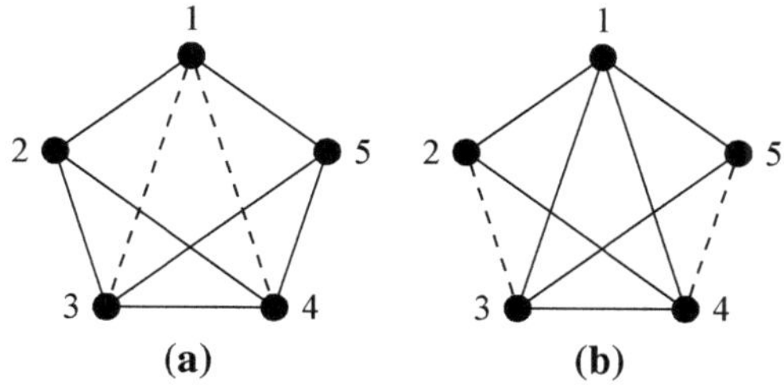

Fig. 4.2 Decomposition of a rigid graph into a minimally infinitesimally rigid graph and other parts. The graph composed by the solid lines is the minimally rigid part and the graph corresponding to the dashed lines is the remaining part

than $2n - 3$. It is possible to select a subgraph $\underline{\mathcal{G}} = (\mathcal{V}, \underline{\mathcal{E}})$ of $\mathcal{G}$ such that it includes $2n - 3$ edges. Specially, let $\underline{\mathcal{G}}$ be a minimally infinitesimally rigid graph. Define an index set $\mathcal{I}$ for the edges as $\mathcal{I} = \{1, \ldots, |\underline{\mathcal{E}}|, |\underline{\mathcal{E}}| + 1, \ldots, |\mathcal{E}|\}$. Here, the indices $\{1, \ldots, |\underline{\mathcal{E}}|\}$ are denoted by the set $\underline{\mathcal{I}}$ and the indices $\{|\underline{\mathcal{E}}| + 1, \ldots, |\mathcal{E}|\}$ are denoted by the set $\tilde{\mathcal{I}}$. For example, the graph of Fig. 4.4 can be decomposed as a minimally rigid part and another part, in various ways. Figure 4.2 shows two possible decompositions.

Let us denote the error of the kth edge corresponding to the index $k \in \mathcal{I}$ as $\bar{e}_k = \|z_k\|^2 - \bar{d}_k^*$. Next let $\underline{e} = (\bar{e}_1, \ldots, \bar{e}_{|\underline{\mathcal{E}}|})^T$ and $\tilde{e} = (\bar{e}_{|\underline{\mathcal{E}}|+1}, \ldots, \bar{e}_{|\mathcal{E}|})^T$, and $\bar{e} = (\underline{e}^T, \tilde{e}^T)^T$. Similarly to the previous chapter, the following potential function is considered:

$$V(p) = \frac{1}{4}\bar{e}^T \bar{e}$$

from which the gradient control law is obtained as

$$\dot{p} = u = -\left[\frac{\partial V}{\partial p}\right]^T = -\mathbb{R}_{\mathcal{G}}^T \bar{e} \tag{4.16}$$

where $\mathbb{R}_{\mathcal{G}}$ is the rigidity matrix, which can be also calculated as $\mathbb{R}_{\mathcal{G}} = \frac{1}{2}[\partial \bar{e}/\partial p]$. From (4.16), the errors can be decomposed as

$$\dot{p} = -\underline{\mathbb{R}}_{\mathcal{G}}^T \underline{e} - \tilde{\mathbb{R}}_{\mathcal{G}}^T \tilde{e} \tag{4.17}$$

where $\underline{\mathbb{R}}_{\mathcal{G}}$ and $\tilde{\mathbb{R}}_{\mathcal{G}}$ are computed as $\underline{\mathbb{R}}_{\mathcal{G}} = \frac{1}{2}\frac{\partial \underline{e}}{\partial p}$ and $\tilde{\mathbb{R}}_{\mathcal{G}} = \frac{1}{2}\frac{\partial \tilde{e}}{\partial p}$. The main idea underlying in this section is to ensure an exponential convergence of the minimally infinitesimally rigid part graph $\underline{\mathcal{G}}$, and then due to the rigidity property, the remaining part will be also exponentially convergent to the desired inter-agent distances. With this motivation, the following modified control law is used to guarantee an exponential convergence:

$$\dot{p} = -k\underline{\mathbb{R}}_{\mathcal{G}}^T\underline{e} - \tilde{\mathbb{R}}_{\mathcal{G}}^T\tilde{e},\ k > 0 \tag{4.18}$$

Then, the error dynamics can be obtained as

$$\begin{aligned}
\dot{\bar{e}} &= \frac{\partial \bar{e}}{\partial p}\dot{p} \\
&= 2\mathbb{R}_{\mathcal{G}}(-k\underline{\mathbb{R}_{\mathcal{G}}}^T\underline{e} - \tilde{\mathbb{R}_{\mathcal{G}}}^T\tilde{e}) \\
&= -2\begin{bmatrix} \underline{\mathbb{R}}_{\mathcal{G}} \\ \tilde{\mathbb{R}}_{\mathcal{G}} \end{bmatrix}\begin{bmatrix} k\underline{\mathbb{R}}_{\mathcal{G}}^T & \tilde{\mathbb{R}}_{\mathcal{G}}^T \end{bmatrix}\begin{bmatrix} \underline{e} \\ \tilde{e} \end{bmatrix} \\
&= -2\begin{bmatrix} k\underline{\mathbb{R}}_{\mathcal{G}}\underline{\mathbb{R}}_{\mathcal{G}}^T\underline{e} + \underline{\mathbb{R}}_{\mathcal{G}}\tilde{\mathbb{R}}_{\mathcal{G}}^T\tilde{e} \\ k\tilde{\mathbb{R}}_{\mathcal{G}}\underline{\mathbb{R}}_{\mathcal{G}}^T\underline{e} + \tilde{\mathbb{R}}_{\mathcal{G}}\tilde{\mathbb{R}}_{\mathcal{G}}^T\tilde{e} \end{bmatrix}
\end{aligned} \tag{4.19}$$

With the above relationship, it is shown that the errors are governed by

$$\dot{\underline{e}} = -2k\underline{\mathbb{R}}_{\mathcal{G}}\underline{\mathbb{R}}_{\mathcal{G}}^T\underline{e} - 2\underline{\mathbb{R}}_{\mathcal{G}}\tilde{\mathbb{R}}_{\mathcal{G}}^T\tilde{e} \tag{4.20}$$

and

$$\dot{\tilde{e}} = -2k\tilde{\mathbb{R}}_{\mathcal{G}}\underline{\mathbb{R}}_{\mathcal{G}}^T\underline{e} - 2\tilde{\mathbb{R}}_{\mathcal{G}}\tilde{\mathbb{R}}_{\mathcal{G}}^T\tilde{e} \tag{4.21}$$

Example 4.1 For the case of Fig. 4.2a, let us assign the index of edges as $(1,2)^e = ed_1, (2,3)^e = ed_2, (3,4)^e = ed_3, (4,5)^e = ed_4, (1,5)^e = ed_5, (1,3)^e = ed_6, (1,4)^e = ed_7, (2,4)^e = ed_8, (3,5)^e = ed_9$. Then, the rigidity matrix is obtained as

$$\mathbb{R}_{\mathcal{G}} = \begin{bmatrix}
p_1^T - p_2^T & p_2^T - p_1^T & 0 & 0 & 0 \\
0 & p_2^T - p_3^T & p_3^T - p_2^T & 0 & 0 \\
0 & 0 & p_3^T - p_4^T & p_4^T - p_3^T & 0 \\
0 & 0 & 0 & p_4^T - p_5^T & p_5^T - p_4^T \\
p_1^T - p_5^T & 0 & 0 & 0 & p_5^T - p_1^T \\
p_1^T - p_3^T & 0 & p_3^T - p_1^T & 0 & 0 \\
p_1^T - p_4^T & 0 & 0 & p_4^T - p_1^T & 0 \\
0 & p_2^T - p_4^T & 0 & p_4^T - p_2^T & 0 \\
0 & 0 & p_3^T - p_5^T & 0 & p_5^T - p_3^T
\end{bmatrix}$$

which, according to Fig. 4.2a, is decomposed as

$$\underline{\mathbb{R}}_{\mathcal{G}} = \begin{bmatrix} p_1^T - p_2^T & p_2^T - p_1^T & 0 & 0 & 0 \\ 0 & p_2^T - p_3^T & p_3^T - p_2^T & 0 & 0 \\ 0 & 0 & p_3^T - p_4^T & p_4^T - p_3^T & 0 \\ 0 & 0 & 0 & p_4^T - p_5^T & p_5^T - p_4^T \\ p_1^T - p_5^T & 0 & 0 & 0 & p_5^T - p_1^T \\ 0 & p_2^T - p_4^T & 0 & p_4^T - p_2^T & 0 \\ 0 & 0 & p_3^T - p_5^T & 0 & p_5^T - p_3^T \end{bmatrix}$$

and

$$\tilde{\mathbb{R}}_{\mathcal{G}} = \begin{bmatrix} p_1^T - p_3^T & 0 & p_3^T - p_1^T & 0 & 0 \\ p_1^T - p_4^T & 0 & 0 & p_4^T - p_1^T & 0 \end{bmatrix}$$

In this case, we can see that $\bar{e}$ is in $\mathbb{R}^7$ and $\tilde{e}$ is in $\mathbb{R}^2$.

Let us first analyze the convergence of (4.20). For a simplicity of presentation, let us denote the first and second terms on the right-hand side of (4.20) as $f(\underline{e}) = -2k\underline{\mathbb{R}}_{\mathcal{G}}\underline{\mathbb{R}}_{\mathcal{G}}^T\underline{e}$ and $\Delta f(\bar{e}) = -2\underline{\mathbb{R}}_{\mathcal{G}}\tilde{\mathbb{R}}_{\mathcal{G}}^T\tilde{e}$. Let $\mathbf{h}_{\mathcal{G}}(p)$ be the edge function corresponding to the graph $\mathcal{G}$ and $\underline{\mathbf{h}}_{\mathcal{G}}(p)$ be the edge function corresponding to the subgraph $\underline{\mathcal{G}}$. Let $\mathbf{h}_K(p)$ be the edge function of $(K_{\mathcal{G}}, p)$, where $K_{\mathcal{G}}$ is the complete graph with the same vertex set. Then, from the rigidity theory (see Sect. 2.2.1), there certainly exists a neighborhood $\mathcal{U}_{edge}$ in the line space of $\underline{\mathbf{h}}_{\mathcal{G}}(p)$ such that $\mathbf{h}_K^{-1}(\mathbf{h}_K(p)) = \underline{\mathbf{h}}_{\mathcal{G}}^{-1}(\underline{\mathbf{h}}_{\mathcal{G}}(p))$ for all $p \in \mathcal{U}_{edge}$.

Note that since the graph considered in this work is rigid, $-2\underline{\mathbb{R}}_{\mathcal{G}}\tilde{\mathbb{R}}_{\mathcal{G}}^T\tilde{e}$ can be expressed as a function of $\underline{e}$ in a local neighborhood $\mathcal{U}_{edge}$ of the realization corresponding to $\underline{e}$. That is, in a local neighborhood $\mathcal{U}_{edge}$, since the equivalence implies congruence of graph, it can be considered as globally rigid in a very small neighborhood of p, $p \in \mathcal{U}_{edge}$. Now, we can obtain an exponential convergence for the minimally rigid subgraph [12].

Lemma 4.4 *Suppose that* $\Delta f = 0$. *Then, the origin of the nominal system* $\dot{\underline{e}} = f(\underline{e})$ *is exponentially stable.*

Proof Consider $\mathcal{D} = \{\underline{e} \in \mathbb{R}^{|\underline{\mathcal{E}}|} : \|\underline{e}\| < \rho, \underline{\mathbf{h}}_{\mathcal{G}}(p) \in \mathcal{U}_{edge}\}$. Using the potential function $\underline{V} = \frac{1}{4}\underline{e}^T\underline{e}$, it leads

$$\begin{aligned} \dot{\underline{V}} &= \frac{\partial \underline{V}}{\partial \underline{e}} f(\underline{e}) \\ &= \frac{1}{2}\underline{e}^T(-2k\underline{\mathbb{R}}_{\mathcal{G}}\underline{\mathbb{R}}_{\mathcal{G}}^T\underline{e}) \\ &\leq -4k\left[\min_{\underline{e}\in\mathcal{D}} \lambda_{min}(\underline{\mathbb{R}}_{\mathcal{G}}\underline{\mathbb{R}}_{\mathcal{G}}^T)\right]\underline{V} \end{aligned} \tag{4.22}$$

As shown in the previous section (see Theorem 4.1), since we consider only the motions on the plane, $\underline{\mathbb{R}}_{\mathcal{G}}\underline{\mathbb{R}}_{\mathcal{G}}^T$ is positive definite for all $\underline{e} \in \mathcal{D}$, which means that

$\lambda_{min}(\underline{\mathbb{R}}_{\mathcal{G}}\underline{\mathbb{R}}_{\mathcal{G}}^T) > 0$ holds in $\mathcal{D}$. Therefore, $\|\underline{e}\|$ converges to 0 locally exponentially fast.

Next, for the convergence analysis of (4.21), let $g(\tilde{e}) = -2\tilde{\mathbb{R}}_{\mathcal{G}}\tilde{\mathbb{R}}_{\mathcal{G}}^T\tilde{e}$ and $\Delta g(\tilde{e}) = -2k\tilde{\mathbb{R}}_{\mathcal{G}}\underline{\mathbb{R}}_{\mathcal{G}}^T\underline{e}$.

Lemma 4.5 *Suppose that $\Delta g(\bar{e}) = 0$. Then, the origin of the nominal system $\dot{\tilde{e}} = g(\tilde{e})$ is exponentially stable.*

Proof Let us show that the matrix $\tilde{\mathbb{R}}_{\mathcal{G}}\tilde{\mathbb{R}}_{\mathcal{G}}^T$ is positive definite in $\mathcal{D}$. Take a point $p^\dagger \in \mathcal{U}_{edge}$. Select a small neighborhood of $p^\dagger$ as $\mathcal{U}_{p^\dagger}$ and consider a point p in $\mathcal{U}_{p^\dagger}$ as $p \in \mathcal{U}_{p^\dagger}$. Then, choosing a potential function $V(p)$ as $V(p) = \frac{k_p}{4}\sum_{i\in\underline{\mathcal{I}}}\tilde{e}_i^2$, and taking a gradient of $V(p)$, by Lojasiewicz lemma, we can see that there exist constants $k_{p^\dagger} > 0$ and $0 < \rho_{p^\dagger} < 1$ such that

$$\|\nabla V(p)\| = k_p\|\tilde{\mathbb{R}}_{\mathcal{G}}^T\tilde{e}\| \geq k_{p^\dagger}\|V(p) - V(p^\dagger)\|^{\rho_{p^\dagger}} = k_{p^\dagger}\|V(p)\|^{\rho_{p^\dagger}} > 0 \quad (4.23)$$

due to $V(p^\dagger) = 0$. Moreover, since the graph is rigid, in a local neighborhood $\mathcal{U}_{edge}$, the motions of edges corresponding to the subgraph $\tilde{\mathbb{R}}_{\mathcal{G}}$ have the same property as in a rigid graph (for more detailed discussions, it is recommended to refer to [17]). Thus, by following the same procedure as Lemma 4.4, it can be shown that $\tilde{\mathbb{R}}_{\mathcal{G}}\tilde{\mathbb{R}}_{\mathcal{G}}^T$ is positive definite within $\mathcal{D}$.

Lemma 4.6 ([7]) *Consider a perturbed system $\dot{x} = f(x) + g(x)$ with $\|g(x)\| \leq \gamma\|x\|$, where γ is a positive constant, $\forall x \in D$. Suppose that the nominal system $\dot{x} = f(x)$ is locally Lipschitz in x on D and the domain $D \subseteq \mathbb{R}^n$ contains the origin $x = 0$. Further suppose that for the nominal system $\dot{x} = f(x)$, there exists a Lyapunov function $V(x)$ such that*

$$c_1\|x\|^2 \leq V(x) \leq c_2\|x\|^2$$
$$\frac{\partial V}{\partial x}f(x) \leq -c_3\|x\|^2$$
$$\left\|\frac{\partial V}{\partial x}\right\| \leq c_4\|x\|$$

in the domain D with positive constants c_1, c_2, c_3, and c_4. If $g(x)$ is also Lipschitz in D and $\gamma < \frac{c_3}{c_4}$, then the origin of $\dot{x} = f(x) + g(x)$ is exponentially stable.

Lemma 4.7 *There exists a k that ensures an exponential convergence of (4.20) and (4.21).*

Proof For the case of (4.20), denote $\gamma = \sum_{\underline{e}\in\mathcal{D},\underline{e}\neq 0}\frac{\|\Delta f(\underline{e})\|}{\|\underline{e}\|}$ and choose k so that $\frac{\gamma}{2\lambda_{min}(\underline{\mathbb{R}}_{\mathcal{G}}\underline{\mathbb{R}}_{\mathcal{G}}^T)} < k$. Then, by Lemma 4.6, since $\left\|\frac{\partial V}{\partial \underline{e}}\right\| = \frac{1}{2}\|\underline{e}\|$, the origin of (4.20) is exponentially stable with $c_1 = c_2 = \frac{1}{4}$, $c_3 = k\min_{\underline{e}\in\mathbb{D}}\lambda_{min}(\underline{\mathbb{R}}_{\mathcal{G}}\underline{\mathbb{R}}_{\mathcal{G}}^T)$ and $c_4 = \frac{1}{2}$. For the case (4.21), the potential function $\tilde{V}(p) = \frac{k_p}{4}\|\tilde{e}\|^2$ is used. Then, from $\dot{\tilde{V}} = \frac{k_p}{2}\tilde{e}^T\dot{\tilde{e}} = -k_p\tilde{e}^T\tilde{\mathbb{R}}_{\mathcal{G}}\tilde{\mathbb{R}}_{\mathcal{G}}^T\tilde{e}$, we also select $c_3 = k_p\min_{\tilde{e}\in\mathbb{D}}\lambda_{min}(\tilde{\mathbb{R}}_{\mathcal{G}}\tilde{\mathbb{R}}_{\mathcal{G}}^T)$ and $c_4 =$

$\frac{k_p}{2}$. Now, due to the fact that $\underline{e} \to 0$ exponentially fast, we can see that $\|\Delta g\| = \|2k\tilde{\mathbb{R}}_{\mathcal{G}}\underline{\mathbb{R}}_{\mathcal{G}}^T\underline{e}\| \to 0$ exponentially fast. Thus, we can infer that $\gamma < \frac{c_3}{c_4} = \frac{1}{2} \min_{\tilde{e}\in\mathbb{D}} \lambda_{min}(\tilde{\mathbb{R}}_{\mathcal{G}}\tilde{\mathbb{R}}_{\mathcal{G}}^T)$ as $t \to \infty$. Consequently, (4.21) is also exponentially stable.

Theorem 4.3 *The dynamics* (4.18) *converges to a desired configuration locally exponentially fast.*

Proof The proof is direct from Lemmas 4.4, 4.5, and 4.7.

Theorem 4.3 has a drawback over Theorem 4.1, because it needs to satisfy the condition $\frac{\gamma}{2\lambda_{min}(\underline{\mathbb{R}}_{\mathcal{G}}\underline{\mathbb{R}}_{\mathcal{G}}^T)} < k$ that may require a piece of global information. Also, the new controller (4.18) needs to separate the graph into two parts $\underline{\mathbb{R}}_{\mathcal{G}}$ and $\tilde{\mathbb{R}}_{\mathcal{G}}$; but for this separation, a centralized coordinator may be necessary. It is remarkable that, from a further analysis of (4.20), it was also shown that a gradient control law, with general potential functions, ensures a locally exponential convergence to a desired configuration via linearization [16, 17]. The convergence speed is determined by $\lambda_{min}(\underline{\mathbb{R}}_{\mathcal{G}}\underline{\mathbb{R}}_{\mathcal{G}}^T)$. The nonzero smallest eigenvalue $\lambda_{min}(\underline{\mathbb{R}}_{\mathcal{G}}\underline{\mathbb{R}}_{\mathcal{G}}^T)$ has been called the *index of rigidity* [19]. It means that if the graph tends to be more rigid, then the convergence speed becomes faster.

4.3 Local Asymptotic Stability in d-Dimensional Space

In this section, we study a local stability of formation control systems in general d-dimensional space under gradient control law. For this, let us define a local potential function ϕ_i as follows:

$$\phi_i(p) \triangleq \frac{k_p}{2} \sum_{j\in\mathcal{N}_i} \gamma(\|z_{ji}\|^2 - \|p_j^* - p_i^*\|^2) \tag{4.24}$$

where $p_i \in \mathbb{R}^d$, $k_p > 0$ and $\gamma : \mathbb{R} \to \mathbb{R}_+$ is a function satisfying the following conditions:

- $\gamma(x) \geq 0$ for any $x \in \mathbb{R}$ and $\gamma(x) = 0$ if and only if $x = 0$,
- $\gamma(x)$ is analytic in a neighborhood of 0.

Note that in (4.24), p_i^* and p_j^* are desired positions of agents i and j. In the position-based formation control, they may be given in a position. But, in displacement-based formation control, p_i^* and p_j^* could be any ones satisfying $p_j^* - p_i^* = (p_j - p_i)^*$, where $(p_j - p_i)^*$ is the desired displacement vector. In distance-based formation control systems, they could be any ones satisfying $\|p_j^* - p_i^*\| = d_{ij}^*$. The summation of all local potential functions ϕ_i is given as

$$\phi(p) \triangleq \frac{k_p}{2} \sum_{(i,j)^e \in \mathcal{E}_+} \gamma(\|z_{ji}\|^2 - \|p_j^* - p_i^*\|^2) = \frac{k_p}{2} \sum_{(i,j)^e \in \mathcal{E}_+} \gamma(\|z_{ij}\|^2 - \|p_j^* - p_i^*\|^2) \tag{4.25}$$

Note that without notational confusion, in undirected graphs, since $\|z_{ij}\| = \|z_{ji}\|$, we use the right-most equation although the sensing measurements are denoted as z_{ji}. With the potential function given in (4.24), the control input for agent i can be designed as

$$\begin{aligned} u_i &= -\nabla_{p_i}\phi_i \\ &= -\left[\frac{\partial \phi_i}{\partial p_i}\right]^T \\ &= -\left[\frac{k_p}{2}\sum_{j\in\mathcal{N}_i}\frac{\partial\gamma(\bar{e}_{ij})}{\partial \bar{e}_{ij}}\frac{\partial \bar{e}_{ij}}{\partial p_i}\right]^T \\ &= k_p \sum_{j\in\mathcal{N}_i}\frac{\partial\gamma(\bar{e}_{ij})}{\partial \bar{e}_{ij}}(p_j - p_i) \end{aligned} \tag{4.26}$$

where $\bar{e}_{ij} = \|z_{ij}\|^2 - \|p_i^* - p_j^*\|^2 = \|z_{ij}\|^2 - \|z_{ij}^*\|^2$. The local stability of the control law (4.26) was analyzed in [10]. The potential function is more general compared to the traditional gradient control law. The traditional gradient law is a form of $\gamma = \sum(\bar{d}_{ij} - \bar{d}_{ij}^*)^2 = \sum \bar{e}_{ij}^2$; so the control input is $u_i = \sum_{j\in\mathcal{N}_i} \bar{e}_{ij}(p_j - p_i)$. Hence, if the error is bigger, the agent generates more control efforts since the input u_i is proportional to the magnitude of the error. However, depending upon the sensing characteristics of distributed agents, the control efforts may be inverse-proportional to the magnitude of the error, which is described in the following example.

Example 4.2 The inverse-proportional potential function may be designed as

$$\gamma_i(\bar{e}_{ij}) = \frac{c\bar{e}_{ij}^2}{k_1\bar{e}_{ij}^4 + k_2\bar{e}_{ij}^2 + k_3}$$

where $c > 0, k_1 > 0, k_2 > 0$, and $k_3 > 0$. Clearly, it is analytic and $\gamma_i(\bar{e}_{ij}) = 0$ if and only if $\bar{e}_{ij} = 0$. Since the control input is computed as

$$u_i = \sum_{j\in\mathcal{N}_i} \frac{(2k_1 - 4ck_1)\bar{e}_{ij}^5 + (2k_2 - 2ck_2)\bar{e}_{ij}^3 + 2k_3\bar{e}_{ij}}{(k_1\bar{e}_{ij}^4 + k_2\bar{e}_{ij}^2 + k_3)^2}(p_j - p_i)$$

we have $u_i \to 0$ when $\bar{e}_{ij} \to \infty$.

In the remaining part of this section, let us summarize the results of [10]. Define the stacked vectors $z = (z_1^T, \ldots, z_m^T)^T = (\mathbb{H}_+ \otimes \mathbb{I}_d)p$ and $z^* = (z_1^{*T}, \ldots, z_m^{*T})^T = (\mathbb{H}_+ \otimes \mathbb{I}_d)p^*$, where $\mathbb{H}_+$ is the incidence matrix corresponding to the edge set $\mathcal{E}_+$.

Then, the control inputs (4.26) can be arranged in a vector form as

$$\begin{aligned} u &= -\nabla\phi(p) \\ &= -k_p(\mathbb{H}_+ \otimes \mathbb{I}_d)^T \mathrm{diag}(z)\Gamma(\bar{e}) \end{aligned} \tag{4.27}$$

where $\Gamma(\bar{e})$ is computed as

$$\Gamma(\bar{e}) = \left(\frac{\partial\gamma(\bar{e}_1)}{\partial\bar{e}_1}, \ldots, \frac{\partial\gamma(\bar{e}_m)}{\partial\bar{e}_m}\right)^T \tag{4.28}$$

For a given desired realization $p^* = (p_1^{*T}, \ldots, p_n^{*T})^T \in \mathbb{R}^{nd}$, the set of desired frameworks, which are congruent to $(\mathcal{G}, p^*)$, is defined as

$$E_{p^*} \triangleq \{p \in \mathbb{R}^{nd} : \|p_i - p_j\| = \|p_i^* - p_j^*\|, \forall i, j \in \mathcal{V}\} \tag{4.29}$$

and the set of realizations that are equivalent to $(\mathcal{G}, p^*)$ is defined as

$$E'_{p^*} \triangleq \{p \in \mathbb{R}^{nd} : \|p_i - p_j\| = \|p_i^* - p_j^*\|, \forall (i, j)^e \in \mathcal{E}_+\} \tag{4.30}$$

It is clear that $E_{p^*} \subseteq E'_{p^*}$, in general, and $E_{p^*} = E'_{p^*}$ if the graph is globally rigid. Furthermore, if the graph is rigid, then there exists a neighborhood U_{p^*} of p^* such that $E_{p^*} \cap U_{p^*} = E'_{p^*} \cap U_{p^*}$. Note that the sets E_{p^*} and E'_{p^*} are not compact (not bounded). To conduct stability analysis as done in the previous sections, we use the link space $z = (z_1^T, \ldots, z_m^T)^T = (\mathbb{H}_+ \otimes \mathbb{I}_d)p$ [5]. Thus, z is in the image of $\mathbb{H}_+ \otimes \mathbb{I}_d$, i.e., $z \in \mathrm{Im}(\mathbb{H}_+ \otimes \mathbb{I}_d)$. To proceed, the following edge function $\mathbf{h}_\mathcal{G}(z)$ is used:

$$\mathbf{h}_\mathcal{G}(z) = \frac{1}{2}(\|z_1\|^2, \ldots, \|z_m\|^2)^T \tag{4.31}$$

The edge function can be also parameterized by position vectors as $\mathbf{h}_\mathcal{G}(z) = \mathbf{h}_\mathcal{G}(p)$. Then, the following relationship can be obtained:

$$\begin{aligned} \frac{\partial\mathbf{h}_\mathcal{G}(p)}{\partial p} &= \frac{\partial\mathbf{h}_\mathcal{G}(z)}{\partial z}\frac{\partial z}{\partial p} \\ &= \mathrm{diag}(z)^T(\mathbb{H}_+ \otimes \mathbb{I}_d) \end{aligned} \tag{4.32}$$

Thus, the gradient system (4.27) can be expressed in the link space as

$$\begin{aligned} \dot{z} &= (\mathbb{H}_+ \otimes \mathbb{I}_d)\dot{p} \\ &= -k_p(\mathbb{H}_+ \otimes \mathbb{I}_d)(\mathbb{H}_+ \otimes \mathbb{I}_d)^T \mathrm{diag}(z)\Gamma(\bar{e}) \end{aligned} \tag{4.33}$$

Then, the desired equilibrium set can be defined as a compact set, which is given by

$$E'_{z^*} \triangleq \{z \in \mathrm{Im}(\mathbb{H}_+ \otimes \mathbb{I}_d) : \|z_i\| = \|z_i^*\|, \forall i = 1, \ldots, m\} \tag{4.34}$$

The local stability of the set E'_{z^*} can be easily checked using the following Lyapunov candidate:

$$V(z) = \frac{1}{2}\sum_{i=1}^{m} \gamma(\|z_i\|^2 - \|z_i^*\|^2) \tag{4.35}$$

The derivative of $V(z)$ is obtained as

$$\begin{aligned}
\dot{V}(z) &= \frac{\partial V(z)}{\partial z}\dot{z} \\
&= -k_p(\nabla V(z))^T(\mathbb{H}_+ \otimes \mathbb{I}_d)(\mathbb{H}_+ \otimes \mathbb{I}_d)^T \mathrm{diag}(z)\Gamma(\bar{e}) \\
&= -k_p[\mathrm{diag}(z)\Gamma(\bar{e})]^T(\mathbb{H}_+ \otimes \mathbb{I}_d)(\mathbb{H}_+ \otimes \mathbb{I}_d)^T \mathrm{diag}(z)\Gamma(\bar{e}) \\
&= -k_p\|\nabla\phi(p)\|^2 \\
&\leq 0
\end{aligned} \tag{4.36}$$

where the relationships $\nabla V(z) = \mathrm{diag}(z)\Gamma(\bar{e})$ and $\|\nabla\phi(p)\| = (\mathbb{H}_+ \otimes \mathbb{I}_d)^T \mathrm{diag}(z)$ $\Gamma(\bar{e})$ are used. For the local asymptotic stability, it is required to show that there exists a neighborhood $U_{E'_{z^*}}$ of E'_{z^*} such that $\dot{V}(z) < 0$ for any $z \in U_{E'_{z^*}} \setminus E'_{z^*}$.

Now using the Lojasiewicz's inequality Lemma 4.2, the following result can be obtained:

Lemma 4.8 *For any $p^\dagger \in E'_{p^*}$, there exists a neighborhood $U_{p^\dagger}$ such that for all $\xi \in U_{p^\dagger} \setminus E'_{p^*}$, $\|\nabla\phi(\xi)\| > 0$.*

Proof Since γ is analytic, the global potential function is analytic. From the Lojasiewicz's inequality Lemma 4.2, there exist $k_\phi > 0$ and $\rho_\phi \in [0, 1)$, and a neighborhood $U_{p^\dagger}$ of $p^\dagger$, where $p^\dagger \in E'_{p^*}$ such that

$$\|\nabla\phi(\xi)\| \geq k_\phi\|\phi(\xi) - \phi(p^\dagger)\|^{\rho_\phi} = k_\phi\|\phi(\xi)\|^{\rho_\phi} \tag{4.37}$$

for all $\xi \in U_{p^\dagger}$. Also, we have $\phi(\xi) = 0$ only when $\xi \in E'_{p^*}$. Thus, for any $\xi \in U_{p^\dagger} \setminus E'_{p^*}$, $\|\nabla\phi(\xi)\| > 0$.

Let $\sigma_{min}(\mathbb{H}_+ \otimes \mathbb{I}_d)$ and $\sigma_{max}(\mathbb{H}_+ \otimes \mathbb{I}_d)$ denote the nonzero minimum and maximum singular values of $\mathbb{H}_+ \otimes \mathbb{I}_d$, respectively. Then, we can have the following lemma.

Lemma 4.9 *The following inequalities are satisfied.*

$$\sigma_{min}(\mathbb{H}_+ \otimes \mathbb{I}_d)\|p - p^\dagger\| \leq \|z - \bar{z}\| \leq \sigma_{max}(\mathbb{H}_+ \otimes \mathbb{I}_d)\|p - p^\dagger\|$$

Proof Let us select $\bar{z} \in E'_{z^*}$, then $\inf_{\eta \in E'_{z^*}} \|z - \eta\| = \|z - \bar{z}\|$ holds because E'_{z^*} is compact and the function $f_z(\eta) \triangleq \|z - \eta\|$ is continuous [13]. Also, for a given z, there exists a $p \in \mathbb{R}^{nd}$ such that $p \in \mathrm{Im}(\mathbb{H}_+ \otimes \mathbb{I}_n)$ and $(\mathbb{H}_+ \otimes \mathbb{I}_n)p = z$. Similarly for a given z^*, there exists a $p^\dagger \in E'_{p^*}$ such that $p^\dagger \in \text{Im}(\mathbb{H}_+ \otimes \mathbb{I}_d)$ and $(\mathbb{H}_+ \otimes \mathbb{I}_d)p^\dagger = \bar{z}$ due to the realizability of the desired configuration. Thus, the following inequalities are obtained:

$$\begin{aligned}\sigma_{min}(\mathbb{H}_+ \otimes \mathbb{I}_d)\|p - p^\dagger\| &\leq \|z - \bar{z}\| \\ &= \|(\mathbb{H}_+ \otimes \mathbb{I}_d)(p - p^\dagger)\| \\ &\leq \sigma_{max}(\mathbb{H}_+ \otimes \mathbb{I}_d)\|p - p^\dagger\| \end{aligned} \tag{4.38}$$

which completes the proof.

With the above lemmas, the asymptotic stability of E'_{p^*} can be ensured as follows [10].

Theorem 4.4 *With the gradient control law (4.26), the desired equilibrium set E'_{p^*} is locally asymptotically stable.*

Proof From Lemma 4.8, for any $p^\dagger \in E'_{p^*}$, there exists a neighborhood $U_{p^\dagger}$ of $p^\dagger$ such that $\|\nabla\phi(\xi)\| > 0$ for all $\xi \in U_{p^\dagger} \setminus E'_{p^*}$. Let us denote the neighborhood of $p^\dagger$ where $\|\nabla\phi(\xi)\| > 0$ holds by the set

$$U_{p^\dagger} = \{p \in \mathbb{R}^{nd} : \|p - p^\dagger\| < \epsilon_p^*\} \tag{4.39}$$

Further define

$$U_{E'_{z^*}}(\epsilon_z) = \{z \in \text{Im}(\mathbb{H}_+ \otimes \mathbb{I}_d) : \inf_{\eta \in E'_{z^*}} \|z - \eta\| < \epsilon_z\} \tag{4.40}$$

Let $\epsilon_z^* = \sigma_{min}(\mathbb{H}_+ \otimes \mathbb{I}_d)\epsilon_p^*$. Then, for any $U_{E'_{z^*}}(\epsilon_z^*)$, there exists $p \in \mathbb{R}^{nd}$ such that $(\mathbb{H}_+ \otimes \mathbb{I}_d)p = z$ and

$$\|(\mathbb{H}_+ \otimes \mathbb{I}_d)^T \nabla V(z)\| = \|\nabla\phi(p)\| > 0 \tag{4.41}$$

Define a level set Ω such as $\Omega(c) \triangleq \{z \in \text{Im}(\mathbb{H}_+ \otimes \mathbb{I}_d) : \|V(z)\| < c\}$. Then, for a sufficiently small $c^* > 0$, it holds $\Omega(c^*) \subset U_{E'_{z^*}}(\epsilon_z)$. Thus, for any $z \in \Omega(c^*) \setminus E'_{z^*}$, there exists $p \in \mathbb{R}^{nd}$ such that $\dot{V}(z) = -k_p\|\nabla\phi(p)\|^2 < 0$, which implies the local asymptotic stability of the set E'_{z^*}. Thus, from Lemma 4.9, since $\|z - \bar{z}\|$ converges to zero, $\|p - p^\dagger\|$ converges to zero, which means that the set E'_{p^*} is locally asymptotically stable.

Theorem 4.5 *With the gradient control law (4.26), if the desired framework $(\mathcal{G}, p^*)$ is infinitesimally rigid, then E_{p^*} is locally asymptotically stable.*

Proof From the definition of the graph rigidity, it is clear that there exists a neighborhood $U_{p^\dagger}$ of $p^\dagger$ where $p^\dagger \in E_{p^*}$, such that $E_{p^*} \cap U_{p^\dagger} = E'_{p^*} \cap U_{p^\dagger}$. Thus, E_{p^*} is locally asymptotically stable.

With the property of $\gamma(x) = 0$ if and only if $x = 0$, since $\bar{e}_{ij}$ converges to zero, the control input u_i also converges to zero. Thus, it can be observed that the final formation will be stationary.

4.4 Summary and Simulations

This chapter has focused on local convergences of the formation to desired configurations. The idea of Sect. 4.1 is quite intuitive since it controls the edges of graphs such that the desired lengths could be achieved in a least square sense. After computing control inputs by the least square approach, it divides the control inputs to the incident vertices. The difficulty of this analysis lies in filling a mathematical gap between the link space and the non-compact equilibrium set. We conducted all the analysis in the link space, which is a compact set, and then the computed control laws are sent back to the unbounded Euclidean space for the control of motions of agents. In Sect. 4.2, a local exponential convergence has been ensured for a gradient control law. But, the control law requires a global coordination for reassigning the network of the agents. Section 4.3, which is reused from [10], is dedicated to a generalization of gradient control law in terms of the applicable dimensions and potential functions. We have provided a general potential function that needs to satisfy some conditions and analyzed the convergence in general d-dimensional space. The extension to double integrator dynamics was conducted in [10] and extension to a formation under leader–follower setups, where a leader is considered to move in a low velocity, was carried out in [11]. It is noticeable that the extensions to double integrator and leader–follower formations require a measurement of velocity, but it is assumed to be available in local coordinate frame. For further extensions to double integrator dynamics, it is recommendable to refer to [4, 18] also.

It is still tough to analytically prove that the gradient control law could not assure an almost global convergence for general graphs. Let us consider Fig. 4.3 that depicts a $K(4)$ formation in 2-D. Initially, the four agents are in a desired configuration satisfying all the desired inter-agent distances (see Fig. 4.3a). Let us suppose that the desired distance between agents 2 and 4 is changed as shown in Fig. 4.3b. Then, from the traditional gradient control law, the agent 2 is controlled as $\dot{p}_2 = -\sum_{j=\mathcal{N}_2} \bar{e}_{2j} z_{2j}$. So, at initial time, all the errors except e_{24}, which is positive, are zero. Thus, the agent 2 is forced to move toward the agent 4 as shown in Fig. 4.3c. But, shortly after the movement, the errors e_{12} and e_{23} become negative. Thus, the agent 2 is forced back to the opposite direction as shown in Fig. 4.3d. Thus, it looks that the gradient control law cannot escape from a local minimum or undesired equilibrium points from general initial configurations. However, unlike the intuition, for any initial location of agent 2, the desired configuration has been readily achieved from numerical simulations. It is due to the fact that when the desired distance between agents 2 and 4 has been changed, the gradient control law divides the control efforts to all agents evenly and not only agent 2, but also all other agents work to achieve the desired configuration up to translations and rotations. Furthermore, since a flip of graph is allowed in the

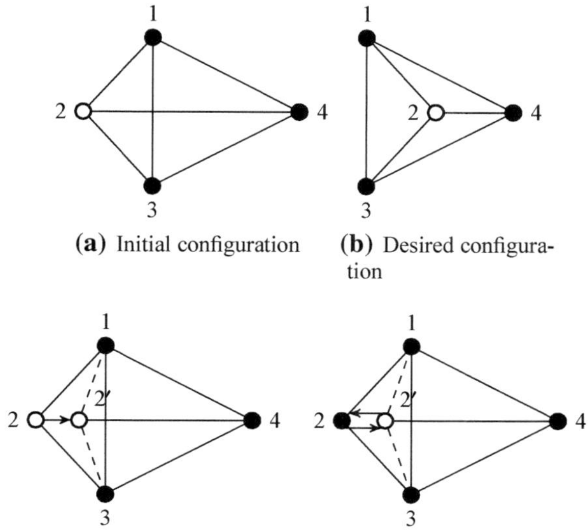

Fig. 4.3 Intuition for a global stabilization of $K(4)$ (**a**) Initial configuration (**b**) Desired configuration (**c**) Move towards to 4 (**d**) Move back to 2

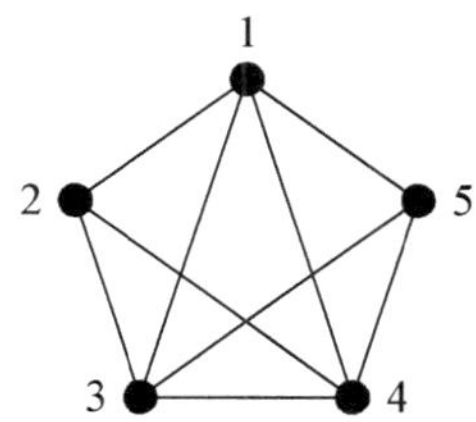

Fig. 4.4 Five agents for simulation

formation control, the desired configuration would be achieved for any initial cases for $K(4)$. Thus, it is still not easy to show a counterexample for the $K(4)$ numerically or analytically.

To find a convergence to an undesired equilibrium point, let us consider five agents as depicted in Fig. 4.4. Agents are connected to all other agents except the edge between agents 2 and 5; and hence, it is a $K(5) - 1$ graph with 9 edges. The initial positions of agents are $(1, 0)^T$, $(3, 0)^T$, $(4, 2)^T$, $(2, 4)^T$, and $(0, 2)^T$. When the desired reference positions are given as $(4, 2)^T$, $(0, 3)^T$, $(1, 0)^T$, $(2, -3)^T$, and $(0, 2)^T$, the desired configuration is not achieved. Figure 4.5 shows the simulation results for this case. As shown in the plots of the squared distance errors, the formation is not stabilized to the desired configuration. But, when the desired reference positions are given as $(1.5, 2)^T$, $(4, -1)^T$, $(1, 0)^T$, $(2, -3)^T$, and $(0, 1)^T$, the desired configuration is achieved as shown in Fig. 4.6. Thus, it is clear that the gradient control law can stabilize the formation depending upon agents' initial positions. Hence, in this sense, the gradient control law cannot stabilize formations up to isomorphism in general.

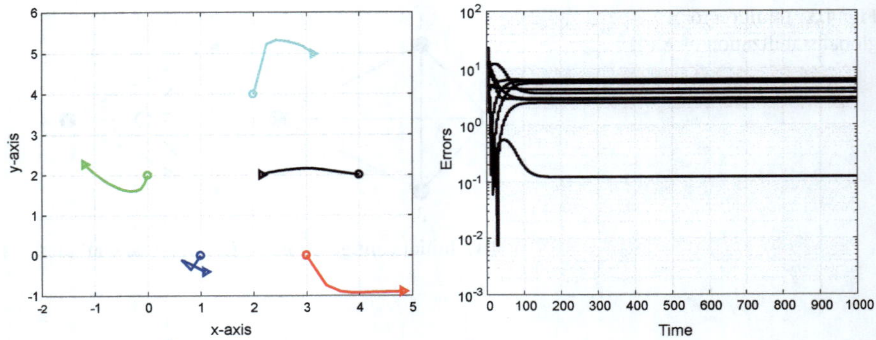

Fig. 4.5 Left—Trajectories of agents: Agents do not converge to the desired configuration. Right—Errors of squared distances

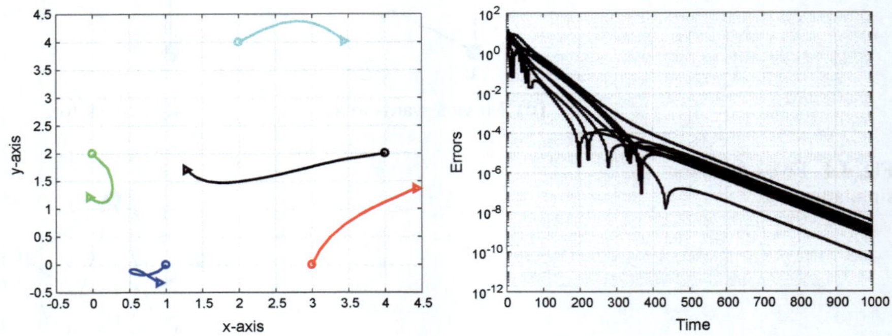

Fig. 4.6 Left—Trajectories of agents: Agents converge to the desired configuration. Right—Errors of squared distances

4.5 Notes

This chapter has been dedicated to local stabilization of gradient-based formation control laws with desired distance constraints. There have been some efforts to ensure a global stabilization with additional constraints. For example, let us consider Fig. 4.7 which shows a triangular formation in 2-dimensional space. The local coordinate frames of agents in the 2-dimensional space are not aligned, i.e., x- and y-axes of the agents' local coordinate frames are not aligned. But, if we can define the virtual z-axis of agents, then we may assume that the z-axes of agents are aligned, as shown in Fig. 4.7.[1] Then, when agent 3 senses the neighboring agents 1 and 2 in its own local coordinate frame, it can see them in counterclockwise or clockwise directions. Depending upon the counterclockwise or clockwise directions, the area of triangular

[1] However, it is hard to image an aligned virtual axis in 3- or higher dimensional spaces. In higher dimensional spaces, we may have to define an aligned virtual manifold rather than an axis.

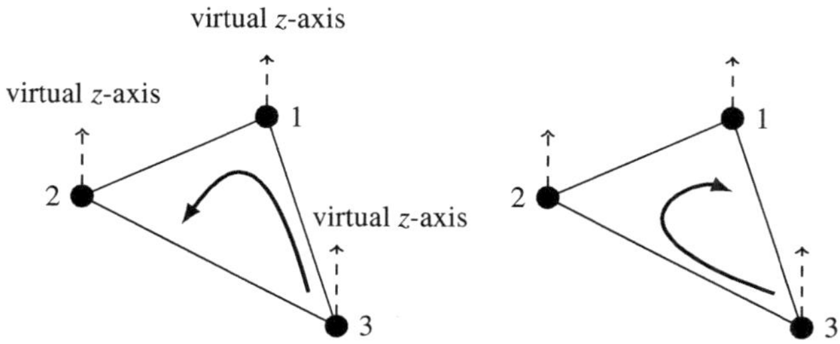

(a) Counterclockwise direction **(b)** Clockwise direction

Fig. 4.7 Additional constraints with the aligned virtual z-axes (virtual z axes are aligned): area constraints or desired order of neighbors

may have different signs (i.e., negative area or positive area). In [3], the signed area of the triangular has been used to avoid the flip ambiguity, with an extension to n agents for triangulated Laman graphs [15], and in [6], the desired order of neighboring agents was used to have a unique formation. It will be worthwhile to extend these results to the cases of 3-dimensional space.

The results of Sect. 4.1 are reused and reproduced from [2, 9]. For more discussions on the realizability of formations in Euclidean space, it is recommended to refer to [2]. The main results of Sect. 4.2 are reused from [12] and Sect. 4.3 is reused from [10]. The following copyright and permission notices are acknowledged.

- The text of Sect. 4.1 is © [2010] IEEE. Reprinted, with permission, from [8].
- The statements of Lemmas 4.4, 4.5, 4.7, and Theorem 4.3 with the proofs are © [2014] IEEE. Reprinted, with permission, from [12].
- The statements of Lemmas 4.1, 4.3, Theorems 4.1, and 4.2 with the proofs are © Reprinted from [9], with permission from Elsevier.
- The statements of Lemmas 4.8, 4.9, Theorems 4.4, and 4.5 with the proofs are © Reproduced from [10] by permission of John Wiley and Sons.

References

1. Absil, P.-A., Kurdyka, K.: On the stable equilibrium points of gradient systems. Syst. Control Lett. **55**(7), 573–577 (2006)
2. Ahn, H.-S., Oh, K.-K.: Command coordination in multi-agent formation: Euclidean distance matrix approaches. In: Proceedings of the International Conference on Control, Automation and Systems, pp. 1592–1597 (2010)
3. Anderson, B.D.O., Sun, Z., Sugie, T., Azuma, S.-I., Sakurama, K.: Formation shape control with distance and area constraints. IFAC J. Syst. Control **1**, 2–12 (2017)
4. Cai, X., de Queiroz, M.: Rigidity-based stabilization of multi-agent formations. ASME J. Dyn. Syst. Meas. Control **136**(1), 014502-1–014502-7 (2014)
5. Dörfler, F., Francis, B.: Geometric analysis of the formation problem for autonomous robots. IEEE Trans. Autom. Control **55**(10), 2379–2384 (2010)

6. Kang, S.-M., Park, M.-C., Ahn, H.-S.: Distance-based cycle-free persistent formation: global convergence and experimental test with a group of quadcopters. IEEE Trans. Ind. Electron. **64**(1), 380–389 (2017)
7. Khalil, H.K.: Nonlinear Systems. Prentice Hall, New Jersey (2002)
8. Oh, K.-K., Ahn, H.-S.: Distance-based formation control using Euclidean distance dynamics matrix: general cases. In: Proceedings of the American Control Conference, pp. 4816–4821 (2011)
9. Oh, K.-K., Ahn, H.-S.: Formation control of mobile agents based on inter-agent distance dynamics. Automatica **47**(10), 2306–2312 (2011)
10. Oh, K.-K., Ahn, H.-S.: Distance-based undirected formations of single- and double-integrator modeled agents in n-dimensional space. Int. J. Robust Nonlinear Control **24**(12), 1809–1820 (2014)
11. Oh, K.-K., Ahn, H.-S.: Leader-follower type distance-based formation control of a group of autonomous agents. Int. J. Control Autom. Syst. **15**(4), 1738–1745 (2017)
12. Park, M.-C., Ahn, H.-S.: Exponential stabilization of infinitesimally rigid formations. In: Proceedings of the International Conference on Control, Automation and Information Sciences (ICCAIS), pp. 36–40 (2014)
13. Rudin, W.: Principles of Mathematical Analysis, 3rd edn. McGraw-Hill, New York (1976)
14. Stoll, M.: Introduction to Real Analysis. Addition Wesley Higher Mathematics (2001)
15. Sugie, T., Anderson, B.D.O., Sun, Z., Dong, H.: On a hierarchical control strategy for multi-agent formation without reflection. In: Proceedings of the 57th IEEE Conference on Decision and Control, pp. 2023–2028 (2018)
16. Sun, Z.: Cooperative coordination and formation control for multi-agent systems. Ph.D. thesis, Australian National University (2017)
17. Sun, Z., Mou, S., Anderson, B.D.O., Cao, M.: Exponential stability for formation control systems with generalized controllers: a unified approach. Syst. Control Lett. **93**(7), 50–57 (2016)
18. Sun, Z., Anderson, B.D.O., Deghat, M., Ahn, H.-S.: Rigid formation control of double-integrator systems. Int. J. Control **90**(7), 1403–1419 (2017)
19. Trinh, M.-H., Park, M.-C., Sun, Z., Anderson, B.D.O., Pham, V.H., Ahn, H.-S.: Further analysis on graph rigidity. In: Proceedings of the 55th IEEE Conference on Decision and Control, pp. 922–927 (2016)

Chapter 5
Persistent Formations

Abstract As studied in Chaps. 3 and 4, the goal of formation control is to achieve a unique desired configuration by satisfying distance constraints between agents. To achieve a desired distance constraint between two neighboring agents, if both agents work cooperatively in terms of sensing, communications, and actuation, then it may be considered as an undirected case. But, if only one of two neighboring agents connected by an edge works for this task, it may be considered as a directed case. In Chaps. 3 and 4, we studied global stabilization and local stabilization for undirected cases except Sect. 3.3. In Sect. 3.3, we studied a polygon formation under directed sensing and control topologies. However, even though we could satisfy the desired constraints, the formation would not be unique. The main goal of Sect. 3.3 was to achieve a desired polygon formation. That is, although the desired distance constraints between neighboring agents are satisfied, the realized configuration may not be unique in polygon graphs (cycle graphs). It may be realistic to suppose that only an agent can control an edge; so each edge of a graph may have a direction in the general setup (for sensings, communications, and actuations). By adding a requirement of achieving a unique configuration into directed graph setup, we may define a unique formation under directed graphs. This chapter is dedicated to address formation control problems under directed sensing and actuation topologies.

5.1 Acyclic Minimally Persistent Formations in 2-Dimensional Space

In directed formation control problems, there could be two different cases according to the existence of cycles in the underlying topology. As studied in Sect. 3.3, although it may be possible to achieve a polygon formation when we do not consider a unique configuration, it is not clear how to realize a formation into a specific dimensional space under directed graph setup with multiple cycles [1]. That is why the problem of unique realization without cycles has been focused in formation control literature. Thus, the two cases should be distinguished according to the existence of cycles in directed graphs, i.e., directed formation with the existence of cycles vs. directed formation without cycles. The former one has been studied under the name of *leader-*

H.-S. Ahn, *Formation Control*, Studies in Systems, Decision and Control 205, https://doi.org/10.1007/978-3-030-15187-4_5

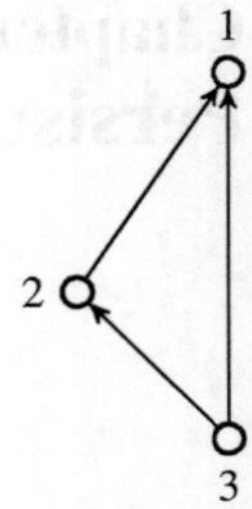

Fig. 5.1 Triangular acyclic minimally persistent formation

remote-follower and formation with *three coleaders* [10], and the latter one has been studied under the name of *cycle-free (acyclic) persistence* [3]. This chapter focuses on cycle-free persistent formations. To be a persistent formation, a necessary condition for the graph is that the underlying topology, without considering the edge directions, should be rigid. But, this condition is not sufficient to imply persistence; so another necessary condition should be given, which is to assign edge directions appropriately such that the direction constraints should be realizable and consistent among agents. Thus, the persistence concept is a combination of rigidity of underlying topology and consistence of direction assignments. For more details, refer to Sect. 2.2.2.

As one of the simplest problems in persistent formation control, in this section, let us first consider acyclic minimally persistent formations in $\mathbb{R}^2$. This section is a summary of [4, 5]. Consider the three-agent case, with the graph topology $\overrightarrow{\mathcal{G}} = (\mathcal{V}, \overrightarrow{\mathcal{E}})$, where $\overrightarrow{\mathcal{E}} = \{(2, 1)^{\bar{e}}, (3, 1)^{\bar{e}}, (3, 2)^{\bar{e}}\}$. This problem is called acyclic triangular formation [2], and is depicted in Fig. 5.1.

Let $\{d_{12}^*, d_{13}^*, d_{23}^*\}$ be the set of desired distances of the edges, with the following realizability constraint:

$$d_{ij}^* < d_{ik}^* + d_{jk}^*, i, j, k \in \{1, 2, 3\} \tag{5.1}$$

Also let $(\overrightarrow{\mathcal{G}}, p^*)$ be a framework with a realization p^* that satisfies the constraints of the desired distances $\{d_{12}^*, d_{13}^*, d_{23}^*\}$. Then, the objective of formation control is to achieve a framework $(\overrightarrow{\mathcal{G}}, p)$ such that $(\overrightarrow{\mathcal{G}}, p^*)$ and $(\overrightarrow{\mathcal{G}}, p)$ become congruent in the 2-dimensional space. To design the control laws, the potential functions $\phi_2 = \frac{1}{4}\bar{e}_{12}^2$ and $\phi_3 = \frac{1}{4}(\bar{e}_{13}^2 + \bar{e}_{23}^2)$ are used. It is supposed that the leader agent, i.e., agent 1, is stationary. So, for the pair of agents 1 and 2, only the agent 2 controls the edge $(2, 1)^{\bar{e}}$. By taking the gradient of the potential function $\phi_2 = \frac{1}{4}\bar{e}_{12}^2$, the control law for agent 2 is designed as

$$\dot{p}_2 = u_2 = -\left[\frac{\partial \phi_2}{\partial p_2}\right]^T = \bar{e}_{12} z_{12} = -\bar{e}_{12} z_{21} \tag{5.2}$$

For the agent 3, it has two constraints $(3, 1)^{\bar{e}}$ and $(3, 2)^{\bar{e}}$. Also, taking the gradient, its control law can be designed as

$$\dot{p}_3 = u_3 = -Q\left[\frac{\partial \phi_3}{\partial p_3}\right]^T = Q(\bar{e}_{13}z_{13} + \bar{e}_{23}z_{23}) \tag{5.3}$$

where Q is the orientation distortion matrix defined as [4]:

$$Q = \begin{bmatrix} \cos\theta & -\sin\theta \\ \sin\theta & \cos\theta \end{bmatrix} \tag{5.4}$$

with $0 < \cos\theta < 1$. Note that the control laws (5.2) and (5.3) can be implemented in local coordinate frames as

$$\dot{p}_2^2 = u_2^2 = \bar{e}_{12}z_{12}^2 \tag{5.5}$$

$$\dot{p}_3^3 = u_3^3 = Q\bar{e}_{13}z_{13}^3 + Q\bar{e}_{23}z_{23}^3 \tag{5.6}$$

Let us assume that there is no communication between agents. Also, for the realization of the above control laws, both the sensing and actuation topologies are assumed *acyclic minimally persistent* graphs. Thus, the sensing and actuation graphs need to be coincident, i.e., $\mathcal{G} = (\mathcal{V}, \overrightarrow{\mathcal{E}}^s) = (\mathcal{V}, \overrightarrow{\mathcal{E}}^a)$. Since $z_{ij} = p_i - p_j$, the above control laws can be expressed in the link space as

$$\dot{z}_{23} = \bar{e}_{12}z_{12} - Q(\bar{e}_{13}z_{13} + \bar{e}_{23}z_{23}) \tag{5.7}$$

$$\dot{z}_{13} = -Q(\bar{e}_{13}z_{13} + \bar{e}_{23}z_{23}) \tag{5.8}$$

$$\dot{z}_{12} = -\bar{e}_{12}z_{12} \tag{5.9}$$

Note that the control law (5.5) for agent 2 is similar to the law (3.29), where the control effort is provided along the edge vector z_{jk}^k. Also the control law (5.6) for agent 3 is also similar to (3.26). Actually, the acyclic triangular formation can be stabilized using the approach developed in Chap. 3. But, for a more rigorous analysis of undesired or degenerate equilibrium points, we develop a linearization-based approach in this chapter. With the above setups, we can show a convergence of agents to a desired configuration under some initial conditions.

Lemma 5.1 *Let us consider a triangular formation depicted in Fig. 5.1 and let the sensing and actuation graphs,* $(\mathcal{V}, \overrightarrow{\mathcal{E}}^s)$ *and* $(\mathcal{V}, \overrightarrow{\mathcal{E}}^a)$*, be acyclic minimally persistent. Under the gradient control laws* (5.2) *and* (5.3)*, the errors* $\bar{e}_{12}$*,* $\bar{e}_{13}$ *and* $\bar{e}_{23}$ *converge to zero exponentially fast if they are not collinear at* $t = t_0$.

Proof The proof is given in [2].

It is remarkable that the proof of Lemma 5.1 can be also conducted by modifying the idea given in Example 3.1. Let us replace $i = 1$, $j = 2$, and $k = 3$. Since the leader, i.e., agent 1, is assumed stationary, we have $u_1 = 0$. Since the agent 2 is updated similar to (3.29) as

$$u_2^2 = -\frac{k_o \bar{e}_{12}}{4\|p_1^2\|^2} p_1^2$$

Thus, it is the same form as (5.2). For agent 3, the control input is computed from

$$\begin{bmatrix} (p_3 - p_1)^T \\ (p_3 - p_2)^T \end{bmatrix} u_3 = -\frac{k_o}{4} \begin{bmatrix} \bar{e}_{13} \\ \bar{e}_{23} \end{bmatrix}$$

which results in a similar form as (5.3). Under the assumption of collinearity of initial configuration, the convergence to the desired inter-agent distances can be shown quite straightforwardly by following the analysis given in Sect. 3.2.

In what follows, based on Lemma 5.1, we would like to develop further analysis, taking account of degenerate configurations (refer to Sect. 12.1.1 for the definition of the degenerate configurations). Mainly, we would like to divide the collinear case into several equilibrium sets. To obtain such equilibrium points, equalizing $\dot{p}_2 = 0$ and $\dot{p}_3 = 0$, we decompose the equilibrium set as $\mathcal{E} = \mathcal{D} \cup \mathcal{U}$, where the desired equilibrium set $\mathcal{D} \triangleq \{p : e = 0\}$ and undesired equilibrium set $\mathcal{U} = \mathcal{U}_1 \cup \mathcal{U}_2 \cup \mathcal{U}_3$ are given as [5]:

$$\mathcal{U}_1 \triangleq \{p : z = 0\}$$
$$\mathcal{U}_2 \triangleq \{p : z_{12} = 0, \bar{e}_{13} = \bar{e}_{23} = 0\}$$
$$\mathcal{U}_3 \triangleq \{p : \bar{e}_{12} = 0, \bar{e}_{13} z_{13} + \bar{e}_{23} z_{23} = 0, \bar{e}_{13} \neq 0 \text{ or } \bar{e}_{23} \neq 0\}$$

For the main results, we use the following lemmas.

Lemma 5.2 *It is true that* $\bar{e}_{13} + \bar{e}_{23} < 0$ *for all* $p \in \mathcal{U}_3$.

Proof The proof is inferred from the proof of Lemma 5.7, which is a more general one for the 3-D case; so it is omitted. The direct proof can be also found in [2].

Lemma 5.3 *With the control law (5.5) for the agent 2,* e_{12} *converges to zero exponentially fast and* p *does not converge to the cases of* $z = 0$.

Proof From $\dot{\bar{e}}_{12} = 2 z_{12}^T \dot{z}_{12}$, with (5.9), it is shown that $\dot{\bar{e}}_{12} = -2\bar{e}_{12}(\bar{e}_{12} + \bar{d}_{12}^*)$, which results in

$$\bar{e}_{12}(t) = \frac{\gamma(t_0) e^{-2\bar{d}_{12}^* t}}{1 - \gamma(t_0) e^{-2\bar{d}_{12}^* t}} \tag{5.10}$$

where $\gamma(t_0) = \frac{\bar{e}_{12}(t_0)}{\bar{e}_{12}(t_0) + \bar{d}_{12}^*(t_0)}$ at $t = t_0$. Thus, $\bar{e}_{12}(t)$ converges to zero exponentially fast as $t \to \infty$, which implies $e_{12}(t)$, where $e_{12} = d_{12} - d_{12}^*$, also converges to zero exponentially fast. Hence, since both $z_{12} = 0$ and $e_{12} = 0$ cannot be true at the same configuration, p does not converge to the cases of $z = 0$.

The above lemma implies that the undesired equilibrium sets $\mathcal{U}_1$ and $\mathcal{U}_2$ are not stable. For the analysis of $\mathcal{U}_3$, the following lemma is developed [4]:

Lemma 5.4 *The undesired equilibrium set $\mathcal{U}_3$ is unstable.*

Proof See the appendix.

Example 5.1 (*Unstability of $\mathcal{U}_3$*) Assume that agents 1, 2, and 3 are collinear with coordinate values as $p_1 = (2, 0)^T$, $p_2 = (1, 0)^T$, and $p_3 = (1.5, 0)^T$. The desired inter-agent distances are given as $d_{12}^* = d_{23}^* = d_{13}^* = 1$. Then, we have $\bar{e}_{12} = 0$ and $\bar{e}_{13}z_{13} + \bar{e}_{23}z_{23} = 0$, with $\bar{e}_{13} = -0.75$ and $\bar{e}_{23} = -0.75$. Thus, we have

$$\begin{bmatrix} \dot{z}_{13} \\ \dot{z}_{23} \end{bmatrix} = -Q \otimes \begin{bmatrix} -0.75 & -0.75 \\ -0.75 & -0.75 \end{bmatrix} \begin{bmatrix} z_{13} \\ z_{23} \end{bmatrix} \tag{5.11}$$

The eigenvalues of the orientation distortion matrix Q are $\cos\theta \pm \mathrm{i}\sin\theta$ and eigenvalues of $\begin{bmatrix} -0.75 & -0.75 \\ -0.75 & -0.75 \end{bmatrix}$ are $-0.75, -0.75$. Thus, the eigenvalues are computed as $0.75(\cos\theta + \mathrm{i}\sin\theta)$, $0.75(\cos\theta + \mathrm{i}\sin\theta)$, $0.75(\cos\theta - \mathrm{i}\sin\theta)$ and $0.75(\cos\theta - \mathrm{i}\sin\theta)$. Therefore, the system (5.11) is unstable.

Next, let us consider the convergence of formation. For this, define the following sets:

$$\begin{aligned} \mathcal{C}_1 &\triangleq \{p : \|z_{12}\| = 0\} \\ \mathcal{C}_2 &\triangleq \{p : \mathrm{rank}[z_{13}, z_{23}] < 2, \|z_{12}\| \neq 0\} \\ \mathcal{C} &\triangleq \mathcal{C}_1 \cup \mathcal{C}_2 \end{aligned}$$

Since the equality $\|z_{12}\| = 0$ implies $\mathrm{rank}[z_{13}, z_{23}] < 2$, it is true that $\mathcal{C} = \{p : \mathrm{rank}[z_{13}, z_{23}] < 2\}$, and $\mathcal{C}_1$ and $\mathcal{C}_2$ are disjoint. Also it is obvious that $\mathcal{U}_2 \subset \mathcal{C}_1$ and $\mathcal{U}_3 \subset \mathcal{C}_2$. From (5.9), it is concluded that if $p(t_0) \in \mathcal{C}_1$, then $p(t) \in \mathcal{C}_1$ forever.

Theorem 5.1 *For all $p \in \mathcal{C}_2 \backslash \mathcal{U}_3$, p does not stay in $\mathcal{C}$.*

Proof Since $\|z_{12}\|$ converges to d_{12}^* unless $\|z_{12}(t_0)\| = 0$, p does not approach $\mathcal{C}_1$. From the definition of $\mathcal{U}_3$, if $p \notin \mathcal{U}_3$, then $\bar{e}_{12} \neq 0$ or $\bar{e}_{13}z_{13} + \bar{e}_{23}z_{23} \neq 0$. Thus, for all $p \in \mathcal{C}_2 \backslash \mathcal{U}_3$, if (i) $\bar{e}_{12} = 0$ and $\bar{e}_{13}z_{13} + \bar{e}_{23}z_{23} \neq 0$, $\dot{p}_3 \neq 0$. Also z_{13} and z_{23} are not parallel each other. Thus, agent 3 does not stay in $\mathcal{C}$. On the other hand, (ii) if $\bar{e}_{12} \neq 0$ and $\bar{e}_{13}z_{13} + \bar{e}_{23}z_{23} = 0$, agent 2 moves such that $\bar{e}_{12}$ becomes zero. Thus, $\bar{e}_{13}z_{13} + \bar{e}_{23}z_{23}$ becomes a nonzero vector, which completes the proof.

Theorem 5.2 *If $p(t_0) \notin \mathcal{C}_1 \cup \mathcal{U}_3$, then e_{12}, e_{13} and e_{23} exponentially converge to zero.*

Proof From the following relationship

$$\forall p \in \mathcal{C}, p \notin \mathcal{C}_1 \cup \mathcal{U}_3 \Longleftrightarrow \forall p \in \mathcal{C}, p \in \mathcal{C}_2 \cap \mathcal{U}_3^c = \mathcal{C}_2 \backslash \mathcal{U}_3 \tag{5.12}$$

and from Theorem 5.1, it is clear that every p in $\mathcal{C}$, but not in $\mathcal{C}_1 \cup \mathcal{U}_3$, escapes from the collinear configurations. Therefore, e_{12}, e_{13} and e_{23} exponentially converge to zero by Lemmas 5.1, 5.3 and 5.4.

To extend the above results to n-agent cases, we use Theorem 2.8. Let us suppose that both $(\mathcal{V}, \overrightarrow{\mathcal{E}}^s)$ and $(\mathcal{V}, \overrightarrow{\mathcal{E}}^a)$ are acyclic minimally persistent, and they are coincident $(\mathcal{V}, \overrightarrow{\mathcal{E}}^s) = (\mathcal{V}, \overrightarrow{\mathcal{E}}^a)$. In such a case, the first three agents are exactly same as depicted in Fig. 5.1. Thus, when the agents 2 and 3 are controlled by (5.2) and (5.3), the triangular shape composed of agents 1, 2 and 3 is globally exponentially convergent to the desired configuration. For any agents after the agent 3, it has two neighbors (i.e., out-degree 2). For $i \in \mathbb{N}$ and $i \geq 3$, denote the neighboring agents as i_α and i_β. Then, $i_\alpha, i_\beta < i$. Similarly to Lemma 5.4, let us use the potential function $\phi_i = \frac{1}{4}(\bar{e}_{i_\alpha i}^2 + \bar{e}_{i_\beta i}^2)$. Then, by taking a derivative, it is obtained as

$$\dot{\phi}_i = -\cos\theta \|\bar{e}_{i_\alpha i} z_{i_\alpha i} + \bar{e}_{i_\beta i} z_{i_\beta i}\|^2 + \bar{e}_{i_\alpha i} z_{i_\alpha i}^T u_{i_\alpha} + \bar{e}_{i_\beta i} z_{i_\beta i}^T u_{i_\beta} \tag{5.13}$$

where u_{i_α} and u_{i_β} are control inputs for the agents i_α and i_β. To study the convergence of the overall formation, consider a formation produced by a subgraph $\overrightarrow{\mathcal{G}}_s$ such that $\overrightarrow{\mathcal{G}}_s = (\mathcal{V}_s, \overrightarrow{\mathcal{E}}_s)$, $\mathcal{V}_s = \{1, \ldots, s\} \subset \mathcal{V}$, $\overrightarrow{\mathcal{E}}_s = \{(j, i)^e \in \overrightarrow{\mathcal{E}} : j, i \in \mathcal{V}_s\} \subset \overrightarrow{\mathcal{E}}$. As aforementioned, when $s = 3$, it is globally exponentially convergent. To use an induction idea on s, let us assume that the length of ed_{ji}, $j < i$ is bounded by an exponentially decaying function for all $(j, i)^e \in \overrightarrow{\mathcal{E}}_s$ for $s = k - 1$. For $s = k$, $\dot{\phi}_i$ is given by (5.13) with $i = k$. We assumed that $k_\alpha < k$ and $k_\beta < k$. Thus, u_{k_α} and u_{k_β} are bounded by exponentially decaying functions. Thus, the lengths of ed_{ji} for all $ed_{ji} \in \overrightarrow{\mathcal{E}}_s$ are bounded for all $s \in \{3, \ldots, n\}$. With the argument thus far, in the case of general acyclic minimally persistent formations, it can be claimed that the boundedness of errors is ensured, which is summarized in the following theorem:

Theorem 5.3 *For acyclic minimally persistent formations, when the agents are updated by gradient control laws simultaneously, the errors of all the agents are bounded.*

Proof For agents $\mathcal{V} = \{1, 2, 3\}$, the errors exponentially converge to zero and u_2 and u_3 are bounded by exponentially decaying functions. For agent 4, $\dot{\phi}_i$ is bounded since when $\bar{e}_{4_\alpha 4}$ or $\bar{e}_{4_\beta 4}$ are large, it is true that $\cos\theta \|\bar{e}_{i_\alpha i} z_{i_\alpha i} + \bar{e}_{i_\beta i} z_{i_\beta i}\|^2$ is much bigger that $\bar{e}_{i_\alpha i} z_{i_\alpha i}^T u_{i_\alpha} + \bar{e}_{i_\beta i} z_{i_\beta i}^T u_{i_\beta}$. Thus, the error of agent 4 is bounded, which implies that the control effort u_4 is also bounded. By following the same argument to the agent 5, it can be shown that the control effort u_5 is also bounded. Thus, by the induction, all the states of the agents are bounded.

Example 5.2 (*Escaping the collinearity*) Assume that agents 1, 2, and 3 are collinear initially, with coordinate values as $p_1 = (2, 0)^T$, $p_2 = (1, 0)^T$, and $p_3 = (0, 0)^T$. Let $\bar{e}_{12} = 0$ and $\bar{e}_{13} z_{13} + \bar{e}_{23} z_{23} \neq 0$, and let the desired inter-agent distances be $d_{12}^* = d_{23}^* = d_{13}^* = 1$. Then, it can be shown that the agents escape the set $\mathcal{C}_2$. Due to $\bar{e}_{12} = 0$, we have $\dot{z}_{12} = 0$, which implies $\dot{p}_2 = 0$. Furthermore, from

$$\bar{e}_{13}z_{13} + \bar{e}_{23}z_{23} = 4\begin{bmatrix} 2 \\ 0 \end{bmatrix} + 0.0\begin{bmatrix} 1 \\ 0 \end{bmatrix} \neq 0$$

we have

$$\dot{z}_{13} = \dot{z}_{23} = -\dot{p}_3 = -4Q\begin{bmatrix} 2 \\ 0 \end{bmatrix}$$

Thus, since Q is a rotation matrix, the y-axis component of $\dot{p}_3$ is not zero if $\theta \neq 0$. Consequently, the agents escape the collinearity.

5.2 Acyclic Minimally Persistent Formations in 3-Dimensional Space

This section focuses on acyclic minimally persistent formation in 3-dimensional space, as an extension of the previous section. Similarly to the previous section, in this section, we also assume that $(\mathcal{V}, \vec{\mathcal{E}}^s)$ and $(\mathcal{V}, \vec{\mathcal{E}}^a)$ are acyclic minimally persistent and they are coincident; but we extend the problem to 3-dimensional space by way of reproducing the results of [6]. To be acyclic minimally persistent formation in 3-D, a follower agent needs to have three outgoing edges toward the leader agents. Let $\phi(x, d^*)$ be a function that satisfies the following assumption:

Assumption 5.2.1 If d^* is a constant, then

- $\phi(x, d^*)$ is nonnegative and $g(x, d^*) = \frac{\partial \phi(x,d^*)}{\partial x}$ is strictly monotonically increasing
- $\phi(x, d^*)$ and $g(x, d^*)$ are continuously differentiable on $x \in (-{d^*}^2, \infty)$ and equal zero if and only if $x = 0$
- $\phi(x, d^*)$ is analytic in a neighborhood of 0

Denote $z_{ij} = p_i - p_j$, $\bar{e}_{ij} = \|z_{ij}\|^2 - {d_{ij}^*}^2$, and $g_{ij} = g(\bar{e}_{ij}, d_{ij}^*)$. Since the function $g(x, d^*)$ is strictly monotonically increasing, its derivative is positive as $\rho_x = \frac{\partial g(x,d^*)}{\partial x} > 0$.

Let us define a local potential for each agent as $V_i : \mathbb{R}^{3(|\mathcal{N}_i|+1)} \rightarrow \mathbb{R}_+$ such as

$$V_i(\ldots, p_j^i, \ldots) = \frac{1}{2}\sum_{j \in \mathcal{N}_i} \phi(\bar{e}_{ij}, d_{ij}^*) \tag{5.14}$$

The gradient control law is proposed as

$$\dot{p}_i = u_i = -\nabla_{p_i} V_i = -\sum_{j \in \mathcal{N}_i} g_{ij} z_{ij} = \sum_{j \in \mathcal{N}_i} g_{ij} z_{ji} \tag{5.15}$$

Since the agent 1 has no neighbors, we have $\mathcal{N}_1 = \emptyset$. Agents 2 and 3 are interacted as $\mathcal{N}_2 = \{1\}$ and $\mathcal{N}_3 = \{1, 2\}$. For the agents $i \geq 4$, there are three neighbors such as $\mathcal{N}_i = \{i_1, i_2, i_3\}$, where $i > i_3 > i_2 > i_1$ without loss of generality. Then the dynamics for the agents can be expressed as

$$\dot{p}_1 = 0 \tag{5.16}$$

$$\dot{p}_2 = g_{21}z_{12} \tag{5.17}$$

$$\dot{p}_3 = g_{31}z_{13} + g_{32}z_{23} \tag{5.18}$$

$$\dot{p}_i = g_{ii_1}z_{i_1i} + g_{ii_2}z_{i_2i} + g_{ii_3}z_{i_3i}, \ \forall i \geq 4 \tag{5.19}$$

As commented in the previous section, control laws for agents 2 and 3 can be also designed by using the inter-agent dynamics presented in Sect. 3.2. It is still quite straightforward to design a control law for agent i by the inter-agent dynamics. However, for agent i, $i \geq 4$, which may not place in the plane composed by the leader agents 1, 2, and 3, it may be difficult to ensure a realizability in 3-D along with noncoplanarity. That is why we propose the above gradient-based control laws (5.16)–(5.19) which are quite intuitive. For this, on the base of Sect. 5.1, we begin with the reexpression of Lemma 5.1.

Corollary 5.1 *Under the control laws (5.16), (5.17), and (5.18), the agents* 1, 2 *and* 3 *exponentially converge to the equilibrium with* $g_{21} = g_{31} = g_{32} = 0$ *if three agents* 1, 2 *and* 3 *are not collinear at the initial time.*

As already mentioned, it is supposed that agent 1 is stationary. Given the fixed agent 1, if agent 2 converges to the desired position, then agent 3 can consider the leader agents (i.e., agents 1 and 2) are fixed. Thus, under this setup, we can make the following assumption.

Assumption 5.2.2 Given agent i and for agents $i_j \in \mathcal{N}_i$, there exist fixed points $p^*_{i_1}$, $p^*_{i_2}$, and $p^*_{i_3}$ and the time T such that $p_{i_j} = p^*_{i_j}$, $\dot{p}_{i_j}(t) = 0$ for all $t > T$, $j = 1, 2, 3$.

Note that the above assumption is not rigorous mathematically though it can be rationalized by using the concept of input-to-state stability (see Sect. 5.3). Since the asymptotic convergence is only ensured as $t \rightarrow \infty$, it may take a long time before a convergence. So, we provide a finite-time convergence in Sect. 5.4 to overcome the drawbacks of this section.

To proceed under the Assumption 5.2.2, with $\frac{d\phi_{ij}}{d\bar{e}_{ij}} = g_{ij}$, $\frac{d\bar{e}_{ij}}{dt} = \frac{d\|z_{ij}\|^2}{dt} = z_{ij}^T(\dot{p}_i - \dot{p}_j)$, and $\dot{p}_i = -\sum_{j \in \mathcal{N}_i} g_{ij}z_{ij} = -\nabla_{p_i} V_i$ for agents $i \geq 4$, the derivative of the potential function is used as

$$\begin{aligned}
\dot{V}_i &= \sum_{j \in \mathcal{N}_i} \frac{d\phi_{ij}}{d\bar{e}_{ij}} \frac{d\bar{e}_{ij}}{dt} \\
&= \left(\sum_{j \in \mathcal{N}_i} g_{ij} z_{ij}^T \right) \dot{p}_i - \sum_{j \in \mathcal{N}_i} g_{ij} z_{ij}^T \dot{p}_j
\end{aligned}$$

$$= -\|\nabla_{p_i} V_i\|^2 - \sum_{j \in \mathcal{N}_i} g_{ij} z_{ij}^T \dot{p}_j \tag{5.20}$$

From the above equation, it follows that $\dot{V}_i + \sum_{j \in \mathcal{N}_i} g_{ij} z_{ij}^T \dot{p}_j = -\|\nabla_{p_i} V_i\|^2 \leq 0$. Thus, the term $V_i + \int_{t_0}^{\infty} \sum_{j \in \mathcal{N}_i} g_{ij} z_{ij}^T \dot{p}_j dt$ is nonincreasing. Further, from the Assumption 5.2.2, since $\dot{p}_j$ is an exponentially decaying function, with $V_{p_i} = \frac{1}{2} p_i^T p_i$, we have

$$\dot{V}_{p_i} = -\sum_{j \in \mathcal{N}_i} g_{ij} (\|p_i\|^2 - p_i^T p_j) = -\sum_{j \in \mathcal{N}_i} g(\bar{e}_{ij}, d_{ij}^*)(\|p_i\|^2 - p_i^T p_j) \tag{5.21}$$

Thus, since p_j is finite, when $p_i \to \infty$, $\dot{V}_{p_i}$ will be negative, which means that g_{ij} and z_{ij} are also bounded. Consequently, we can see that $\int_{t_0}^{\infty} \sum_{j \in \mathcal{N}_i} g_{ij} z_{ij}^T \dot{p}_j dt$ is bounded, which further implies $V_i(t)$ is bounded (i.e., exists and finite). Thus, by Barbalat's lemma and Theorem 2.23, both $\dot{V}_i$ and $\sum_{j \in \mathcal{N}_i} g_{ij} z_{ij}^T \dot{p}_j$ will converge to 0, from which it follows that the points $p_i(t)$ will converge to the equilibrium set

$$\mathcal{U}_{i,eq} \triangleq \{p_i \in \mathbb{R}^d : \nabla_{p_i} V_i = \sum_{j \in \mathcal{N}_i} g_{ij} z_{ji} = 0\} \tag{5.22}$$

as $t \to \infty$. Divide the above set into the correct equilibrium $\mathcal{U}_{i,eq}^C$ and incorrect equilibrium set $\mathcal{U}_{i,eq}^I$ respectively as

$$\mathcal{U}_{i,eq}^C = \{p_i \in \mathbb{R}^d : g_{ii_1} = g_{ii_2} = g_{ii_3} = 0\} \tag{5.23}$$

$$\mathcal{U}_{i,eq}^I = \{p_i \in \mathbb{R}^d : \sum_{j \in \mathcal{N}_i} g_{ij} z_{ji} = 0;\ \exists g_{ij} \neq 0,\ j \in \mathcal{N}_i\} \tag{5.24}$$

Note that $\mathcal{U}_{i,eq} = \mathcal{U}_{i,eq}^C \cup \mathcal{U}_{i,eq}^I$ and $\mathcal{U}_{i,eq}^C \cap \mathcal{U}_{i,eq}^I = \emptyset$. The undesired equilibrium set $\mathcal{U}_{i,eq}^I$ satisfies $\sum_{j \in \mathcal{N}_i} g_{ij} z_{ji} = 0$; so the vectors z_{ij} should be on the same plane, which further implies four points are of coplanar. To investigate the repulsiveness of undesired equilibrium set, we use the following notations, with an abuse of notation. A triangle $v_1 v_2 v_3$ is composed of three vertices v_1, v_1, and v_3. The vertices have their coordinate values in Euclidean space. Denote $z_{v_i v_j}$ as the vector $z_{v_i v_j} = v_i - v_j$. Also denote $\|z_{v_i v_j}\|$ as the length of the vector $z_{v_i v_j}$ and $\alpha_{v_1 v_3}^{v_2}$ as the subtended angle between two vectors $z_{v_1 v_2}$ and $z_{v_3 v_2}$. A tetrahedron $v_1 v_2 v_3 v_4$ is composed of four vertices v_1, v_2, v_3, and v_4. The three subtended angles at node v_1 in a tetrahedron are denoted as θ_1, θ_2 and θ_3 as depicted in Fig. 5.2. These angles can be also written as $\theta_1 = \alpha_{v_2 v_3}^{v_1}$, $\theta_2 = \alpha_{v_3 v_4}^{v_1}$ and $\theta_3 = \alpha_{v_2 v_4}^{v_1}$. Then, the following lemmas are utilized.

Lemma 5.5 ([11]) *Let us consider two triangles $v_1 v_2 v_3$ and $w_1 w_2 w_3$. If $\|z_{v_1 v_2}\| < \|z_{w_1 w_2}\|$, $\|z_{v_1 v_3}\| < \|z_{w_1 w_3}\|$, and $\|z_{v_2 v_3}\| > \|z_{w_2 w_3}\|$, then $\alpha_{v_2 v_3}^{v_1} > \alpha_{w_2 w_3}^{w_1}$.*

Lemma 5.6 *Given a tetrahedron $v_1 v_2 v_3 v_4$, it is true that $\alpha_{v_2 v_3}^{v_1} + \alpha_{v_3 v_4}^{v_1} + \alpha_{v_2 v_4}^{v_1} < 2\pi$ and $\alpha_{v_2 v_3}^{v_1} + \alpha_{v_3 v_4}^{v_1} > \alpha_{v_2 v_4}^{v_1}$, $\alpha_{v_2 v_3}^{v_1} + \alpha_{v_2 v_4}^{v_1} > \alpha_{v_3 v_4}^{v_1}$, and $\alpha_{v_3 v_4}^{v_1} + \alpha_{v_2 v_4}^{v_1} > \alpha_{v_2 v_3}^{v_1}$.*

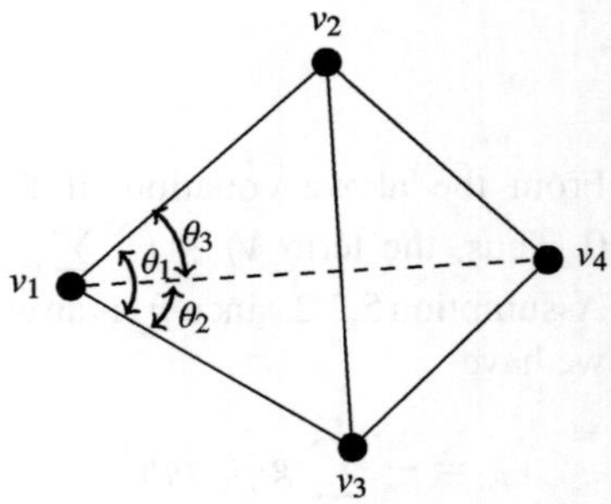

Fig. 5.2 The tetrahedral with vertices v_1, v_2, v_3 and v_4, and subtended angles θ_1, θ_2 and θ_3. © [2017] IEEE. Reprinted, with permission, from [6]

Proof The proof is clear from geometrical relationships and triangular inequalities.

Next, to examine a repulsiveness of undesired configurations, we make the following lemma [6]:

Lemma 5.7 *If* $p_i \in \mathcal{U}^I_{i,eq}$, *then* $g_{ii_1} + g_{ii_2} + g_{ii_3} < 0$.

Proof See the appendix.

Example 5.3 (*Example of Lemma* 5.7) Let the leader agents 1, 2, and 3 be located at $p_1 = (-2, -1, 0)^T$, $p_2 = (0, \sqrt{5}, 0)^T$, and $p_3 = (2, -1, 0)^T$. Let a follower agent, agent 4, be located at the origin, i.e., $p_4 = (0, 0, 0)^T$. It can be shown that $g_{41} + g_{42} + g_{43} < 0$, when the desired configuration of the set of agents 1, 2, 3, and 4 is not of coplanar. If the desired configuration is not of coplanar, the reference position of agent 4 can be given as $p_4^* = (0, 0, \epsilon)^T$ where $\epsilon \neq 0$. Then, for example since $g_{4i} = g(\bar{e}_{41})$ and it strictly monotonically increases as a function of $\bar{e}_{41}$, and $\bar{e}_{41} = \bar{d}_{41} - \bar{d}^*_{41} = \sqrt{5} - \sqrt{5 + \epsilon^2} < 0$, we can know that $g_{41} < 0$. Similarly, we know that $g_{42} < 0$ and $g_{43} < 0$. Thus, $g_{41} + g_{42} + g_{43} < 0$ if $\epsilon \neq 0$.

Based on Lemma 5.7, for the stability analysis, let us define the following set

$$\mathcal{C}_i = \{p_i : \text{rank}[z_{ii_1}, z_{ii_2}, z_{ii_3}] < 3\} \tag{5.25}$$

With the above set, we can now make the following theorem:

Theorem 5.4 *If* $p_{i_j} = p^*_{i_j}$ *and* $\dot{p}_{i_j} = 0$, *for* $j = 1, 2, 3$ *at time* T, *and agent* i *is outside* $\mathcal{C}_i$, *where* $i_j \in \mathcal{N}_i$, *then the trajectory* $p_i(t)$ *does not converge to* $\mathcal{U}^I_{i,eq}$.

Proof Let h be the distance from agent i to the plane containing agents i_1, i_2, i_3. Then from the control law (5.19), it follows:

$$\dot{h}(t) = -(g_{ii_1} + g_{ii_2} + g_{ii_3})h(t), \ t > T \tag{5.26}$$

which can be rewritten as $h(t) = h(T)e^{-\int_T^t (g_{ii_1} + g_{ii_2} + g_{ii_3})d\tau}$. Define an open set

$$\Sigma_{p_i} \triangleq \{p_i : g_{ii_1} + g_{ii_2} + g_{ii_3} < 0\} \tag{5.27}$$

Then, it is obvious that $\mathcal{U}_{i,eq}^{I} \subset \Sigma_{p_i}$ and $\Sigma_{p_i} \cap \mathcal{U}_{i,eq}^{C} = \emptyset$. Assume that p_i converges to an equilibrium in $\mathcal{U}_{i,eq}^{I}$, which means there exists a finite time T_i large enough such that $p_i(T_i) \in \Sigma_{p_i}$ or $g_{ii_1} + g_{ii_2} + g_{ii_3} < 0$ for all $t \geq T_i$. It implies $e^{-\int_{T_i}^{t} g_{ii_1}+g_{ii_2}+g_{ii_3} d\tau} > 0$. Since V_i is bounded, it is true that $g_{ii_1} + g_{ii_2} + g_{ii_3}$ is also bounded. Thus, $h(T)e^{-\int_{T}^{T_i} g_{ii_1}+g_{ii_2}+g_{ii_3} dt} > 0$. So, the following inequality holds

$$h(t) = h(T)e^{-\int_{T}^{T_i} g_{ii_1}+g_{ii_2}+g_{ii_3} dt} e^{-\int_{T_i}^{t} g_{ii_1}+g_{ii_2}+g_{ii_3} dt} > 0 \tag{5.28}$$

This is a contradiction to the assumption that $p_i(t)$ converges to $\mathcal{U}_{i,eq}^{I}$.

Theorem 5.5 *Let agent i be not in $\mathcal{C}_i$, i.e., $p_i(T) \notin \mathcal{C}_i$. Then, the trajectory of $p_i(t)$ globally asymptotically converges to the desired equilibrium $\mathcal{U}_{i,eq}^{C}$.*

Proof To investigate the convergence to the desired equilibrium, consider a neighborhood $U_{\bar{p}_i}$ of $\bar{p}_i$ such that $U_{\bar{p}_i} \cap \mathcal{C}_i = \emptyset$, where $\bar{p}_i \in \mathcal{U}_{i,eq}^{C}$. Let $U_{\bar{p}_i}^{o} = U_{\bar{p}_i} \setminus \{\bar{p}_i\}$. So, if $p_i \in U_{\bar{p}_i}^{o}$, then $p_i \notin \mathcal{U}_{i,eq}^{C}$. There exist constants $k_i > 0$, $\rho_i \in [0, 1)$ such that $\|\nabla_{p_i} V_i\| \geq k_i \|V_i(\bar{p}_i) - V_i(p_i)\|^{\rho_i} > 0$ for all $p_i \in U_{\bar{p}_i}^{o}$ since V_i is a real analytic function. Let $e_{p_i} = p_i - \bar{p}_i$; then it follows

$$\dot{e}_{p_i} = \dot{p}_i = -\nabla_{p_i} V_i = -\nabla_{e_{p_i}} V_i \tag{5.29}$$

Let T be the time when $p_j = \bar{p}_j$, $\dot{p}_j = 0$, $j = i_1, i_2, i_3$ for all $t \geq T$. Then, from (5.29), since $\dot{V}_i = -\|\nabla_{p_i} V_i\|^2 < 0$, the agent i locally asymptotically converges to $\bar{p}_i$. But, due to the repulsiveness property of an undesired equilibrium as confirmed in Theorem 5.4, it is clear that the trajectory of $p_i(t)$ almost globally asymptotically converges to the desired equilibrium.[1]

An exponential convergence can be shown using the property of $g(x, d^*)$ as follows:

Theorem 5.6 *Let $p(t_0)$ be in the neighborhood of $\mathcal{U}_{i,eq}^{C}$. Then, $p_i(t)$ converges to the desired equilibrium $\mathcal{U}_{i,eq}^{C}$ exponentially fast.*

Proof The derivative of the function $g(\bar{e}_{ij}, d_{ij}^*)$ with respect to $\bar{e}_{ij}$ is positive and bounded. Let $\rho_{max} = \max_{p_i \in U_{\bar{p}_i}^{o}} \{\frac{\partial g_{ii_1}}{\partial \bar{e}_{ii_1}}, \frac{\partial g_{ii_2}}{\partial \bar{e}_{ii_2}}, \frac{\partial g_{ii_3}}{\partial \bar{e}_{ii_3}}\}$ and $\rho_{min} = \min_{p_i \in U_{\bar{p}_i}^{o}} \{\frac{\partial g_{ii_1}}{\partial \bar{e}_{ii_1}}, \frac{\partial g_{ii_2}}{\partial \bar{e}_{ii_2}}, \frac{\partial g_{ii_3}}{\partial \bar{e}_{ii_3}}\}$. Since $\phi(\bar{e}_{ij}, d_{ij}^*) = g_{ij} = 0$ when $\bar{e}_{ij} = 0$, it follows

$$\phi(\bar{e}_{ij}, d_{ij}^*) = \int_0^{\bar{e}_{ij}} \int_0^{\bar{e}_{ij}} \frac{\partial g_{ij}}{\partial \bar{e}_{ij}} d\bar{e}_{ij} d\bar{e}_{ij} \leq \rho_{max} \|\bar{e}_{ij}\|^2 \tag{5.30}$$

$$\rho_{min} \|\bar{e}_{ij}\| \leq \|g_{ij}\| = \left\| \int_0^{\bar{e}_{ij}} \frac{\partial g_{ij}}{\partial \bar{e}_{ij}} d\bar{e}_{ij} \right\| \tag{5.31}$$

[1] In this monograph, when the undesired or incorrect equilibrium sets are unstable, while the desired equilibrium sets are stable, then without loss of generality, we call the desired sets almost globally asymptotically stable.

for all $p_i \in U^o_{\bar{p}_i}$, $j \in \{1, 2, 3\}$. Moreover, for any $p_i \in U^o_{\bar{p}_i}$, z_{ii_1}, z_{ii_2} and z_{ii_3} are not collinear and g_{ii_1}, g_{ii_2} and g_{ii_3} are not zero. Thus, $g_{ii_1} z_{ii_1} + g_{ii_2} z_{ii_2} + g_{ii_3} z_{ii_3} \neq 0$, which means that there exists $\delta > 0$ such that $\|g_{ii_1} z_{ii_1} + g_{ii_2} z_{ii_2} + g_{ii_3} z_{ii_3}\|^2 > \delta(\|g_{ii_1}\|^2 + \|g_{ii_2}\|^2 + \|g_{ii_3}\|^2)$. Then, considering $p_i \in U^o_{\bar{p}_i}$, we have

$$
\begin{aligned}
V_i &= \sum_{j=1}^{3} \phi(\bar{e}_{ii_j}, d^*_{ii_j}) \\
&\leq \rho_{max}(\|\bar{e}_{ii_1}\|^2 + \|\bar{e}_{ii_2}\|^2 + \|\bar{e}_{ii_3}\|^2) \\
&\leq \frac{\rho_{max}}{\rho^2_{min}}(\|g_{ii_1}\|^2 + \|g_{ii_2}\|^2 + \|g_{ii_3}\|^2)
\end{aligned} \tag{5.32}
$$

Thus, we have

$$
\dot{V}_i = -\|g_{ii_1} z_{ii_1} + g_{ii_2} z_{ii_2} + g_{ii_3} z_{ii_3}\|^2 < -\delta \frac{\rho^2_{min}}{\rho_{max}} V_i \tag{5.33}
$$

which implies V_i converges to zero exponentially fast in the neighborhood of $\bar{p}_i \in \mathcal{U}^C_{i,eq}$.

5.3 Acyclic Persistent Formations in 2-Dimensional Space

In Sects. 5.1 and 5.2, it was shown that acyclic minimally persistent formations could be achieved in 2- and 3-dimensional spaces respectively by gradient control laws. However, it is noticeable that the minimally persistent formation does not ensure a stabilization of formation to a desired configuration even though the desired inter-agent distances are satisfied. It is required to have at least three constraints to have a unique realization in 2-dimensional space (2-D) and four constraints in 3-dimensional space (3-D). In this section, we would like to use the results of Sect. 5.2 to ensure a stabilization of formation to a desired configuration in 2-D. The idea utilized in this section is similar to the approach of Sect. 12.1. Since a convergence of an agent with 3 constraints can be achieved in 3-D, we consider one axis components in 3-D as virtual values in 2-D. That is, in 2-D, we assume that the follower agents have 3 constraints to its leaders; but in such a case, since it is not minimally persistent, it is difficult to prove the convergence. So, we use the formation control law developed for 3-D in Sect. 5.2 and apply it to 2-D by setting one of the components (for example, z-axis component) as a virtual value. Then, in 2-D, each agent computes its control input by using the gradient control law applied for 3-D; but takes only x- and y-axis components for implementations in 2-D.

Let us consider the same setup as Sect. 5.2; so a follower agent i has three neighboring leaders as $\mathcal{N}_i = \{i_1, i_2, i_3\}$. We suppose that given an agent i, there exists a virtual agent i^v associated with the agent i moving in 3-D. Figure 5.3 depicts the

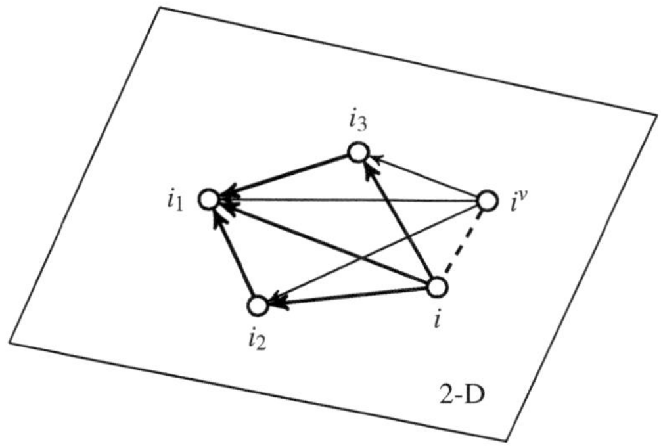

Fig. 5.3 The vertices i, i_1, i_2 and i_3 are on the same plane, while the virtual vertex i^v is not in the same plane. © [2018] IEEE. Reprinted, with permission, from [8]

virtual agent moving in 3-D that corresponds to the agent i in 2-D. In this case, it can be considered that the agent i's position is a projection of agent i^v's position onto the plane, which can be expressed as

$$p_{i^v} = \begin{bmatrix} 1 & 0 \\ 0 & 1 \\ 0 & 0 \end{bmatrix} p_i + \begin{bmatrix} 0 \\ 0 \\ 1 \end{bmatrix} h_i \tag{5.34}$$

where p_{i^v} denotes the virtual position in 3-D of the virtual agent i^v, p_i denotes the position of the agent i in 2-D, and h_i is the distance from p_{i^v} to p_i (i.e., $h_i = \| p_{i^v} - p_i \|$. Note that the actual position of agent i is p_i; the virtual position information p_{i^v} and relevant dynamics are used just for a computational purpose to drive agents in 2-D under the setup in 3-D. When the agent i has three desired squared distance constraints to its neighboring leaders (they are denoted as i_1, i_2 and i_3) as $\bar{d}^*_{ij}$, where $j = i_1, i_2, i_3$, the virtual agent i^v has desired squared distances to the leaders as $\bar{d}^*_{i^v j} = \bar{d}^*_{ij} + (h^*_i)^2$ where h^*_i is the desired distance between i and i^v. We select the initial virtual distance of h_i simply as $h_i(t_0) > 0$. Without notational confusion, given a virtual agent i^v, we define $z_{i^v j} = p_{i^v} - p_j$, $\bar{e}_{i^v j} = \|z_{i^v j}\|^2 - \bar{d}^*_{i^v j}$, and $g_{i^v j} = g(\bar{e}_{i^v j}, \bar{d}^*_{i^v j})$ and $\rho_{i^v j} = \rho(\bar{e}_{i^v j}, \bar{d}^*_{i^v j})$ where the functions $g(\cdot,\cdot)$ and $\rho(\cdot,\cdot)$ are defined in Assumption 5.2.1. It is remarkable that when an agent has three constraints to other agents in 3-D, there are two points that satisfy the distance constraints. However when the two points are projected onto the plane, their projections are merged at a same point. So, the ambiguity could disappear in 2-D. Similar to the control laws (5.16), (5.17), (5.18), and (5.19), we use the following laws:

$$\dot{p}_1 = 0 \tag{5.35}$$

$$\dot{p}_2 = g_{21} z_{12} \tag{5.36}$$

$$\dot{p}_3 = g_{31} z_{13} + g_{32} z_{23} \tag{5.37}$$

$$\dot{p}_i = \begin{bmatrix} 1 & 0 & 0 \\ 0 & 1 & 0 \end{bmatrix} \dot{p}_{i^v}, \; i \geq 4 \tag{5.38}$$

$$\dot{p}_{i^v} = g_{i^v i_1} z_{i_1 i^v} + g_{i^v i_2} z_{i_2 i^v} + g_{i^v i_3} z_{i_3 i^v} \tag{5.39}$$

It is clear that the control laws given in (5.38) and (5.39) can be implemented in a distributed way. The agent i can compute z_{ji^v} and $g_{i^v j}$ from the measurements z_{ji} and g_{ij}. Let us define the desired configuration set as

$$\mathcal{U}^D \triangleq \{p \in \mathbb{R}^{2n} : g_{ij} = 0, \forall (i, j)^{\bar{e}} \in \overrightarrow{\mathcal{E}}\} \tag{5.40}$$

Also, given a desired realization p^* as $p^* \in \mathcal{U}^D$, we define the following vectors:

$$e_{p[1:i-1]} = \begin{bmatrix} p_1 - p_1^* \\ p_2 - p_2^* \\ \vdots \\ p_{i-1} - p_{i-1}^* \end{bmatrix} \text{ and } e_{p[i]} = \begin{bmatrix} p_{i^v} - p_{i^v}^* \\ p_i - p_i^* \end{bmatrix}$$

For the overall stability analysis, with the dynamics of agents $1, \ldots, i$, we use the following cascade system:

$$\dot{e}_{p[i]} = f_1(e_{p[i]}, e_{p[1:i-1]}), i \geq 4 \tag{5.41}$$

$$\dot{e}_{p[1:i-1]} = f_2(e_{p[1:i-1]}) \tag{5.42}$$

By the input-to-state stability (see Lemma 2.11), if the origin of (5.42) and origin of the unforced dynamics of (5.41) are asymptotically stable, the cascade systems (5.41)–(5.42) are also asymptotically stable. Without loss of generality, let us suppose that the leaders i_1, i_2 and i_3 are moving in 2-D, while the virtual follower i^v is moving in a virtual 3-D. Then similarly to the sets (5.23)–(5.24), we can define the following sets:

$$\mathcal{U}^C_{i^v,eq} = \{p_{i^v} \in \mathbb{R}^3 : g_{i^v i_1} = g_{i^v i_2} = g_{i^v i_3} = 0\} \tag{5.43}$$

$$\mathcal{U}^I_{i^v,eq} = \{p_{i^v} \in \mathbb{R}^3 : \sum_{j \in \mathcal{N}_i} g_{i^v j} z_{i^v j} = 0;\ \exists g_{i^v j} \neq 0, j \in \mathcal{N}_i\} \tag{5.44}$$

Now, following the same process as conducted in Sect. 5.2 with Assumption 5.2.2, it can be claimed that the set $\mathcal{U}^C_{i^v,eq}$ is locally exponentially stable, and the set $\mathcal{U}^I_{i^v,eq}$ is unstable (repulsive). Then, denoting a realization p as $p = (p[1:, i-1]^T, p_{i^v})^T$, it is shown that the Lebesgue measure for $p \in \mathcal{U}^I_{i^v,eq}$ is zero since the algebraic realization of four points such that they are collinear rarely occurs. Thus, borrowing *almost everywhere* concept from Lebesgue measure zero [9], we say that the desired equilibrium set $\mathcal{U}^C_{i^v,eq}$ is *almost* globally asymptotically stable. Therefore, since the set $\mathcal{U}^C_{i^v,eq}$ is almost globally asymptotically stable, it is obvious that the set $\mathcal{U}^C_{i,eq} = \{p_i \in \mathbb{R}^2 : g_{ii_1} = g_{ii_2} = g_{ii_3} = 0\}$ is also almost globally asymptotically stable. It is remarkable that the control laws (5.35)–(5.39) can be further generalized to the case when a follower agent has more than three constraints. For example, the control law

(5.39) can be used for agent 4. For the agent 5, it may have four leaders, and agent 6 may have five leader agents. For agent 5, the control laws can be modified as

$$\dot{p}_5 = \begin{bmatrix} 1\,0\,0\,0 \\ 0\,1\,0\,0 \end{bmatrix} \dot{p}_{5^v}$$

$$\dot{p}_{5^v} = g_{5^v i_1} z_{i_1 5^v} + g_{5^v i_2} z_{i_2 5^v} + g_{5^v i_3} z_{i_3 5^v} + g_{5^v i_4} z_{i_4 5^v}$$

where $i_1 = 1, i_2 = 2, i_3 = 3$, and $i_4 = 4^v$. For the case of agent 4, it has one virtual axis, while the agent 5 has two virtual axes. For the case of agent 6 that has five leader constraints, it needs to have three virtual axes. The systematic idea and approach can be borrowed from Sect. 12.1.

5.4 Finite-Time Convergence of Formations

In Sect. 5.3, almost globally asymptotic convergence has been ensured using a global input-to-state stability concept, with the Assumption 5.2.2. However, since it is an asymptotic stability and the agents are connected in a cascade way, the asymptotic convergence also will be ensured in a cascade way. Thus, it may take a long time until a convergence. In this section, we would like to ensure the convergence of the set of leaders in finite-time so that the overall convergence time can be of finite and can be anticipated. To achieve a finite-time convergence of this section, we extract a result from [8]. For the finite-time convergence, the gains g_{ij} given in the control laws (5.35)–(5.39) are replaced by g_{ij}^f which are defined as one of the following laws:

$$g_{ij}^{f_1} = \frac{g_{ij}}{\| \sum_{j \in \mathcal{N}_i} g_{ij} z_{ij} \|} \tag{5.45}$$

$$\text{or } g_{ij}^{f_2} = |g_{ij}|^{\alpha} \text{sign}(g_{ij}) \triangleq \text{sig}(g_{ij})^{\alpha}, \; \alpha \in (0, 1) \tag{5.46}$$

Let us consider the four-agent case which is composed of a leader, the first follower, the second follower, and an ordinary follower. The leader does not have any neighboring agent, while the first follower has the leader as its single neighboring agent and the second follower has the leader and the first follower as its two neighboring agents. The ordinary follower has three neighboring leaders. Without loss of generality, we denote the leader by agent 1, the first follower by agent 2, and the second follower by agent 3.

The finite-time convergence of the agent 2 is summarized in the following theorem:

Theorem 5.7 *When the agent 2 is controlled by (5.36) with replaced g_{21} by $g_{21}^{f_1}$ or by $g_{21}^{f_2}$, it converges to the desired equilibrium in a finite time.*

Proof For the agent 2, the dynamics can be written as

$$\dot{p}_2 = \frac{g_{21}}{\|g_{21}z_{21}\|}z_{12} \tag{5.47}$$

$$\text{or } \dot{p}_2 = \text{sig}(g_{21})^{\alpha} z_{12} \tag{5.48}$$

Denoting $\omega \triangleq d_{21} - d_{21}^*$, by Assumption 5.2.1, it is true that $\omega > 0$ ($<$ or $=$, respectively) if and only if $g_{21} > 0$ ($<$ or $=$, respectively). Now, when the agent 2 is updated by (5.47), it follows

$$\dot{\omega} = \frac{\partial \|z_{21}\|}{\partial z_{21}}\dot{z}_{21} = \frac{1}{\|z_{21}\|}z_{21}^T\dot{z}_{21} = -\frac{1}{\|z_{21}\|}z_{21}^T\frac{g_{21}}{\|g_{21}z_{21}\|}z_{21} = -\frac{g_{21}}{\|g_{21}\|} \tag{5.49}$$

The above equality can be rewritten as $\dot{\omega} = -\text{sign}(g_{21}) = -\text{sign}(\omega)$. Taking the potential function $V_\omega = \omega^2 \geq 0$ and its derivative $\dot{V}_\omega = -2\omega\text{sign}(\omega)$, we have $\dot{V}_\omega < 0$ for all $\omega \neq 0$ and $\dot{V}_\omega = 0$ if and only if $\omega = 0$. Moreover $\dot{V}_\omega + (V_\omega)^{0.5} = -2|\omega| + |\omega| = -|\omega| \leq 0$. Thus, according to Lemma 2.15, the signal $\omega(t)$ converges to zero in finite time which implies $d_{21} \to d_{21}^*$ in a finite time.

Next, when it is updated by (5.48), we use the potential function $V = \frac{1}{2}|g_{21}|^{\alpha+1}$, which has the following derivative:

$$\begin{aligned}
\dot{V} &= \frac{1}{2}\frac{\partial |g_{21}|^{\alpha+1}}{\partial g_{21}}\frac{\partial g_{21}}{\partial t} \\
&= \frac{1}{2}(\alpha+1)|g_{21}|^{\alpha}\frac{g_{21}}{|g_{21}|}\frac{\partial g_{21}}{\partial t} \\
&= \frac{1}{2}(\alpha+1)|g_{21}|^{\alpha}\frac{g_{21}}{|g_{21}|}\frac{\partial g(\bar{e}_{21}, \bar{d}_{21}^*)}{\partial \bar{e}_{12}}\frac{\partial \bar{e}_{12}}{\partial t} \\
&= (\alpha+1)\text{sig}(g_{21})^{\alpha}\rho_{21}z_{21}^T\dot{z}_{21}
\end{aligned} \tag{5.50}$$

where $\rho_{21} = \frac{\partial g(\bar{e}_{21}, \bar{d}_{21}^*)}{\partial \bar{e}_{12}} > 0$. Inserting (5.48) into the right-hand side of (5.50) yields

$$\begin{aligned}
\dot{V} &= -(\alpha+1)\rho_{21}\left[\text{sig}(g_{21})^{\alpha}\right]^2\|z_{21}\|^2 \\
&= -(\alpha+1)\rho_{21}|g_{21}|^{2\alpha}\|z_{21}\|^2 \\
&\leq 0
\end{aligned} \tag{5.51}$$

Thus, we know that $\dot{V} = 0$ if $g_{21} = 0$ or $z_{21} = 0$. Thus, by LaSalle's invariance principle (see Theorem 2.22), p_2 converges to either the position corresponding to $g_{21} = 0$ or $z_{21} = 0$. Let us suppose that p_2 converges to a point $p_2^\dagger$ when $\|z_{21}\| = 0$. Then, there must exist a finite time T such that $-\bar{d}_{21}^* < \bar{e}_{21}(T) < 0$ holds, which implies $g_{21}(p_2^\dagger) < g_{21}(T) < 0$. Thus, we have $V(T) < V(p_2^\dagger)$, which is a contradiction to the fact that $\dot{V} \leq 0$. Additionally, in the vicinity of $p_2^\ddagger$ (denote it as $U_{p_2^\ddagger}$) defined by

$g_{21}(p_2^{\ddagger}) = 0$, since $\dot{V} \leq -c_2 V^{\frac{2\alpha}{\alpha+1}}$ where $c_2 = \min_{p_2 \in U_{p_2^{\ddagger}}}((\alpha+1)\rho_{21}\|z_{21}\|^2)^{\frac{2\alpha}{\alpha+1}}$, p_2 converges to the desired equilibrium in a finite time by Lemma 2.15.

The finite-time controller of agent 2 is quite intuitive. The following example compares the gradient law and finite-time gradient law.

Example 5.4 Let the agent 2, which is the first follower, be supposed to approach toward the agent 1, which is the leader. The control law (5.48) is a finite-time controller, while the traditional gradient control law is $\dot{p}_2 = g_{21}z_{12}$. When $|g_{21}| \gg 1$, we have the inequality $|g_{21}| > \text{sig}(g_{21})^{\alpha}$. Thus, when the error is big, the traditional gradient control law provides more control efforts. But, when $|g_{21}| \ll 1$, we have $|g_{21}| < \text{sig}(g_{21})^{\alpha}$. Thus, when the error is small, the finite-time controller provides more control efforts. In fact, as the error becomes smaller and smaller, the control effort of the finite-time control law is more significant. For example, with $\alpha = 0.5$, when $g_{21} = 0.01$, we have $\dot{p}_2 = 0.01z_{12}$ in the traditional gradient law, but have $\dot{p}_2 = 0.1z_{12}$ in the finite-time control law. But, when $\alpha = 0.1$, we have $\dot{p}_2 = 0.6309z_{12}$ for the finite-time control law. Consequently, when the magnitude of error is big, i.e., $|g_{ij}| \geq 1$, the traditional gradient law may be used for speeding up the convergence. But, when the error becomes smaller such as $|g_{ij}| < 1$, we can change the control law to finite-time one, and gradually we can reduce α toward zero for improving the convergence speed. Finally, when α is set to zero, the finite-time control law with a `sig` function reduces to the sign function control law.

Next, let us study the motions of other agents. For a convenience of notation, let us assume that the neighbor nodes of agent i have been converged to desired points at T_i. First we investigate the convergence of formation when agents are updated by (5.45). Let us consider the virtual node of agent i as i^v and its potential function $V_{i^v} = \sum_{j \in \mathcal{N}_{i^v}} \phi(\bar{e}_{i^v j}, d^*_{i^v j})$. Let us suppose that the virtual agent i^v has three neighbor nodes as i_1, i_2 and i_3. Then the dynamics of i^v can be obtained as

$$\begin{aligned}\dot{p}_{i^v} &= -\frac{\nabla_{p_{i^v}} V_{i^v}}{\|\nabla_{p_{i^v}} V_{i^v}\|} \\ &= -\frac{g_{i^v i_1} z_{i^v i_1} + g_{i^v i_2} z_{i^v i_2} + g_{i^v i_3} z_{i^v i_3}}{\|g_{i^v i_1} z_{i^v i_1} + g_{i^v i_2} z_{i^v i_2} + g_{i^v i_3} z_{i^v i_3}\|}\end{aligned} \tag{5.52}$$

The derivative of the potential function is given as

$$\begin{aligned}\dot{V}_{i^v} &= \sum_{j \in \mathcal{N}_{i^v}} \frac{\partial \phi(\bar{e}_{i^v j}, d^*_{i^v j})}{\partial \bar{e}_{i^v j}} \frac{\partial \bar{e}_{i^v j}}{\partial t} \\ &= \sum_{j \in \mathcal{N}_{i^v}} g_{i^v j} z^T_{i^v j} (\dot{p}_{i^v} - \dot{p}_j) \\ &= -\sum_{j \in \mathcal{N}_{i^v}} g_{i^v j} z^T_{i^v j} \left(\frac{g_{i^v i_1} z_{i^v i_1} + g_{i^v i_2} z_{i^v i_2} + g_{i^v i_3} z_{i^v i_3}}{\|g_{i^v i_1} z_{i^v i_1} + g_{i^v i_2} z_{i^v i_2} + g_{i^v i_3} z_{i^v i_3}\|} \right) - \sum_{j \in \mathcal{N}_{i^v}} g_{i^v j} z^T_{i^v j} \dot{p}_j\end{aligned}$$

$$= -\|g_{i^v i_1} z_{i^v i_1} + g_{i^v i_2} z_{i^v i_2} + g_{i^v i_3} z_{i^v i_3}\| - \sum_{j \in \mathcal{N}_{i^v}} g_{i^v j} z_{i^v j}^T \dot{p}_j$$

Now, by following the same argument of Theorem 5.3, the control efforts of the leader agents, i.e., $\dot{p}_j$, are bounded. Thus, if i^v diverges (i.e., i diverges), then $\dot{V}_{i^v}$ will be negative, which further implies that the errors of all the agents are bounded. Therefore, since the leader agents converge to the desired points in a finite time T_i, we can equalize $\dot{V}_{i^v}$ as

$$\dot{V}_{i^v} = -\|g_{i^v i_1} z_{i^v i_1} + g_{i^v i_2} z_{i^v i_2} + g_{i^v i_3} z_{i^v i_3}\| \leq 0 \tag{5.53}$$

at $t \geq T_i$. Consequently, the virtual agent i^v converges to the set $\mathcal{U}_{i^v} \triangleq \{p_{i^v} | g_{i^v i_1} z_{i^v i_1} + g_{i^v i_2} z_{i^v i_2} + g_{i^v i_3} z_{i^v i_3} = 0\}$. Let $h_i(t)$ be the distance from the virtual agent i^v to the plane containing agents i_1, i_2, i_3. Then, from (5.39), we can obtain

$$\dot{h}_i(t) = -(g_{i^v i_1} + g_{i^v i_2} + g_{i^v i_3}) h_i(t)$$

Let us divide the equilibrium set as the desired one $\mathcal{U}_{i^v}^C \triangleq \{p_{i^v} \in \mathbb{R}^3 | g_{i^v i_1} = g_{i^v i_2} = g_{i^v i_3} = 0\}$ and undesired one $\mathcal{U}_{i^v}^I = \mathcal{U}_{i^v} \backslash \mathcal{U}_{i^v}^C$.

Theorem 5.5 *When the follower agent is updated by either (5.52) or (5.46), it converges to desired equilibrium $\mathcal{U}_{i^v}^C$ in a finite time.*

Proof Let us first consider the control law (5.52). Define two sets $\Sigma^+ \triangleq \{p_{i^v} \in \mathbb{R}^3 | g_{i^v i_1} + g_{i^v i_2} + g_{i^v i_3} > 0\}$ and $\Sigma^- \triangleq \{p_{i^v} \in \mathbb{R}^3 | g_{i^v i_1} + g_{i^v i_2} + g_{i^v i_3} < 0\}$. Let us suppose that the sets Σ^+ and Σ^- include the undesired set $\mathcal{U}_{i^v}^I$ such as $\mathcal{U}_{i^v}^I \subset \Sigma^+$ or $\mathcal{U}_{i^v}^I \subset \Sigma^-$. First let us assume that the agent i^v is in the set Σ^+. Then, it converges to the plane; but in this case, according to the results in Sects. 5.3 and 5.2, the agent converges to the desired equilibrium. Next, consider the set Σ^-. Then, there exists a finite time $T > T_i$ large enough such that $p_{i^v}(T) \in \Sigma^-$ and $g_{i^v i_1} + g_{i^v i_2} + g_{i^v i_3} < 0$ for all $t \geq T$. So, for all $t \geq T$, we have $\dot{h}_i(t) > 0$. But, this is a contradiction to the convergence to the set $\mathcal{U}_{i^v}$, which implies that the virtual agent converges to the desired equilibrium.

It is now required to show that the convergence could be achieved in a finite time. From $\frac{\partial V_{i^v}}{\partial p_{i^v}^T} = \sum_{j \in \mathcal{N}_{i^v}} (g_{i^v j} z_{i^v j})^T$ and $\frac{\partial^2 V_{i^v}}{\partial p_{i^v}^2} = \sum_{j \in \mathcal{N}_{i^v}} \left(\frac{\partial g_{i^v j}}{\partial \bar{e}_{ij}} \frac{\partial \bar{e}_{ij}}{\partial p_{i^v}} z_{i^v j}^T + g_{i^v j} \frac{\partial z_{i^v j}^T}{\partial p_{i^v}} \right)$, the Hessian of the potential function can be obtained as

$$H(V_{i^v}) = \sum_{j \in \mathcal{N}_{i^v}} (\rho_{i^v j} z_{i^v j} z_{i^v j}^T + g_{i^v j} \otimes \mathbb{I}_2) \tag{5.54}$$

Using the following denotation

$$A_{i^v j} \triangleq z_{i^v j} z_{i^v j}^T = \rho_{i^v j} \begin{bmatrix} (x_{i^v} - x_j)^2 & (x_{i^v} - x_j)(y_{i^v} - y_j) & (x_{i^v} - x_j) h_i \\ (x_{i^v} - x_j)(y_{i^v} - y_j) & (y_{i^v} - y_j)^2 & (y_{i^v} - y_j) h_i \\ (x_{i^v} - x_j) h_i & (y_{i^v} - y_j) h_i & h_i^2 \end{bmatrix}$$

and using the vector $v = (v_x, v_y, v_z)^T \neq 0$, we have

$$v^T \left(\sum_{j \in \mathcal{N}_{i^v}} \rho_{i^v j} A_{i^v j} \right) v = \sum_{j \in \mathcal{N}_{i^v}} \rho_{i^v j} [v_x(x_i - x_j) + v_y(y_i - y_j) + v_z h_i]^2$$

Let $U_{p_{i^v}}$ be a neighborhood of the desired equilibrium $p^*_{i^v}$ such that $U_{p_{i^v}} \cap \mathcal{U}^I_{i^v} = \emptyset$. Since $p_{i^v} - p_j$ are linearly independent for all $p_{i^v} \in U_{p_{i^v}}$, it is true that $v^T \left(\sum_{j \in \mathcal{N}_{i^v}} \rho_{i^v j} A_{i^v j} \right) v > 0$. Thus, there exists a neighborhood $\hat{U}_{p_{i^v}} \in U_{p_{i^v}}$ such that the Hessian $H(V_{i^v})$ is positive definite for all $p_{i^v} \in \hat{U}_{p_{i^v}}$. Therefore, using Lemma 2.14, the virtual agent p_{i^v} converges to the desired equilibrium in a finite time, which also leads the convergence of p_i to the desired equilibrium.

Now, consider the control law (5.46). Let us select the potential function of agent i^v as $V_{i^v} = \sum_{j \in \mathcal{N}_{i^v}} |g_{i^v j}|^{\alpha+1}$, which leads the following derivative:

$$\begin{aligned} \dot{V}_{i^v} &= \sum_{j \in \mathcal{N}_{i^v}} ((\alpha + 1)\text{sig}(g_{i^v j})^{\alpha} \rho_{i^v j} z^T_{i^v j} \dot{z}_{i^v j}) \\ &\leq -\rho_{min}(\alpha + 1) \|\text{sig}(g_{i^v j})^{\alpha} z_{i^v j}\|^2 \leq 0 \end{aligned}$$

where $\rho_{min} = \min_{j \in \mathcal{N}_{i^v}} \rho_{i^v j}$. In the neighborhood $U_{p_{i^v}}$ of the desired equilibrium of p_{i^v}, $p_{i^v} - p_j$, $j \in \mathcal{N}_{i^v}$, are linearly independent. Let $c_i = \min_{p_{i^v} \in U_{p_{i^v}}, j \in \mathcal{N}_{i^v}} ((\alpha + 1)\rho_{i^v j} \|p_{i^v} - p_j\|^2)^{2\alpha/(\alpha+1)}$, then following the similar process as (5.53), we can have

$$\dot{V}_{i^v} \leq -c_i(|g_{i^v i_1}|^{2\alpha} + |g_{i^v i_2}|^{2\alpha} + |g_{i^v i_3}|^{2\alpha})$$

Since $0 < \alpha < 1$ and $0 < \frac{2\alpha}{\alpha+1} < 1$, it holds that $(|g_{i^v i_1}|^{\alpha+1} + |g_{i^v i_2}|^{\alpha+1} + |g_{i^v i_3}|^{\alpha+1})^{\frac{2\alpha}{\alpha+1}} \leq |g_{i^v i_1}|^{2\alpha} + |g_{i^v i_2}|^{2\alpha} + |g_{i^v i_3}|^{2\alpha}$. Thus, we can have $\dot{V}_{i^v} \leq -c_i V_{i^v}^{\frac{2\alpha}{\alpha+1}}$. According to Lemma 2.15, p_{i^v} converges to the desired equilibrium in a finite time. Hence using the similar analysis as the control law (5.52), it is shown that $\mathcal{U}^I_{i^v}$ is repulsive, which implies $\mathcal{U}^C_{i^v}$ is almost globally stable in a finite time.

5.5 Summary and Simulations

For ensuring a global stabilization, it was required to have a graph that does not include a cycle. Thus, the basic underlying topology is acyclic persistence. The sensing and control variables in acyclic minimally persistent formations for 2-D and 3-D studied in Sects. 5.1 and 5.2 are the relative displacements and edge distances, respectively. However, when an agent has more than 2 constraints in 2-D, and more than 3 constraints in 3-D, the traditional gradient control laws cannot stabilize the formation. Thus, we strictly require the number of constraints for each agent to 2 in

Table 5.1 Variables and network properties of acyclic (minimally) persistent formations (AMPF or APF)

	AMPF	APF
Sensing variables	p^i_{ji}	p^i_{ji}
Control variables	$\|z_{ji}\|$	$\|z_{ji}\|$
Computational variables	None	$\dot{p}_{i^v}$
Configuration	Not unique	Unique

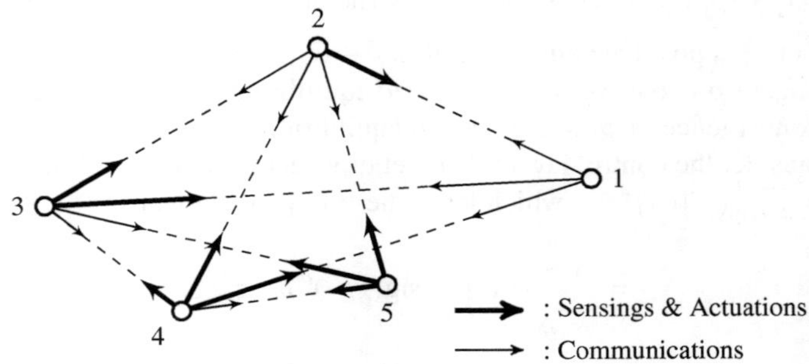

Fig. 5.4 Sensing/actuation and communication topologies for simulations. The sensing and actuation directions are coincident; but the communication direction is opposite

2-D or 3 in 3-D, respectively. Hence, the underlying topology for sensing and control is acyclic minimally persistent. Since there should be no cycle in the topology, the edges are considered directed toward the leaders. Table 5.1 summarizes these points for the acyclic minimally persistent formations of Sects. 5.1 and 5.2. In Sect. 5.3, we added one more constraint to each agent to change the formation to acyclic persistent formations.

Thus, we could remove the *minimally* property in the underlying topology. But, to add one more constraint, each node needs to have a virtual computation such as (5.39); so the computational load of each agent increases which is a weakness of the algorithm. But, unlike Sect. 12.1, since it is acyclic persistence graph, it is not necessary to have communications between agents. By this way, since three constraints can be imposed to each agent, the position of a follower can be uniquely determined with respect to the three leaders. Thus, the achieved formation could be unique. Table 5.1 also shows these points. It is also noticeable that, if some more edges are added such that the out-degree becomes more than 4, the proposed control laws in (5.35)–(5.39) should be further modified. The generalization could be done using the approach of Sect. 12.1 via using more virtual variables.

Let us evaluate the convergence characteristics of acyclic minimally persistent formations in $\mathbb{R}^3$ of Sect. 5.2. For the simulation, we consider the network topology depicted in Fig. 5.4. The potential function needs to satisfy the conditions of Assumption 5.2.1. We consider the following three potential functions $\phi^1 = \frac{1}{2}\bar{e}_{ij}$,

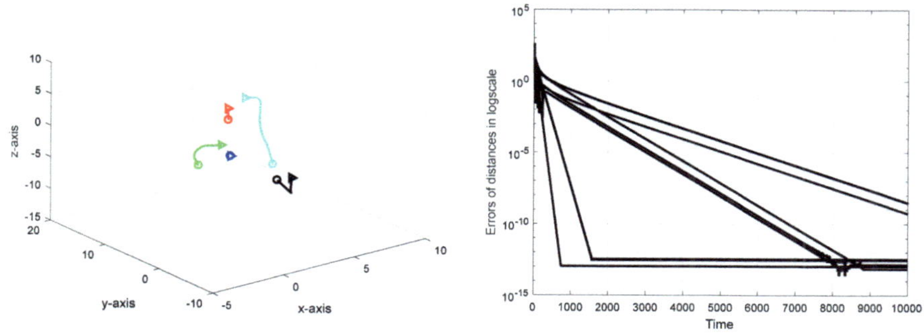

Fig. 5.5 Left: Trajectories of positions of agents, with potential function ϕ^1. The initial positions are marked by $\circ$ and the final positions are marked by $\triangleright$. Right: Errors of distances, with potential function ϕ^1

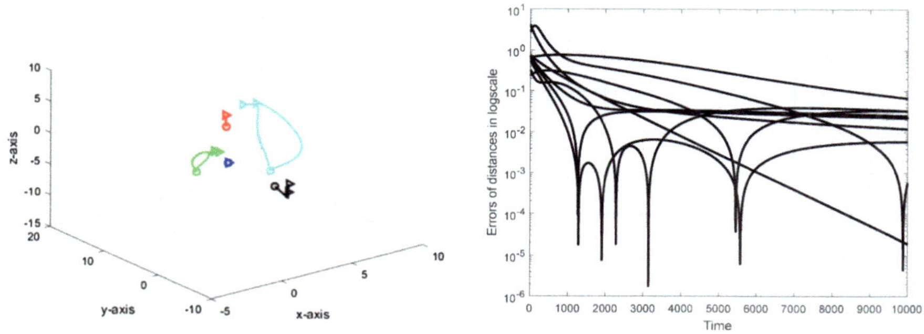

Fig. 5.6 Left: Trajectories of positions of agents, with potential function ϕ^2. The initial positions are marked by $\circ$ and the final positions are marked by $\triangleright$. Right: Errors of distances, with potential function ϕ^2

$\phi^2 = \bar{e}_{ij} - \ln(\bar{e}_{ij} + \bar{d}^*_{ij})/\bar{d}^*_{ij}$, and $\phi^3 = \bar{e}^2_{ij}/(\bar{e}_{ij} + \bar{d}^*_{ij})^2$, which respectively result in control functions as follows:

$$
\begin{aligned}
g^1_{ij} &= \bar{e}_{ij} \\
g^2_{ij} &= 1 - \frac{\bar{d}^*_{ij}}{\bar{e}_{ij} + \bar{d}^*_{ij}} \\
g^3_{ij} &= 1 - \frac{(\bar{d}^*_{ij})^2}{(\bar{e}_{ij} + \bar{d}^*_{ij})^2}
\end{aligned}
$$

Note that the potential function $\phi^1 = \frac{1}{2}\bar{e}_{ij}$ is a typical one used in traditional gradient control laws. The left plots of Figs. 5.5, 5.6, and 5.7 show the trajectories of agents according to the potential functions, while the right plots of Figs. 5.5, 5.6, and 5.7 show the errors in the distances. It is observed that the traditional potential function

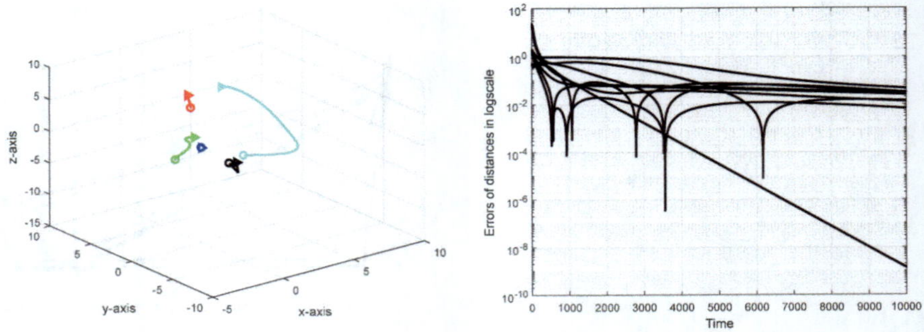

Fig. 5.7 Left: Trajectories of positions of agents, with potential function ϕ^3. The initial positions are marked by $\circ$ and the final positions are marked by $\triangleright$. Right: Errors of distances, with potential function ϕ^3

has the best performance in terms of the convergence speed than others as shown in the errors. The other potential functions do not have significant difference in terms of the convergence speed.

5.6 Notes

The results of Sect. 5.1 are reused from [4, 5] and the results of Sect. 5.2 are reused from [6]. For a generalized acyclic persistent formation in 2-D using distance and displacement constraints in a finite-time convergence, it is recommended to refer to [7]. In this chapter, the leader agent was assumed stationary. When the leader moves from a point to another target point, with the help of the input-to-state stability, we can have a similar stability result. For the analysis of a moving agent along with experiments, it is referred to [4]. The results of Sect. 5.4 have been reproduced from [8]. In this section, we have combined the control law for a finite-time convergence and control law for acyclic persistent formations, to make a finite-time convergence of acyclic persistent formations in $\mathbb{R}^2$; it can be considered as a combination of Sect. 5.3 and [7]. It is also remarkable that the formations studied in [10] under the name of *leader-remote-follower* and *three coleaders* include cycles in the underlying topology; but, the formations are still minimally persistent. It might be possible to use the virtual variables to enhance the control schemes of [10] to general persistent formations. The following copyright and permission notices are acknowledged.

- The text of Sect. 5.1 is © [2012] IEEE. Reprinted, with permission, from [5].
- The text of Sect. 5.2 is © [2017] IEEE. Reprinted, with permission, from [6].
- The statements of Theorems 5.7 and 5.5 with the proofs are © [2018] IEEE. Reprinted, with permission, from [8].
- The statements of Lemmas 5.3, 5.4, Theorems 5.1, 5.2, and 5.3 with the proofs are © Reprinted by permission from Springer-Nature: [4] (2016).

References

1. Belabbas, M.A.: On global stability of planar formations. IEEE Trans. Autom. Control **58**(8), 2148–2153 (2013)
2. Cao, M., Anderson, B.D.O., Morse, A.S., Yu, C.: Control of acyclic formations of mobile autonomous agents. In: Proceedings of the 47th IEEE Conference on Decision and Control, pp. 1187–1192 (2008)
3. Hendrickx, J.M., Anderson, B.D.O., Delvenne, J.-C., Blondel, V.D.: Directed graphs for the analysis of rigidity and persistence in autonomous agent systems. Int. J. Robust Nonlinear Control **17**(10–11), 960–981 (2007)
4. Park, M.-C., Ahn, H.-S.: Distance-based acyclic minimally persistent formations with non-steepest descent control. Int. J. Control Autom. Syst. **14**(1), 163–173 (2016)
5. Park, M.-C., Oh, K.-K., Ahn, H.-S.: Modified gradient control for acyclic minimally persistent formations to escape from collinear position. In: Proceedings of the IEEE Conference on Decision and Control, pp. 1423–1427 (2012)
6. Pham, V.H., Trinh, M.H., Ahn, H.-S.: Distance-based directed formation control in 3-dimensional space. In: Proceedings of the SICE Annual Conference, pp. 886–891 (2017)
7. Pham, V.H., Trinh, M.H., Ahn, H.-S.: A finite-time convergence of generalized acyclic persistent formation in 3-d using relative position measurements. In: Proceedings of the 17th International Conference on Control, Automation and Systems, pp. 230–235 (2017)
8. Pham, V.H., Trinh, M.H., Ahn, H.-S.: Finite-time convergence of acyclic generically persistent formation. In: Proceedings of the American Control Conference, pp. 3642–3647 (2018)
9. Schröder, B.S.W.: Mathematical Analysis - A Concise Introduction. Wiley, USA (2008)
10. Summers, T.H., Yu, C., Dasgupta, S., Anderson, B.D.O.: Control of minimally persistent leader-remote-follower and coleader formations in the plane. IEEE Trans. Autom. Control **56**(12), 2778–2792 (2011)
11. Summers, T.H., Yu, C., Anderson, B.D.O., Dasgupta, S.: Formation shape control: global asymptotic stability of a four-agent formation. In: Proceedings of the 48th IEEE Conference on Decision and Control and the 28th Chinese Control Conference, pp. 3002–3007 (2009)

Part III
Orientation Alignment-based Approaches

Chapter 6
Formation Control via Orientation Alignment

Abstract Based on gradient control laws, the global stabilization is assured for some specific formations in Chap. 3, and the local stabilization has been assured for general graphs in Chap. 4. It is noticeable that the traditional gradient control laws, or modified gradient control laws, only rely upon error signals measured at local coordinate frames. Thus, it is purely based on sensing information without any communications with neighboring agents. However, as commented in Chap. 1 and as shown in Fig. 4.5, although the gradient control law that does not use any communication is computationally light, it may not ensure a global convergence for general graphs. In this chapter, we assume that the neighboring agents could communicate with each other. The agents can sense each other relatively; then, the sensing variables and/or computational variables may be exchanged between neighboring agents. For example, it may be assumed that bearing measurements in misaligned coordinate frames can be exchanged between neighboring agents. With the help of communications, it can then be shown that a (quasi-) global convergence under more generalized initial conditions could be assured. Here, the key problem we would like to treat for a global convergence is to align the directions of agents, which also may be a key issue in collective behaviors in nature [2, 3].

6.1 Formation Control via Orientation Estimation

In *orientation alignment*, there are two different problems. The first problem is to estimate the orientation angles of distributed agents. From Fig. 6.1, when the orientation angles θ_i and θ_j are estimated with respect to a common coordinate frame, it is called *orientation estimation*. The second problem is to control the orientation angles of agents. If the angles θ_i and θ_j in Fig. 6.1 are controlled to achieve $\theta_i \rightarrow \theta_j$, it is called *orientation control*. Thus, by the orientation control, all the directions of axes of coordinate frames of agents are aligned. Hence, by orientation alignment, we mean either orientation estimation or orientation control. By making the orientations of coordinate frames of distributed agents be aligned, the formation control problem in distance-based setup is transformed to a formation control problem in consensus-based setup (displacement-based setup with aligned orientation).

H.-S. Ahn, *Formation Control*, Studies in Systems,
Decision and Control 205, https://doi.org/10.1007/978-3-030-15187-4_6

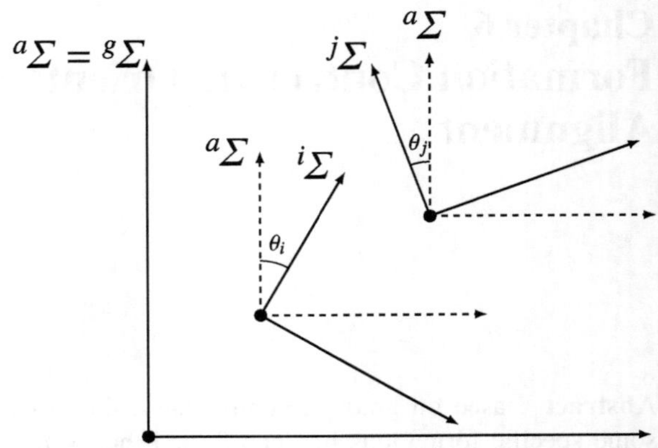

Fig. 6.1 Global coordinate frame (${}^g\Sigma$), aligned coordinate frame (${}^a\Sigma$), and local coordinate frames (${}^i\Sigma$ and ${}^j\Sigma$), and orientation angles (θ_i and θ_j). © [2014] IEEE. Reprinted, with permission, from [8]

A formation control scheme via orientation estimation was developed under the setups of local communications between neighboring agents in 2-dimensional space in [6–8]. This section extracts some key results from these works. Consider n agents in the plane:

$$\dot{p}_i = u_i, i = 1, \ldots, n \tag{6.1}$$

where $p_i \in \mathbb{R}^2$ and $u_i \in \mathbb{R}^2$ denote the position and the control input of agent i with respect to a global coordinate frame ${}^g\Sigma$. Although the expressions of the dynamics are given in a global coordinate frame for an analysis in a common frame, the implementations of the control laws are conducted in local coordinate frames ${}^i\Sigma$. If the coordinate transformation matrix from the global coordinate frame to the ith local coordinate frame is denoted as R_i^g, the control input u_i with respect to the global coordinate frame can be transformed into the local frame as $u_i^i = R_i^g u_i$. Here, the coordinate transformation matrix R_i^g is a function of the orientation θ_i of the agent i as depicted in Fig. 6.1. Note that the orientation of agent i in 2-D can be formally defined as the angle θ_i satisfying $v_i = R_i^g(\theta_i)v^g$, where v_i is a vector expressed in ${}^i\Sigma$ and v^g is a vector expressed in ${}^g\Sigma$. Thus, for calculating the orientation angles of distributed agents with respect to a common direction, it is necessary to have a common reference frame. By multiplying $R_i^g(\theta_i)$ to the both sides of (6.1), the agent dynamics can be expressed in the local frame as

$$\dot{p}_i^i = u_i^i, i = 1, \ldots, n \tag{6.2}$$

For two agents i and j, respectively, it is noticeable that the control inputs u_i^i and u_j^j, where $i \neq j$ and $\theta_i \neq \theta_j$, are expressed at different coordinate frames. Thus, the basis vectors of u_i^i and u_j^j are different, which means that the input vectors u_i^i and u_j^j cannot be compared with each other because the underlying reference frames for the signals u_i^i and u_j^j are different.

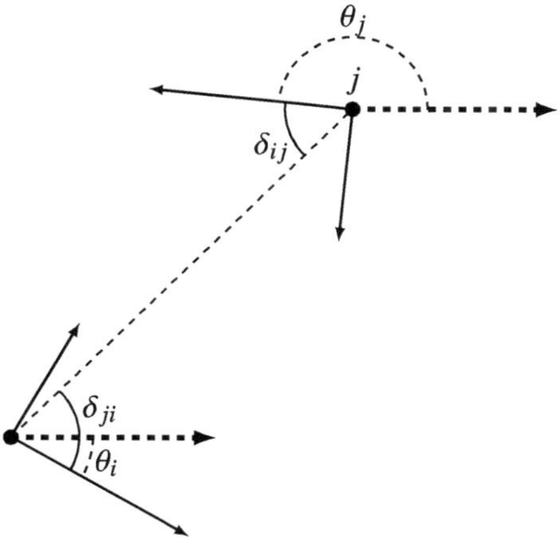

Fig. 6.2 Bearing angles between neighboring agents

In distributed formation control, it is of interest to design the control input u_i only using local information. It is particularly assumed that agent i can measure the distance d_{ji} and the bearing angle δ_{ji} of the neighboring agent j with respect to the local coordinate frame ${}^i\Sigma$. The bearing angle can be considered as the direction of the vector, connecting two neighboring agents as shown in Fig. 6.2, measured in ${}^i\Sigma$. Thus, in bearing angles, it is not necessary to have a common global reference frame different from the orientation angles. It is remarkable that the bearing angles and distances measured in misaligned local coordinate frames are scalar components. By combining the measurements d_{ji} and δ_{ji}, the relative position of the agent j expressed with respect to ${}^i\Sigma$ can be written as

$$p^i_{ji} \triangleq p^i_j - p^i_i = p^i_j \tag{6.3}$$

As aforementioned, the control inputs u^i_i and u^j_j cannot be compared, which means that formation control for distributed agents, whose directions are not aligned, cannot be executed in a common frame. The control inputs, however, would have a common coordinate frame if orientations are aligned, i.e., $\theta_i = \theta_j$, for all $i, j \in \mathcal{V}$. Of course, if all the agents can measure a common direction, for example, true north, then the problem becomes simple. They can rotate their coordinate frames such that the direction of the frame would be aligned to the true north. But, in such a case, it is considered that the agent can sense some global information. In this chapter, we would like to have the agents be aligned to a common direction with only local measurements.

In order to get local measurements, let us suppose that agents could sense the bearing angles of their neighboring agents δ_{ji} as depicted in Fig. 6.2. Observe that the bearing angles are expressed in local coordinate frames. Thus, we need to find a relationship between the bearing angles and orientation angles in order to align the orientations of agents. This can be simply obtained from the prime value operator PV, which is defined as

$$\theta_{ji} \triangleq \mathrm{PV}(\theta_j - \theta_i) = [(\theta_j - \theta_i + \pi)\mathrm{modulo}(2\pi)] - \pi, \ j \in \mathcal{N}_i \tag{6.4}$$

Example 6.1 Let us consider four cases $(\theta_i, \theta_j) = (\frac{\pi}{6}, \frac{4}{3}\pi), (\frac{\pi}{6}, \frac{2}{3}\pi), (\frac{4}{3}\pi, \frac{\pi}{6}), (\frac{2}{3}\pi, \frac{\pi}{6})$. From the definition of $\theta_{ji} = \theta_j - \theta_i$, we compute $\theta_{ji} = \frac{7}{6}\pi, \frac{1}{2}\pi, -\frac{7}{6}\pi, -\frac{1}{2}\pi$ respectively. From the prime value operation (6.4), they are computed as $-\frac{5}{6}\pi, \frac{1}{2}\pi, \frac{5}{6}\pi, -\frac{1}{2}\pi$. Thus, we can see that the prime value operation computes the difference of orientation angles toward the short-distance direction. Also, the clockwise direction is considered as positive, and the counterclockwise direction is considered as negative.

In the Eq. (6.4), the orientation angles θ_i and θ_j are of global information, which cannot be available at the local coordinate frame. However, the bearing angles δ_{ij} and δ_{ji} are obtainable at the local frames as already mentioned. Suppose that δ_{ji} is available at the ith agent, while δ_{ij} is available at the jth agent as shown in Fig. 6.2. The availability of δ_{ji} at i, and availability of δ_{ij} at j are defined by the sensing topology. Further, assuming that agents i and j can exchange the bearing measurements, which is defined by the communication topology, agent i can have both bearing measurements δ_{ij} and δ_{ji}. Then, from Fig. 6.2, we can find the relationship $\delta_{ji} + \theta_i + \pi = \delta_{ij} + \theta_j$, where $+$ is along the counterclockwise direction. This relationship can be changed as $\theta_j - \theta_i = \delta_{ji} - \delta_{ij} + \pi$. If we take a short-distance direction, $\mathrm{PV}(\theta_j - \theta_i)$ can be computed as

$$\mathrm{PV}(\theta_j - \theta_i) = \mathrm{PV}(\delta_{ji} - \delta_{ij} + \pi) = [(\delta_{ji} - \delta_{ij})\mathrm{modulo}(2\pi)] - \pi \tag{6.5}$$

The above relationship, however, only provides diffusive couplings $\theta_j - \theta_i$ between the orientations of two neighboring agents. Thus, it still does not provide the orientation angle of each agent explicitly, which means that each agent needs to compute its orientation with the information of (6.5). To this aim, it is supposed that there are orientation estimators inside the agents, which is possible since agents have distributed computational capability. The input to the estimator is the value of the right-hand side of (6.5), and the output is the estimated orientation $\hat{\theta}_i$. Then, with the orientation estimator, it can be considered that the estimated orientation $\hat{\theta}_i$ is available to the agent i. Also, assuming that neighboring agents can exchange $\hat{\theta}_i$ and $\hat{\theta}_j$, agent i can get $\hat{\theta}_{ji}$, which is defined as follows:

$$\hat{\theta}_{ji} \triangleq \mathrm{PV}(\hat{\theta}_j - \hat{\theta}_i) \tag{6.6}$$

Thus, for the orientation alignment, each agent senses the bearing angles δ_{ji} and communicate with neighboring agents to exchange δ_{ji} and $\hat{\theta}_i$. Then, with these exchanged information, each agent updates its orientation $\hat{\theta}_i$ according to estimation law (see (6.8)).

Table 6.1 Comparison between gradient-based formation control laws and orientation estimation-based formation control laws

	Gradient-based formation laws	Orientation estimation-based formation laws
Sensing	p^i_{ji}	p^i_{ji}
Control	$\|z_{ij}\|$	p^i_{ji}
Communications	None	$\delta_{ij}, \hat{\theta}_i$
Computation	None	$\hat{\theta}_i$

The main idea of using (6.5) and (6.6) is to ensure the convergence of $\text{PV}(\hat{\theta}_j - \hat{\theta}_i)$ to $\text{PV}(\theta_j - \theta_i)$, that is, it is attempted to achieve

$$\text{PV}(\hat{\theta}_j - \hat{\theta}_i) \longrightarrow \text{PV}(\theta_j - \theta_i) \tag{6.7}$$

where $\hat{\theta}_i$ is estimated as a function of $\hat{\theta}_{ji}$, δ_{ji}, and δ_{ij}. If $\hat{\theta}_i$ is estimated such that (6.7) does hold, then it is true that $\hat{\theta}_i \to \theta_i + \Delta\theta_\infty$, where $\Delta\theta_\infty$ is a common offset orientation angle from the orientations defined with respect to the global coordinate frame ${}^g\Sigma$.

It may be important to emphasize the differences between the gradient-based formation control laws and orientation estimation-based formation control laws, which are summarized in Table 6.1. In gradient-based approaches, there is no communication and computation, while in orientation estimation approaches, there are additional communications between neighboring agents and additional computation at each agent. Also, the control variables should be distinguished in these two approaches.

For the orientation estimation, the following consensus-type estimation law on a circle space is proposed

$$\dot{\hat{\theta}}_i = k_{\hat{\theta}} \sum_{j \in \mathcal{N}_i} w_{ij}(\hat{\theta}_{ji} - \theta_{ji}) \tag{6.8}$$

where $k_{\hat{\theta}} > 0$ and $\hat{\theta}_i(t_0) = 0$. Since the Eq. (6.8) is a typical consensus protocol, it will reach a consensus for connected undirected graphs, strongly connected digraphs, and for a graph having directed rooted tree (arborescence). The weighting coefficients w_{ij} determine the network topology. As shown in Table 6.1, according to sensing, control, communications, and computation topologies, the variables treated in the network are distinguished. Also as explained in Sect. 1.3, according to sensing, control, communications, and computation, the graph topology and edge directions can be differently characterized. For the estimation law (6.8), agent i needs to have information of δ_{ij} and $\hat{\theta}_j$ delivered from the neighboring agent j. For the computation θ_{ji}, it is required to use δ_{ij} and δ_{ji} which are measured by both agents. Thus, the sensing topology should be of bidirectional. Here, it is important to note that δ_{ij} and δ_{ji} are only needed to be measured initially and only once since the orientations of agents are fixed. To understand the communication topology, we need to pay a little attention. Since δ_{ij} and $\hat{\theta}_j$ are delivered from agent j to agent i, the agent j is one of incoming agents[1] in terms of communications, i.e., $j \in \mathcal{N}_i^I$. Of course, if agent i delivers information of δ_{ji} and $\hat{\theta}_i$ to agent j, then the agent j is one of outgoing agents of agent i, i.e., $j \in \mathcal{N}_i^O$. But, as mentioned already, the dynamics (6.8) reaches a consensus when the topology is a connected undirected graph (average consensus), a strongly connected digraph (consensus), a strongly connected and balanced digraph (average consensus), and for a graph having directed rooted tree, which is called arborescence (consensus to a root node). Thus, if the communication topology has an arborescence, then an alignment in $\hat{\theta}_i$ could be achieved. Thus, the sensing graph $\mathcal{G} = (\mathcal{V}, \mathcal{E}^s)$ is an undirected one, while the communication graph $\mathcal{G} = (\mathcal{V}, \mathcal{E}^c)$ could be a directed one. Also since the actuation topology $\mathcal{E}^a$ could be different from these topologies, without notational confusion, when we mention *graph* in this chapter, we mean $\mathcal{G} = (\mathcal{V}, \mathcal{E}^s \cup \mathcal{E}^c \cup \mathcal{E}^a)$. In the estimation (6.8), the agent i needs to keep its updated value $\hat{\theta}_i$ and needs to deliver it to outgoing-neighboring agents j to compute $\hat{\theta}_{ij}$. The diffusive orientation angle θ_{ji} is computed by (6.5) with measurements δ_{ij} and δ_{ji} initially once.

As noted in [10], on the circle, there exists anti-consensus phenomenon depending upon initial orientation angles of agents. Hence, to avoid the anti-consensus in the estimation, it is supposed that the initial angles satisfy $\max_{i \in \mathcal{V}} \theta_i(t_0) - \min_{i \in \mathcal{V}} \theta_i(t_0) < \pi$. Then, defining $\tilde{\theta}_i = \hat{\theta}_i - \theta_i$, it is true that $\max_{i \mathcal{V}} \tilde{\theta}_i(t_0) - \min_{i \mathcal{V}} \tilde{\theta}_i(t_0) < \pi$. Using the fact that $\hat{\theta}_{ji} - \theta_{ji} = \hat{\theta}_j - \theta_j - (\hat{\theta}_i - \theta_i)$, the error dynamics of the orientation estimation can be written as

$$\dot{\tilde{\theta}} = -k_{\hat{\theta}} \mathbb{L} \tilde{\theta} \tag{6.9}$$

where $\mathbb{L}$ is the weighted Laplacian matrix and $\tilde{\theta} = (\tilde{\theta}_1, \ldots, \tilde{\theta}_n)^T$. It is clear that (6.9) is a Laplacian dynamics. Thus, it will converge to a vector $\tilde{\theta}_\infty \mathbf{1}_n$ exponentially fast, where $\tilde{\theta}_\infty$ is a constant, if the graph has a spanning tree. As aforementioned, the bearing angles δ_{ji} and δ_{ij} only need to be measured and exchanged once between neighboring agents. Then, the agents only need to exchange $\hat{\theta}_i$ and $\hat{\theta}_j$ for the computation $\hat{\theta}_{ji}$. Thus, we can increase the gain w_{ij} in (6.8) for speeding up the convergence speed as far as there is a communication between neighboring agents. That is, the sensing frequency and communication frequency could be different. Although in such a circumstance, if the communication speed is faster than the sensing speed, the consensus of (6.8) still can be achieved quickly.

Upon consensus, it is true that $\hat{\theta}_i \rightarrow \theta_i + \tilde{\theta}_\infty$ as $t \rightarrow \infty$, which implies $\hat{\theta}_j - \hat{\theta}_i \rightarrow \theta_j - \theta_i$. Since the estimated orientation angles have a common offset angle $\tilde{\theta}_\infty$, it can be considered that the estimated orientation angles $\hat{\theta}_i$, for all $i \in \mathcal{V}$, are expressed with respect to a virtual common frame ${}^c\Sigma$ that is rotated by the angle $\tilde{\theta}_\infty$ from the global coordinate frame ${}^g\Sigma$. Here, note that the angle $\tilde{\theta}_\infty$ is not known; in fact, it is

[1] For a directed edge $(i, j)^{\bar{e}}$, the node i is an incoming node to node j. But, for node i, the node j is an outgoing node.

not necessary to know this value since we ignore the rotations in the desired formation configuration. Using the estimated angle $\hat{\theta}_i$ from (6.8) and with the desired relative position $p_j^* - p_i^*$ in a global coordinate frame, the formation control law can be proposed as

$$u_i^i = k_p \sum_{j \in \mathcal{N}_i} w_{ij}[(p_j^i - p_i^i) - R^i(-\hat{\theta}_i)(p_j^* - p_i^*)], \forall i \in \mathcal{V} \tag{6.10}$$

where $R^i = (R_i^g)^{-1}$. Note that $R^i(-\hat{\theta}_i)(p_j^* - p_i^*) = R_i(\hat{\theta}_i)(p_j^* - p_i^*)$ is expressed with respect to ${}^i\Sigma$, while the desired relative displacement $p_j^* - p_i^*$ is expressed with respect to ${}^g\Sigma$. In (6.10), the agent i uses the information of p_j^i and $p_j^* - p_i^*$. The control topology, or called actuation topology, is determined by the weighting w_{ij} and the given desired relative positions $p_j^* - p_i^*$. If it is assumed $w_{ij} = w_{ji}$ for all $(i, j)^e \in \mathcal{E}^a$ and $p_j^* - p_i^*$ is given to agent i, and $p_i^* - p_j^*$ is given to agent j, it is an undirected one. However, the actuation topology of (6.10) could be different from the communication topology; even it can be a directed one. In any case, if the actuation topology ensures a consensus, as will be clear in what follows, the desired formation defined by displacement vectors will be achieved. For a further analysis, define $e_{\theta_i} = \tilde{\theta}_i - \tilde{\theta}_\infty$ and $e_{p_i^c} = p_i^c - p_i^*$. Then the dynamics of agents can be written as

$$\begin{aligned}
\dot{p}_i^c &= u_i^c \\
&= R^i(\theta_i + \tilde{\theta}_\infty)u_i^i \\
&= k_p \sum_{j \in \mathcal{N}_i} w_{ij}[R^i(\theta_i + \tilde{\theta}_\infty)(p_j^i - p_i^i) - R^i(\theta_i + \tilde{\theta}_\infty)R_i(\hat{\theta}_i)(p_j^* - p_i^*)] \\
&= k_p \sum_{j \in \mathcal{N}_i} w_{ij}[(p_j^c - p_i^c) - R^i(-e_{\theta_i})(p_j^* - p_i^*)]
\end{aligned} \tag{6.11}$$

where $R^i(-e_{\theta_i}) = R_i^g(e_{\theta_i})$. From $\dot{e}_{p_i^c} = \dot{p}_i^c - \dot{p}_i^*$, the error dynamics of position vectors can be obtained as

$$\dot{e}_{p_i^c} = k_p \sum_{j \in \mathcal{N}_i} w_{ij}(e_{p_j^c} - e_{p_i^c}) + k_p \sum_{j \in \mathcal{N}_i} w_{ij}[\mathbb{I}_2 - R^i(-e_{\theta_i})](p_j^* - p_i^*) \tag{6.12}$$

Then, with the notations $e_{p^c} = (e_{p_1^c}, \ldots, e_{p_n^c})^T$, $e_\theta = (e_{\theta_1}, \ldots, e_{\theta_n})^T$, and $D(R_e^i) = \text{blkdg}(R^i(e_{\theta_1}), \ldots, R^i(e_{\theta_n}))$, the overall error dynamics can be obtained as

$$\dot{e}_{p^c} = -k_p(\mathbb{L} \otimes I_2)e_{p^c} - k_p[\mathbb{I}_{2n} - D(R_e^i)^{-1}](\mathbb{L} \otimes \mathbb{I}_2)p^* \tag{6.13}$$

$$\dot{e}_\theta = -k_{\hat{\theta}}\mathbb{L}e_\theta \tag{6.14}$$

Remark 6.1 Note that, the Laplacian matrix $\mathbb{L}$ in (6.13) is determined by actuation topology, while the Laplacian matrix $\mathbb{L}$ in (6.14) is determined by communication topology. Thus, if the actuation topology and the communication topology are different, the Laplacian matrices given in (6.13) and (6.14) would be different.

Lemma 6.1 *If the graph $\mathcal{G} = (\mathcal{V}, \mathcal{E}^s)$ has a spanning tree and initial orientation angles satisfy the inequality* $\max_{i\in\mathcal{V}} \theta_i(t_0) - \min_{i\in\mathcal{V}} \theta_i(t_0) < \pi$, *there exist positive constants k_γ and λ_γ such that* $\|\mathbb{I}_{2n} - D(R_e^i)\| \leq k_\gamma e^{-\lambda_\gamma t}\|e_\theta(t_0)\|$ *and* $\|\mathbb{I}_{2n} - D(R_e^i)^{-1}\| \leq k_\gamma e^{-\lambda_\gamma t}\|e_\theta(t_0)\|$ *hold for all $t \geq t_0$.*

Proof It is clear that there exist positive constants k_{e_θ} and λ_{e_θ} such that

$$\|e_\theta(t)\| \leq k_{e_\theta} e^{-\lambda_{e_\theta}(t-t_0)}\|e_\theta(t_0)\| \tag{6.15}$$

From the property of consensus, it is well known that $\max_{i\in\mathcal{V}} e_{\theta_i}$ is a nonincreasing function, and $\min_{i\in\mathcal{V}} e_{\theta_i}$ does not decrease [5]. Thus, if $\max_{i\in\mathcal{V}} \theta_i(t_0) - \min_{i\in\mathcal{V}} \theta_i(t_0) < \pi$ does hold, then $\max_{i\in\mathcal{V}} \theta_i(t) - \min_{i\in\mathcal{V}} \theta_i(t) < \pi$ holds. Thus the property $\|\mathbb{I}_{2n} - D(R_e^i)\| = \max_{i=1,\ldots,n} \|\mathbb{I}_2 - R^i(e_{\theta_i})\|$ and $\|\mathbb{I}_2 - R^i(e_{\theta_i})\| = \sqrt{2}\sqrt{1-\cos e_{\theta_i}} \leq \sqrt{2}|e_{\theta_i}|$ will be satisfied because the condition $\max_{i\in\mathcal{V}} \theta_i(t) - \min_{i\in\mathcal{V}} \theta_i(t) < \pi$ is satisfied. Then, the following inequality can be obtained:

$$\begin{aligned}\|\mathbb{I}_{2n} - D(R_e^i(t))\| &\leq \sqrt{2} \max_{i=1,\ldots,n} |e_{\theta_i}| \\ &\leq \sqrt{2}\|e_\theta(t)\|\end{aligned} \tag{6.16}$$

From (6.16), it is direct to obtain $\|\mathbb{I}_{2n} - D(R_e^i(t))\| \leq k_\gamma e^{-\lambda_\gamma t}\|e_\theta(t_0)\|$ with positive constants k_γ and λ_γ. Furthermore, since $R^i(e_{\theta_i})$ is a SO(2) matrix, it is also true that $\|\mathbb{I}_{2n} - D(R_e^i(t))^{-1}\| \leq k_\gamma e^{-\lambda_\gamma t}\|e_\theta(t_0)\|$.

From the above lemma, it is obvious that $\|k_p(\mathbb{I}_{2n} - D(R_e^i(t))^{-1})(\mathbb{L}\otimes\mathbb{I}_2)p^*\| \leq k'_\gamma e^{-\lambda_{\gamma'} t}\|e_\theta(t_0)\|$ where $k'_\gamma > 0$ and $\gamma' > 0$. Thus, $\|k_p(\mathbb{I}_{2n} - D(R_e^i(t))^{-1})(\mathbb{L}\otimes\mathbb{I}_2) p^*\|$ exponentially converges to zero. For the main result of this section, let us define a consensus manifold $\mathcal{E}_{2n} \triangleq \{[\xi_1^T, \ldots, \xi_n^T]^T \in \mathbb{R}^{2n} : \xi_i = \xi_j, \forall i, j \in \{1, \ldots, n\}\}$ and define a metric $\text{dist}(x, \mathcal{E}_{2n}) \triangleq \inf_{\xi\in\mathcal{E}_{2n}} \|x - \xi\|$. Then, we can have the following main result of this section [8]:

Theorem 6.1 *If the graph has a spanning tree and initial orientation angles satisfy the condition* $\max_{i\in\mathcal{V}} \theta_i(t_0) - \min_{i\in\mathcal{V}} \theta_i(t_0) < \pi$, *then $e_{p^c}(t)$ exponentially converges to a point $e_{p^c}^\infty \in \mathcal{E}_{2n}$.*

Proof See the appendix.

It is worth noting that the right-hand side of (A.21) is composed of two parts. The first one is due to the position error dynamics (6.13) and the second part is due to the orientation error (6.14), where the orientation error acts as an input to the position error dynamics. As shown in the right-hand side of (A.21), the two error dynamics are combined, with upper boundaries by exponential convergence. As concluded in (A.25), the combined dynamics is still exponentially converging to zero. Thus, we can see that when both the dynamics (6.13) and (6.14) are globally

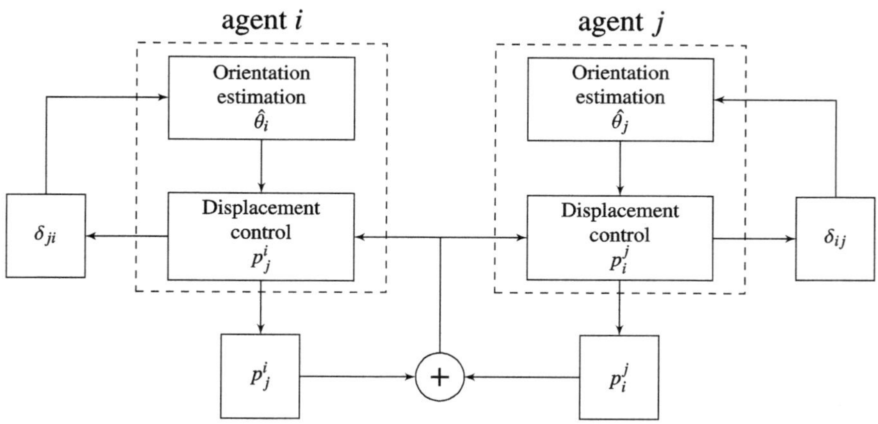

Fig. 6.3 The overall structure of the orientation estimation-based formation control

exponentially stable, with one of them being the input to the other dynamics, then the combined dynamics is also globally exponentially stable. This result may be also achieved by the input-to-state stability (see Sect. 2.4.3). However, the analysis conducted in the proof of Theorem 6.1 ensures an exponential convergence, which is stronger than asymptotical stability that can be ensured by the input-to-state stability. Also note that Theorem 6.1 ensures a stabilization to the desired formation under the circumstance of $\max_{i\in\mathcal{V}} \theta_i(t_0) - \min_{i\in\mathcal{V}} \theta_i(t_0) < \pi$, which may be considered as a region of convergence. Since the condition $\max_{i\in\mathcal{V}} \theta_i(t_0) - \min_{i\in\mathcal{V}} \theta_i(t_0) < \pi$ means that the agents are looking toward the same half circle initially, the condition is not too strict from a sense of collective behaviors. We will call the results of Theorem 6.1 as *quasi-global* stability. Also, depending on application scenarios, the condition may not be necessary or can be relaxed (see Example 6.2). Figure 6.3 shows the overall structure of the orientation estimation-based formation control scheme presented in this section. Agents conduct the orientation estimation with distributed sensings and communications. Then, these values are used for the displacement-based control in local coordinate frames. The actuations result in the variations of the relative displacements (p_j^i and p_i^j) and the bearing angles (δ_{ji} and δ_{ij}).

Example 6.2 Consider two agents as depicted in Fig. 6.2. First, let $\delta_{21} = \frac{\pi}{3}$ and $\delta_{12} = \frac{\pi}{4}$. Then, we have $\theta_{21} = \text{PV}(\theta_2 - \theta_1) = \text{PV}(\delta_{21} - \delta_{12} + \pi) = -\frac{11}{12}\pi$, and $\theta_{12} = \frac{11}{12}\pi$. Then, we can have

$$
\begin{aligned}
\begin{bmatrix} \dot{\hat{\theta}}_1 \\ \dot{\hat{\theta}}_2 \end{bmatrix} &= \begin{bmatrix} 1 & -1 \\ -1 & 1 \end{bmatrix} \begin{bmatrix} \hat{\theta}_1 \\ \hat{\theta}_2 \end{bmatrix} - \begin{bmatrix} \theta_{21} \\ \theta_{12} \end{bmatrix} \\
&= \begin{bmatrix} 1 & -1 \\ -1 & 1 \end{bmatrix} \begin{bmatrix} \hat{\theta}_1 \\ \hat{\theta}_2 \end{bmatrix} - \begin{bmatrix} -\frac{11}{12}\pi \\ \frac{11}{12}\pi \end{bmatrix}
\end{aligned}
$$

Hence, $\hat{\theta}_1$ and $\hat{\theta}_2$ are estimated as $\hat{\theta}_1 = \theta^* - \frac{11}{12}\pi$ and $\hat{\theta}_2 = \theta^*$ where θ^* is the common offset orientation angle. Next, let the axes of agent 2 be rotated by π angle further. Then, we would have $\delta_{21} = \frac{\pi}{3}$ and $\delta_{12} = \frac{\pi}{4} + \pi = \frac{5}{4}\pi$. In such case, we have $\theta_{21} = \frac{1}{12}\pi$ and $\theta_{12} = -\frac{1}{12}\pi$. Thus, we would have $\hat{\theta}_1 = \theta^* - \frac{1}{12}\pi$ and $\hat{\theta}_2 = \theta^*$ as the estimations of the orientation angles. Consequently, for the case of two agents, the initial angle requirement $\max_{i\in\mathcal{V}} \theta_i(t_0) - \min_{i\in\mathcal{V}} \theta_i(t_0) < \pi$ is not necessary. Thus, for a general graph, if we conduct the orientation estimation sequentially, almost for any initial condition, the orientation alignment can be conducted. But, this process will be only valid in a sequential process; it cannot be done in a simultaneous estimation.

6.2 Switching Topology

In, the formation control problems studied so far, the underlying topologies of initial graph and desired graph are identical. But, the realizations of the initial and desired may be different since $p(t_0) \neq p^*$. Also, the graphs during the transition from the initial configuration to final configuration are required to be isomorphic to the initial graph. But, in certain cases, the network topology may vary as time propagates. For example, as the interagent distances are varying, the communication or sensing topology may change. In such cases, the initial graph and the graphs during the transition are not isomorphic any more (i.e., the topologies could be different), although the actuation topology still needs to be isomorphic (i.e., the actuation topology does not change). In this section, we assume that the underlying network topology is switching. Then, the results of previous section can be extended to a switching topology as far as the network is uniformly connected. Let a directed graph with time-varying weights be denoted as $\vec{\mathcal{G}}(t) = (\mathcal{V}, \vec{\mathcal{E}}(t), \mathcal{W}(t))$. The digraph $\vec{\mathcal{G}}(t)$ is called uniformly connected if, for any $t \geq t_0$, there exist a finite time T and a vertex $i \in \mathcal{V}$ such that i is the root of a spanning tree of the graph $(\mathcal{V}, \cup_{\tau\in[t,T]}\mathcal{E}(t), \int_t^T \mathcal{W}(\tau)d\tau)$. Let the Laplacian matrix be denoted as $\mathbb{L}(t)$ for the time-varying graph $\mathcal{G}(t)$. Formation control via orientation alignment for uniformly connected networks was developed in [8]. Due to similarity of overall derivations when compared to the equations in Sect. 6.1, in this section, we summarize the results of [8] in a concise way. As a basic lemma for the uniformly connected networks, let us adopt the following lemma first.

Lemma 6.2 [10] *From the time-varying graph $\mathcal{G}(t)$, consider the following Laplacian dynamics*

$$\dot{x}(t) = -(\mathbb{L}(t) \otimes \mathbb{I}_n)x(t) \tag{6.17}$$

where $x(t) \in \mathbb{R}^n$. If the graph is uniformly connected, then $x(t)$ exponentially converges to a point x such that $x_1 = \cdots = x_n$.

For the uniformly connected network, an average consensus is no more ensured although a consensus is always ensured. Under the time-varying topology, the overall error dynamics can be changed from (6.13) and (6.14) as

$$\dot{e}_{p^c} = -k_p(\mathbb{L}(t) \otimes \mathbb{I}_2)e_{p^c} - k_p[\mathbb{I}_{2n} - D(R_e^i(t))^{-1}](\mathbb{L}(t) \otimes \mathbb{I}_2)p^* \tag{6.18}$$

$$\dot{e}_\theta = -k_{\hat{\theta}}\mathbb{L}(t)e_\theta \tag{6.19}$$

where $\mathbb{L}(t)$ is the time-varying Laplacian of the graph $\mathcal{G}(t)$. Using the Lemma 6.2, the following result is direct.

Lemma 6.3 *If the time-varying graph $\mathcal{G}(t)$ is uniformly connected and the initial orientation angles satisfy the inequality $\max_{i \in \mathcal{V}} \theta_i(0) - \min_{i \in \mathcal{V}} \theta_i(0) < \pi$, then by the orientation estimation law (6.8), the error of the estimated orientation exponentially converges to $\tilde{\theta}_\infty \mathbf{1}_n$ where $\tilde{\theta}_\infty$ is a constant.*

With the above lemma, it is shown that as $t \to \infty$, $R^i(e_{\theta_i}) \to \mathbb{I}_2$. Thus, both $\|\mathbb{I}_{2n} - D(R_e^i(t))\|$ and $\|\mathbb{I}_{2n} - D(R_e^i(t))^{-1}\|$ exponentially converge to zero. Now, following the same procedure as the proof of Theorem 6.1, we can obtain the following theorem.

Theorem 6.2 *If the time-varying graph $\mathcal{G}(t)$ is uniformly connected and the initial orientation angles satisfy the condition $\max_{i \in \mathcal{V}} \theta_i(0) - \min_{i \in \mathcal{V}} \theta_i(0) < \pi$, then $e_{p^c}(t)$ exponentially converges to a point $e_{p^c}^\infty \in \mathcal{E}_{2n}$.*

Proof From Lemmas 6.2 and 6.3, denoting $w(t) \triangleq k_p[\mathbb{I}_{2n} - D(R_e^i(t))^{-1}](\mathbb{L}(t) \otimes \mathbb{I}_2)p^*$, it is obvious that if $w(t) = 0$, then $\dot{e}_{p^c} = -k_p(\mathbb{L}(t) \otimes \mathbb{I}_2)e_{p^c}$ is uniformly exponentially stable. The solution of (6.18) is obtained as $e_{p^c}(t) = \phi_{e_{p^c}}(t, t_0)e_{p^c}(t_0) + \int_{t_0}^t \phi_{e_{p^c}}(t, \tau)w(\tau)d\tau$, where $\phi_{e_{p^c}}(t, t_0)$ is the state transition matrix from t_0 to t. Now with the metric $\text{dist}(\cdot, \mathcal{E}_{2n})$, it can be shown that

$$\begin{aligned} \text{dist}(e_{p^c}(t), \mathcal{E}_{2n}) &\le \text{dist}(\phi_{e_{p^c}}(t, t_0)e_{p^c}(t_0), \mathcal{E}_{2n}) + \text{dist}\Big(\int_{t_0}^t \phi_{e_{p^c}}(t, \tau)w(\tau)d\tau, \mathcal{E}_{2n}\Big) \\ &\le k_{\mathcal{E}} e^{-\lambda_{\mathcal{E}}(t-t_0)}\text{dist}(e_{p^c}(t_0), \mathcal{E}_{2n}) + \int_{t_0}^t \text{dist}(\phi_{e_{p^c}}(t, \tau)w(\tau), \mathcal{E}_{2n})d\tau \end{aligned} \tag{6.20}$$

Next, denoting $\eta(t) \triangleq \int_{t_0}^t \text{dist}(\phi_{e_{p^c}}(t, \tau)w(\tau), \mathcal{E}_{2n})d\tau$, by the same idea of (A.22), it is shown that $\eta(t) \le \frac{k_{\mathcal{E}}}{\lambda_{\mathcal{E}}} \sup_{t_0 \le \tau \le t}\{\text{dist}(w(\tau), \mathcal{E}_{2n})\}$. Since $\text{dist}(x, \mathcal{E}_{2n}) \le \|x\|$, it follows that $\sup_{t_0 \le \tau \le t}\{\text{dist}(w(\tau), \mathcal{E}_{2n})\} \le \sup_{t_0 \le \tau \le t}\{\|w(\tau)\|\}$. Thus, (6.20) can be changed as

$$\text{dist}(e_{p^c}(t), \mathcal{E}_{2n}) \le k_{\mathcal{E}} e^{-\lambda_{\mathcal{E}}(t-t_0)}\text{dist}(e_{p^c}(t_0), \mathcal{E}_{2n}) + \frac{k_{\mathcal{E}}}{\lambda_{\mathcal{E}}} \sup_{t_0 \le \tau \le t}\{\|w(\tau)\|\} \tag{6.21}$$

With the fact that $\|\mathbb{I}_{2n} - D(R_e^i(t))\| \to 0$ as $t \to \infty$, it is clear that $w(t) \to 0$ as $t \to \infty$. Then, by replacing t_0 in the above equation by $(t + t_0)/2$, it is shown that

$$\text{dist}(e_{p^c}(t), \mathcal{E}_{2n}) \leq k_{\mathcal{E}} e^{-\lambda_{\mathcal{E}}(t-t_0)/2} \text{dist}(e_{p^c}((t+t_0)/2), \mathcal{E}_{2n}) + \frac{k_{\mathcal{E}}}{\lambda_{\mathcal{E}}} \sup_{(t+t_0)/2 \leq \tau \leq t} \{\|w(\tau)\|\} \tag{6.22}$$

Now, with Lemma 6.5, we can see that the term $\|w(\tau)\|$ is upper bounded as $\|w(\tau)\| \leq ke^{-\lambda_k(\tau-t_0)}\|e_\theta(t_0)\|$, where $k, \lambda_k > 0$. Hence, as $t \to \infty$, the first term and second term of the right-hand side of the above inequality converge to zero exponentially, which implies that $e_{p^c}(t)$ exponentially converges to the set $\mathcal{E}_{2n}$. Due to $\dot{e}_{p^c} \to 0$, it will converge to a point $e_{p^c}^{\infty} \in \mathcal{E}_{2n}$.

6.3 Formation Control via Orientation Control

In the previous sections, the orientations of agents have been estimated up to a common rotation. Then, using these estimations, we can find orientations of agents with respect to a common frame. Since, in these approaches, we did not control the orientation of agents, the formation control is conducted in Euclidean spaces. It may be necessary to actively control the orientations of agents for some applications, which means that a formation control is conducted in SE(2) $= \mathbb{R}^2 \times$ SO(2) space. As an example, an array of cameras mounted on mobile agents may need to control their directions toward a same direction. Motivated from this, in this section, by adopting the control scheme of [7], the orientation angles are further controlled such that the orientations are physically aligned to a common direction. Since the relative orientation angle θ_{ji} can be obtained via communications between neighboring agents in which the relative bearing angles δ_{ij} and δ_{ji} are exchanged, the following orientation control law can be implemented in a distributed way:

$$\dot{\theta}_i(t) = u_{\theta_i} = k_\theta \sum_{j \in \mathcal{N}_i} a_{ij}(t)\theta_{ji} \tag{6.23}$$

where $k_\theta > 0$, θ_i are in the circle space $\mathbb{S}$, i.e., $\theta_i \in \mathbb{S}$, and $a_{ij}(t)$ can be considered as the strength in communication couplings between neighboring nodes. The consensus or synchronization problem on the circle has a non-convex configuration characteristics due to $\theta_i + 2\pi = \theta_i$; thus, a simple linear consensus algorithm does not work properly [11]. In order not to include the periodicity of 2π and to make a contraction along a part of the circle, we need to have a constraint for initial angles. Under the condition $\max_{i \in \mathcal{V}} \theta_i(t_0) - \min_{i \in \mathcal{V}} \theta_i(t_0) < \pi$, it is true that $\theta_j(t) - \theta_i(t) = \text{PV}(\theta_j(t) - \theta_i(t))$. Then, the consensus strategy (6.23) can be concisely written as

$$\dot{\theta} = -k_\theta \mathbb{L}(t)\theta(t) \tag{6.24}$$

Note that unlike (6.8), in (6.23), the relative orientations θ_{ji} need to be measured every sampling instants for the consensus update. It means that the agents need to sense the bearing vectors every sampling instants. Thus, the overall consensus speed of (6.24) may be slower than the consensus of (6.9).

Now, under the setup of the typical consensus problem, the following lemma can be generated for the dynamics (6.23).

Lemma 6.4 *If the time-varying graph, $\mathcal{G}(t)$ is uniformly connected and the initial orientation angles satisfy the inequality* $\max_{i \in \mathcal{V}} \theta_i(t_0) - \min_{i \in \mathcal{V}} \theta_i(t_0) < \pi$*, then by the orientation control law (6.23), the orientations $\theta_i(t)$ converge to a common value θ_∞ exponentially fast.*

In the above lemma, the condition $\max_{i \in \mathcal{V}} \theta_i(t_0) - \min_{i \in \mathcal{V}} \theta_i(t_0) < \pi$ is critical. The following example shows a case when the condition is not satisfied.

Example 6.3 Let us consider a graph topology as depicted in Fig. 3.6. Assume that the initial angles are given as $\theta_1(t_0) = \frac{1}{6}\pi$, $\theta_2(t_0) = \frac{5}{6}\pi$, $\theta_3(t_0) = \frac{7}{6}\pi$, and $\theta_4(t_0) = \frac{5}{3}\pi$. It is clear that the initial condition is not satisfied. Figure 6.4 depicts the trajectories of orientation angles. It is noticeable that θ_{ji} is not computed by $\theta_j - \theta_i$, which is an expression in the linear space. Rather than computing by $\theta_j - \theta_i$, it is computed by $\mathrm{PV}(\theta_j - \theta_i)$ as defined in (6.4). If the relative angles are computed by $\theta_j - \theta_i$, a consensus will be always achieved.

Note that the orientation angles θ_i are defined with respect to the global coordinate frame ${}^g\Sigma$. In the above lemma, orientation angles converge to a common value

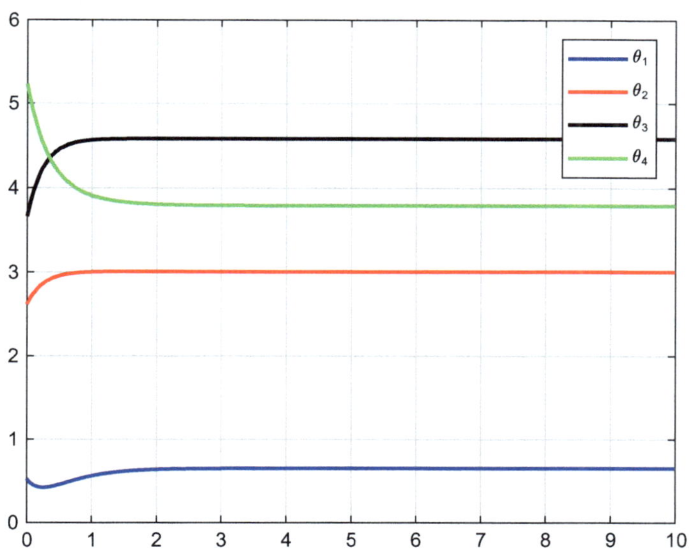

Fig. 6.4 Trajectories of orientation angles when the initial condition is not satisfied

θ_∞. The converged value θ_∞ also needs to be defined with respect to ${}^g\Sigma$. Let the coordinate frame aligned along the direction of θ_∞ be called the aligned coordinate frame, which is denoted as ${}^a\Sigma$. For the formation control of agents, the following distributed control law for the agent i is proposed:

$$u_i^i(t) = k_p \sum_{j\in\mathcal{N}_i} a_{ij}(t)[(p_j^i(t) - p_i^i(t)) - (p_j^* - p_i^*)] \tag{6.25}$$

where $k_p > 0$. Define the misaligned orientation angle as $e_{\theta_i}(t) = \theta_i(t) - \theta_\infty$. Then, the control law (6.25) can be expressed with respect to ${}^a\Sigma$ as

$$u_i^i(t) = k_p \sum_{j\in\mathcal{N}_i} a_{ij}(t)[(R^i_{e_{\theta_i}(t)})^{-1}(p_j^a(t) - p_i^a(t)) - (p_j^* - p_i^*)] \tag{6.26}$$

where $R^i_{e_{\theta_i}(t)}$ is the rotation matrix from ${}^i\Sigma$ to ${}^a\Sigma$, and $p_i^a(t)$ is the position expressed in ${}^a\Sigma$. Note that in fact, we do not need to know the aligned coordinate frame in (6.26) since the control law (6.25) is implemented in the local coordinate frame. Then, the position is redefined in ${}^a\Sigma$ as

$$\begin{aligned}
\dot{p}_i^a(t) &= u_i^a \\
&= R^i_{e_{\theta_i}(t)} u_i^i(t) \\
&= k_p R^i_{e_{\theta_i}(t)} \sum_{j\in\mathcal{N}_i} a_{ij}(t)[(R^i_{e_{\theta_i}(t)})^{-1}(p_j^a(t) - p_i^a(t)) - (p_j^* - p_i^*)] \\
&= k_p \sum_{j\in\mathcal{N}_i} a_{ij}(t)[(p_j^a(t) - p_i^a(t)) - R^i_{e_{\theta_i}(t)}(p_j^* - p_i^*)]
\end{aligned} \tag{6.27}$$

where $u_i^a(t)$ is the virtual control input expressed in ${}^a\Sigma$. For a further analysis, define $e^a_{p_i}(t) \triangleq p_i^a(t) - p_i^*$. Then, the position error in ${}^a\Sigma$ can be obtained as

$$\dot{e}^a_{p_i}(t) = k_p \sum_{j\in\mathcal{N}_i} a_{ij}(t)(e^a_{p_j}(t) - e^a_{p_i}(t)) + k_p \sum_{j\in\mathcal{N}_i} a_{ij}(t)(\mathbb{I}_2 - R^i_{e_{\theta_i}(t)})(p_j^* - p_i^*) \tag{6.28}$$

Denoting $w_i(t) \triangleq k_p \sum_{j\in\mathcal{N}_i} a_{ij}(t)(\mathbb{I}_2 - R^i_{e_{\theta_i}(t)})(p_j^* - p_i^*)$, $e_p^a(t) = (e^a_{p_1}, \ldots, e^a_{p_n})^T$, and $w(t) = (w_1(t), \ldots, w_n(t))^T$, the above equation can be written in a vector form as

$$\dot{e}_p^a(t) = -k_p(\mathbb{L}(t) \otimes \mathbb{I}_2)e_p^a(t) + w(t) \tag{6.29}$$

Remark 6.2 A comment similar to Remark 6.1 can be provided here. The Laplacian matrix $\mathbb{L}$ in (6.24) is determined by communication topology, while the Laplacian matrix $\mathbb{L}$ in (6.29) is determined by actuation topology. Due to the reason similar to Remark 6.1, the Laplacian matrix $\mathbb{L}$ in (6.29) could be generated from directed graph. Also, the Laplacian matrices in (6.24) and (6.29) may be different.

To investigate the convergence of $w(t)$, the following lemma is developed with the definition $e_\theta(t) = (e_{\theta_1}, \ldots, e_{\theta_n})^T$.

Lemma 6.5 *The additional perturbation term $w(t)$ decays exponentially in the sense of $\|w(t)\| \leq k_w e^{-\lambda_w(t-t_0)}\|e_\theta(t_0)\|$ where $k_w, \lambda_w > 0$, if $\max_{i\in\mathcal{V}}\theta_i(t_0) - \min_{i\in\mathcal{V}}\theta_i(t_0) < \pi$ and the time-varying graph $\mathcal{G}(t)$ is uniformly connected.*

Proof Since $R^i_{e_{\theta_i}(t)}$ is a rotation matrix, it is true that $\|\mathbb{I}_2 - R^i_{e_{\theta_i}(t)}\| < \sqrt{2}|e_{\theta_i}(t)|$ under the condition $\max_{i\in\mathcal{V}}\theta_i(t) - \min_{i\in\mathcal{V}}\theta_i(t) < \pi$, which is satisfied due to the property of consensus algorithm. Then $w(t)$ is upper bounded as

$$\begin{aligned}
\|w(t)\| &\leq \sum_{i=1}^{n}\|w_i\| \\
&\leq k_p \sum_{i=1}^{n}\sum_{j\in\mathcal{N}_i} a_{ij}(t)\|\mathbb{I}_2 - R^i_{e_{\theta_i}(t)}\|\|(p_j^* - p_i^*)\| \\
&\leq \sqrt{2}k_p \sum_{i=1}^{n}\sum_{j\in\mathcal{N}_i} a_{ij}(t)|e_{\theta_i}(t)|\|(p_j^* - p_i^*)\|
\end{aligned} \tag{6.30}$$

Using the properties of norm of vector $w(t)$, $\|w(t)\| \leq \|w(t)\|_1 \leq \sqrt{n}\|w(t)\|$, $w(t) \in \mathbb{R}^n$, it is clear that there exists a constant k_{w,e_θ} such that $\sqrt{2}k_p\sum_{i=1}^n \sum_{j\in\mathcal{N}_i} a_{ij}(t)|e_{\theta_i}(t)|\|(p_j^* - p_i^*)\| \leq k_{w,e_\theta}\|e_\theta(t)\|$. Also since $\|e_\theta(t)\| \leq k_{e_\theta}e^{-\lambda_{e_\theta}(t-t_0)}$ $\|e_\theta(t_0)\|$ for all $t \geq t_0$, the inequality can be further changed as

$$\|w(t)\| \leq k_{w,e_\theta}k_{e_\theta}e^{-\lambda_{e_\theta}(t-t_0)}\|e_\theta(t_0)\| \tag{6.31}$$

which completes the proof.

Let the solution of (6.29) be denoted as $e_{p^a}(t) = \phi_{e_{p^a}}(t, t_0)e_{p^a}(t_0) + \int_{t_0}^{t}\phi_{e_{p^a}}(t, \tau)$ $w(\tau)d\tau$, where $\phi_{e_{p^a}}(t, t_0)$ is the state transition matrix from t_0 to t. For the main result of this section, define $E_{p^*} \triangleq \{p_i | p_j^a - p_i^a = p_j^* - p_i^*, \forall (i, j)^e \in \mathcal{E}\}$. Then, we can obtain the following convergence property [7]:

Theorem 6.3 *Under the distributed control law (6.25), $p^a(t) = (p_1^a(t), \ldots, p_n^a(t))^T$ exponentially converges to E_{p^*} if $\max_{i\in\mathcal{V}}\theta_i(t_0) - \min_{i\in\mathcal{V}}\theta_i(t_0) < \pi$ and the time-varying graph $\mathcal{G}(t)$ is uniformly connected.*

Proof In the proof, it will be shown that $\text{dist}(e_{p^a}(t), \mathcal{E}_{2n})$ converges to zero exponentially fast. The proof is similar to the proof of Theorem 6.2. Using the property of the metric $\text{dist}(\cdot)$, since there exist $k_{e_p}, \lambda_{e_p} > 0$ such that $\text{dist}(\phi_{e_{p^a}}(t, t_0)e_{p^a}(t_0), \mathcal{E}_{2n}) \leq k_{e_p}e^{-\lambda_{e_p}(t-t_0)}\text{dist}(e_{p^a}(t_0), \mathcal{E}_{2n})$, the following inequality holds

$$\begin{aligned}\mathrm{dist}(e_{p^a}(t), \mathcal{E}_{2n}) &\le \mathrm{dist}(\phi_{e_{p^a}}(t, t_0)e_{p^a}(t_0), \mathcal{E}_{2n}) + \mathrm{dist}\left(\int_{t_0}^{t} \phi_{e_{p^a}}(t, \tau)w(\tau)d\tau, \mathcal{E}_{2n}\right) \\ &\le k_{e_p} e^{-\lambda_{e_p}(t-t_0)}\mathrm{dist}(e_{p^a}(t_0), \mathcal{E}_{2n}) \\ &\quad + \int_{t_0}^{t} \mathrm{dist}(\phi_{e_{p^a}}(t, \tau)w(\tau), \mathcal{E}_{2n})d\tau \end{aligned} \tag{6.32}$$

Repeating the same process as the proof of Theorem 6.2, the term $\int_{t_0}^{t} \mathrm{dist}(\phi_{e_{p^a}}(t, \tau) w(\tau), \mathcal{E}_{2n})d\tau$ is upper bounded as $\int_{t_0}^{t} \mathrm{dist}(\phi_{e_{p^a}}(t, \tau)w(\tau), \mathcal{E}_{2n})d\tau \le k'_{\mathcal{E}} \sup_{t_0 \le \tau \le t} \{\|w(\tau)\|\}$, with a positive constant $k'_{\mathcal{E}}$. Thus, by combining (6.32) with (6.31), the inequality (6.32) can be changed as

$$\mathrm{dist}(e_{p^a}(t), \mathcal{E}_{2n}) \le k_{e_p} e^{-\lambda_{e_p}(t-t_0)}\mathrm{dist}(e_{p^a}(t_0), \mathcal{E}_{2n}) + k_{e_p, e_\theta} e^{-\lambda_{e_\theta}(t-t_0)} \|e_\theta(t_0)\| \tag{6.33}$$

where k_{e_p, e_θ} is a positive constant. Then, by replacing t_0 in the above equation by $(t + t_0)/2$, it is shown that

$$\begin{aligned}\mathrm{dist}(e_{p^a}(t), \mathcal{E}_{2n}) &\le k_{e_p} e^{-\lambda_{e_p}(t-t_0)/2}\mathrm{dist}(e_{p^a}((t + t_0)/2), \mathcal{E}_{2n}) \\ &\quad + k_{e_p, e_\theta} e^{-\lambda_{e_\theta}(t-t_0)/2} \|e_\theta((t + t_0)/2)\| \end{aligned} \tag{6.34}$$

which implies as $t \to \infty$, the first term and second term of the right-hand side of the above inequality converge to zero exponentially. Thus, $e_{p^a}(t)$ exponentially converges to the set $\mathcal{E}_{2n}$. Since $\dot{e}_{p^a} \to 0$, it converges to a point $e_{p^a}^{\infty} \in \mathcal{E}_{2n}$. Thus, the set E_{p^*} is exponentially stable.

Figure 6.5 shows the overall structure of the orientation control-based formation control system. Different from Fig. 6.3, the active orientation control results in the

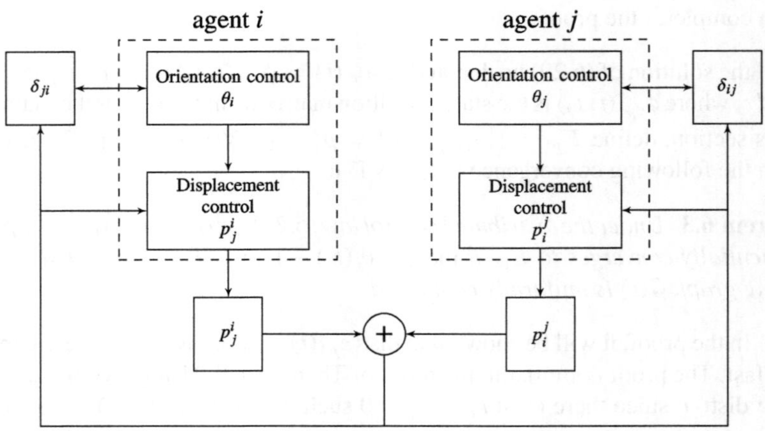

Fig. 6.5 The overall structure of the orientation control-based formation control

variation of the bearing angles. Thus, the bearing angles are changed as functions of orientation control and position control of agents.

The formation control scheme presented in this section is a combination of distributed orientation control and displacement-based formation control. Although these two processes are conducted in a simultaneous way, the key idea is to transform the distance-based approach to the displacement-based approach only using local measurements. The following example shows the significance of the orientation control.

Example 6.4 Consider two agents and suppose that their orientations have not been aligned perfectly due to some uncertainties or biases. Let the orientation of agent 1 be deviated from the orientation of the agent 2 by an amount of θ. Then, the distributed formation control laws for agents 1 and 2 can be formulated as

$$\begin{aligned}\dot{p}_1 &= R(\theta)(p_2 - p_1) - (p_2^* - p_1^*)\\ \dot{p}_2 &= (p_1 - p_2) - (p_1^* - p_2^*)\end{aligned}$$

which can be combined as

$$\begin{bmatrix}\dot{p}_1\\ \dot{p}_2\end{bmatrix} = \begin{bmatrix}-R(\theta) & R(\theta)\\ \mathbb{I}_2 & -\mathbb{I}_2\end{bmatrix}\begin{bmatrix}p_1\\ p_2\end{bmatrix} - \begin{bmatrix}-\mathbb{I}_2 & \mathbb{I}_2\\ \mathbb{I}_2 & -\mathbb{I}_2\end{bmatrix}\begin{bmatrix}p_1^*\\ p_2^*\end{bmatrix} \tag{6.35}$$

The desired equilibrium point is given as

$$p_{eq}^* = \begin{bmatrix}p_1^* + p^c\\ p_2^* + p^c\end{bmatrix} \tag{6.36}$$

By inserting (6.36) to the right-hand side of (6.35), we have

$$\begin{bmatrix}-R(\theta) & R(\theta)\\ \mathbb{I}_2 & -\mathbb{I}_2\end{bmatrix}\begin{bmatrix}p_1^* + p^c\\ p_2^* + p^c\end{bmatrix} - \begin{bmatrix}-\mathbb{I}_2 & \mathbb{I}_2\\ \mathbb{I}_2 & -\mathbb{I}_2\end{bmatrix}\begin{bmatrix}p_1^*\\ p_2^*\end{bmatrix} = \begin{bmatrix}\mathbb{I}_2 - R(\theta) & R(\theta) - \mathbb{I}_2\\ \mathbf{0}_{2\times 2} & \mathbf{0}_{2\times 2}\end{bmatrix}\begin{bmatrix}p_1^*\\ p_2^*\end{bmatrix} \neq \mathbf{0}_{4\times 1}$$

Consequently, we can see that if $\theta \neq 0$, then the desired formation cannot be achieved. Thus, without orientation alignment, the displacement-based formation control would always have some errors in the formation configuration.

6.4 Summary and Simulations

Tables 6.2 and 6.3 provide a comparison between the orientation estimation-based formation control law studied in Sect. 6.1 and orientation control-based formation control studied in the previous section. As shown in these tables, the orientation control-based approach does not require a computation of orientation estimation; but it has an additional control input u_{θ_i}.

Table 6.2 Variables and network properties of orientation estimation-based formation systems

	Variables	Topology	Edge direction
Sensing	p^i_{ji}	Spanning tree	Undirected
Control	p^i_{ji}	Spanning tree (strongly connected)	Undirected (directed)
Communications	$\delta_{ij}, \hat{\theta}_i$	Spanning tree (arborescence)	Undirected (directed)
Computation	$\hat{\theta}_i, u^i_i$	None	N/A

Table 6.3 Variables and network properties of orientation control-based formation systems

	Variables	Topology	Edge direction
Sensing	p^i_{ji}	Spanning tree	Undirected
Control	p^i_{ji}, θ_i	Spanning tree (strongly connected)	Undirected (directed)
Communications	δ_{ij}	Spanning tree (arborescence)	Undirected (directed)
Computation	u^i_i	None	N/A

In the main theorems of Sects. 6.1 and 6.3, we required a spanning tree for the underlying topology. But, since the orientation angles are estimated on the basis of consensus in Sect. 6.1 and the orientation angles can be reached to a common value by also consensus in Sect. 6.3, the communications could be a directed one as long as a consensus can be obtained. Thus, under the directed graph topology, since a consensus can be reached if the network is strongly connected, or if there exists an arborescence (i.e., there exists a directed path from a root to for any other node; a rooted-out directed tree), the communication topology could be directed. Consequently, the sensing topology $(\mathcal{V}, \mathcal{E}^s)$ is undirected, while the basic communication topology $(\mathcal{V}, \mathcal{E}^c)$ could be directed one, in both the orientation estimation-based and orientation control-based approaches. Also, as commented in Remarks 6.1 and 6.2, the actuation topology could be a directed one as far as a convergence to a consensus manifold is ensured.

An agent needs to send an information to neighboring agents, which are defined by a communication topology. Thus, to deliver information to specific agents, it may be required to encode the identification number (or address) in communication protocol. In this sense, the communication may have to be implemented to a network where agents are assigned with specific addresses or identification numbers. In such case, to achieve a desired formation, agents of a network should have identification numbers, which is disadvantageous over the formation control laws presented in Part II. To overcome such a drawback, an index-free formation control law may be designed; for example, see [9].

Let us consider five agents shown in Fig. 6.6 for numerical verifications of Sect. 6.3. The topologies for communications and actuations are equivalent; but

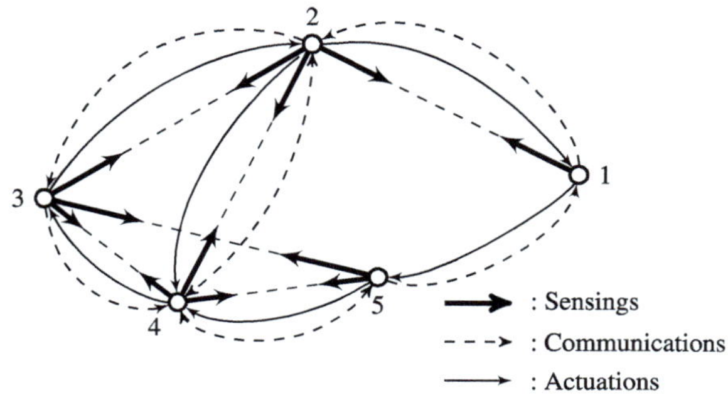

Fig. 6.6 Sensing and communication/actuation topologies for simulations

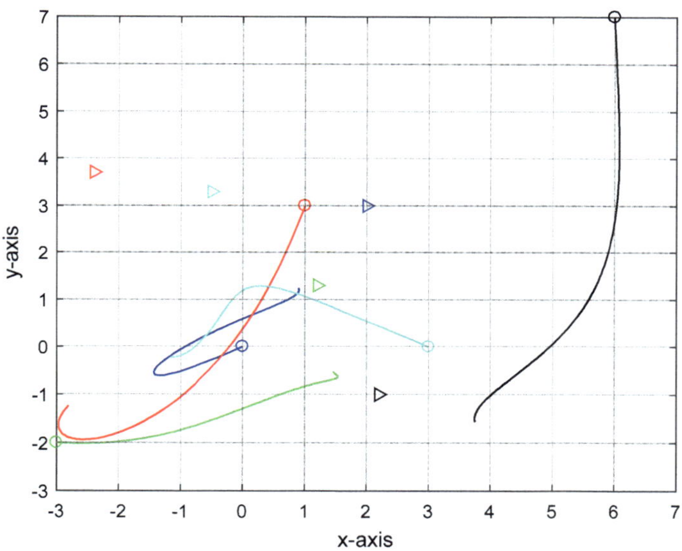

Fig. 6.7 Trajectories of positions of five agents, without compensations of the rotations and translations. The trajectories of agents are distinguished by the colors. The initial positions are marked by ∘ and the desired positions are marked by ▹

the directions are opposite. The sensing topology is different from the communication and actuation topologies. The simulation results in Figs. 6.7 to 6.8 are obtained with initial orientation angles as $\theta_1(t_0) = \pi/3, \theta_2(t_0) = \pi/3, \theta_3(t_0) = \pi/20, \theta_4(t_0) = \pi/2.1$, and $\theta_5(t_0) = 1.9\pi$. These initial angles satisfy the condition for the alignment. Figure 6.7 shows the initial positions and desired positions of agents, and trajectories of agents. Since the positions converge to the desired configuration up to translations and rotations, the converged positions do not match the desired positions. The left plots in Fig. 6.8 show the initial positions and desired positions of agents, and trajectories of agents after the compensation of translations and rotations. It shows that

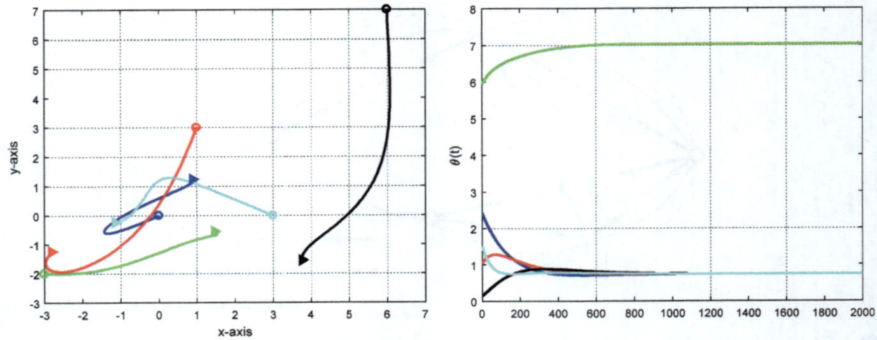

Fig. 6.8 Left: Trajectories of positions of five agents, after compensations of the rotations and translations. Right: Trajectories of orientations of five agents (θ_i). The trajectories of agents are distinguished by the colors

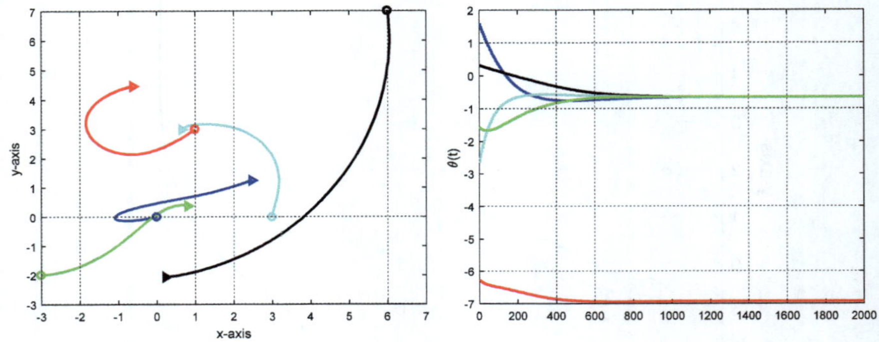

Fig. 6.9 Left: Trajectories of positions of agents with initial orientation angles $\theta_1(t_0) = \pi/2$, $\theta_2(t_0) = -2\pi$, $\theta_3(t_0) = \pi/10$, $\theta_4(t_0) = -\pi/1.2$, and $\theta_5(t_0) = -\pi/2$. Right: Trajectories of orientation angles of agents with initial orientation angles $\theta_1(t_0) = \pi/2$, $\theta_2(t_0) = -2\pi$, $\theta_3(t_0) = \pi/10$, $\theta_4(t_0) = -\pi/1.2$, and $\theta_5(t_0) = -\pi/2$

the agents have converged to the desired configuration. The right plots in Fig. 6.8 depict the trajectories of orientation angles of agents, which shows that the angles have reached a consensus.

Figure 6.9 shows simulation results with initial orientation angles $\theta_1(t_0) = \pi/2$, $\theta_2(t_0) = -2\pi$, $\theta_3(t_0) = \pi/10$, $\theta_4(t_0) = -\pi/1.2$, and $\theta_5(t_0) = -\pi/2$. Although the alignment condition is not satisfied, the agents converge to desired configuration and orientation angles have reached a consensus. Although the underlying topologies for sensings, communications, and actuations are different, the desired formation configuration and orientation alignment have been achieved. So, in orientation alignment-based formation control, there are certain degrees of freedom for designing the different network topologies for distributed sensings, communications, and actuations.

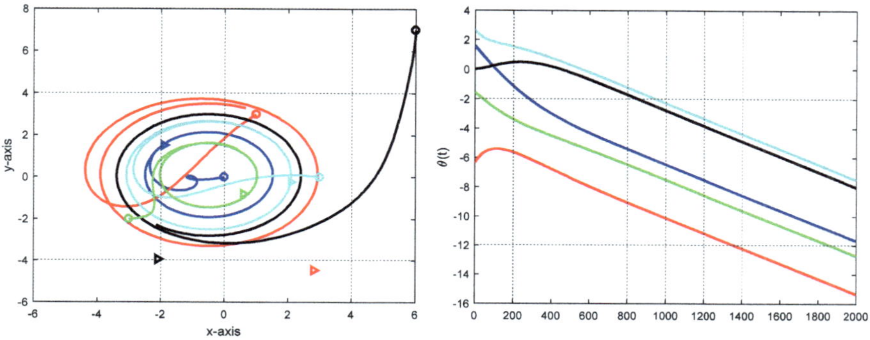

Fig. 6.10 Left: Trajectories of positions of agents with initial orientation angles $\theta_1(t_0) = \pi/2.01$, $\theta_2(t_0) = -2\pi$, $\theta_3(t_0) = 0.01\pi$, $\theta_4(t_0) = \pi/1.2$, and $\theta_5(t_0) = -\pi/2.001$. Right: Trajectories of orientation angles of agents with initial orientation angles $\theta_1(t_0) = \pi/2.01, \theta_2(t_0) = -2\pi, \theta_3(t_0) = 0.01\pi, \theta_4(t_0) = \pi/1.2$, and $\theta_5(t_0) = -\pi/2.001$

On the other hand, with slightly different initial orientation angles $\theta_1(t_0) = \pi/2.01$, $\theta_2(t_0) = -2\pi$, $\theta_3(t_0) = 0.01\pi$, $\theta_4(t_0) = \pi/1.2$, and $\theta_5(t_0) = -\pi/2.001$, Fig. 6.10 shows that the agents do not converge to the desired configuration and orientation angles diverge. From the numerical test, we can see that the condition $\max_{i\in\mathcal{V}} \theta_i(t_0) - \min_{i\in\mathcal{V}} \theta_i(t_0) < \pi$ is a sufficient one. The trajectories of orientations in Fig. 6.10 show that the alignment is not achieved; but the agents rotate continuously with a constant speed and with the some relative offset angles.

6.5 Notes

The results of this chapter have been mainly reused and reproduced from [6–8]. The orientation estimation- based formation control could guarantee a quasi-global asymptotic stability while it requires communications between neighboring agents. It is an obvious advantage over gradient-based control laws to achieve the quasi-global stability, while it is a disadvantage to have communications between neighboring agents. Also, it was assumed that the agents control their motions only for formation configuration, without considering a movement of dynamic formations. When agents move in a constant velocity, using a concept of Heaviside step function, the convergence of formation also can be ensured [1]. Some analyses for formation stabilization with mismatched alignments, but without active estimation or control, have been conducted in [4]. The following copyright and permission notices are acknowledged.

- The statement of Theorem 6.2 with the proof is © [2012] IEEE. Reprinted, with permission, from [7].

References

1. Ahn, H.-S., Lee, B.-H., Oh, K.-K., Jeong, K.: Orientation alignment-based formation control with reference node. In: Proceedings of the 34th Chinese Control Conference and SICE Annual Conference, pp. 6942–6947 (2015)
2. Cavagna, A., Cimarelli, A., Giardina, I., Parisi, G., Santagati, R., Stefanini, F., Tavarone, R.: From empirical data to inter-individual interactions: unveiling the rules of collective animal behavior. Math. Model. Methods Appl. Sci. **20**(supp01), 1491–1510 (2010)
3. Couzin, I.D., Krause, J., Franks, N.R., Levin, S.A.: Effective leadership and decision-making in animal groups on the move. Nature **433**, 513–516 (2005)
4. Meng, Z., Anderson, B.D.O., Hirche, S.: Formation control with mismatched compasses. Automatica **69**, 232–241 (2016)
5. Moreau, L.: Stability of multiagent systems with time-dependent communication links. IEEE Trans. Autom. Control **50**(2), 169–182 (2005)
6. Oh, K.-K., Ahn, H.-S.: Formation control of mobile agents without an initial common sense of orientation. In: Proceedings of the IEEE Conference on Decision and Control, pp. 1428–1432 (2012)
7. Oh, K.-K., Ahn, H.-S.: Orientation alignment-based formation control in multi-agent systems. In: Proceedings of the IEEE/ASME International Conference Mechatronics and Embedded Systems and Applications, pp. 42–45 (2012)
8. Oh, K.-K., Ahn, H.-S.: Formation control and network localization via orientation alignment. IEEE Trans. Autom. Control **59**(2), 540–545 (2014)
9. Sakurama, K., Ahn, H.-S.: Index-free assignment formation of networked multi-agent systems. In: Proceedings of the 2018 Annual American Control Conference (ACC), pp. 466–471 (2018)
10. Sarlette, A., Sepulchre, R.: Consensus optimization on manifolds. SIAM J. Control Optim. **48**(1), 56–76 (2009)
11. Sarlette, A., Sepulchre, R.: Synchronization on the circle, pp. 1–21 (2009). arXiv:0901.2408 [math.OC]

Chapter 7
Formation Control via Orientation and Position Estimation

Abstract In the previous chapter, the orientation angles θ_i are estimated or controlled such that $\hat{\theta}_i - \hat{\theta}_j = \theta_i - \theta_j$. Then, after the orientation alignment, the agents with aligned orientation could be controlled in a displacement-based setup. That is, given a desired configuration p^*, the formation control has been achieved up to rotation and translation in the sense of $p_j - p_i \to p_j^* - p_i^*$ with respect to a common coordinate frame. Thus, the relative displacements between neighboring agents have been controlled. But, if we could estimate the positions of agents and the estimated positions are used to control the motions of agents, then the desired formation might be achieved more rapidly. In this chapter, we would like to estimate the positions of agents, up to a common offset in positions, using only relative measurements.

7.1 Formation Control via Orientation Control and Position Estimation in 2-Dimensional Space

The dynamics for the position estimation and formation control are combined into a single form for simultaneously updating the estimation and control. Let a global coordinate frame be denoted by ${}^g\Sigma$, local coordinate frame by ${}^i\Sigma$, and aligned coordinate frame by ${}^a\Sigma$. Under distance-based sensing topology, the positions of agents are estimated in the sense of $\hat{p}_i \to p_i + p^\dagger$, where $p^\dagger$ is a constant vector, with respect to ${}^a\Sigma$. Then by directly controlling $\hat{p}_i$ that is the estimate of p_i, we would like to ensure that $\hat{p}_i \to p_i^* + p^\ddagger$, where $p^\ddagger$ is also a constant vector, in a common coordinate frame ${}^c\Sigma$. Note that the aligned coordinate frame ${}^a\Sigma$ could be a common coordinate frame. But, the common coordinate frame is not necessarily the aligned coordinate frame ${}^a\Sigma$ since ${}^c\Sigma$ can be any arbitrary coordinate frame, while ${}^a\Sigma$ is the coordinate frame after the orientation synchronization.

The goal of the formation control via orientation control and position estimation in 2-dimensional space is to transform the problem set up in the distance-based one to the position-based one by using relative position sensings [13]. The distance-based setup has been transformed into the displacement-based setup in the previous chapter. This chapter further transforms the displacement-based setup into the position-based setup. Consider the following n agents on the plane:

H.-S. Ahn, *Formation Control*, Studies in Systems, Decision and Control 205, https://doi.org/10.1007/978-3-030-15187-4_7

$$\dot{p}_i = v_i \tag{7.1}$$

$$\dot{\theta}_i = u_{\theta_i} \tag{7.2}$$

It is noticeable that the dynamics (7.1)–(7.2) is equivalent to the dynamics (6.1). But, in (6.1), we do not explicitly represent the dynamics for orientation although it was used in defining relative measurements, orientation alignment, and control. When the dynamics (6.1) is used for formation control, or for consensus, it is usually supposed that the orientations of agents are aligned in existing works. In this sense, it would be clear to write the dynamics of distributed agents as (7.1)–(7.2).

Similarly to Chap. 6, with $\theta_{ji} \triangleq \text{PV}(\theta_j - \theta_i)$, where $\text{PV}(\theta_j - \theta_i)$ can be calculated by $\text{PV}(\theta_j - \theta_i) = \text{PV}(\delta_{ji} - \delta_{ij} + \pi)$, the orientations can be aligned by the angular rate control as

$$u_{\theta_i} = k_\theta \sum_{j \in \mathcal{N}_i} a_{ij} \theta_{ji} \tag{7.3}$$

where θ_{ji} is assumed to be available to agent i by bidirectional communications between neighboring agents i and j. Then the orientation dynamics can be concisely written as

$$\dot{\theta} = -k_\theta \mathbb{L} \theta \tag{7.4}$$

which implies all orientation angles converge to a common value, i.e., $\theta_i \to \theta_\infty$, for all $i \in \mathcal{V}$ as $t \to \infty$. The coordinate frame with the direction that is aligned to θ_∞ is denoted by ${}^a\Sigma$. Let the positions of agents be estimated with respect to ${}^a\Sigma$ as follows:

$$\dot{\hat{p}}_i^a = k_{\hat{p}} \sum_{j \in \mathcal{N}_i} a_{ij} [(\hat{p}_j^a - \hat{p}_i^a) - p_j^i] + v_i^i \tag{7.5}$$

It is assumed that the estimated positions $\hat{p}_j^a$ and $\hat{p}_i^a$ between neighboring agents can be exchanged. The update (7.5) is conducted in each agent; so it is a distributed computation work of agents. Since the update is done with a virtual coordinate frame, it is not necessary to have a knowledge for the aligned coordinate frame. The linear velocity input v_i^i is designed as

$$v_i^i = k_p (p_i^* - \hat{p}_i^a) \tag{7.6}$$

The above position control law ignores an offset up to a common rotation and common translation. Let the difference between an arbitrarily aligned orientation angle θ_∞ and orientation θ_i be denoted as $e_{\theta_i} = \theta_\infty - \theta_i$. For a simplicity of presentation, denoting $R(e_{\theta_i}) = R_i^a(e_{\theta_i})$, where R_i^a is the orientation transformation matrix from ${}^a\Sigma$ to ${}^i\Sigma$, we have the relationships $v_i^a = R^{-1}(e_{\theta_i}) v_i^i$ and $p_{ji}^i = R(e_{\theta_i})(p_j^a - p_i^a)$. Then, (7.1), (7.2), (7.4), (7.5), and (7.6) can be combined to obtain the following equations [13]:

$$\dot{p}^a = k_p(p^* - \hat{p}^a) - k_p[\mathbb{I}_{2n} - D(R_{e_\theta})^{-1}](p^* - \hat{p}^a) \tag{7.7}$$

$$\dot{\hat{p}}^a = -k_{\hat{p}}(\mathbb{L} \otimes \mathbb{I}_2)(\hat{p}^a - p^a) + k_p(p^* - \hat{p}^a) - k_{\hat{p}}[\mathbb{I}_{2n} - D(R_{e_\theta})](\mathbb{L} \otimes \mathbb{I}_2)p^a \tag{7.8}$$

$$\dot{e}_\theta = -k_\theta \mathbb{L} e_\theta \tag{7.9}$$

where $D(R_{e_\theta}) = \text{blkdg}(R(e_{\theta_i}))$. For the stability analysis, further define $e_{p^a} \triangleq p^* - p^a$ and $e_{\hat{p}^a} \triangleq p^a - \hat{p}^a$, and $e_s \triangleq [e_{p^a}{}^T, e_{\hat{p}^a}^T]^T$. Then the following error dynamics can be obtained:

$$\dot{e}_s = A_s e_s + \Delta A_s(e_\theta)e_s + D_s(e_\theta) \tag{7.10}$$

$$\dot{e}_\theta = -k_\theta \mathbb{L} e_\theta \tag{7.11}$$

where

$$A_s = \left[\begin{array}{c|c} -k_p\mathbb{I}_{2n} & -k_p\mathbb{I}_{2n} \\ \hline \mathbf{0}_{2n\times 2n} & -k_{\hat{p}}(\mathbb{L}\otimes\mathbb{I}_2) \end{array}\right]$$

$$\Delta A_s(e_\theta) = \left[\mathbb{I}_2 \otimes (\mathbb{I}_{2n} - D(R_{e_\theta})^{-1})\right]\left[\begin{array}{c|c} k_p\mathbb{I}_{2n} & k_p\mathbb{I}_{2n} \\ \hline -k_p\mathbb{I}_{2n} & -k_p\mathbb{I}_{2n} \end{array}\right] - \left[\mathbb{I}_2 \otimes (\mathbb{I}_{2n} - D(R_{e_\theta}))\right]\left[\begin{array}{c|c} \mathbf{0}_{2n\times 2n} & \mathbf{0}_{2n\times 2n} \\ \hline k_{\hat{p}}(\mathbb{L}\otimes\mathbb{I}_2) & \mathbf{0}_{2n\times 2n} \end{array}\right]$$

$$D_s(e_\theta) = [\mathbb{I}_{2n} - D(R_{e_\theta})]\left[\begin{array}{c} \mathbf{0}_{2n\times 2n} \\ k_{\hat{p}}(\mathbb{L}\otimes\mathbb{I}_2)p^* \end{array}\right]$$

Lemma 7.1 *If the time-varying graph $\mathcal{G}(t)$ is uniformly connected and initial orientation angles satisfy* $\max_{i\in\mathcal{V}} \theta_i(t_0) - \min_{i\in\mathcal{V}} \theta_i(t_0) < \pi$, *there exist k_γ and $\lambda_\gamma > 0$ such that $\|\mathbb{I}_{2n} - D(R_{e_\theta})(t)\| \leq k_\gamma e^{-\lambda_\gamma(t-t_0)}\|e_\theta(t_0)\|$ and $\|\mathbb{I}_{2n} - D(R_{e_\theta})^{-1}\| \leq k_\gamma e^{-\lambda_\gamma(t-t_0)}\|e_\theta(t_0)\|$ for $t \geq t_0$.*

Proof From (7.11), since there exist positive constants k_{e_θ} and λ_{e_θ} such that $\|e_\theta(t)\| \leq k_{e_\theta} e^{-\lambda_{e_\theta}(t-t_0)}\|e_\theta(t_0)\|$ is satisfied, the proof is exactly the same to the proof of Lemma 6.1.

Lemma 7.2 *If the time-varying graph $\mathcal{G}(t)$ is uniformly connected, the Laplacian matrix $\mathbb{L}(t)$ is piecewise continuous and initial orientation angles satisfy* $\max_{i\in\mathcal{V}} \theta_i(t_0) - \min_{i\in\mathcal{V}} \theta_i(t_0) < \pi$, *then $e_s(t)$ is bounded for $t \geq t_0$.*

Proof From (7.10), for an analysis purpose, the terms $\Delta A_s(e_\theta)e_s + D_s(e_\theta)$ are considered input or dissipating terms. So, the state transition matrix ϕ_{A_s} can be defined from the matrix A_s. Since $\mathbb{L}$ is the Laplacian matrix, the matrix A_s has eigenvalues that are not positive, which means there exists a finite constant M_{A_s} such that $\|\phi_{A_s}(t, t_0)\| \leq M_{A_s}$ holds. The solution of (7.10) can be expressed as

$$e_s(t) = \phi_{A_s}(t, t_0)e_s(t_0) + \int_{t_0}^{t} \phi_{A_s}(t, \tau)[\Delta A_s(e_\theta(\tau))e_s(\tau) + D_s(e_\theta(\tau))]d\tau \tag{7.12}$$

With Lemma 7.1, it is obvious that there exist $k_{D_s}, \lambda_{D_s} > 0$ such that

$$\|D_s(e_\theta(t))\| \leq k_{D_s} e^{-\lambda_{D_s}(t-t_0)} \|D_s(e_\theta(t_0))\| \tag{7.13}$$

Thus, it is true that

$$\begin{aligned}
\|e_s(t)\| &\leq \|\phi_{A_s}(t, t_0)\| \|e_s(t_0)\| + \int_{t_0}^{t} \|\phi_{A_s}(t, \tau)\| \|\Delta A_s(e_\theta(\tau))\| \|e_s(\tau)\| d\tau \\
&\quad + \int_{t_0}^{t} \|\phi_{A_s}(t, \tau)\| \|D_s(e_\theta(\tau))\| d\tau \\
&\leq M_{A_s} \|e_s(t_0)\| + M_{A_s} \int_{t_0}^{t} \|\Delta A_s(e_\theta(\tau))\| \|e_s(\tau)\| d\tau \\
&\quad + \frac{k_{D_s}}{\lambda_{D_s}} M_{A_s} \|D_s(e_\theta(t_0))\|
\end{aligned} \tag{7.14}$$

It follows from the Gronwall–Bellman lemma[1] that

$$\begin{aligned}
\|e_s(t)\| &\leq M_{A_s} \|e_s(t_0)\| e^{\int_{t_0}^{t} M_{A_s} \|\Delta A_s(e_\theta(\tau))\| d\tau} \\
&\quad + \frac{k_{D_s}}{\lambda_{D_s}} M_{A_s} \|D_s(e_\theta(t_0))\| e^{\int_{t_0}^{t} M_{A_s} \|\Delta A_s(e_\theta(\tau))\| d\tau}
\end{aligned} \tag{7.15}$$

The Lemma 7.1 implies there exist $k_{\Delta A_s}$ and $\lambda_{\Delta A_s} > 0$ such that $\|\Delta A_s(e_\theta(\tau))\| \leq k_{\Delta A_s} e^{-\lambda_{\Delta A_s}(t-t_0)} \|e_\theta(t_0)\|$. Thus, we can see that $e^{\int_{t_0}^{t} M_{A_s} \|\Delta A_s(e_\theta(\tau))\| d\tau}$ is bounded for $t \geq t_0$, which further implies that $e_s(t)$ is bounded for $t \geq t_0$.

Lemma 7.3 *With the orientation alignment law (7.3), position estimation law (7.5), and position control law (7.6), if the time-varying graph $\mathcal{G}(t)$ is uniformly connected and initial orientation angles satisfy* $\max_{i \in \mathcal{V}} \theta_i(t_0) - \min_{i \in \mathcal{V}} \theta_i(t_0) < \pi$, *then there exists a constant vector $e_{\hat{p}^a}^{\infty}$ such that $e_{\hat{p}^a}(t)$ asymptotically converges to $\mathbf{1}_n \otimes e_{\hat{p}^a}^{\infty}$ as $t \to \infty$.*

Proof See the appendix.

Lemma 7.4 *Let us suppose that the underlying assumptions of Lemma 7.3 are still satisfied. Then, there exists a constant vector $e_{\hat{p}^a}^{\infty}$ such that $e_{p^a}(t)$ asymptotically converges to $-\mathbf{1}_n \otimes e_{\hat{p}^a}^{\infty}$, respectively, as $t \to \infty$.*

[1]Let the functions $x(t)$, $\lambda(t)$, and $u(t)$ be continuous functions, and additionally let $u(t)$ be non-negative, satisfying $x(t) \leq \lambda(t) + \int_a^t u(s)x(s)ds$. Then, the following inequality holds [8]:

$$x(t) \leq \lambda(t) + \int_a^t \lambda(s)u(s)e^{\int_s^t u(\tau)d\tau} ds.$$

Proof Let $\xi_s \triangleq e_{p^a} + \mathbf{1}_n \otimes e^{\infty}_{\hat{p}^a}$ and $v_s \triangleq k_p[\mathbb{I}_{2n} - D(R_{e_\theta})^{-1}](e_{p^a} + e_{\hat{p}^a})$. Then, from $\xi_s(t) = p^* - p^a + \mathbf{1}_n \otimes e^{\infty}_{\hat{p}^a}$, it can be shown that

$$\begin{aligned}\dot{\xi}_s(t) &= -\dot{p}^a(t)\\ &= -k_p(p^* - \hat{p}^a) + k_p[\mathbb{I}_{2n} - D(R_{e_\theta})^{-1}](p^* - \hat{p}^a)\\ &= -k_p(p^* - p^a + \mathbf{1}_n \otimes e^{\infty}_{\hat{p}^a}) + k_p\hat{p}^a - k_p(p^a - \mathbf{1}_n \otimes e^{\infty}_{\hat{p}^a})\\ &\quad + k_p[\mathbb{I}_{2n} - D(R_{e_\theta})^{-1}](p^* - \hat{p}^a)\\ &= -k_p\xi_s(t) - k_p e_{\hat{p}^a} + k_p\mathbf{1}_n \otimes e^{\infty}_{\hat{p}^a} + k_p[\mathbb{I}_{2n} - D(R_{e_\theta})^{-1}](e_{p^*} + e_{\hat{p}^a})\end{aligned} \tag{7.16}$$

Thus,

$$\begin{aligned}\xi_s(t) = e^{-k_p(t-t_0)}\xi_s(t_0) - k_p\int_{t_0}^{t} e^{-k_p(t-\tau)}(e_{\hat{p}^a}(\tau) - \mathbf{1}_n \otimes e^{\infty}_{\hat{p}^a})d\tau\\ + \int_{t_0}^{t} e^{-k_p(t-\tau)}v_s(\tau)d\tau\end{aligned} \tag{7.17}$$

which results in

$$\|\xi_s(t)\| \le e^{-k_p(t-t_0)}\|\xi_s(t_0)\| + \sup_{t_0\le\tau\le t}\left\{\|(e_{\hat{p}^a}(\tau) - \mathbf{1}_n \otimes e^{\infty}_{\hat{p}^a})\| + \frac{1}{k_p}\|v_s(\tau)\|\right\} \tag{7.18}$$

Without loss of generality, by replacing t_0 by $t/2$, the above inequality can be rewritten as

$$\|\xi_s(t)\| \le e^{-k_p(t/2)}\|\xi_s(t/2)\| + \sup_{t/2\le\tau\le t}\left\{\|(e_{\hat{p}^a}(\tau) - \mathbf{1}_n \otimes e^{\infty}_{\hat{p}^a})\| + \frac{1}{k_p}\|v_s(\tau)\|\right\} \tag{7.19}$$

Therefore, since $e_{\hat{p}^a}(t/2) \to \mathbf{1}_n \otimes e^{\infty}_{\hat{p}^a}$ and $v_s(t/2) \to 0$, the right-hand side of (7.19) goes to zero as $t/2 \to \infty$. Thus, it is concluded that $\xi_s(t) \to 0$, which implies $e_{p^a}(t) \to -\mathbf{1}_n \otimes e^{\infty}_{\hat{p}^a}$.

Now, with the above lemmas, we are able to make the following main result for both orientation control and position estimation in 2-D [13]:

Theorem 7.1 *With the orientation alignment law (7.3), position estimation law (7.5), and position control law (7.6), if the time-varying graph $\mathcal{G}(t)$ is uniformly connected and initial orientation angles satisfy* $\max_{i\in\mathcal{V}}\theta_i(t_0) - \min_{i\in\mathcal{V}}\theta_i(t_0) < \pi$, *then the orientations of agents are aligned, the positions of agents are estimated up to a common position offset, and the desired formation is achieved up to also a common position offset.*

Proof From Lemmas 7.3 and 7.4, it was shown that $p^a \to p^* - \mathbf{1}_n \otimes e^{\infty}_{\hat{p}^a}$ and $\hat{p}^a \to p^a + \mathbf{1}_n \otimes e^{\infty}_{\hat{p}^a}$, which completes the proof.

Table 7.1 Characteristics of formation control via orientation control and position estimation

	Variables	Topology	Edge direction
Sensing	p^i_{ji}	Connected tree	Undirected
Control	p^i_{ji}, θ_i	Connected tree	Undirected
Communications	$\delta_{ij}, \hat{p}^a_i$	Connected tree (Arborescence)	Undirected (Directed)
Computation	$v^i_i, \hat{p}^a_i$	None	N/A

Table 7.1 shows the characteristics of formation control via orientation control and position estimation. When comparing with Table 6.3, it is clear that the formation control via orientation control and position estimation requires more communications and computation; but since it controls the position directly, it may have a rapid convergence speed. Also, since it handles the positions directly, the collision between agents may be avoided. Due to the same reason as shown in Table 6.3, a consensus can be reached if there exists an arborescence in the directed graph setup. Thus, the communication topology can be relaxed as a directed one. If a consensus, on the basis of initial values of orientation angles of agents, is supposed to occur, then there should be a path from any node to any other node. Thus, a strongly connected graph is required. Furthermore, if all the nodes are balanced in terms of in- and outdegrees, then we may be able to use an average consensus value for the alignment [15]. Figure 7.1 shows the overall structure of the formation control algorithm via orientation control and position estimation. It shows that the orientation control, position estimation, and position control are conducted in a closed loop. Note that, the analysis for the convergence in this block diagram is carried out simultaneously.

It is noticeable that if all the orientations of agents are aligned, then the orientation control in Fig. 7.1 is not necessary. In such a case, the formation control problem is to transfer the displacement-based problem to the position-based problem, via a position estimation [11]. The following example provides a detailed process for this problem.

Example 7.1 ([11]) Let us suppose that the orientations of agents are aligned. Then, the positions of agents can be estimated and the control law can be designed based on the estimated positions. Since the orientations of agents are aligned, we can simply ignore the axes of the local coordinate frames. The positions of agents are estimated, based on (7.5), such as $\dot{\hat{p}}_i = k_{\hat{p}} \sum_{j \in \mathcal{N}_i} a_{ij}[(\hat{p}_j - \hat{p}_i) - p^i_j] + v_i$ and the positions are controlled as $\dot{p}_i = v_i = k_p(p^*_i - \hat{p}_i)$. Then, the overall dynamics can be written as

$$\dot{p} = k_p(p^* - \hat{p}) \tag{7.20}$$

$$\dot{\hat{p}} = -k_{\hat{p}}(\mathbb{L} \otimes \mathbb{I}_2)(\hat{p} - p) + k_p(p^* - \hat{p}) \tag{7.21}$$

Let $e_{\hat{p}} = p - \hat{p}$ and $e_p = p - p^*$. Then, from (7.20) and by subtracting (7.21) from (7.20), we can have

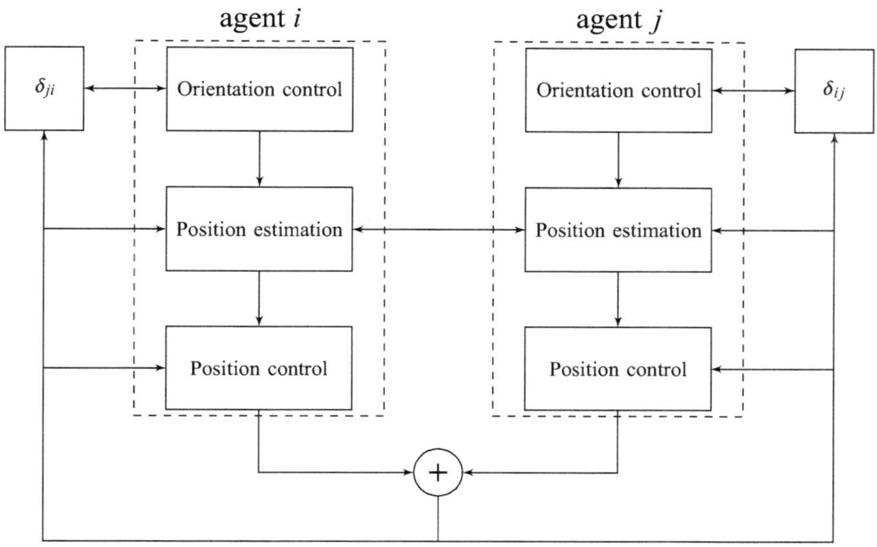

Fig. 7.1 The overall structure of formation control via orientation control and position estimation

$$\dot{e}_p = -k_p e_p + k_p e_{\hat{p}} \tag{7.22}$$
$$\dot{e}_{\hat{p}} = -k_{\hat{p}}(\mathbb{L} \otimes \mathbb{I}_2) e_{\hat{p}} \tag{7.23}$$

From the above Eq. (7.23), it is clear that $e_{\hat{p}}$ converges to a consensus point $p^{\infty} = \mathbb{I}_2 \otimes p^* \in \mathbb{R}^{2n}$, where $p^* \in \mathbb{R}^2$ is a common position offset, exponentially fast. Thus, if we denote $\zeta_{\hat{p}} = e_{\hat{p}} - p^{\infty}$ and $\zeta_p = e_p - p^{\infty}$, we can have

$$\dot{\zeta}_p = -k_p \zeta_p + k_p \zeta_{\hat{p}} \tag{7.24}$$
$$\dot{\zeta}_{\hat{p}} = -k_{\hat{p}}(\mathbb{L} \otimes \mathbb{I}_2) \zeta_{\hat{p}} \tag{7.25}$$

Then, due to the same reasons as the proof of Theorem 6.1, $e_{\hat{p}}$ and e_p converge to p^{∞} globally exponentially, which means that $p \to p^* + p^{\infty}$.

7.2 Formation Control via Orientation Control and Position Estimation in 3-Dimensional Space

The results of the previous section can be extended to 3-dimensional space by replacing the orientation angle by the rotation matrix. For this extension, this section introduces results of [12]. Let us consider a rigid body in 3-dimensional space (3-D) that can be expressed with a position vector with three elements and three Euler angles. The position is calculated by integrating linear translation motions while the Euler angles are calculated by integrating angular motions. If the three Euler angles are aligned to a common reference frame, we call it orientation alignment in 3-D. Thus, the orientation alignment in 3-D means that the Euler angles of agents are coincident. Let $u_i \in \mathbb{R}^3$ and $\omega_i \in \mathbb{R}^3$ denote the linear and angular velocities of agent i with respect to ${}^g\Sigma$. To describe the linear and angular motions, with notation $\omega_i = (\omega_{1i}, \omega_{2i}, \omega_{3i})^T$ for angular rate, it is convenient to use the following skew-symmetric matrix:

$$(\omega_i)^\wedge = S(\omega_i) \triangleq \begin{bmatrix} 0 & -\omega_{3i} & \omega_{2i} \\ \omega_{3i} & 0 & -\omega_{1i} \\ -\omega_{2i} & \omega_{1i} & 0 \end{bmatrix} \triangleq \Omega_i \tag{7.26}$$

Inversely, by the symbol $\vee$, the inverse operation of (7.26) is defined as

$$S(\omega_i)^\vee = \omega_i \tag{7.27}$$

With the above definitions, the n-agent systems can be described as [6]:

$$\dot{p}_i = u_i \tag{7.28}$$
$$\dot{R}_i = R_i(\omega_i)^\wedge = -(\omega_i)^\wedge R_i \tag{7.29}$$

where $R_i = R_i^g$. Since we consider a group of single-integrator dynamical systems, the linear velocities u_i, and angular velocities ω_i are the control inputs. From relative displacement measurements, it is possible to suppose that the relative positions from neighboring agents of agent i can be measured as

$$p_j^i = R_i(p_j - p_i), \ j \in \mathcal{N}_i \tag{7.30}$$

For the orientation alignment, it is further supposed that the agent i can measure the orientation of agent j with respect to ${}^i\Sigma$, which implies that the following measurement is also available to the agent i (see Fig. 1.3):

$$R_{ji} = R_j R_i^{-1}, \ j \in \mathcal{N}_i \tag{7.31}$$

Let the estimated position be denoted by $\hat{p}_i$. Then, similarly to the previous section, the goal of formation control is to ensure $R_j(t)R_i^{-1}(t) \to \mathbb{I}_3$, $\hat{p}_j(t) - \hat{p}_i(t) \to p_j(t) - p_i(t)$, and $p_j(t) - p_i(t) \to p_j^*(t) - p_i^*(t)$ as $t \to \infty$.

For the orientation alignment of the agents, the following alignment law, which is adopted from [5, 7], is used

$$\omega_i = -k_o \sum_{j \in \mathcal{N}_i} \frac{1}{2} a_{ij} (R_j R_i^{-1} - R_i R_j^{-1})^\vee \tag{7.32}$$

where $k_o > 0$. In the above alignment law, it is required to have both $R_j R_i^{-1}$ and $R_i R_j^{-1}$ to update the angular velocity of agent i. For the agent i, it can measure $R_j R_i^{-1}$; but $R_i R_j^{-1}$ is measured by neighboring agent j. Thus, the information $R_i R_j^{-1}$ should be delivered from agent j to agent i. From this observation, it may be important to distinguish the sensing topology and communication topology. In the update (7.32), for an edge shared by two neighboring agents i and j, the sensing should be done mutually. So, it can be considered as of bidirectional, which means that the sensing topology is undirected. However, for the orientation alignment between two neighboring agents, it is not necessary to update the orientations of two agents in a mutual way. In fact, if a graph is balanced and strongly connected, then the orientation alignment can be achieved using an average consensus although the underlying graph is directed. Note if a leader is stationary and other agents need to align their directions to the orientation of the leader, then an arborescence tree is required for alignment as discussed in Sect. 7.1. In such a case, a strongly connected graph is not required.

Continuing from (7.32), for an instance, the sensing and communication topologies are depicted for a three-agent case in Fig. 7.2. The sensing graph $\mathcal{G} = (\mathcal{V}, \mathcal{E}^s)$ is undirected and the communication graph $\mathcal{G} = (\mathcal{V}, \mathcal{E}^c)$ is strongly connected. Assuming that the sensing topology is undirected, the information $R_j R_i^{-1} - R_i R_j^{-1}$ is available at the agent i. Thus, the alignment law (7.32) can be implemented in a distributed way. That is, with the orientation alignment law (7.32), the orientation transformation matrix R_i is updated as

$$\dot{R}_i = k_o \sum_{j \in \mathcal{N}_i} \frac{1}{2} a_{ij} (R_j R_i^{-1} - R_i R_j^{-1}) R_i \tag{7.33}$$

It is worth mentioning that that the orientation control law (7.32) is implemented via local coordinate frames of agents, and Eq. (7.33) is only used for analysis purpose. To examine the convergence of (7.33), the following Lyapunov candidate is utilized.

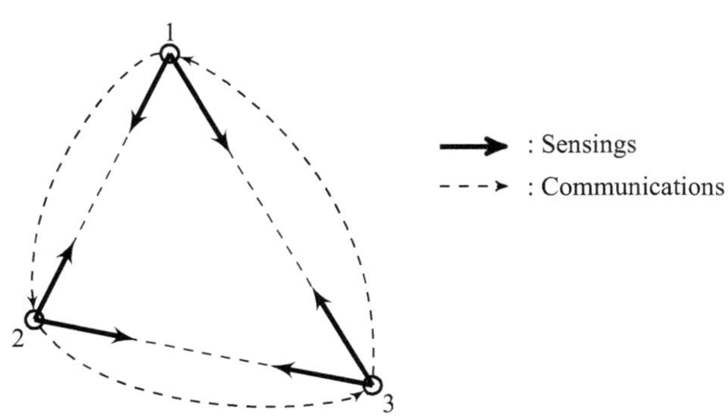

Fig. 7.2 An example for the sensing and communication topologies

$$U(R_1, \ldots, R_n) = 2\sum_{i=1}^{n}\sum_{j=1}^{n} \text{trace}(\mathbb{I}_3 - R_j R_i^{-1}) \tag{7.34}$$

From [7], the following lemma can be obtained:

Lemma 7.5 *Assume that the sensing graph $\mathcal{G} = (\mathcal{V}, \mathcal{E}^s)$ is connected and the communication graph $\mathcal{G} = (\mathcal{V}, \mathcal{E}^c)$ is strongly connected, and $R_j(t_0)R_i^{-1}(t_0)$, for all $(i, j)^e \in \mathcal{E}^s$, are positive definite. Then, there exist $k_U, \lambda_U > 0$ such that $U(t) \leq k_U U(t_0)e^{-\lambda_U(t-t_0)}$ for all $t \geq t_0$.*

The above lemma implies that there exists $R_\infty \in \mathbb{R}^{3\times 3}$ such that $R_i(t)$ asymptotically converges to R_∞, i.e.,

$$\lim_{t\to\infty} R_i(t) = R_\infty, i = 1, \ldots, n \tag{7.35}$$

It is possible to further elaborate to show an exponential convergence. To show this, it is helpful to use the concept of rotation vector in 3-dimensional space, i.e., an axis of rotation [10]. Given two rotation matrices $R_1(t)$ and $R_2(t)$, which provide the orientations of coordinate frames ${}^1\Sigma$ and ${}^2\Sigma$, respectively, there always exists a unit vector r_{12} defined with respect to ${}^g\Sigma$ that can relate the two rotation matrices. By rotating the axis around the direction r_{12} by an amount ϑ_{12}, the matrix $R_1(t)$ can become $R_2(t)$. So, there always exists a coordinate frame ${}^{ij}\Sigma$ on which an orientation transformation matrix $R_i(t)$ can be transformed to another orientation transformation matrix $R_j(t)$ by only one single rotation ϑ_{ij}.[2] With the notation $\omega_{ij} \triangleq \frac{1}{2}(R_j R_i^{-1} - R_i R_j^{-1})^\vee$, the following lemma can be developed for an exponential convergence between neighboring agents.

[2] The rotation is expressed by an exponential formula, which is also called Rodrigues' formula [10]. Let the rotation from ${}^1\Sigma$ to ${}^2\Sigma$ be computed as $R_2 R_1^{-1}$. Then, the Rodrigues' formula is given as

$$e^{(r_{12})^\wedge \vartheta_{12}} = \mathbb{I}_3 + (r_{12})^\wedge \sin\vartheta_{12} + [(r_{12})^\wedge]^2(1 - \sin\vartheta_{12})$$

which is equivalent to $R_{21} = R_2 R_1^{-1}$. Let the elements of R_{21} be given as

$$R_{21} = \begin{bmatrix} s_{11} & s_{12} & s_{13} \\ s_{21} & s_{22} & s_{23} \\ s_{31} & s_{32} & s_{33} \end{bmatrix}$$

Then, the angle ϑ_{12} and the unit vector r_{12} are computed as [10]:

$$\vartheta_{12} = \cos^{-1}\left(\frac{\text{trace}(R_{21}) - 1}{2}\right) \tag{7.36}$$

$$r_{12} = \frac{1}{2\sin\vartheta_{12}} \begin{bmatrix} s_{32} - s_{23} \\ s_{13} - s_{31} \\ s_{21} - s_{12} \end{bmatrix} \tag{7.37}$$

Lemma 7.6 *Let us consider two agents with orientation angles R_i and R_j. When their orientations are updated by ω_i in* (7.32)*, the orientations are aligned exponentially fast as $t \to \infty$.*

Proof Since the two orientations of two agents can be coincident by a single rotation around a unit axis, without loss of generality, let us suppose that two agents i and j are moving on arbitrary x-y plane. Then, ω_{ij} can be obtained as

$$\omega_{ij} = \begin{bmatrix} 0 \\ 0 \\ \sin(\vartheta_j - \vartheta_i) \end{bmatrix} \tag{7.38}$$

Due to the property of oscillators [2], the angles between two neighboring nodes, i.e., ϑ_j and ϑ_i for all $(i, j)^e \in \mathcal{E}^s$, are attracted. To be more specific, in (7.38), two points ϑ_i and ϑ_j can be considered as points on the surface of a sphere (the two points define a great circle on the sphere). Then, agent i is updated as $\dot{\vartheta}_i = \sin(\vartheta_j - \vartheta_i)$ and agent j is updated as $\dot{\vartheta}_j = \sin(\vartheta_i - \vartheta_j)$, when taking account of the interactions only between the neighboring agents i and j. Thus, due to the contraction property between two phases ϑ_i and ϑ_j, the spherical distance between the two points ϑ_i and ϑ_j is contracted exponentially fast.

With the above lemma, we can now develop the following theorem (note that the same conclusion can be found in [7]).

Theorem 7.2 *The convergence of $R_i(t)$ to R_∞ is exponentially fast, if $R_j(t_0)R_i^{-1}(t_0)$, for all $(i, j)^e \in \mathcal{E}^s$, are positive definite.*

Proof Let us first interpret the physical meaning of $\omega_{ij} \triangleq \frac{1}{2}(R_j R_i^{-1} - R_i R_j^{-1})^\vee$ in 3-dimensional space. For R_i, there exists an axis which can be rotated by a certain angle to transform ${}^g\Sigma$ to ${}^i\Sigma$ by a single rotation. Let us denote this angular rotation as ϕ_i. For R_j, there also exists an axis which can be rotated by a certain angle to transform ${}^g\Sigma$ to ${}^j\Sigma$ by a single rotation. Let us denote this angular rotation as ϕ_j. Thus, $\omega_{ij} \triangleq \frac{1}{2}(R_j R_i^{-1} - R_i R_j^{-1})^\vee$ can be considered to be achievable by two angular rotations ϕ_i and ϕ_j, which means that the relative orientation of two agents in 3-dimensional space is defined by two angular components. Therefore, the relative orientation of two agents are represented by the locations of two points on the sphere. Let us denote these two points as ϑ_i and ϑ_j on the sphere. The two points are also attracted due to Lemma 7.6. Let us select another neighboring node k of agent i. Then the two points corresponding to the relative orientation define another great circle on the same sphere. In this case, it is also true that the spherical distance between the two points ϑ_i and ϑ_k are also contracted exponentially fast by Lemma 7.6. Thus, extending to general n agents, we define a geodesic convex set $\mathcal{C}_i(t)$ as the area of surface of sphere that contains all the locations corresponding to the orientations of neighboring agents of agent i at $t \geq t_0$. Then, it is clear that $\mathcal{C}_i(t_2) \subseteq \mathcal{C}_i(t_1) \subseteq \mathcal{C}_i(t_0)$, $t_2 \geq t_1 \geq t_0$, which further implies $\sum_{i=1}^n \mathcal{C}_i(t_2) \subseteq \sum_{i=1}^n \mathcal{C}_i(t_1) \subseteq \sum_{i=1}^n \mathcal{C}_i(t_0)$, $t_2 \geq t_1 \geq t_0$. Hence, with Lemma 7.5, it is shown that $R_i(t) \to R_\infty$ exponentially as $t \to \infty$ if $R_j(t_0)R_i^{-1}(t_0)$, for all $(i, j)^e \in \mathcal{E}^s$, are positive definite.

The physical meaning of the positive definiteness of $R_j(t_0)R_i^{-1}(t_0)$ is quite intuitive. The following lemma provides a necessary and sufficient condition for this.

Lemma 7.7 *The matrix $R_j(t_0)R_i^{-1}(t_0)$ is positive definite if and only if the rotation angle ϑ_{ji} defined in (7.36) is less than $\frac{\pi}{2}$, i.e., $|\vartheta_{ji}| \leq \frac{\pi}{2}$.*

Proof Let the orientation transformation matrices R_i and R_j be expressed as $R_i^{ij\Sigma}$ and $R_j^{ij\Sigma}$ with respect to an arbitrarily coordinate frame ${}^{ij}\Sigma$. Then, we can use the fact that $R_i^{ij\Sigma}$ can be transformed to $R_j^{ij\Sigma}$ by a single rotation on the coordinate frame ${}^{ij}\Sigma$. It is possible to consider that the rotated angle of $R_i^{ij\Sigma}$ from ${}^{ij}\Sigma$ is $\vartheta_i^{ij\Sigma}$, and the rotated angle of $R_j^{ij\Sigma}$ from ${}^{ij}\Sigma$ is $\vartheta_j^{ij\Sigma}$. In this setup, it is possible to transform $R_i^{ij\Sigma}$ to $R_j^{ij\Sigma}$ by rotating $\vartheta_j^{ij\Sigma} - \vartheta_i^{ij\Sigma}$. Thus, without loss of generality, assuming that R_i and R_j, and ϑ_i and ϑ_j are expressed with respect to ${}^{ij}\Sigma$, $R_j(t_0)R_i^{-1}(t_0)$ can be simplified as

$$R_jR_i^{-1} = \begin{bmatrix} \cos(\vartheta_j - \vartheta_i) & -\sin(\vartheta_j - \vartheta_i) & 0 \\ \sin(\vartheta_j - \vartheta_i) & \cos(\vartheta_j - \vartheta_i) & 0 \\ 0 & 0 & 1 \end{bmatrix} \tag{7.39}$$

Thus, to be positive definite, it is necessary and sufficient to have $|\vartheta_j - \vartheta_i| < \frac{\pi}{2}$.

From the existence of R_∞, let an arbitrary coordinate frame be denoted as ${}^c\Sigma$ whose basis vectors are aligned to the axes of R_∞. For the position estimation of agents with respect to ${}^c\Sigma$, the following position estimation law is proposed

$$\dot{\hat{p}}_i^c = k_{\hat{p}} \sum_{j\in\mathcal{N}_i} a_{ij}[(\hat{p}_j^c - \hat{p}_i^c) - (p_j^i - p_i^i)] + u_i^i \tag{7.40}$$

The position control input u_i^i defined with respect to ${}^i\Sigma$ is designed as

$$u_i^i = k_p(p_i^* - \hat{p}_i^c) \tag{7.41}$$

The overall dynamics of the system is described in ${}^c\Sigma$. Consider the relative orientation of ${}^i\Sigma$ with respect to ${}^c\Sigma$, which can be written as $R_iR_\infty^{-1}$. Denoting $R_i^c \triangleq -R_iR_\infty^{-1}$, the following update rule can be obtained:

$$\begin{aligned} \dot{R}_i^c &= -\dot{R}_iR_\infty^{-1} \\ &= -k_o \sum_{j\in\mathcal{N}_i} \frac{1}{2}a_{ij}(R_jR_i^{-1} - R_iR_j^{-1})R_iR_\infty^{-1} \\ &= k_o \sum_{j\in\mathcal{N}_i} \frac{1}{2}a_{ij}(R_j^c(R_i^c)^{-1} - R_i^c(R_j^c)^{-1})R_i^c \end{aligned} \tag{7.42}$$

When comparing (7.42) with (7.33), we can see that they are equivalent. Thus, by Theorem 7.2, we can see that R_i^c, for all $i \in \mathcal{V}$, converges to a constant rotation matrix

R^c_∞. Using the relationships $u^c_i = R^i_c u^i_i$ and $p^c_j - p^c_i = R^i_c(p^i_j - p^i_i)$, it can be shown that

$$\begin{aligned}\dot{p}^c_i &= u^c_i \\ &= R^i_c k_p(p^*_i - \hat{p}^c_i) \qquad (7.43)\\ \dot{\hat{p}}^c_i &= k_{\hat{p}} \sum_{j\in\mathcal{N}_i} a_{ij}[(\hat{p}^c_j - \hat{p}^c_i) - (R^i_c)^{-1}(p^c_j - p^c_i)] + k_p(p^*_i - \hat{p}^c_i) \qquad (7.44)\end{aligned}$$

Letting $p^c = (p^{cT}_1, \ldots, p^{cT}_n)^T$, $\hat{p}^c = ((\hat{p}^c_1)^T, \ldots, (\hat{p}^c_n)^T)^T$, and $\Gamma = \text{blkdg}(R^1_c, \ldots, R^n_c)$, the overall dynamics can be expressed as

$$\begin{aligned}\dot{p}^c &= k_p(p^* - \hat{p}^c) - k_p[\mathbb{I}_{3n} - \Gamma](p^* - \hat{p}^c) \qquad (7.45)\\ \dot{\hat{p}}^c &= -k_{\hat{p}}(\mathbb{L}\otimes\mathbb{I}_3)(\hat{p}^c - p^c) + k_p(p^* - \hat{p}^c) - k_{\hat{p}}[\mathbb{I}_{3n} - \Gamma^{-1}](\mathbb{L}\otimes\mathbb{I}_3)p^c \qquad (7.46)\end{aligned}$$

where $\mathbb{L}$ is the Laplacian matrix corresponding to the communication topology $\mathcal{G} = (\mathcal{V}, \mathcal{E}^c)$. It is noticeable that a consensus is achieved if the graph is strongly connected although the consensus value is not an average of initial values [15]. As aforementioned, for an average consensus, it is required to have the balanced topology for the underlying graph. Now by defining error variables $e_{p^c} \triangleq p^* - p^c$ and $e_{\hat{p}^c} \triangleq p^c - \hat{p}^c$, the following equations can be obtained

$$\begin{aligned}\dot{e}_{p^c} &= -\dot{p}^c \\ &= -k_p(p^* - \hat{p}^c) + k_p[\mathbb{I}_{3n} - \Gamma](p^* - \hat{p}^c) \qquad (7.47)\\ \dot{e}_{\hat{p}^c} &= -k_{\hat{p}}(\mathbb{L}\otimes\mathbb{I}_3)e_{\hat{p}^c} - k_p[\mathbb{I}_{3n} - \Gamma](e_{p^c} + e_{\hat{p}^c}) \\ &\quad - k_{\hat{p}}[\mathbb{I}_{3n} - \Gamma^{-1}](\mathbb{L}\otimes\mathbb{I}_3)e_{p^c} + k_{\hat{p}}[\mathbb{I}_{3n} - \Gamma^{-1}](\mathbb{L}\otimes\mathbb{I}_3)p^* \qquad (7.48)\end{aligned}$$

With a new vector $e = ((e_{p^c})^T, (e_{\hat{p}^c})^T)^T$, the following error dynamics finally can be generated:

$$\dot{e} = Ae + \Delta A(\Gamma)e + D(\Gamma) \qquad (7.49)$$

where

$$\begin{aligned}A &\triangleq \begin{bmatrix} -k_p\mathbb{I}_{3n} & -k_p\mathbb{I}_{3n} \\ \mathbf{0}_{3n\times 3n} & -k_{\hat{p}}(\mathbb{L}\otimes\mathbb{I}_3)\end{bmatrix}\\ \Delta A(\Gamma) &\triangleq k_p\begin{bmatrix} (\mathbb{I}_{3n} - \Gamma) & \mathbf{0}_{3n\times 3n} \\ \mathbf{0}_{3n\times 3n} & (\mathbb{I}_{3n} - \Gamma)\end{bmatrix}\begin{bmatrix} \mathbb{I}_{3n} & \mathbb{I}_{3n} \\ -\mathbb{I}_{3n} & -\mathbb{I}_{3n}\end{bmatrix} \\ &\quad - \begin{bmatrix} (\mathbb{I}_{3n} - \Gamma^{-1}) & \mathbf{0}_{3n\times 3n} \\ \mathbf{0}_{3n\times 3n} & (\mathbb{I}_{3n} - \Gamma^{-1})\end{bmatrix}\begin{bmatrix} \mathbf{0}_{3n\times 3n} & \mathbf{0}_{3n\times 3n} \\ k_{\hat{p}}(\mathbb{L}\otimes\mathbb{I}_3) & \mathbf{0}_{3n\times 3n}\end{bmatrix}\\ D(\Gamma) &\triangleq \begin{bmatrix} (\mathbb{I}_{3n} - \Gamma^{-1}) & \mathbf{0}_{3n\times 3n} \\ \mathbf{0}_{3n\times 3n} & (\mathbb{I}_{3n} - \Gamma^{-1})\end{bmatrix}\begin{bmatrix} \mathbf{0}_{3n\times 3n} \\ k_{\hat{p}}(\mathbb{L}\otimes\mathbb{I}_3)p^*\end{bmatrix}\end{aligned}$$

Note that ω_i of (7.32) is expressed with respect to ${}^g\Sigma$, although $R_j R_i^{-1} - R_i R_j^{-1}$ can be available from local relative measurements. Since $R_i(t)$ is updated from (7.33) in agent i, using $R_i(t)$, it is possible to implement ω_i in agent i as follows:

$$\omega_i^i = k_o \sum_{j \in \mathcal{N}_i} \frac{1}{2} a_{ij} R_i (R_j R_i^{-1} - R_i R_j^{-1})^{\vee} \tag{7.50}$$

For the main results, the following lemmas are developed first:

Lemma 7.8 *For the orientation alignment law (7.33), if the sensing graph $\mathcal{G} = (\mathcal{V}, \mathcal{E}^s)$ has a spanning tree and the communication graph $\mathcal{G} = (\mathcal{V}, \mathcal{E}^c)$ is strongly connected, and $R_j(t_0) R_i^{-1}(t_0)$ for all $(i, j)^e \in \mathcal{E}^s$ are positive definite, then there exist $k_\omega, \lambda_\Gamma > 0$ such that*

$$\begin{aligned} \|\mathbb{I}_{3n} - \Gamma(t)\| &\leq k_\omega U(t_0) e^{-\lambda_\Gamma t} \\ \|\mathbb{I}_{3n} - \Gamma^{-1}(t)\| &\leq k_\omega U(t_0) e^{-\lambda_\Gamma t} \end{aligned} \tag{7.51}$$

for all $t \geq t_0$.

Proof From the definition of the Frobenius norm, it is true that

$$\begin{aligned} \|\mathbb{I}_3 - R_j R_i^{-1}\|_F &= \sqrt{\operatorname{tr}((\mathbb{I}_3 - R_j R_i^{-1})^T (\mathbb{I}_3 - R_j R_i^{-1}))} \\ &= \sqrt{\operatorname{tr}(\mathbb{I}_3 - R_j R_i^{-1} + \mathbb{I}_3 - R_i R_j^{-1})} \\ &= \sqrt{2\operatorname{tr}(\mathbb{I}_3 - R_j R_i^{-1})} \end{aligned}$$

which, with (7.34), leads to

$$U(R_1, \ldots, R_n) = \|\mathbb{I}_3 - R_j R_i^{-1}\|_F^2$$

From Theorem 7.2, it is clear that $\|\mathbb{I}_3 - R_j(t) R_i^{-1}(t)\|_F$ exponentially converges to zero as $t \to \infty$. Thus, using the fact that $\|\mathbb{I}_3 - R_j R_i^{-1}\| \leq \|\mathbb{I}_3 - R_j R_i^{-1}\|_F$, it is obvious that $\|\mathbb{I}_3 - R_j(t) R_i^{-1}(t)\|$ exponentially converges to zero as $t \to \infty$. Thus, there exist k_R and $\lambda_R > 0$ such that $\|\mathbb{I}_3 - R_i^{-1} R_\infty\| \leq k_R U(t_0) e^{-\lambda_R t}$ and $\|\mathbb{I}_3 - R_\infty R_i^{-1}\| \leq k_R U(t_0) e^{-\lambda_R t}$. Now, combining the above discussions with (7.42), we can obtain:

$$\begin{aligned} \|\mathbb{I}_3 - (R_i^c)^{-1} R_\infty^c\| &\leq k_R U(t_0) e^{-\lambda_R t} \\ \|\mathbb{I}_3 - R_\infty^c (R_i^c)^{-1}\| &\leq k_R U(t_0) e^{-\lambda_R t} \end{aligned}$$

Thus, the above inequalities imply the following inequalities:

$$\|\mathbb{I}_{3n} - \Gamma(t)\| \leq k_\omega U(t_0)e^{-\lambda_\Gamma t} \tag{7.52}$$

$$\|\mathbb{I}_{3n} - \Gamma^{-1}(t)\| \leq k_\omega U(t_0)e^{-\lambda_\Gamma t} \tag{7.53}$$

which completes the proof.

Lemma 7.9 *For the orientation alignment law (7.33), if the interaction graph $\mathcal{G}$ has a spanning tree and $R_j(t_0)R_i^{-1}(t_0)$ for all $(i, j)^e \in \mathcal{E}^s$ are positive definite, then $e = ((e_{p^c})^T, (e_{\hat{p}^c})^T)^T$ is bounded for all $t \geq t_0$.*

Proof See the appendix.

Theorem 7.3 *Suppose the orientation transformation matrix $R_i(t)$ is updated by (7.33) and the orientation alignment is conducted by (7.50) with position estimation law (7.40) and position control law (7.41). If the sensing graph $\mathcal{G} = (\mathcal{V}, \mathcal{E}^s)$ is connected and the communication graph $\mathcal{G} = (\mathcal{V}, \mathcal{E}^c)$ is strongly connected, and $R_j(t_0)R_i^{-1}(t_0)$ for all $(i, j)^e \in \mathcal{E}^s$ are positive definite, then there exists $e_{\hat{p}^c}^\infty \in \mathbb{R}^3$ such that e_{p^c} and $e_{\hat{p}^c}$ asymptotically converge to $\mathbf{1}_n \otimes e_{\hat{p}^c}^\infty$ as $t \to \infty$.*

Proof See the appendix.

If the orientations of agents are aligned, then similarly to Example 7.1, we can also design a position-estimation-based formation control under the displacement-based setup. All the equations in Example 7.1 are exactly applied to the case of 3-dimensional space.

7.3 Summary and Simulations

In Sect. 7.1, we have used distance-based setup in terms of relative sensings, but we controlled the estimated position directly; so in terms of actuation, it is a position-based control. So, we have combined the distance-based displacement control and displacement-based position control. In Sect. 7.2, it was extended to 3-D, assuming that the relative orientations, $R_jR_i^{-1}$, between neighboring agents are measured. It is important to distinguish $\theta_{ji} = \theta_j - \theta_i$ in 2-D and $R_{ji} = R_jR_i^{-1}$ in 3-D. In fact, in 2-D, it is intuitive and feasible to obtain θ_{ji} by exchanging δ_{ji} and δ_{ij}. But it is hard to suppose that the bearing measurements of neighboring agents in 3-D can be measured in local coordinate frames. Thus, the assumption that the relative orientations R_{ji} are available to individual agents is not so realistic. In the literature, several results are available on how to obtain R_{ji}. For example, in [4, 5], a visual motion observer was introduced to estimate the relative orientations. A vision camera attached to agent i recognizes feature points of orthogonal axes of coordinate frame of agent j. But, since the feature points should be attached to the rigid body axes, it may be difficult to apply the algorithm for agents that do not have any features. In [9], it is supposed that two neighboring agents commonly measure some features of environment; but the solution for a relative rotation matrix is not unique. In [14], a

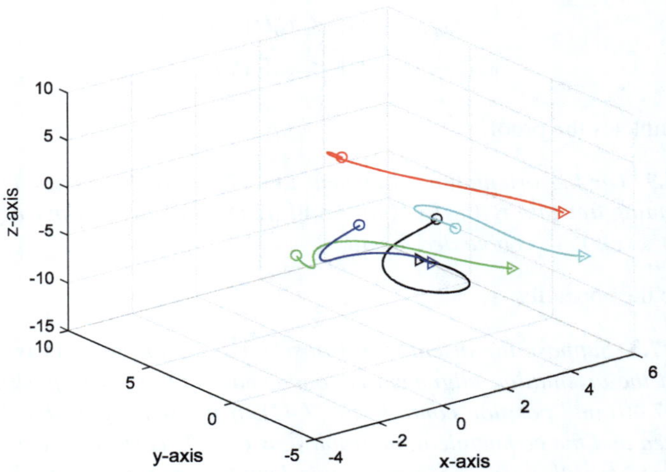

Fig. 7.3 Trajectories of positions of agents with the initial orientation condition satisfying the requirements

least squared optimization approach was proposed to obtain relative orientations via exploiting some kinematic relationships between neighboring agents. Although they have produced some necessary and sufficient conditions for an orientation localization of triangular sensing networks, a distributed solution was not provided. In [16], without using any features, but only using direction-only measurements, they have designed a distributed method for computing relative orientations between neighboring agents, assuming that they have a common neighboring agent. Additionally, they also presented a relative orientation localization algorithm as well as absolute orientation localization algorithm that can be performed in a distributed way, by way of Henneberg-style triangular extensions. Thus, it is recommendable to use the algorithms introduced in [16] for obtaining R_{ji} for use in distributed formation control.

We test the theoretical results of Sect. 7.2. Figures 7.3 and 7.4 show the simulation results of five agents under the same topology as Fig. 6.6. Let us select the initial orientation angles of agent as: $\phi_1(t_0) = 0$, $\phi_2(t_0) = \pi/4$, $\phi_3(t_0) = \pi/2$, $\phi_4(t_0) = 3\pi/4$, $\phi_5(t_0) = 0.95\pi$; $\theta_1(t_0) = -0.69\pi$, $\theta_2(t_0) = -3\pi/4$, $\theta_3(t_0) = -\pi/2$, $\theta_4(t_0) = -\pi/4$, $\theta_5(t_0) = -0.1\pi$; and $\psi_1(t_0)=\pi$, $\psi_2(t_0)=4\pi/5$, $\psi_3(t_0)=\pi/2$, $\psi_4(t_0) = \pi/3$, $\psi_5(t_0) = 0.04\pi$. These initial angles satisfy the initial requirement of Theorem 7.3. As shown in these figures, the positions and orientation converge to the desired configuration since all the required conditions are satisfied. Figures 7.5 and 7.6 are simulation results with initial conditions of $\phi_1(t_0) = \pi/2.4$, $\phi_2(t_0) = 1.5\pi$, $\phi_3(t_0) = -1.3\pi$, $\phi_4(t_0) = \pi/5$, $\phi_5(t_0) = -\pi/8$; $\theta_1(t_0) = \pi/1.9$, $\theta_2(t_0) = -\pi/3$, $\theta_3(t_0) = -\pi/3$, $\theta_4(t_0) = \pi/2.1$, $\theta_5(t_0) = 1.9\pi$; and $\psi_1(t_0) = 1.8\pi$, $\psi_2(t_0) = -1.2\pi$, $\psi_3(t_0) = 1.2\pi$, $\psi_4(t_0) = 0.4\pi$, $\psi_5(t_0) = -0.8\pi$. Note that these initial orientations do not satisfy the requirements of Theorem 7.3. However, although the required condition is not satisfied, the positions and orientations of agents con-

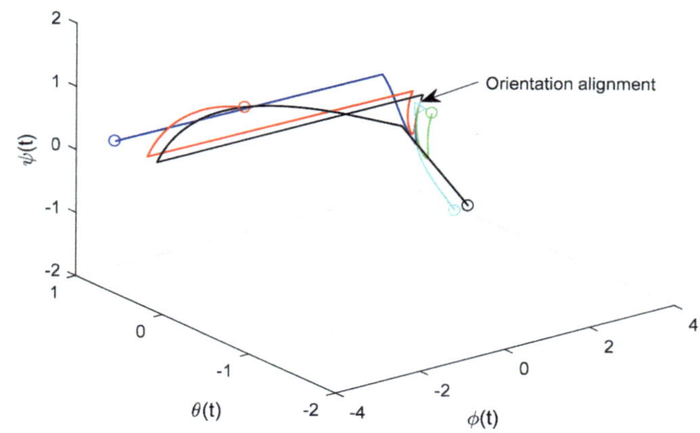

Fig. 7.4 Trajectories of orientations of agents with the initial orientation condition satisfying the requirements

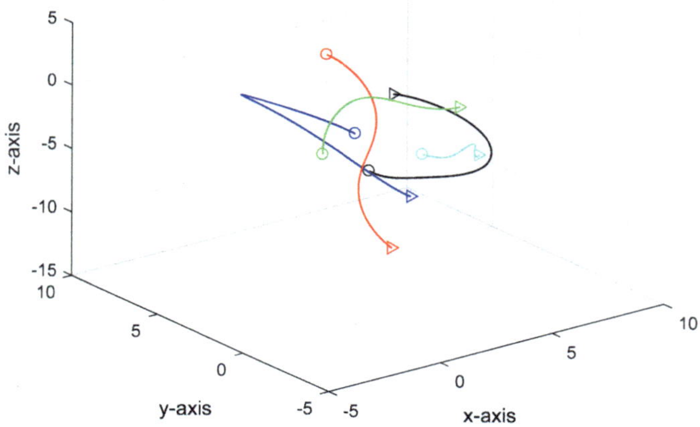

Fig. 7.5 Trajectories of positions of agents with the initial orientation condition that does not satisfy the requirements

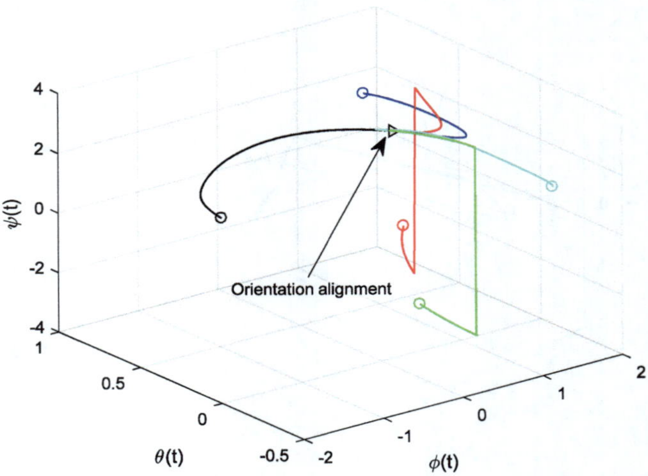

Fig. 7.6 Trajectories of orientations of agents with the initial orientation condition that does not satisfy the requirements

verge to the desired configuration. From a number of simulation tests, we have observed that although the condition $R_j(t_0)R_i^{-1}(t_0) > 0, \forall (i, j)^e \in \mathcal{E}^s$ is not satisfied, the agents easily converge to the desired configuration, although there are some cases that do not converge to the desired configuration. From the comparison of Figs. 7.3, 7.4 and 7.5, 7.6, however, we can observe that the case with the initial condition satisfying the requirements shows a quick convergence to the desired configuration and to the orientation synchronization. Thus, the results of this section can be considered also as a sufficient condition.

7.4 Notes

This chapter introduced a position estimation scheme on the base of local relative displacement measurements. It seems difficult to estimate the positions of agents in a distributed way when the measurements are pure distances, although a sequential approach [3] or centralized approaches for globally rigid graphs [1] look feasible. For more technical results on network localization, it is recommended to refer to Chap. 13. The formation control problem formulated in this chapter is conceptually analogous to output feedback control of linear systems. In output feedback control, the states of the system are estimated from the outputs; then the estimated states are used for a feedback control. In the formation control via position estimation, the positions of agents are estimated from relative local information; then, the estimated positions are directly controlled. From this observation, we can see that there is a duality between observability of linear systems and localizability of multi-agent systems.

The results of Sect. 7.1 are reused from [13] and the results of Sect. 7.2 are reused from [12]. The following copyright and permission notices are acknowledged.

- The statements of Lemma 7.6, Theorem 7.2, Lemmas 7.8, 7.9, and Theorem 7.3 with the proofs are © [2014] IEEE. Reprinted, with permission, from [12].
- The text of Sect. 7.1 is © Reprinted by permission from Springer-Nature: [13] (2018).

References

1. Aspnes, J., Eren, T., Goldenberg, D.K., Morse, A.S., Whiteley, W., Yang, Y.R., Anderson, B.D.O., Belhumeur, P.N.: A theory of network localization. IEEE Trans. Mobile Comput. **5**(12), 1663–1678 (2006)
2. Dörfler, F., Bullo, F.: Synchronization in complex networks of phase oscillators: a survey. Automatica **50**(6), 1539–1564 (2014)
3. Fang, J., Cao, M., Morse, A.S., Anderson, B.D.O.: Sequential localization of sensor networks. SIAM J. Control Optim. **48**(1), 321–350 (2009)
4. Fujita, M., Hatanaka, T., Kobayashi, N., Ibuki, T., Spong, M.W.: Visual motion observer-based pose synchronization: A passivity approach. In: Proceedings of the Joint 48th IEEE Conference on Decision and Control and 28th Chinese Control Conference, pp. 2402–2407 (2009)
5. Hatanaka, T., Igarashi, Y., Fujita, M., Spong, M.W.: Passivity-based pose synchronization in three dimensions. IEEE Trans. Autom. Control **57**(2), 360–375 (2012)
6. Hughes, P.C.: Spacecraft Attitude Dynamics. Dover Publications, INC., Mineola (1986)
7. Igarashi, Y., Hatanaka, T., Fujita, M., Spong, M.W.: Passivity-based attitude synchronization in SE(3). IEEE Trans. Control Syst. Technol. **17**(5), 1119–1134 (2009)
8. Khalil, H.K.: Nonlinear Systems. Prentice Hall, Upper Saddle River (2002)
9. Montijano, E., Cristofalo, E., Zhou, D., Schwager, M., Sagüés: Vision-based distributed formation control without an external positioning system. IEEE Trans. Robot. **32**(2), 339–351 (2016)
10. Murray, R.M., Li, Z., Sastry, S.S.: A Mathematical Introduction to Robotic Manipulation. CRC Press, Boca Raton (1994)
11. Oh, K.-K., Ahn, H.-S.: Formation control of mobile agents based on distributed position estimation. IEEE Trans. Autom. Control **58**(3), 737–742 (2013)
12. Oh, K.-K., Ahn, H.-S.: Formation control of rigid bodies based on orientation alignment and position estimation. In: Proceedings of 14th International Conference on Control, Automation and Systems, pp. 740–744 (2014)
13. Oh, K.-K., Ahn, H.-S.: Distributed formation control based on orientation alignment and position estimation. Int. J. Control Autom. Syst. **16**(3), 1112–1119 (2018)
14. Piovan, G., Shames, I., Fidan, B., Bullo, F., Anderson, B.D.O.: On frame and orientation localization for relative sensing networks. Automatica **49**, 206–213 (2013)
15. Saber, R.O., Murray, R.M.: Agreement problems in networks with directed graphs and switching topology. Technical Report, CIT-CDS 03005, Control and Dynamical Systems, California Institute of Technology (2003)
16. Tran, Q.V., Ahn, H.-S., Anderson, B.D.O.: Distributed orientation localization of multi-agent systems in 3-dimensional space with direction-only measurements. In: Proceedings of the 57th IEEE Conference on Decision and Control, pp. 2883–2889 (2018)

Chapter 8
Formation Control via Global Orientation Alignment

Abstract In Chap. 6, formation controls via orientation alignment, including orientation estimation and orientation control, were presented. However, the algorithms are valid only for a quasi-global convergence when the initial orientation angles are restricted to the condition $\max_{i\in\mathcal{V}} \theta_i(t_0) - \min_{i\in\mathcal{V}} \theta_i(t_0) < \pi$. In this chapter, we would like to remove this restriction such that the orientation alignment could be achieved for almost all initial conditions. It will be shown that the orientation alignment could be done for almost any initial orientation angles with more information exchanges between neighboring nodes, and with more computational load in each node. Thus, there is a cost in implementing a global orientation alignment algorithm. But, since all the measurements are relative and information exchanges take place between neighboring agents, it is still a distributed control law. For a global convergence, we utilize virtual variables that transform a non-convex circle or a sphere to the linear Euclidean space. The global orientation alignment problem is defined in a non-convex circle or a sphere; for the convergence analysis, we conduct analysis in the Euclidean space by way of using virtual variables. Then, after analyzing and designing the control law in the Euclidean space, we again transform the control law into the circle or the spherical space.

8.1 Formation Control in 2-Dimensional Space

The global orientation alignment scheme in 2-dimensional space can be achieved by way of exchanging a virtual variable between neighboring agents. This section introduces a formation control scheme via global orientation alignment developed in [6]. Consider n single-integrator agents in 2-D. Assume that $\theta_{ji} \triangleq \text{PV}(\theta_j - \theta_i)$ is available to agent i. Note that $\text{PV}(\theta_j - \theta_i)$ can be obtained as $\text{PV}(\theta_j - \theta_i) = \text{PV}(\delta_{ji} - \delta_{ij} + \pi)$ (see (6.4)). The goal of the global orientation estimation is to estimate $\theta_i,\ i \in \mathcal{V}$ from θ_{ji} for any $\theta_i(t_0)$ in a distributed way. Here, by *for any* $\theta_i(t_0)$, it means that we would like to estimate the angles without the condition $\max_{i\in\mathcal{V}} \theta_i(t_0) - \min_{i\in\mathcal{V}} \theta_i(t_0) < \pi$. The main idea is to map θ_i that is defined on a circle space $\mathbb{S}$ to a point in complex space $\mathbb{C} \setminus \{0\}$, i.e., $\mathbb{T}(\theta_i) \mapsto z_i$, where $\theta_i \in \mathbb{S}, z_i \in \mathbb{C} \setminus \{0\}$, and $\mathbb{T}$ is one-to-one and onto mapping from $\mathbb{S}$ to $\mathbb{C} \setminus \{0\}$. Then, we attempt to

H.-S. Ahn, *Formation Control*, Studies in Systems, Decision and Control 205, https://doi.org/10.1007/978-3-030-15187-4_8

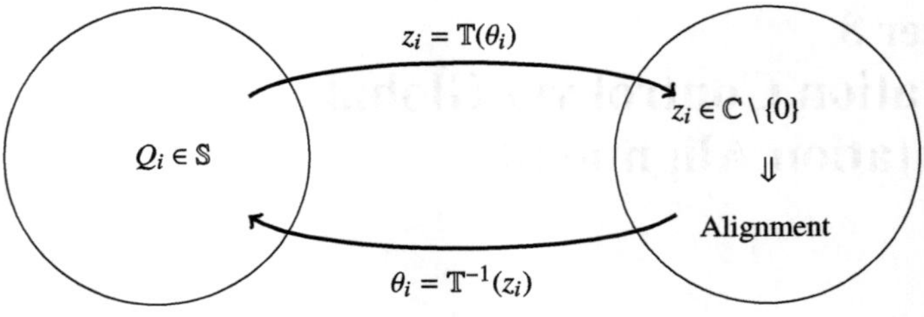

Fig. 8.1 The main idea of global orientation alignment

achieve an alignment in the complex space, and then we map the auxiliary alignment variables back to the circle space, i.e., $\mathbb{T}(z_i)^{-1} \mapsto \theta_i$. Figure 8.1 describes the idea of the global orientation alignment procedure. The idea described in this figure can be generalized to general d-dimensional space (See Sect. 8.2 for 3-dimensional case).

Let us map a true orientation angle $\theta_i \in \mathbb{S}$ to a complex number z_i^* as $z_i^* = \mathbb{T}(\theta_i) \triangleq e^{\mathrm{i}\theta_i}$, where the text i denotes the imaginary number. Let $z^* = (z_1^*, \ldots, z_n^*)^T$. If the point $z_i^* \in \mathbb{C}$ corresponding to the true orientation θ_i is exactly estimated, then we can obtain the true θ_i by solving $z_i^* = \cos(\theta_i) + \mathrm{i}\sin(\theta_i)$. Let the estimated complex number be denoted as $\hat{z}_i^* \in \mathbb{C}$. Then, the objective of the distributed orientation estimation problem is to ensure $\angle \hat{z}_i^* \to \angle z_i^* + \angle \alpha$ as $t \to \infty$ where α is a constant complex number, which is a common offset. The following estimation law is proposed:

$$\dot{\hat{z}}_i(t) = \sum_{j \in \mathcal{N}_i} (e^{-\mathrm{i}\theta_{ji}} \hat{z}_j(t) - \hat{z}_i(t)), \forall i \in \mathcal{V} \tag{8.1}$$

For the above update, the neighboring agents i and j need to exchange the estimated $\hat{z}_i(t)$ and the bearing angles δ_{ji}. Thus, similar to the discussions given in Sect. 6.4, the sensing topology needs to be bidirectional in order to have θ_{ji}, while the update (8.1) could be of directional as far as an alignment would be achieved, which means that the communication topology could be of directed; i.e., strongly connected or a graph with an arborescence.

Table 8.1 shows the variables and topologies required in formation control via global orientation alignment. Comparing to Table 6.2, we can see that the variable $\hat{\theta}_i$ in communications and computation is replaced by the virtual variable $\hat{z}_i(t)$. The Eq. (8.1) can be combined into a concise form as

$$\dot{\hat{z}}(t) = \hat{\mathbb{L}} \hat{z}(t) \tag{8.2}$$

where $\hat{\mathbb{L}}$, which is a kind of complex Laplacian matrix, is defined as

Table 8.1 Variables and network properties of formation via global orientation alignment

	Variables	Topology	Edge direction
Sensing	p_{ji}^i	Connected tree	Undirected
Control	p_{ji}^i, θ_i	Spanning tree (strongly connected)	Undirected (Directed)
Communications	$\delta_{ij}, \hat{z}_i$	Spanning tree (arborescence)	Undirected (Directed)
Computation	$u_i^i, \dot{\hat{z}}_i$	None	N/A
Initialization	$\hat{z}_i(t_0)$	None	N/A

$$[\hat{\mathbb{L}}]_{ij} = \begin{cases} -|\mathcal{N}_i|, & i = j \\ e^{-\mathrm{i}\theta_{ji}}, & j \in \mathcal{N}_i \\ 0, & j \notin \mathcal{N}_i \end{cases} \tag{8.3}$$

Then, the algebraic property of (8.2) can be checked as follows:

Lemma 8.1 *Let the sensing graph have a spanning tree, i.e., $\mathcal{G} = (\mathcal{V}, \mathcal{E}^s)$ is connected, and the communication topology $\mathcal{G} = (\mathcal{V}, \mathcal{E}^c)$ is directed and strongly connected or it has an arborescence. Then, for $\hat{\mathbb{L}}$ in (8.2), zero is a simple eigenvalue with the corresponding eigenvector z^*. Also, every eigenvalue except for zero eigenvalue has strictly negative real part.*

Proof With a diagonal matrix $D_z = \text{diag}[z_i^*]$, which is nonsingular, the matrix $\hat{\mathbb{L}}$ can be similarly transformed as $\bar{\mathbb{L}} = D_z^{-1}\hat{\mathbb{L}}D_z$. The entries of $\bar{\mathbb{L}}$ are represented such as

$$[\bar{\mathbb{L}}]_{ij} = [D_z^{-1}\hat{\mathbb{L}}D_z]_{ij} = [D_z^{-1}]_{ii}[\hat{\mathbb{L}}]_{ij}[D_z]_{jj} \tag{8.4}$$

So, the diagonal entries of $\bar{\mathbb{L}}$ are same as the ones of $\hat{\mathbb{L}}$, while off-diagonal entries are computed as $[\bar{\mathbb{L}}]_{ij} = e^{-\mathrm{i}\theta_i} e^{-\mathrm{i}\theta_{ji}} e^{\mathrm{i}\theta_j} = 1$, which means that the matrix $\bar{\mathbb{L}}$ can be considered as a Laplacian matrix with weights of 1. Thus, all eigenvalues of $\bar{\mathbb{L}}$ have strictly negative real parts except for a simple zero eigenvalue with the corresponding eigenvector $\xi = (1, 1, \ldots, 1)^T$. Hence from the linear transformation $\bar{\mathbb{L}}\xi = D_z^{-1}\hat{\mathbb{L}}D_z\xi = D_z^{-1}\hat{\mathbb{L}}z^* = 0$, we can see that cz^*, where c is a constant, is the eigenvector corresponding to the zero eigenvalue of $\hat{\mathbb{L}}$.

Example 8.1 (*Eigenvalues of Laplacian* $\hat{\mathbb{L}}$) Consider three agents with orientation angles $\theta_1 = \frac{3}{2}\pi$, $\theta_2 = \frac{2}{3}\pi$, and $\theta_3 = \frac{7}{6}\pi$. Then, we have $\theta_{21} = -\theta_{12} = -\frac{5}{6}\pi$, $\theta_{31} = -\theta_{13} = -\frac{1}{3}\pi$, and $\theta_{32} = -\theta_{23} = \frac{1}{2}\pi$. Then, we calculate the Laplacian matrix $\hat{\mathbb{L}}$ as

$$\hat{\mathbb{L}} = \begin{bmatrix} -2 & \cos\frac{5}{6}\pi + \mathrm{i}\sin\frac{5}{6}\pi & \cos\frac{1}{3}\pi + \mathrm{i}\sin\frac{1}{3}\pi \\ \cos\frac{5}{6}\pi - \mathrm{i}\sin\frac{5}{6}\pi & -2 & \cos\frac{1}{2}\pi - \mathrm{i}\sin\frac{1}{2}\pi \\ \cos\frac{1}{3}\pi - \mathrm{i}\sin\frac{1}{3}\pi & \cos\frac{1}{2}\pi + \mathrm{i}\sin\frac{1}{2}\pi & -2 \end{bmatrix} \tag{8.5}$$

From the above Laplacian, we compute the eigenvalues and the corresponding eigenvectors as $0, -3, -3$, and $(0.2887 + \mathrm{i}0.5, -\mathrm{i}0.5774, 0.5774)^T$, $(-0.6908 - \mathrm{i}0.2894, -0.4535 - \mathrm{i}0.4650, 0.131)^T$, $(-0.0945 - \mathrm{i}0.3112, 0.0737 + \mathrm{i}0.4892, 0.8059)^T$. Thus, from $(0.2887 + \mathrm{i}0.5, -\mathrm{i}0.5774, 0.5774)^T$, we compute $\theta_1 = \frac{1}{3}\pi$, $\theta_2 = -\frac{1}{2}\pi$, and $\theta_3 = 0$, which implies that the eigenvector corresponding to the zero eigenvalue of $\hat{\mathbb{L}}$ represents the true orientation angles of agents up to a common rotation. The diagonal matrix D_z is computed as

$$D_z = \begin{bmatrix} \cos(\frac{3}{2}\pi) + \mathrm{i}\sin(\frac{3}{2}\pi) & 0 & 0 \\ 0 & \cos(\frac{2}{3}\pi) + \mathrm{i}\sin(\frac{2}{3}\pi) & 0 \\ 0 & 0 & \cos(\frac{7}{6}\pi) + \mathrm{i}\sin(\frac{7}{6}\pi) \end{bmatrix} \quad (8.6)$$

From $D_z^{-1}\hat{\mathbb{L}}D_z$, we compute

$$\bar{\mathbb{L}} = \begin{bmatrix} -2 & 1 & 1 \\ 1 & -2 & 1 \\ 1 & 1 & -2 \end{bmatrix} \quad (8.7)$$

which is the typical Laplacian of weighting 1 for each edges.

The results of Lemma 8.1 imply that the equilibrium set of (8.2) is $\mathcal{E} \triangleq \{\mathbf{q} : \mathbf{q} \in \{\mathbb{C}^n \cap \mathrm{span}\{z^*\}\}\}$. It is not desirable to have $\hat{z}$ to converge to the origin. Therefore, the desired equilibrium set is defined as $\mathcal{E}' \triangleq \mathcal{E} \setminus \{0\}$. For a square matrix A, define the column space of matrix A as $C(A)$. Then, it is true that $C(A) = \mathit{null}(A^T)^{\perp}$ where $\mathit{null}(A^T)^{\perp}$ denotes the orthogonal space of null space of A^T. Now, the convergence of (8.2) is analyzed as follows:

Theorem 8.1 *For the dynamics (8.2), there exists a finite point $\hat{z}_\infty \in \mathcal{E}$ such that $\hat{z}(t)$ globally exponentially converges if and only if $\mathcal{G}$ has a spanning tree. Further $\hat{z}_\infty \in \mathcal{E}'$ if and only if an initial value $\hat{z}(t_0) \notin C(\hat{\mathbb{L}})$.*

Proof Using the coordinate transformation $\hat{z}(t) = D_z v(t)$, (8.2) can be written as $\dot{v}(t) = D_z^{-1}\hat{\mathbb{L}}D_z v(t) = \bar{\mathbb{L}}v(t)$. Since $\bar{\mathbb{L}}$ is a Laplacian matrix, there exists an equilibrium set $\mathcal{E}_v \triangleq \{\xi_i \in \mathbb{C} : \xi_1 = \cdots = \xi_n\}$ such that $v(t)$ globally exponentially converges. Then there exist a finite point $v_\infty \in \mathcal{E}_v$ and constants $k_v, \lambda_v > 0$ such that

$$\|v(t) - v_\infty\| \le k_v e^{-\lambda_v(t-t_0)}\|v(t_0) - v_\infty\| \quad (8.8)$$

By using the coordinate transformation, (8.8) can be written as

$$\|D_z^{-1}(\hat{z}(t) - \hat{z}_\infty)\| \le k_v e^{-\lambda_v(t-t_0)}\|D_z^{-1}(\hat{z}(t_0) - \hat{z}_\infty)\| \quad (8.9)$$

where $\hat{z}_\infty = D_z v_\infty$. Using the fact that $\|D_z^{-1}(\hat{z}(t) - \hat{z}_\infty)\|^2 = (D_z^{-1}(\hat{z}(t) - \hat{z}_\infty))^T (D_z^{-1}(\hat{z}(t) - \hat{z}_\infty)) = \|(\hat{z}(t) - \hat{z}_\infty)\|^2$, (8.8) can be further rewritten as

$$\|(\hat{z}(t) - \hat{z}_\infty)\| \leq k_v e^{-\lambda_v(t-t_0)} \|(\hat{z}(t_0) - \hat{z}_\infty)\| \tag{8.10}$$

The solution of (8.2), i.e., $\hat{z}(t)$, is given as $\hat{z}(t) = e^{\hat{\mathbb{L}}(t-t_0)}\hat{z}(t_0)$. Let the right and left eigenvectors corresponding to zero eigenvalue be denoted as p_n and w_n, respectively. Then, since all eigenvalues except for zero eigenvalue are negative real, the steady-state solution of (8.2) can be obtained as

$$\lim_{t\to\infty} \hat{z}(t) = p_n w_n \hat{z}(t_0) \tag{8.11}$$

This shows that $\hat{z}(t)$ converges to the origin if and only if $\hat{z}(t_0)$ is perpendicular to w_n, which is the left eigenvector of the zero eigenvalue. Thus, it follows that $w_n\hat{z}(t_0) = 0$ if and only if $\hat{z}(t_0)$ is in $null(\hat{\mathbb{L}}^T)^\perp = C(\hat{\mathbb{L}})$, which completes the proof.

From (8.10), it is clear that $\hat{z}_i$ converges to $\hat{z}^i_\infty$. Let the angle of agent i be denoted as $\theta_i = \angle z_i^*$, and the angle of agent i with a constant offset $\alpha \in \mathbb{C}$ be denoted as $\theta_i^* = \theta_i + \angle\alpha$. Also, let the estimated angle be denoted as $\hat{\theta}_i$ (i.e., $\hat{\theta}_i = \angle\hat{z}_i$). Let us denote the error in the estimated angle as $e_{\theta_i}(t) \triangleq \hat{\theta}_i(t) - \theta_i^*$. Then, from (8.10), there exist constants $k_\theta, \lambda_\theta > 0$ such that

$$\|e_\theta(t)\| \leq k_\theta e^{-\lambda_\theta(t-t_0)} \|e_\theta(t_0)\| \tag{8.12}$$

where $e_\theta(t) = (e_{\theta_1}(t), \ldots, e_{\theta_n}(t))^T$. The following control law is used:

$$u_i^i(t) = k_u \sum_{j\in\mathcal{N}_i} a_{ij}((p_j^i - p_i^i) - (R^i)^{-1}(\hat{\theta}_i(t))(p_j^* - p_i^*)) \tag{8.13}$$

where $R^i(\hat{\theta}_i(t))$ is the rotation matrix with the rotation angle $\hat{\theta}_i(t)$ from ${}^i\Sigma$ to ${}^g\Sigma$, and $(R^i)^{-1}(\hat{\theta}_i(t))$ can be simply written as $R_i(\hat{\theta}_i(t))$. The control law uses the information of p_j^i and $p_j^* - p_i^*$; thus the control topology (actuation topology) is determined by the given desired relative positions. In the given desired position, if j is a neighbor of agent i, and agent i is a neighbor of agent j, then the actuation topology will be bidirectional. Otherwise, it could be directed one as far as the underlying actuation topology ensures a convergence to a consensus manifold. Thus, for (8.2) and (8.13), we can also provide exactly the same comments as given in Remark 6.1.

The dynamics of agents with respect to a common reference frame ${}^a\Sigma$ can be expressed as

$$\begin{aligned}
\dot{p}_i^a &= u_i^a \\
&= R^i(\theta_i^*)u_i^i(t) \\
&= k_u \sum_{j\in\mathcal{N}_i} a_{ij}((p_j^* - p_i^*) - R^i(\theta_i^* - \hat{\theta}_i(t))(p_j^* - p_i^*)) \\
&= k_u \sum_{j\in\mathcal{N}_i} a_{ij}((p_j^* - p_i^*) - R_i(e_{\theta_i})(p_j^* - p_i^*))
\end{aligned} \tag{8.14}$$

where u_i^a denotes the control input of agent i with respect to a virtual coordinate frame ${}^a\Sigma$. Defining the error of position as $e_{p_i}^a(t) \triangleq p_i^a(t) - p_i^*$, the error dynamics can be obtained as follows:

$$\dot{e}_{p_i}^a(t) = k_u \sum_{j\in\mathcal{N}_i} a_{ij}((e_{p_j}^a(t) - e_{p_i}^a(t)) + (\mathbb{I}_2 - R_i(e_{\theta_i}))(p_j^* - p_i^*)) \tag{8.15}$$

Denoting $w_i(t) \triangleq k_u \sum_{j\in\mathcal{N}_i} a_{ij}((\mathbb{I}_2 - R_i(e_{\theta_i}))(p_j^* - p_i^*))$, the above equation can be rewritten as

$$\dot{e}_p^a(t) = -k_u(\mathbb{L} \otimes \mathbb{I}_2)e_p^a(t) + w(t) \tag{8.16}$$

Based on the definition of $w(t)$, and using the fact that $\|\mathbb{I}_2 - R_i(e_\theta)\| \leq \sqrt{2}\|e_\theta(t)\|$, there exist $k_w, k_{e_\theta}, \lambda_{e_\theta} > 0$ such that

$$\begin{aligned}
\|w(t)\| &\leq \sum_{i\in\mathcal{V}} \|w_i(t)\| \\
&\leq k_u \sum_{i\in\mathcal{V}} \sum_{j\in\mathcal{N}_i} a_{ij}\|\mathbb{I}_2 - R_i(e_{\theta_i})\|\|p_j^* - p_i^*\| \\
&\leq \sqrt{2}k_u \sum_{i\in\mathcal{V}} \sum_{j\in\mathcal{N}_i} a_{ij}\|e_\theta(t)\|\|p_j^* - p_i^*\| \\
&\leq k_w\|e_\theta(t)\| \leq k_w k_{e_\theta} e^{-\lambda_{e_\theta} t}\|e_\theta(t_0)\|
\end{aligned} \tag{8.17}$$

From (8.16), the solution of $e_p^a(t)$ is given as

$$e_p^a(t) = \phi(t, t_0)e_p^a(t_0) + \int_{t_0}^{t} \phi(t, \tau)w(\tau)d\tau \tag{8.18}$$

where $\phi(t, t_0) = e^{-k_u(\mathbb{L}\otimes\mathbb{I}_2)(t-t_0)}$.

Theorem 8.2 *Under the distributed estimation law (8.1) and control law (8.13), if the underlying graph $\mathcal{G} = (\mathcal{V}, \mathcal{E}^s \cup \mathcal{E}^c \cup \mathcal{E}^a)$ has a spanning tree and $\hat{z}(t_0) \notin C(\hat{\mathbb{L}})$, then the desired formation is globally exponentially achieved up to translation and rotation, i.e., $p_i(t) \to p_i^* + p_\infty$, where $p_\infty \in \mathbb{R}^2$ is a constant.*

Proof The proof can be completed by following the proof of Theorem 6.1.

Example 8.2 (*Trivial consensus*) Let us consider the same case as Example 8.1 with the Laplacian matrix (8.5). To achieve the desired convergence, the initial condition needs to satisfy $\hat{z}(t_0) \notin C(\hat{\mathbb{L}})$. Thus, it is required to have $\hat{z}(t_0) \notin \text{span}\{v_1, v_2, v_3\}$, where $v_1 = (-2, -\frac{\sqrt{3}}{2} - \mathrm{i}\frac{1}{2}, \frac{1}{2} - \mathrm{i}\frac{\sqrt{3}}{2})^T$, $v_2 = (-\frac{\sqrt{3}}{2} + \mathrm{i}\frac{1}{2}, -2, \mathrm{i})^T$, and $v_3 = (\frac{1}{2} + \mathrm{i}\frac{\sqrt{3}}{2}\pi, -\mathrm{i}, -2)^T$. It means that, if $\hat{z}(t_0) = c_1v_1 + c_2v_2 + c_3v_3$, where c_1, c_2, and c_3 are constants, a consensus to the trivial state will be achieved.

8.2 Formation Control in 3-Dimensional Space

The results of the previous section can be extended to 3-dimensional space by using the rotation matrix that can be parameterized with two virtual variables. This section introduces these extensions based on the results developed in [4, 8, 9]. Let R_i denote the orientation transformation from ${}^g\Sigma$ to ${}^i\Sigma$, from which it can be defined as $p_j - p_i = R_i^{-1}(p_j^i - p_i^i)$. The goal of formation control is to ensure $p_j^a - p_i^a \rightarrow p_j^* - p_i^*$ as $t \rightarrow \infty$ for all $(i, j)^e \in \mathcal{E}$ for an arbitrary coordinate frame ${}^a\Sigma$. Since the desired displacements in the formation control are specified by desired relative displacements, we ignore an ambiguity by a common rotation. Thus, the formation goal needs to be achieved up to an arbitrary coordinate frame ${}^a\Sigma$. The orientation in the d-dimensional space is defined by an $d \times d$ proper orthogonal matrix R which contains, as column vectors, the unit vectors r_i identifying the directions of axes of ${}^i\Sigma$ with respect to ${}^g\Sigma$. The rotation matrix R, which is also called orientation transformation matrix, is an element of special orthogonal group as $R \in \mathrm{SO}(d)$. Thus, the orientation transformation matrix R in a general d-dimensional space has the properties such as $R^T R = \mathbb{I}_d$ and $det(R) = +1$.

When the global reference frame ${}^g\Sigma$ is the Cartesian coordinate frame with standard basis e_i, the directions of axes of local frame are represented as $r_i = Re_i$. Let R_{ji} denote the relative orientation of jth local coordinate frame with respect to the ith local coordinate frame. Thus, we assume that R_{ji} is measured by agent i, while R_{ij} is measured by agent j. It is well known that R_{ji} can be computed as

$$R_{ji} = R_j R_i^{-1} = R_j R_i^T \tag{8.19}$$

The relative orientation is also called relative point on SO(3), which is formally defined as follows:

Definition 8.1 The *left-invariant relative point* on SO(3) of agent j with respect to agent i is $R_{ij}^l = R_i^{-1} R_j$, and the *right-invariant relative point* on SO(3) of j with respect to i is $R_{ji}^r = R_j R_i^{-1}$.

Note that the left- and right-invariant relative points satisfy $R_{ij}^l = R_{ji}^r = \mathbb{I}_3$, $\forall i, j \in \mathcal{V}$ if and only if a consensus on SO(3) is achieved.

The objective of the orientation estimation problem is to get $\hat{R}_i \in \mathrm{SO}(d)$ from the measurements R_{ji} such that $R_i(t)^T \hat{R}_i \rightarrow R^c$ as $t \rightarrow \infty$ for an arbitrary common orientation transformation matrix $R^c \in \mathrm{SO}(d)$. Given a R_i, let us denote $B_i \in \mathrm{SO}(d)$ as

$$B_i \triangleq R_i X,\ X \in \mathrm{SO}(d), \forall i \in \mathcal{V} \tag{8.20}$$

Then, by (8.19), it follows that

$$B_j = R_{ji} B_i, \forall (i, j)^e \in \mathcal{E} \tag{8.21}$$

Thus, finding the steady-state value of $\hat{R}_i$ is equivalent to find B_i satisfying the equality (8.21). It is supposed that $d-1$ auxiliary variables are used to define the orientation of each agent, and the auxiliary variables of the ith agent are denoted as $z_{i,k} \in \mathbb{R}^d,\ k \in \{1,\ldots,d-1\}$. Let B_i be represented as $B_i = [b_{i,1}, b_{i,2}, \ldots, b_{i,d}]$ where $b_{i,k} \in \mathbb{R}^d,\ k \in \{1,\ldots,d\}$ is the kth column vector of B_i. To generate orthonormal column vectors of B_i from $z_{i,k} \in \mathbb{R}^d,\ k \in \{1,\ldots,d-1\}$, the well-known Gram–Schmidt process [10] can be utilized. The Gram–Schmidt process is outlined in Algorithm 1, which generates the orthonormal vectors $b_{1,1},\ldots,b_{i,d-1}$ from $z_{i,1},\ldots,z_{i,d-1}$. In the algorithm, the symbol $\langle\cdot,\cdot\rangle$ denotes the operator of inner product. To make B_i be a member of SO(d), it is further required to ensure that $det(B_i)$

Algorithm 1 Gram–Schmidt orthonormalization process

$j = 1$ and $m = 1$
if $j = 1$ **then**
  $v_{i,1} \triangleq z_{i,1}$
  $b_{i,1} = \frac{v_{i,1}}{\|v_{i,1}\|}$
end if
for $m < d-1$ **do**
  $m = j + 1$
  $v_{i,m} \triangleq z_{i,m} - \sum_{k=1}^{m-1} \langle z_{i,m}, b_{i,k}\rangle b_{i,k}$
  $b_{i,m} \triangleq \frac{v_{i,m}}{\|v_{i,m}\|}$
  $j = j + 1$
end for

be equivalent to 1. For this, since $b_{i,d}$ should be selected such that $det(B_i) = 1$, there could be many ways to make $det(B_i) = 1$. One of the approaches is to consider $b_{i,d}$ as a pseudovector for making $det(B_i)$ to be one. In d-dimensional vector space, a pseudovector can be derived from the element of the $(d-1)$th exterior powers, denoted by the wedge products $\wedge^{d-1}(\mathbb{R}^d)$, which is defined as

$$b_{i,1} \wedge b_{i,2} \wedge \cdots \wedge b_{i,d-1} \tag{8.22}$$

The wedge product represents the signed hyper-volume of d-dimensional parallelogram whose edges are d vectors on the same space [2]. Thus, in 3-dimensional space, the wedge product $b_{i,1} \wedge b_{i,2}$ is a signed volume. For the update of $d-1$ auxiliary variables, the following update law is proposed:

$$\dot{z}_{i,k} = u_i = \sum_{j\in\mathcal{N}_i} a_{ij}(R_{ji}^{-1} z_{j,k} - z_{i,k}),\ \forall i \in \mathcal{V},\ \forall k \in \{1,\ldots,d-1\} \tag{8.23}$$

where $a_{ij} > 0$. In the update (8.23), $z_{i,k}$ is updated independently for each k. Thus, in (8.23), there are $d-1$ independent updates for computing u_i. Defining $z_i \triangleq (z_{1,k}^T, z_{2,k}^T, \ldots, z_{n,k}^T)^T$, (8.23) can be written in a concise form as

$$\dot{z}_i = \hat{\mathbb{L}} z_i, \ \forall k \in \{1, \dots, d-1\} \tag{8.24}$$

where $\hat{\mathbb{L}} \in \mathbb{R}^{dn \times dn}$ is a block matrix with element matrices

$$\hat{\mathbb{L}}_{ij} = \begin{cases} a_{ij} R_{ij}, & j \in \mathcal{N}_i \\ -\sum_{j \in \mathcal{N}_i} a_{ij} \mathbb{I}_d, & i = j \\ \mathbf{0}_{d \times d}, & j \notin \mathcal{N}_i \end{cases} \tag{8.25}$$

The eigenvalues of $\hat{\mathbb{L}}$ are examined for stability analysis of (8.24). Let us define a block diagonal matrix $D \triangleq \text{blkdg}(R_1, R_2, \dots, R_n)$ where $R_i \in \mathbb{R}^{d \times d}, i = \{1, \dots, n\}$. By a similarity transformation as $\bar{\mathbb{L}} = D^{-1} \hat{\mathbb{L}} D$, the (i, j)th element matrix of $\bar{\mathbb{L}}$ is obtained as

$$\bar{\mathbb{L}}_{ij} = R_i^T \hat{\mathbb{L}}_{ij} R_j = \begin{cases} a_{ij} \mathbb{I}_d, & j \in \mathcal{N}_i \\ -\sum_{j \in \mathcal{N}_i} a_{ij} \mathbb{I}_d, & i = j \\ \mathbf{0}_{d \times d}, & j \notin \mathcal{N}_i \end{cases} \tag{8.26}$$

Thus, from (8.26), $\bar{\mathbb{L}}$ is rewritten as $\bar{\mathbb{L}} = -\mathbb{L}_H \otimes \mathbb{I}_d$, where $\mathbb{L}_H \in \mathbb{R}^{n \times n}$ has zero row sum with dominant diagonal entries and is a underlying Laplacian matrix of the transformed $\bar{\mathbb{L}}$. Note that $\mathbb{L}_H$ has strictly positive real eigenvalues except for a simple zero eigenvalue corresponding to the eigenvector $\xi = (1, 1, \dots, 1) \in \mathbb{R}^n$ (it is a direct extension of 2-dimensional case; for example, see Example 8.1). Using the coordinate transformation $z_i = D q_k$, (8.24) can be changed as

$$\dot{q}_i(t) = \bar{\mathbb{L}} q_k(t) = -\mathbb{L}_H \otimes \mathbb{I}_d q_k(t), \ \forall k \in \{1, \dots, d-1\} \tag{8.27}$$

From the above equation, it is clear that q_k converges to the equilibrium set $\mathcal{E}_q = \{q = (q_1^T, q_2^T, \dots, q_n^T)^T \in \mathbb{R}^{dn} | q_1 = q_2 = \cdots = q_n\}$. To avoid a trivial consensus to zero, the desired equilibrium set is defined as $\mathcal{E}_q' = \mathcal{E}_q \setminus \{0\}$. Similar to the previous section, the column space of matrix A is denoted as $C(A)$. The convergence property of (8.27) is now provided as follows [5]:

Lemma 8.2 *Let the underlying graph topology of* (8.24) *and* (8.27) *be directed as* $\vec{\mathcal{G}}$ *with a rooted-out branch (arborescence). Then, for the dynamics (8.27), there exists a finite point* $q_k^\infty \in \mathcal{E}_q'$ *for each* $k \in \{1, \dots, d-1\}$ *such that* q_k *globally exponentially converges if and only if an initial value* $q_k(t_0) \notin C(\mathbb{L}_H \otimes \mathbb{I}_d)$, $\forall k \in \{1, \dots, d-1\}$.

Proof It is clear that q_k exponentially converges to the set $\mathcal{E}_q$. Thus, for $k \in \{1, 2, \dots, d-1\}$, there exists a finite point $q_k^\infty \in \mathcal{E}_q$ and constants $k_q, \lambda_q > 0$ such that

$$\|q_k(t) - q_k^\infty\| \le k_q e^{-\lambda_q (t - t_0)} \|q_k(t_0) - q_k^\infty\| \tag{8.28}$$

The solution of $q_k(t)$ can be written as follows:

$$q_k(t) = e^{-\mathbb{L}_H \otimes \mathbb{I}_d (t-t_0)} q(t_0) \tag{8.29}$$

Since the matrix $\mathbb{L}_H$ has positive real eigenvalues except for a simple zero eigenvalue, we take a similarity transformation as $G^{-1}\mathbb{L}_H G = J_H$, where J_H is the Jordan form, and $G^{-1} = [w_1^T, \ldots, w_n^T]^T$ and $G = [g_1, \ldots, g_n]$ are the left and right eigenvector matrices. Let us denote the left and right eigenvectors corresponding to the zero eigenvalue by w_n and g_n. Then, as $t \to \infty$, from

$$e^{J_H \otimes \mathbb{I}_d (t-t_0)} \to \begin{bmatrix} 0 & & & \\ & 0 & & \\ & & \ddots & \\ & & & 1 \end{bmatrix} \otimes \mathbb{I}_d \tag{8.30}$$

Thus, the steady-state solution of q_k is obtained as follows:

$$\begin{aligned} \lim_{t\to\infty} q_k(t) &= \lim_{t\to\infty} (G \otimes \mathbb{I}_d) e^{J_H \otimes \mathbb{I}_d (t-t_0)} (G^{-1} \otimes \mathbb{I}_d) q_k(t_0) \\ &= (g_n \otimes \mathbb{I}_d^T)(w_n \otimes \mathbb{I}_d) q_k(t_0) \\ &= (g_n w_n) q_k(t_0) \end{aligned} \tag{8.31}$$

This means that $q_k(t)$ converges to zero if and only if $q_k(t_0)$ is perpendicular to the $(w_n \otimes \mathbb{I}_n)$. It follows that $(w_n \otimes \mathbb{I}_n) q_k(t_0) = 0$ if and only if $q_k(t_0) \in null(\mathbb{L}_H^T \otimes \mathbb{I}_n)^{\perp}$, which completes the proof.

Lemma 8.2 implies that $\lim_{t\to\infty} z_i(t) = D q_k^{\infty}$ for each k. Let $B_i \in \mathrm{SO}(d)$ be derived from the Gram–Schmidt process of $z_{i,k}$, $\forall k \in \{1, \ldots, d-1\}$ and from the calculation of $b_{i,d}$. Then, it can be shown that there exists $B_i^{\infty} \in \mathrm{SO}(d)$ to which $B_i(t)$ converges. For this, we need some manipulations. With the relationship $z_{i,k} = R_i q_{i,k}$, by following the process in Algorithm 1, the Gram–Schmidt orthonormalization can be obtained as

$$\begin{aligned} v_{i,1} &= R_i q_{i,1} \\ v_{i,2} &= R_i q_{i,2} - \langle R_i q_{i,2}, \frac{v_{i,1}}{\|v_{i,1}\|} \rangle \frac{v_{i,1}}{\|v_{i,1}\|} \\ &\vdots \\ v_{i,d-1} &= R_i q_{i,d-1} - \sum_{k=1}^{d-2} \langle R_i q_{i,d-1}, \frac{v_{i,k}}{\|v_{i,k}\|} \rangle \frac{v_{i,k}}{\|v_{i,k}\|} \end{aligned}$$

Further, denting $v_{i,m} \triangleq R_i x_{i,m}$, $m = 1, \ldots, d-1$, and replacing $v_{i,m}$ by $x_{i,m}$, $\forall\, m = 1, \ldots, d-1$, $x_{i,m}$ can be written as follows:

$$\begin{aligned} x_{i,1} &= q_{i,1} \\ x_{i,2} &= q_{i,2} - \langle q_{i,2}, \tfrac{x_{i,1}}{\|x_{i,1}\|}\rangle \tfrac{x_{i,1}}{\|x_{i,1}\|} \\ &\vdots \\ x_{i,d-1} &= q_{i,d-1} - \textstyle\sum_{k=1}^{d-2} \langle q_{i,d-1}, \tfrac{x_{i,k}}{\|x_{i,k}\|}\rangle \tfrac{x_{i,k}}{\|x_{i,k}\|} \end{aligned} \tag{8.32}$$

Note that the above relationship is also a Gram–Schmidt procedure with respect to $q_{i,m}$, $\forall m \in \{1, \ldots, d-1\}$. Then, with the above relationships, in the following theorem [5], we can show that $B_i(t) \to R_{ij}B_j(t)$, $\forall (i, j)^e \in \mathcal{E}$ as $t \to \infty$.

Theorem 8.3 *Suppose that the $n \times n$ matrix $B_i(t)$ for the ith agent is derived from the Gram–Schmidt procedure and the pseudovector (8.22) of z_i obtained from (8.24). Then, there exists a common matrix $X \in SO(d)$ such that B_i converges to R_iX as $t \to \infty$ for all $i \in \{1, \ldots, n\}$.*

Proof From Lemma 8.2, since $q_{i,k}$ converges to a finite point $q_k^\infty \in \mathbb{R}^d$ as $t \to \infty$, the relationship (8.32) means that there exists a finite vector $x_k^\infty \in \mathbb{R}^n$ such that $x_{i,k}(t) \to x_k^\infty$ as $t \to \infty$. Consequently, from the relationship $b_{i,k} = R_i \frac{x_{i,k}}{\|x_{i,k}\|}$, $\forall k \in \{1, 2, \ldots, d-1\}$, it is clear that B_i converges to R_iX.

From the above theorem, it is obvious that $B_i(t) \to R_{ij}B_j(t)$. To utilize the results thus far to the formation control in 3-dimensional space, two auxiliary variables $z_{i,1}$ and $z_{i,2}$ for the ith agent are necessary. Then, using the Gram–Schmidt procedure, the first and second column vectors, $b_{i,1}$ and $b_{i,2}$, of $B_i \in \mathbb{R}^3$ are calculated. Then the third column vector can be computed as

$$b_{i,3} = b_{i,1} \times b_{i,2}, \forall i \in \mathcal{V} \tag{8.33}$$

Since B_i relates the orientation transformation matrix R_i with respect to a common offset transformation matrix X as defined in (8.20), it can be considered that $\hat{R}_i(t) = B_i(t) \in \text{SO}(3)$. The following formation control law is proposed

$$u_i^i = k_u \sum_{j \in \mathcal{N}_i} l_{ij}((p_j^i - p_i^i) - \hat{R}_i(p_j^* - p_i^*)) \tag{8.34}$$

where $k_u, l_{ij} > 0$. Using $u_i = R_i^{-1}u_i^i$, the position dynamics of the ith agent can be expressed in a global coordinate frame as

$$\begin{aligned} \dot{p}_i &= k_u \sum_{j \in \mathcal{N}_i} l_{ij} R_i^{-1}((p_j^i - p_i^i) - \hat{R}_i(p_j^* - p_i^*)) \\ &= k_u \sum_{j \in \mathcal{N}_i} l_{ij}((p_j - p_i) - R_i^{-1}\hat{R}_i(p_j^* - p_i^*)) \end{aligned} \tag{8.35}$$

From Theorem 8.3, since $B_i \to R_iX$, it is true that $R_i^{-1}\hat{R}_i = R_i^{-1}R_iX = X$ because $\hat{R}_i$ is replaced by R_iX. Also since X is fixed, it can be considered that $R_i^{-1}\hat{R}_i \to R^c$, where R^c is a common offset rotation matrix. Defining $e_i \triangleq p_i - R^c p_i^*$, the error dynamics can be obtained as

$$\dot{e}_i = k_u \sum_{j\in\mathcal{N}_i} l_{ij}(e_j - e_i) + k_u \sum_{j\in\mathcal{N}_i} l_{ij}(R^c - R_i^{-1}\hat{R}_i)(p_j^* - p_i^*) \tag{8.36}$$

Denoting $w_i(t) \triangleq k_u \sum_{j\in\mathcal{N}_i} l_{ij}(R^c - R_i^{-1}\hat{R}_i)(p_j^* - p_i^*)$, the above equation can be written as

$$\dot{e}(t) = -k_u \mathbb{L} \otimes \mathbb{I}_3 e(t) + w(t) \tag{8.37}$$

where $\mathbb{L}$ is the Laplacian determined by the connectivities l_{ij} of the actuation topology. Since $R_i^{-1}\hat{R}_i$ converges to R^c exponentially fast, there exist $k_w, \lambda_w > 0$ such that

$$\|w(t)\| \leq k_w e^{-\lambda_w(t-t_0)}\|w(t_0)\| \tag{8.38}$$

The solution of (8.37) is given by

$$e(t) = e^{-\mathbb{L}\otimes\mathbb{I}_3(t-t_0)}e(t_0) + \int_{t_0}^{t} e^{-\mathbb{L}\otimes\mathbb{I}_3(t-\tau)}w(\tau)d\tau \tag{8.39}$$

Based on Theorem 8.3, we can now obtain the main result of this section:

Theorem 8.4 *Under the orientation estimation law (8.24) and formation control law (8.34), there exist a finite point p_∞ and constant matrix R^c such that $p_i(t), \forall i \in \mathcal{V}$ globally exponentially converges to $R^c p_i^* + p_\infty$ if the sensing topology $\mathcal{G} = (\mathcal{V}, \mathcal{E}^s)$ has a spanning tree, the communication topology $\mathcal{G} = (\mathcal{V}, \mathcal{E}^c)$ has an arborescence, and the actuation topology $\mathcal{G} = (\mathcal{V}, \mathcal{E}^c)$ has an arborescence, and $\hat{q}_i(t_0)$, $k = 1, 2$ is not in $C(\mathbb{L}_H \otimes \mathbb{I}_3)$.*

Proof See the appendix.

8.3 Orientation Control in 2-Dimensional Space

This section further attempts to achieve a formation control that changes both the orientations and positions of agents. Note that in the previous chapters, the orientations have been estimated and the estimated orientations are used for formation control. Since the orientations of agents are defined on a unit circle, the orientation control of agents in 2-D can be understood as a motion control of nodes in the circle [7]. Consider n agents on the unit circle modeled as

$$\dot{\theta}_i = \omega_i, \; i \in \mathcal{V} \tag{8.40}$$

It is supposed that the geodesic distance between neighboring agents j and i can be measured as θ_{ji} in the local coordinate frame of agent i (for this process, see (6.4)). Then, under the assumption that the interaction topology has an arborescence, we would like to design ω_i such that $\theta_i \to \theta^*$ as $t \to \infty$ only using the relative measurements θ_{ji}. Given an angle $\theta \in \mathbb{S}$, there is one-to-one and onto mapping between the rotation matrix R and θ as

$$R(\theta) = \begin{bmatrix} \cos\theta & -\sin\theta \\ \sin\theta & \cos\theta \end{bmatrix}$$

Since the space $\mathbb{S}$ is not a convex space (not linear space), it is not trivial to achieve a consensus on $\mathbb{S}$. The main idea of achieving a consensus of θ_i on $\mathbb{S}$ is to change θ_i into the complex vector space using the following relationship:

$$z_i = e^{\mathrm{i}\theta_i}$$

We will estimate z_i instead of θ_i using the measurements θ_{ji}. The estimated z_i is denoted as $\hat{z}_i$. Then, using the estimated $\hat{z}_i$, each agent calculates $\hat{\theta}_i$, which further can be used for orientation control as $\omega_i(\hat{\theta}_i)$ such that $\theta_i \to \theta^*$. First, we design an estimation law for a common complex value $\alpha \in \mathbb{C}$ such that $\angle\hat{z}_i \to \angle z_i + \angle\alpha$ as $t \to \infty$ for all $i \in \mathcal{V}$. For the estimation, we use the estimation law (8.1), under which the estimated variables converge to desired values as proved in Theorem 8.1. From $z_i = e^{\mathrm{i}\theta_i}$, when the angle θ_i is updated by ω_i, the dynamics of z_i is updated as, in real part and imaginary part separately:

$$\dot{z}_i = \underbrace{\begin{bmatrix} 0 & -\omega_i \\ \omega_i & 0 \end{bmatrix}}_{\triangleq \Omega_i} z_i \tag{8.41}$$

Since the initial θ_i are not known, the initial values of z_i are not known; thus, the update rule of (8.41) cannot be implemented in local agents. The estimated values $\hat{z}_i$ are also called estimated auxiliary variables. The estimated auxiliary variables are assumed to be communicated between neighboring agents. It is worth noting that the sensing topology is undirected, but the communication topology could be directional. The auxiliary variables z_i are estimated as

$$\dot{\hat{z}}_i = \Omega_i \hat{z}_i + \sum_{j \in \mathcal{N}_i} (R_{ji}(t)^T \hat{z}_j - \hat{z}_i) \tag{8.42}$$

Denoting $z = (z_1, z_2, \ldots, z_n)^T$ and $\hat{z} = (\hat{z}_1, \hat{z}_2, \ldots, \hat{z}_n)^T$, and defining

$$H_{ji}(t) \triangleq \begin{cases} R_{ji}(t)^T & j \in \mathcal{N}_i \\ -\sum_{j\in\mathcal{N}_i} \mathbb{I}_2 & i = j \\ \mathbf{0}_{2\times 2} & j \notin \mathcal{N}_i \end{cases} \tag{8.43}$$

Equations (8.41) and (8.42) can be combined as

$$\dot{\hat{z}} = D_\omega \hat{z} + H(t)\hat{z} \tag{8.44}$$

$$\dot{z} = D_\omega z \tag{8.45}$$

where $H(t) \in \mathbb{R}^{2n\times 2n}$ and D_ω are block matrices defined as

$$H(t) = \begin{bmatrix} H_{11}(t) & \cdots & H_{1n}(t) \\ \vdots & \ddots & \vdots \\ H_{n1}(t) & \cdots & H_{nn}(t) \end{bmatrix} \tag{8.46}$$

and $D_\omega = \text{blkdg}(\Omega_1, \Omega_2, \ldots, \Omega_n)$, respectively. With the time-varying nonsingular matrix $D_r(t) = \text{blkdg}(R_1(t), R_2(t), \ldots, R_n(t)) \in \mathbb{R}^{2n\times 2n}$, we conduct a similarity transformation as follows:

$$H_s(t) = D_r(t)^{-1}H(t)D_r(t) = -(\mathbb{L}_H \otimes \mathbb{I}_2) \tag{8.47}$$

where $\mathbb{L}_H \in \mathbb{R}^{n\times n}$ is the Laplacian matrix corresponding to the graph $\mathcal{G}(\mathcal{V}, \mathcal{E})$. For further simplification, introducing a new vector $\hat{q}$ as $\hat{q} \triangleq D_r(t)^{-1}\hat{z}$ and using $\dot{D}_r(t) = D_\omega D_r(t)$, the dynamics of (8.44) can be changed as

$$\begin{aligned} \dot{\hat{z}} &= \dot{D}_r(t)\hat{q} + D_r(t)\dot{\hat{q}} \\ &= D_\omega D_r(t)\hat{q} + D_r(t)\dot{\hat{q}} \end{aligned} \tag{8.48}$$

Now, with (8.48) and (8.44), we have the following dynamics:

$$\dot{\hat{q}} = -(\mathbb{L}_H \otimes \mathbb{I}_2)\hat{q} \tag{8.49}$$

Since $\mathbb{L}_H$ is a Laplacian matrix, $\hat{q}$ converges to a nontrivial consensus manifold $\mathcal{E}_{2n} \triangleq \{[\xi_1^T, \ldots, \xi_n^T]^T \in \mathbb{R}^{2n} : \xi_i = \xi_j, \forall i, j \in \{1, \ldots, n\}\}$ if $\hat{q}(t_0) \notin C(\mathbb{L}_H \otimes \mathbb{I}_2)$. It is noticeable that $\hat{q}$ reaches a consensus; but $\hat{q}_i(t)$ may reach a limit cycle, a synchronized trajectory or a stationary point depending upon initial values of $\hat{z}_i(t_0)$. Also, if the dynamics for the auxiliary variable (8.42) for the leader is not updated, i.e., when the graph has only an arborescence, it may converge to a trajectory. The same phenomenon may happen in 3-dimensional case in Sect. 8.4. Let the converged value of (8.49) be denoted as $\hat{q}^\infty$. Then, it is clear that $\hat{z}_i$ converges to $R_k(t)\hat{q}^\infty$ which means that the estimator (8.44) achieves $\hat{z}_i$ as

$$\angle\hat{z}_i = \angle z_i e^{\mathrm{i}\Delta\theta} = \angle e^{\mathrm{i}(\theta_i + \Delta\theta(t))} \tag{8.50}$$

where $\Delta\theta(t)$ is an unknown common bias arising from the estimation process. We would like to actively control the orientation of each agent, θ_i, such that they converge to a common value. So, it is required to design the angular velocity ω_i of each agent for (8.40). To achieve this goal, we use the estimator (8.42). From (8.42), let us define a tangential mapping of z_i on a unit circle. Let a smooth mapping $\phi : M \to \mathbb{S}$ be defined as $\phi(z_i) = \angle z_i$ where $z_i \in M \in \mathbb{R}^2$. Then, the differential of ϕ at $z_i \in M$, denoted as $\mathrm{d}\left\{\phi_{z_i}\right\}$, is a linear mapping as follows:

$$\mathrm{d}\left\{\phi_{z_i}\right\} : T_{z_i}(M) \to T_{\phi(z_i)}(\mathbb{S}) \tag{8.51}$$

where $T_{z_i}(M)$ is a tangent space of the manifold M at z_i and $T_{\phi(z_i)}(\mathbb{S})$ is the tangent space of the circle space $\mathbb{S}$ at $\phi(z_i)$. Now, with the differential operator defined in (8.51), on the basis of (8.42), $\hat{\theta}_i$ may be updated as

$$\dot{\hat{\theta}}_i \triangleq \mathrm{d}\left\{\dot{\hat{z}}_i\right\} = \underbrace{\mathrm{d}\left\{\Omega_i \hat{z}_i\right\}}_{\triangleq \hat{\omega}_i} + \underbrace{\mathrm{d}\left\{\sum_{j\in\mathcal{N}_i}(R_{ji}(t)^T \hat{z}_j - \hat{z}_i)\right\}}_{\triangleq f_i} \tag{8.52}$$

Likewise, θ_i may be updated as

$$\dot{\theta}_i = \mathrm{d}\left\{\Omega_i z_i\right\} = \omega_i \tag{8.53}$$

Now, in order to achieve a consensus in orientation angles θ_i for all agents, by slightly modifying (8.52), we propose the following distributed estimator law:

$$\dot{\hat{\theta}}_i = \omega_i + f_i \tag{8.54}$$

Then, from (8.54), with the updated $\hat{\theta}_i$, which is computed in a distributed way, the following alignment law is proposed:

$$\dot{\theta}_i = \omega_i = -\sin(\hat{\theta}_i) \tag{8.55}$$

Defining $\hat{\Theta} \triangleq (\hat{\theta}_1, \hat{\theta}_2, \ldots, \hat{\theta}_n)^T$, $\Theta \triangleq (\theta_1, \theta_2, \ldots, \theta_n)^T$, $\omega \triangleq (\omega_1, \omega_2, \ldots, \omega_n)^T$, and $\mathbf{f} \triangleq (f_1, f_2, \ldots, f_n)^T$, (8.54) and (8.55) can be concisely written as

$$\dot{\hat{\Theta}} = \omega + \mathbf{f} \tag{8.56}$$

$$\dot{\Theta} = \omega \tag{8.57}$$

The overall structure of the estimator-based feedback alignment controller is depicted in Fig. 8.2.

For the dynamics (8.57), by equalizing $\sin(\hat{\theta}_i) = 0$, two equilibrium sets can be defined as

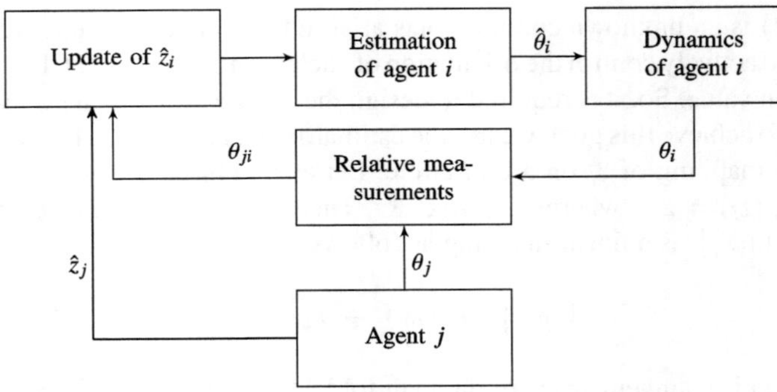

Fig. 8.2 The structure of estimator-based feedback alignment controller

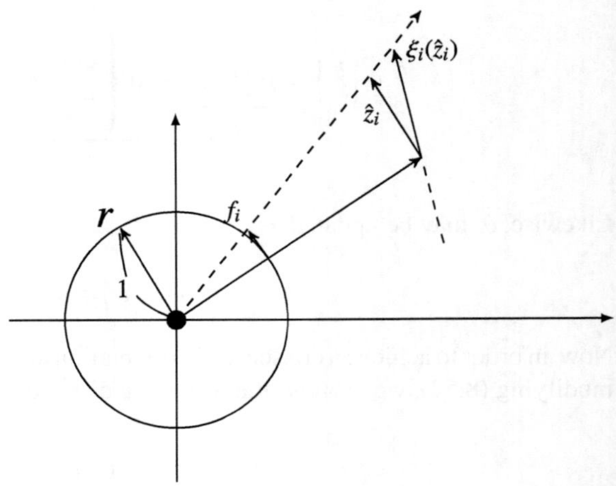

Fig. 8.3 The scaled projection of $\xi_i(\hat{z}_i)$. © [2017] IEEE. Reprinted, with permission, from [7]

$$\mathcal{S}_d \triangleq \{\hat{\theta}_i, \theta_i \in \mathbf{P} : \hat{\theta}_i = 0, \theta_i = \theta^* \in \mathbb{S}, \forall i \in \mathcal{V}\}$$
$$\mathcal{S}_u \triangleq \mathcal{S} \setminus \mathcal{S}_d$$

where $\mathcal{S} \triangleq \{\hat{\theta}_i, \theta_i \in \mathbf{P} : \hat{\theta}_i \in \{0, \pi\}, \theta_i = \theta^* \in \mathbb{S}, \forall i \in \mathcal{V}\}$ and $\mathbf{P} \triangleq [0, 2\pi)$. In (8.52), f_i can be considered as a tangent projection of the vector $\xi_i(\hat{z}_i) \triangleq \sum_{j \in \mathcal{N}_i} (R_{ji}(t)^T \hat{z}_j - \hat{z}_i)$ onto the circle including the point $\hat{z}_i$. As depicted in Fig. 8.3, it can be computed in the following steps. First, having scaled $\hat{z}_i$ as unit, take a rotation by $\pi/2$, which can be computed as $R_{\pi/2} \frac{\hat{z}_i}{\|\hat{z}_i\|}$. Second, we also need to scale $\xi_i(\hat{z}_i)$ as $\frac{\xi_i(\hat{z}_i)}{\|\hat{z}_i\|}$. Then, by taking an inner product, we can obtain the desired projection as [7]:

$$
\begin{aligned}
f_i &= \frac{\xi_i(\hat{z}_i)^T}{\|\hat{z}_i\|} R_{\pi/2} \frac{\hat{z}_i}{\|\hat{z}_i\|} \\
&= \left[\sum_{j \in \mathcal{N}_i} (R_{ji}(t)\hat{z}_j - \hat{z}_i) \right]^T R_{\pi/2} \frac{\hat{z}_i}{\|\hat{z}_i\|^2} \\
&= \sum_{j \in \mathcal{N}_i} \underbrace{\left[(R_{ji}(t)\hat{z}_j - \hat{z}_i)^T R_{\pi/2} \frac{\hat{z}_i}{\|\hat{z}_i\|^2} \right]}_{\triangleq f_{ji}}
\end{aligned}
\tag{8.58}
$$

In the above equation, due to the fact that $\hat{z}_i R_{\pi/2} \hat{z}_i = 0$, f_{ji} is simplified as

$$
\begin{aligned}
f_{ji} &= \hat{z}_j^T R_{ji}(t) R_{\pi/2} \frac{\hat{z}_i}{\|\hat{z}_i\|^2} \\
&= c_{ji} \cos\left(\hat{\theta}_j - \theta_{ji} - \hat{\theta}_i - \frac{\pi}{2}\right) \\
&= c_{ji} \sin(\hat{\theta}_j - \theta_{ji} - \hat{\theta}_i)
\end{aligned}
\tag{8.59}
$$

which can be obtained from $a^T b = \|a\|\|b\| \cos(\angle a - \angle b)$, where $a = \hat{z}_j^T R_{ji}(t)$ and $b = R_{\pi/2} \frac{\hat{z}_i}{\|\hat{z}_i\|^2}$, and $c_{ji} > 0$. Thus, the dynamics of $\hat{\theta}_i$ and θ_i can be expressed as:

$$
\dot{\hat{\theta}}_i = -\sin(\hat{\theta}_i) + \sum_{j \in \mathcal{N}_i} c_{ji} \sin(\hat{\theta}_j - \theta_{ji} - \hat{\theta}_i) \tag{8.60}
$$

$$
\dot{\theta}_i = -\sin(\hat{\theta}_i) \tag{8.61}
$$

Lemma 8.3 *With the feedback alignment law (8.55), the equilibrium point in $\mathcal{S}_u$ is unstable.*

Proof To examine repulsiveness of the set $\mathcal{S}_u$, take a linearization of (8.60)–(8.61) at $\hat{\theta}_i = \pi$ and $\theta_i = \theta^*$ for a specific $i \in \mathcal{V}$. The Jacobian of the system (8.60)–(8.61) is obtained as:

$$
\left[\begin{array}{c|c} -\cos(\hat{\theta}_i) - \alpha_i & \alpha_i \\ \hline -\cos(\hat{\theta}_i) & 0 \end{array} \right] \Bigg|_{\hat{\theta}_i = \pi, \theta_i = \theta^*} = \left[\begin{array}{c|c} 1 - \beta_i & \beta_i \\ \hline 1 & 0 \end{array} \right] \tag{8.62}
$$

where $\alpha_i = \sum_{j \in \mathcal{N}_i} c_{ji} \cos(\hat{\theta}_j - \theta_{ji} - \hat{\theta}_i)$ and $\beta_i = \sum_{j \in \mathcal{N}_i} c_{ji} \cos(\hat{\theta}_j - \pi)$. From the right-hand side of the above equation, we can find eigenvalues as 1 and $-\beta_i$. Thus, since one of the eigenvalues is always positive, it is unstable.

Let us suppose that the estimated orientations are all zero, i.e, $\hat{\theta}_i = 0$. Then, we can have the following linearization:

$$\left[\begin{array}{c|c} -1-\sum_{j\in\mathcal{N}_i} c_{ji}\cos(-\theta_{ji}) & \sum_{j\in\mathcal{N}_i} c_{ji}\cos(-\theta_{ji}) \\ \hline -1 & 0 \end{array}\right] \tag{8.63}$$

which has eigenvalues as $-1, -\sum_{j\in\mathcal{N}_i} c_{ji}\cos(\theta_{ji})$. Thus, depending on θ_{ji}, it may have negative or positive eigenvalues. But, if $\sin(\theta_{ji}) \neq 0$, it does not correspond to the equilibrium from (8.60). The Lemma 8.3 means that when $\theta_i = \theta^*$ for all i, if there exists at least one agent having $\hat{\theta}_i = \pi$, that equilibrium is unstable. Now, it is required to show that the orientation alignment law (8.55) ensures $\theta_i \to \theta^*$ for all i as $t \to \infty$, which is summarized in the following theorems [7]:

Theorem 8.5 *Suppose that the underlying network topology has an arborescence. Then, with the orientation alignment law (8.55), the estimation dynamics (8.56) is almost globally asymptotically stable, i.e.,* $\hat{\Theta} \to 0$ *as* $t \to \infty$.

Proof From (8.59), denoting $\hat{\theta}_{ji} \triangleq \hat{\theta}_j - \hat{\theta}_i$ and $\Delta\hat{\theta}_{ji}(t) \triangleq \hat{\theta}_{ji}(t) - \theta_{ji}(t)$, $f_{ji}(t)$ can be changed as

$$\begin{aligned} f_{ji}(t) &= c_{ji}\sin(\hat{\theta}_j - \theta_{ji} - \hat{\theta}_i) \\ &= c_{ji}\sin(\Delta\hat{\theta}_{ji}(t)) \end{aligned} \tag{8.64}$$

From (8.50), it follows that $\hat{\theta}_i = \theta_i + \Delta\theta(t)$ where $\Delta\theta(t)$ is common to all agents. Thus, $\Delta\hat{\theta}_{ji}(t) \to 0$ as $t \to \infty$. Since $\|\sin(\Delta\hat{\theta}_{ji}(t))\| \leq 1$, there clearly exist constants $k_{ji}, \lambda_{ji} > 0$ such that $\|f_{ji}(t)\| \leq k_{ji}e^{-\lambda_{ji}(t-t_0)}\|f_{ji}(t_0)\|$. Note that we ignore the cases of $\Delta\hat{\theta}_{ji}(t_0) = \pm\pi$, which makes $f_{ji}(t_0) = 0$. In these special cases, although the convergence is not ensured, it is of Lebesgue measure zero.[1] Thus, for the vector $\mathbf{f}$, there exist constants $k_f, \lambda_f > 0$ such that

$$\|\mathbf{f}(t)\| \leq k_f e^{-\lambda_f(t-t_0)}\|\mathbf{f}(t_0)\| \tag{8.65}$$

Let us choose a Lyapunov candidate V_k for $\hat{\theta}_i$ as $V_k = 1 - \cos(\hat{\theta}_i)$. The derivative of the candidate is obtained as

$$\begin{aligned} \dot{V}_k &= \dot{\hat{\theta}}_i \sin(\hat{\theta}_i) = -\sin^2(\hat{\theta}_i) + f_k \sin(\hat{\theta}_i) \\ &\leq -\sin^2(\hat{\theta}_i) + k_f e^{-\lambda_f(t-t_0)}\|\mathbf{f}(t_0)\| \sin(\hat{\theta}_i) \\ &= -V_k(2-V_k) + k_f e^{-\lambda_f(t-t_0)}\|\mathbf{f}(t_0)\|\sqrt{V_k(2-V_k)} \end{aligned} \tag{8.66}$$

where we used $|\sin(\hat{\theta}_i)| = \sqrt{V_k(2-V_k)}$. For a simplicity of analysis, let us denote $\beta(t-t_0) \triangleq k_f e^{-\lambda_f(t-t_0)}\|\mathbf{f}(t_0)\|$. In the above equation, there could be two cases according to the magnitude of $\beta(t-t_0)$. First, if $\beta(t-t_0) \geq \sqrt{V_k(2-V_k)}$, then $\dot{V}_k \geq 0$. But, we know that $\beta(t-t_0)$ decreases to zero exponentially. Also, we know that $0 \leq V_k(2-V_k) = 1 - \cos^2(\hat{\theta}) \leq 1$. Thus, we can see that if $\beta(t-t_0) \geq \sqrt{V_k(2-V_k)}$ for all $t \geq t_0$, then $\sqrt{V_k(2-V_k)} \to 0$ and $\dot{V}_i$ is still bounded, and it also

[1] That is why we call it almost globally asymptotically stable.

converges to zero because $\beta(t - t_0) \to 0$. To satisfy $\sqrt{V_k(2 - V_k)} \to 0$, we need to have either $V_k = 0$ or $2 - V_k = 0$. When $2 - V_k = 0$, we need to have $\hat{\theta}_i = \pm\pi$. But, as shown in Lemma 8.3, it is unstable. Thus, from $V_k = 0$, we have $\hat{\theta}_i = 0$. Consequently, it is clear that as $t \to \infty$, $\beta(t - t_0) \to 0$, $V_k \to 0$, and $\hat{\theta}_i \to 0$, which means that $\hat{\theta}_i = 0$ is a globally asymptotically stable equilibrium point.

Second, on the other hand, if $\beta(t - t_0) < \sqrt{V_k(2 - V_k)}$, then $\dot{V}_k \leq 0$. Let us suppose that this case occurs at $t_0 + T_1 \leq t < t_0 + T_1 + T_2$. Then, in this case, there exists a $0 < \phi < 1$ such that $\phi V_k(2 - V_k) \geq \beta(t - t_0)\sqrt{V_k(2 - V_k)}$. Then, the following is true:

$$\dot{V}_k \leq -(1 - \phi)V_k(2 - V_k) \leq -(1 - \phi)V_k \tag{8.67}$$

By using the comparison lemma,[2] the solution of (8.67) can be upper bounded as

$$V_k(t) \leq V_k(t_0 + T_1)e^{-\int_{t_0+T_1}^{t}(1-\phi)d\tau}$$

From the above inequality, it is clear that $V_k(t)$ converges to zero in an exponential rate. From the relationship $V_k(t) = 1 - \cos\hat{\theta}_i$, since $V_k(t)$ converges to zero exponentially, it is shown that $\hat{\theta}$ also converges to zero exponentially. Thus, there exist constants $k_{\hat{\theta}}, \lambda_{\hat{\theta}} > 0$ such that

$$|\hat{\theta}_i(t)| \leq k_{\hat{\theta}}e^{-\lambda_{\hat{\theta}}(t-t_0-T_1)}|\hat{\theta}_i(t_0 + T_1)|, \; t_0 + T_1 \leq t \leq t_0 + T_1 + T_2 \tag{8.68}$$

Thus, $\hat{\theta}_i = 0$ is a globally exponentially stable equilibrium point when $\beta(t - t_0) < \sqrt{V_k(2 - V_k)}$, i.e., in $t_0 + T_1 \leq t \leq t_0 + T_1 + T_2$.

Now, combining the above two cases and using the fact that $\hat{\theta}_i(t)$ is continuous, we can conclude that $\hat{\theta}_i(t)$ converges to zero asymptotically if $\Delta\hat{\theta}_{ji}(t_0) \neq \pm\pi$, which completes the proof.

Theorem 8.6 *Suppose that the underlying sensing network topology $\mathcal{G} = (\mathcal{V}, \mathcal{E}^s)$ is connected (has a spanning tree), and the communications and actuation topologies, $(\mathcal{V}, \mathcal{E}^c)$ and $(\mathcal{V}, \mathcal{E}^a)$, have an arborescence. Then, with the orientation alignment law (8.55) and estimation law (8.54), all the orientations θ_i of agents converge to a common constant value, i.e., $\theta_i \to \theta^*$.*

Proof From $\hat{\theta}_i = \theta_i + \Delta\theta(t)$ and $\hat{\theta}_i \to 0$, we have $\theta_i \to -\Delta\theta(t)$ as $t \to \infty$. Also since $\dot{\hat{\theta}}_i = -\sin(\hat{\theta}_i) \to 0$ from (8.61), we know that $\Delta\theta(t)$ converges to a constant, which completes the proof.

Similarly to Sect. 6.3, we can use Theorem 8.6 for the formation control of agents with orientation synchronization. We can also use almost global input-to-state stability concept (see Lemmas 2.10 and 2.11) in order to apply the above results to the

[2]Consider a differentiable function $x(t)$, with the inequality $\dot{x}(t) \leq f(t, x(t))$. Let the solution of $\dot{y}(t) = f(t, y(t))$ be given as $y(t)$. Then, $x(t)$ is upper bounded as $x(t) \leq y(t)$. For a general statement, refer to Lemma 3.4 of [3].

formation control. Since it can be derived by following the exactly same procedure as in Sect. 6.3, the detailed derivations are omitted. It is further noticeable that orientation synchronization for a complete graph can be easily designed using existing works [11, 12]. Notice that (8.60) and (8.61) can be combined as:

$$\dot{\vartheta}_i = \sum_{j \in \mathcal{N}_i} c_{ji} \sin(\vartheta_j - \vartheta_i) \tag{8.69}$$

where $\vartheta_i = \hat{\theta}_i - \theta_i$ and $c_{ji} = c > 0$ for all connectivities. As commented in [12], if the network topology is a complete graph, the states of (8.69) will reach a synchronization, i.e., $\vartheta_j = \vartheta_i$, from almost all initial conditions (if they are not in saddle points at initial time). It means that in such a case, with any angular control inputs, the orientation estimation law (8.60) will estimate the orientation angles of agents upto a common offset. On the other hand, if $c_{jk} = c < 0$, under the complete graph topology, then the states ϑ_i will be balanced on the circle space upto a common rotation. For a more general topology, without using any all-to-all communications, a local information algorithm was proposed to ensure a synchronization [11], which uses agent's own angle θ_i and diffusive coupling angles θ_{ji}. Thus, for a complete graph [12], the orientations of agents can be balanced along the circle with the following orientation control law:

$$\dot{\theta}_i = \sum_{j=1}^{n} k_b \sin(\theta_{ji}) \tag{8.70}$$

where $k_b < 0$. Then, the orientation estimation law can be designed as

$$\dot{\hat{\theta}}_i = \sum_{j=1}^{n} k_b \sin(\theta_{ji}) + \sum_{j=1}^{n} k_c \sin(\hat{\theta}_j - \theta_{ji} - \hat{\theta}_i) \tag{8.71}$$

where $k_b < 0$ and $k_c > 0$. Consequently, for a complete graph, a distributed formation control law, along with the orientation control law (8.70) and the orientation estimation law (8.71), will result in a desired formation configuration with orientation angles being balanced.

8.4 Orientation Control in 3-Dimensional Space

The orientation control in 3-D is a problem of controlling the orientation of a rigid body, as a function of Euler angles, which is commonly called special orthogonal group SO(3). The orientation control in $\mathbb{R}^3$ was introduced in [4]. In $\mathbb{R}^3$, a point $Q \in$ SO(3) of the orthogonal group for orientations is characterized by two constraints $Q^T Q = \mathbb{I}_3$ and $det(Q) = 1$. That is, a point Q is a member of SO(3) if and only if $Q^T Q = \mathbb{I}_3$ and $det(Q) = 1$. Given a vector $\omega_i = (\omega_{1i}, \omega_{2i}, \omega_{3i})^T \in \mathbb{R}^3$, the Lie algebra $\mathfrak{so}(3)$ is a set of 3×3 skew-symmetric matrices, $[\omega_i]^\wedge$, defined as

$$\Omega_i \triangleq \begin{pmatrix} 0 & -\omega_{3i} & \omega_{2i} \\ \omega_{3i} & 0 & -\omega_{1i} \\ -\omega_{2i} & \omega_{1i} & 0 \end{pmatrix} \in \mathfrak{so}(3) \underset{[\cdot]^\wedge}{\overset{[\cdot]^\vee}{\rightleftarrows}} \begin{pmatrix} \omega_{1i} \\ \omega_{2i} \\ \omega_{3i} \end{pmatrix} \tag{8.72}$$

In the remaining part of this section, we summarize the results given in [4]. Consider n agents evolving on SO(3). Let $R_i(t) \in \mathrm{SO}(3)$ denote a point of agent i at time t. Assuming that each agent does not share a common reference frame, the ith local reference frame is rotated from the global reference frame with the amount of $R_i(t) \in \mathrm{SO}(3)$. The right-invariant relative point on SO(3) is simply denoted as $R_{ji}(t)$, defined by $R_{ji}(t) = R_j(t)R_i^T(t)$. It is associated with the relative orientation of the jth agent in the view of the ith agent. Further, suppose that two auxiliary variables are generated for each agent. The auxiliary variables are assumed to be communicated between the neighboring agents based on the interaction topology. Note that the sensing and communication graphs could have opposite directions. That is, when an agent i is supposed to sense the relative orientation R_{ji}, an agent j needs to transmit the auxiliary variables of j to the agent i. But, in fact, it is difficult to measure R_{ji} without the sensing of agent j toward the agent i. Thus, the sensing topology would be of bidirectional. Each auxiliary variable at the ith agent is denoted by $\hat{z}_{i,k} \in \mathbb{R}^3$, $k \in \{1, 2\}$. Let $B_i \in \mathrm{SO}(3)$ denote the rotation matrix derived from the auxiliary variables. To generate the orthonormal column vectors of $B_i = [b_{i,1}\ b_{i,2}\ b_{i,3}]$ from $\hat{z}_{i,k}$, we use the Gram–Schmidt procedure and the cross product operation as follows:

$$\begin{aligned} &v_{i,1} := \hat{z}_{i,1} &&, b_{i,1} := \tfrac{v_{i,1}}{\|v_{i,1}\|} \\ &v_{i,2} := \hat{z}_{i,2} - \langle \hat{z}_{i,2}, b_{i,1}\rangle b_{i,1} &&, b_{i,2} := \tfrac{v_{i,2}}{\|v_{i,2}\|} \\ &b_{i,3} := b_{i,1} \times b_{i,2} \end{aligned} \tag{8.73}$$

In general d-dimensional space, we use the wedge vectors to generate orthonormal column vectors as done in (8.22); but in 3-dimensional space, the cross product (8.73) can be used as an alternative of the wedge product. It shows that two auxiliary variables generate the rotation matrix on SO(3). We design the dynamics of auxiliary variable as follows:

$$\dot{\hat{z}}_{i,k} = \sum_{j \in \mathcal{N}_i} a_{ij}(R_{ji}^{-1}\hat{z}_{j,k} - \hat{z}_{i,k}) - \Omega_i \hat{z}_{i,k}, \tag{8.74}$$

$\forall i \in \mathcal{V}$, $k \in \{1, 2\}$, where $a_{ij} \geq 0$ is a weighted value and $\Omega_i = (\omega_i)^\wedge \in \mathfrak{so}(3)$. Each element of ω_i is the angular velocity of the rotating frame. In this chapter, the angular velocity is the control input and has the following relationship with the rotation matrix (see 7.29):

$$\dot{R}_i = R_i \Omega_i = -\Omega_i R_i. \tag{8.75}$$

8.4.1 Convergence Analysis of Auxiliary Variables

By using the coordinate transformation $\hat{z}_{i,k} = R_i \hat{q}_{i,k}$ and $\dot{R}_i = -\Omega_i R_i$, the dynamics of $\hat{z}_{i,k}$, i.e., (8.74), is rewritten in terms of $\hat{q}_{i,k}$ as follows:

$$\dot{\hat{q}}_{i,k} = \sum_{j \in \mathcal{N}_i} a_{ij}(\hat{q}_{j,k} - \hat{q}_{i,k}), \ \forall i \in \mathcal{V}, k \in \{1, 2\} \tag{8.76}$$

Let $\hat{q}_k$ denote a stacked vector formed as $\hat{q}_k := (\hat{q}_{1,k}^T, \hat{q}_{2,k}^T, \ldots, \hat{q}_{n,k}^T)^T$. The dynamics of $\hat{q}_k$ can be written as follows:

$$\dot{\hat{q}}_k = -(\mathbb{L}_H \otimes \mathbb{I}_3)\hat{q}_k, \ k \in \{1, 2\}, \tag{8.77}$$

where $\mathbb{L}_H \in \mathbb{R}^{n \times n}$ is a Laplacian matrix. Then, the convergence of $\hat{q}_{i,k}$ to a consensus value is achieved in an exponential rate. In the problem, convergence of the auxiliary variable $\hat{z}_{i,k}$ to the origin is not desired since the origin is a trivial solution which cannot be transformed to an element of SO(3). Thus, the desired equilibrium set in the orientation estimation is defined by $\mathcal{S}_q = \{x = (x_1^T, x_2^T, \ldots, x_n^T)^T \in \mathbb{R}^{3n} : x_1 = x_2 = \cdots = x_n \neq 0\}$. As mentioned in the previous chapters, the trivial solution lies in the column space of the Laplacian matrix.

Lemma 8.4 *Suppose that the communication and actuation topologies $\mathcal{G} = (\mathcal{V}, \mathcal{E}^c)$ and $\mathcal{G} = (\mathcal{V}, \mathcal{E}^a)$ have a common arborescence. For the dynamics (8.77), there exists a finite point $\hat{q}_k^\infty \in \mathcal{S}_q$, $\hat{q}_k^\infty := (q_k^\infty, q_k^\infty, \ldots, q_k^\infty)^T$, for each $k \in \{1, 2\}$ that $\hat{q}_k$ exponentially converges if and only if an initial value $\hat{q}_k(t_0)$, $\forall k \in \{1, 2\}$ is not in $C(\mathbb{L}_H \otimes \mathbb{I}_3)$.*

Proof The proof is similar to the proof of Lemma 8.2.

The result indicates that for almost all initial values of $\hat{z}_{i,k}$, it converges to the desired equilibrium set regardless of the value of $\Omega_i \in \mathfrak{so}(3)$.

Remark 8.1 The dimension of $C(\mathbb{L}_H)$ is $n - 1$ when $\mathbb{L}_H$ is a $n \times n$ matrix. Then, the Lebesgue measure of $C(\mathbb{L}_H \otimes \mathbb{I}_3)$ is equivalent to zero. This implies that when the initial value of q_i is randomly assigned, the value is almost not in $C(\mathbb{L}_H \otimes \mathbb{I}_3)$. Furthermore, under the dynamics (8.77), the convergence of q_i to $\mathcal{S}_q$ is possible for almost every initial value of $q_i(t_0)$. Thus, it can be said that the proposed approach guarantees *almost global convergence*.

Let us consider the steady-state solution of $B_i = [b_{i,1}\ b_{i,2}\ b_{i,3}] \in \text{SO}(3)$. From the result of Lemma 8.4, the steady-state solutions of $\hat{z}_{i,1}$ and $\hat{z}_{i,2}$ are equivalent to $R_i q_1^\infty$ and $R_i q_2^\infty$, respectively, where $q_1^\infty, q_2^\infty \in \mathbb{R}^3$. Then, the solution of B_i converges to the following value:

$$\lim_{t\to\infty} B_i(t) = \lim_{t\to\infty} R_i(t) \underbrace{[q_1^\infty,\ q_2^\infty,\ q_1^\infty \times q_2^\infty]}_{:=R^c} \in \text{SO}(3) \tag{8.78}$$

where R^c is a common rotation matrix for all agents. With these results, we can make the following theorem:

Theorem 8.7 *Consider n-agents with an underlying sensing network topology* $\mathcal{G} = (\mathcal{V}, \mathcal{E}^s)$ *being connected, and with communications topology* $(\mathcal{V}, \mathcal{E}^c)$ *having an arborescence, in* SO(3). *When the auxilirary variables are updated by* (8.74), *followed by the Gram–Schmidt process* (8.73), *there exists a common rotation matrix* $R^c \in \mathrm{SO}(3)$ *such that* $B_i \to R_i R^c$ *as* $t \to \infty$.

8.4.2 Consensus Protocol on SO(3)

The desired equilibrium set of the dynamics (8.75) is defined by

$$\mathcal{S}_C := \{R_i \in \mathrm{SO}(3), \forall i \in \mathcal{V} : R_1 = R_2 = \cdots = R_n\}$$

Note that the system (8.75) converges to $\mathcal{S}_C$ if and only if the steady-state solution of B_i, $\forall i \in \mathcal{V}$ belongs to $\mathcal{S}_C$. Then, we can consider the dynamics of B_i instead of the dynamics of R_i to make the consensus on SO(3). The derivative of B_i is written as follows:

$$\frac{d}{dt}B_i = \left[\frac{d}{dt}b_{i,1} \ \frac{d}{dt}b_{i,2} \ \frac{d}{dt}b_{i,3}\right]. \tag{8.79}$$

Using the fact that $R_{ji}^{-1} = R_{ij}$, the derivative of each column vector, for $k = 1$, can be obtained as follows.

$$\begin{aligned}
\dot{b}_{i,1} &= \frac{\dot{\hat{z}}_{i,1}}{\|\hat{z}_{i,1}\|} + \left(\frac{d}{dt}(\hat{z}_{i,1}^T\hat{z}_{i,1})^{-\frac{1}{2}}\right)\hat{z}_{i,1} \\
&= \frac{1}{\|\hat{z}_{i,1}\|}\left(\sum_{j\in\mathcal{N}_i} a_{ij}(R_{ij}\hat{z}_{j,1} - \hat{z}_{i,1}) - \Omega_i\hat{z}_{i,1}\right) - \frac{1}{\|\hat{z}_{i,1}\|^3}\left(\sum_{j\in\mathcal{N}_i} a_{ij}\hat{z}_{i,1}^T(R_{ij}\hat{z}_{j,1} - \hat{z}_{i,1})\right. \\
&\quad \left. - \hat{z}_{i,1}^T\Omega_i\hat{z}_{i,1}\right)\hat{z}_{i,1} \\
&= \underbrace{\frac{\sigma_{i,1}}{\|\hat{z}_{i,1}\|} - \frac{\hat{z}_{i,1}^T\sigma_{i,1}}{\|\hat{z}_{i,1}\|^3}\hat{z}_{i,1}}_{\triangleq\rho_{i,1}} - \Omega_i\underbrace{\frac{\hat{z}_{i,1}}{\|\hat{z}_{i,1}\|}}_{=b_{i,1}},
\end{aligned} \tag{8.80}$$

where $\sigma_{i,k}$ denotes the first term in the right equation of (8.74), defined by

$$\sigma_{i,k} := \sum_{j\in\mathcal{N}_i} a_{ij}(R_{ij}\hat{z}_{j,k} - \hat{z}_{i,k}). \tag{8.81}$$

From the Lemma 8.4, we can see that $\hat{z}_{i,k}$ converges to $R_i(t)q_k^{\infty}$ as time goes to infinity. It indicates that the term $\sigma_{i,k}$ in (8.81) vanishes as time goes to infinity. This feature is exploited to analyze the stability of B_i in the latter. The derivative of $b_{i,2}$ is obtained as follows:

$$\dot{b}_{i,2} = \frac{\dot{v}_{i,2}}{\|v_{i,2}\|} + \left(\frac{d}{dt}(v_{i,2}^T v_{i,2})^{-\frac{1}{2}}\right) v_{i,2} \tag{8.82}$$

where $v_{i,2} = \hat{z}_{i,2} - \langle \hat{z}_{i,2}, b_{i,1}\rangle b_{i,1}$. The terms on the right side of (8.82) can be written as follows:

$$\begin{aligned}
\dot{v}_{i,2} =& \dot{\hat{z}}_{i,2} - \frac{d}{dt}\langle \hat{z}_{i,2}, b_{i,1}\rangle b_{i,1} - \langle \hat{z}_{i,2}, b_{i,1}\rangle \frac{d}{dt} b_{i,1} \\
=& \sigma_{i,2} - \Omega_i \hat{z}_{i,2} - \langle \sigma_{i,2}, b_{i,1}\rangle b_{i,1} - \hat{z}_{i,2}^T \left(\frac{\sigma_{i,1}}{\|\hat{z}_{i,1}\|} - \frac{\hat{z}_{i,1}^T \sigma_{i,1}}{\|\hat{z}_{i,1}\|^3} \hat{z}_{i,1} \right) b_{i,1} \\
& - \langle \hat{z}_{i,2}, b_{i,1}\rangle \left(-\Omega_i b_{i,1} + \rho_{i,1}\right) \\
=& - \Omega_i v_{i,2} + \sigma_{i,2} - \langle \sigma_{i,2}, b_{i,1}\rangle b_{i,1} - \langle \hat{z}_{i,2}, \rho_{i,1}\rangle b_{i,1} - \langle \hat{z}_{i,2}, b_{i,1}\rangle \rho_{i,1}
\end{aligned} \tag{8.83}$$

and

$$\left(\frac{d}{dt}(v_{i,2}^T v_{i,2})^{-\frac{1}{2}}\right) v_{i,2} = -\frac{v_{i,2}^T \dot{v}_{i,2}}{\|v_{i,2}\|^3} v_{i,2} \tag{8.84}$$

Then, (8.82) can be rewritten as follows:

$$\begin{aligned}
\dot{b}_{i,2} =& \frac{\dot{v}_{i,2}}{\|v_{i,2}\|} - \frac{v_{i,2}^T \dot{v}_{i,2}}{\|v_{i,2}\|^3} v_{i,2} \\
=& -\Omega_i b_{i,2} + \underbrace{\frac{1}{\|v_{i,2}\|}\zeta + \frac{\langle v_{i,2}, \zeta\rangle}{\|v_{i,2}\|^3} v_{i,2}}_{:=\rho_{i,2}}
\end{aligned} \tag{8.85}$$

where $\zeta = \sigma_{i,2} - \langle \sigma_{i,2}, b_{i,1}\rangle b_{i,1} - \langle \hat{z}_{i,2}, \rho_{i,1}\rangle b_{i,1} - \langle \hat{z}_{i,2}, b_{i,1}\rangle \rho_{i,1}$. Likewise, the derivative of $\dot{b}_{i,3}$ is calculated as follows:

$$\begin{aligned}
\dot{b}_{i,3} =& \frac{d}{dt} b_{i,1} \times b_{i,2} + b_{i,1} \times \frac{d}{dt} b_{i,2} \\
=& (-\Omega_i b_{i,1} + \rho_{i,1}) \times b_{i,2} + b_{i,1} \times (-\Omega_i b_{i,2} + \rho_{i,2}) \\
=& - \Omega_i b_{i,1} \times b_{i,2} - b_{i,1} \times \Omega_i b_{i,2} + \underbrace{\rho_{i,1} \times b_{i,2} + b_{i,1} \times \rho_{i,2}}_{:=\rho_{i,3}} \\
=& (\omega_i \times b_{i,1}) \times b_{i,2} + b_{i,1} \times (\omega_i \times b_{i,2}) + \rho_{i,3}
\end{aligned} \tag{8.86}$$

Consequently, by using the Jacobi identity,[3] (8.86) is rewritten as follows:

$$\begin{aligned}\dot{b}_{i,3} &= -\omega_i \times (b_{i,2} \times b_{i,1}) + \rho_{i,3} \\ &= -\Omega_i (b_{i,1} \times b_{i,2}) + \rho_{i,3} \\ &= -\Omega_i b_{i,3} + \rho_{i,3}\end{aligned} \tag{8.87}$$

Now, we make the following main result of this chapter:

Theorem 8.8 *Suppose that the sensing network topology $\mathcal{G} = (\mathcal{V}, \mathcal{E}^s)$ has a spanning tree and the communication topology $(\mathcal{V}, \mathcal{E}^c)$ has an arborescence. Under the auxiliary update dynamics* (8.74), *the term $\rho_{i,k}$, $\forall i \in \mathcal{V}, k \in \{1, 2, 3\}$, defined in* (8.80), (8.85), *and* (8.86), *satisfies*

$$\|\rho_{i,k}(t)\| \leq k_\rho e^{-\lambda_\rho (t-t_0)} \|\rho_{i,k}(t_0)\|, \ \forall t \geq T + t_0 \tag{8.88}$$

where $k_\rho, \lambda_\rho > 0$ and $T > 0$ are some finite values.

Proof Let us consider the term $\sigma_{i,k}$ in (8.81). From the result of Lemma 8.4, there exist $k_\sigma > 0$ and $\lambda_\sigma > 0$ such that

$$\|\sigma_{i,k}(t)\| \leq k_\sigma e^{-\lambda_\sigma (t-t_0)} \|\sigma_{i,k}(t_0)\| \tag{8.89}$$

for almost every initial value of $\hat{z}_{i,k}$. From the definition of $\rho_{i,k}$ in (8.80), (8.85), and (8.86), it is noted that the term $\rho_{i,k}$ is a combination of $\sigma_{i,1}$ and $\sigma_{i,2}$. Then, the Euclidean norm of $\rho_{i,k}$ satisfies the following inequality:

$$\|\rho_{i,k}(t)\| \leq \|\sigma_{i,1}(t)\| f_{i,1}(\hat{z}_{i,1}, \hat{z}_{i,2}) + \|\sigma_{i,2}(t)\| f_{i,2}(\hat{z}_{i,1}, \hat{z}_{i,2})$$

where $f_1, f_2 : \mathbb{R}^3 \times \mathbb{R}^3 \to \mathbb{R}^+$ has the algebraic fraction whose denominator contains the term such as $\|\hat{z}_{i,k}\|$ or $\|\hat{z}_{i,1}\|^2 \hat{z}_{i,2} - \langle \hat{z}_{i,2}, \hat{z}_{i,1} \rangle \hat{z}_{i,1}$. From the result of Lemma 8.4, the solution of $\hat{z}_{i,k}$ converges to some finite values except for 0 or linearly dependent vector with another auxiliary variable. Then, there exist finite values $T > 0$ and $\delta_1, \delta_2 > 0$ such that

$$\|f_1(\hat{z}_{i,1}, \hat{z}_{i,2}) - \delta_1\| \leq \epsilon_1, \quad \|f_2(\hat{z}_{i,1}, \hat{z}_{i,2}) - \delta_2\| \leq \epsilon_2 \tag{8.90}$$

for all $t \geq T + t_0$, where $\epsilon_1, \epsilon_2 > 0$ are sufficiently small values, which completes the proof.

[3] The Jacobi identity is a property that determines the order of evaluation behaves for the given operation. When a, b, and c are elements of set $\mathbb{R}^3$, the Jacobi identity has a relationship: $a \times (b \times c) + b \times (c \times a) + c \times (a \times b) = 0$

By (8.80), (8.85), and (8.87), the derivative of B_i is calculated as follows:

$$\dot{B}_i B_i^T = -\Omega_i + \rho_i B_i^T, \tag{8.91}$$

where $\rho_i = [\rho_{i,1}, \rho_{i,2}, \rho_{i,3}] \in \mathbb{R}^{3\times 3}$.

Next, we design the control law Ω_i based on the estimated orientation as follows:

$$\Omega_i = \log(B_i^T) = -\log(B_i) = -\frac{\vartheta_i}{2\sin(\vartheta_i)}[B_i^T - B_i] \tag{8.92}$$

where ϑ_i is an angle of rotation axis satisfying $1 + 2\cos(\vartheta_i) = trace(B_i)$ and $-\pi < \vartheta_i < \pi$ [1]. In (8.92), when $trace(B_i) \approx 3$, it means that $\vartheta_i \approx 0$. In such a case, to avoid a possibility of singularity,[4] we simply compute Ω_i as $\Omega_i = -\frac{1}{2}[B_i^T - B_i]$. The inputs to the orientation control are angular rate computed by $\Omega_i^\vee = (\omega_{1i}, \omega_{2i}, \omega_{3i})^T$ where Ω_i is computed from B_i in (8.92). Since the angular rates ω_{1i}, ω_{2i}, and ω_{3i} are inertia forces in single-integrator dynamics, it can be implemented in a local coordinate frame. Now, the behavior of the dynamics for $R_i(t)$ is analyzed as follows:

Theorem 8.9 *Suppose that the sensing network topology $\mathcal{G} = (\mathcal{V}, \mathcal{E}^s)$ has a spanning tree, and the communications and actuation topologies $(\mathcal{V}, \mathcal{E}^c)$ have a common arborescence. Under the dynamics (8.74) and* (8.75) *with control law (8.92), there exists $R^c \in \mathrm{SO}(3)$ such that $R_i(t) \in \mathrm{SO}(3)$ converges to $(R^c)^{-1}$ asymptotically.*

Proof Let us use the following Lyapunov candidate $V_i = 3 - trace(B_i)$, which is a positive semidefinite function. Note that $V_i = 0$ when $B_i = \mathbb{I}_3$. The derivative of V_i is obtained as follows:

$$\begin{aligned}
\dot{V}_i &= -\frac{d}{dt}trace(B_i) = -trace\left(\frac{d}{dt}B_i\right) \\
&= -trace(-\Omega_i B_i + \rho_i) \\
&= trace(\Omega_i B_i) - trace(\rho_i)
\end{aligned} \tag{8.93}$$

When each element of B_i is denoted by $[B_i]_{p,q}$, $\forall p, q \in \{1, 2, 3\}$, the term $trace(\Omega_i B_i)$ is written as follows:

$$trace(\Omega_i B_i) = -\frac{\vartheta_i}{2\sin(\vartheta_i)} trace(\mathbb{I}_3 - B_i B_i) \tag{8.94}$$

Noticing that $trace(\mathbb{I}_3) > trace(B_i B_i)$ as far as $B_i \neq \mathbb{I}_3$, and $\frac{\vartheta_i}{2\sin(\vartheta_i)} > 0$, and denoting $\kappa_i \triangleq trace(\mathbb{I}_3 - B_i B_i)$, we can see that

$$trace(\Omega_i B_i) = -\frac{\vartheta_i}{2\sin(\vartheta_i)}\kappa_i \leq 0$$

[4] From L'Hospital's rule, the term $-\frac{\vartheta_i}{2\sin(\vartheta_i)}$ can be transformed as $-\frac{1}{2}$ as $\theta \to 0$.

Now, let us consider an asymptotic behavior of $trace(\rho_i)$. From the result of Theorem 8.8, there exist a finite value T_1 and a class KL function $\beta(\cdot,\cdot)$ such that

$$|trace(\rho_i(t))| \leq \beta(trace(\rho(t_0)), T_1),\ \forall t \geq T_1 \tag{8.95}$$

Then, the derivative of V_i has the following relations for $t \geq T_1$:

$$\dot{V}_i \leq trace(\Omega_i B_i) + |trace(\rho_i)| \tag{8.96}$$

$$\leq -\frac{\vartheta_i}{2\sin(\vartheta_i)}\kappa_i + \beta(trace(\rho(t_0)), T_1). \tag{8.97}$$

Now, as T_1 goes to infinity, $\beta(trace(\rho(t_0)), T_1)$ converges to zero. Thus, there exist $T_2 > T_1$ and $0 < \phi < 1$ such that

$$\dot{V}_i \leq -(1-\phi)\frac{\vartheta_i}{2\sin(\vartheta_i)}\kappa_i - \phi\frac{\vartheta_i}{2\sin(\vartheta_i)}\kappa_i + \beta(trace(\rho(t_0)), T_1) \tag{8.98}$$

$$\leq -(1-\phi)\frac{\vartheta_i}{2\sin(\vartheta_i)}\kappa_i \leq 0, \tag{8.99}$$

for $t \in (T_1, T_2]$. Consequently, by Barbalat's lemma, i.e., Lemma 2.8, and by Theorem 2.23, there holds $V_i \to 0$, which implies $B_i \to \mathbb{I}_3$ as $t \to \infty$. From the Lemma 8.4, we know that $B_i(t)$ converges to $R_i(t)R^c$, $\forall i \in \mathcal{V}$. With these results, i.e., $B_i(t) \to R_i(t)R^c$ and $B_i(t) \to \mathbb{I}_3$, we obtain the conclusion of $R_i(t) \to (R^c)^{-1}$, $\forall i \in \mathcal{V}$, which completes the proof.

From the definition of R^c in (8.78) and the result of Lemma 8.4, the common rotation matrix R^c is determined from the initial values of auxiliary variables. Thus, it is worth to note that the steady-state solution of $R_i(t)$ is dependent on the initial values of auxiliary variables. The results in this section can be used for the formation control also. For the formation control, we can directly formulate the problem by replacing $\hat{R}_i$ in (8.34) by R_i given in (8.75), which is a function of B_i updated by (8.80), (8.82), and (8.87). If $R_i(t_0)$ is known, then, by propagating (8.75) with an initial value $R_i(t_0)$, each agent can compute $R_i(t)$. However, notice that R_i in (8.75) is not available at local coordinate frame since we have no information of $R_i(t_0)$. So, alternatively, let us replace $\hat{R}_i$ in (8.34) by $B_i(t)$ which makes the following position update:

$$u_i^i = k_u \sum_{j \in \mathcal{N}_i} a_{ij}((p_j^i - p_i^i) - B_i(t)(p_j^* - p_i^*)) \tag{8.100}$$

where $B_i \to R_i R^c$ as $t \to \infty$. Then, from (8.100), the position dynamics with respect to ${}^g\Sigma$ can be obtained as

$$\dot{p}_i = k_u \sum_{j \in \mathcal{N}_i} a_{ij}(p_j - p_i - (R_i)^{-1} B_i(t)(p_j^* - p_i^*)) \tag{8.101}$$

Then, defining the error in the position as $e_i = p_i - R^c p_i^*$, it is updated as

$$\dot{e}_i = k_u \sum_{j \in \mathcal{N}_i} a_{ij}(e_j - e_i) + k_u \sum_{j \in \mathcal{N}_i} a_{ij}(R^c - R_i^{-1} B_i)(p_j^* - p_i^*) \tag{8.102}$$

Denoting $w_i(t) \triangleq k_u \sum_{j \in \mathcal{N}_i} a_{ij}(R^c - R_i^{-1} B_i)(p_j^* - p_i^*)$, we have

$$\dot{e}(t) = -k_u(\mathbb{L} \otimes \mathbb{I}_3)e(t) + w(t) \tag{8.103}$$

where $\mathbb{L}$ is the Laplacian according to the topology a_{ij} for $(i, j)^e \in \mathcal{E}$. Thus, due to the same reason as the proof of Theorem 8.4, we can see that $p_i(t), \forall i \in \mathcal{V}$ globally exponentially converges to $R^c p_i^* + p_\infty$ where R^c and p_∞ are offsets for rotations and translations.

Remark 8.2 Following the similar procedure as (7.40)–(7.46), we can also design position estimation dynamics. After this, the estimated position can be directly utilized as inputs for a position-feedback control. In this case, the positions are estimated via global orientation alignment, from the distance-based setup. Thus, the distance-based setup can be transformed to the position-based setup under global orientation alignment.

For the orientation control in 3-D, we have defined auxiliary variables $\hat{z}_{i,1}$ and $\hat{z}_{i,2}$ for all agents. These variables could be considered as local variables updated on each agent's local coordinate frame. Let us transform these variables to the global frame as $\hat{z}_{i,1}^g = R_g^i \hat{z}_{i,1}$ and $\hat{z}_{i,2}^g = R_g^i \hat{z}_{i,2}$. Then, the following example shows that the average of $\hat{z}_{i,1}^g$ and $\hat{z}_{i,2}^g$ are invariant, which characterizes the dynamic behaviors of (8.74).

Example 8.3 (*Dynamic characteristics of auxiliary variables*) Let us suppose that the underlying graph is connected and undirected. Then, from the consensus property of (8.76), for a given k, we have $\frac{1}{n}\sum_{i=1}^n \hat{q}_{i,k}(t) = \text{const}, \ \forall t \geq t_0$. With the relationship $\hat{q}_{i,k} = R_i^{-1} \hat{z}_{i,k}$, we further have

$$\frac{1}{n}\sum_{i=1}^n R_i^{-1} \hat{z}_{i,k} = \frac{1}{n}\sum_{i=1}^n \hat{z}_{i,k}^g = \text{const}$$

The above relationship implies that when the auxiliary variables are projected onto the global coordinate frame, the average of auxiliary variables does not change. Also, from the above relationship, we have

$$
\begin{aligned}
\frac{d(\sum_{i=1}^{n} R_i^{-1}\hat{z}_{i,k})}{dt} &= \sum_{i=1}^{n} R_i^T \Omega_i^T \hat{z}_{i,k} + \sum_{i=1}^{n} \dot{R}_i^T \left\{ \sum_{j\in\mathcal{N}_i} (R_{ji}^{-1}\hat{z}_{j,k} - \hat{z}_{i,k}) - \Omega_i \hat{z}_{i,k} \right\} \\
&= \sum_{i=1}^{n} \left[\sum_{j\in\mathcal{N}_i} \hat{z}_{j,k}^g - \hat{z}_{i,k}^g \right] \\
&= 0
\end{aligned}
$$

Thus, for a balanced graph, we can obtain the additional constraint $\sum_{i=1}^{n} \hat{z}_{i,k}^g(t) = \sum_{i=1}^{n} R_i^{-1}\hat{z}_{i,k}(t) = 0$ as one of the dynamic characteristics.

8.5 Summary and Simulations

The weakness of Chap. 6, i.e., the quasi-global convergence depending upon initial orientation angles, was remedied in this chapter by exchanging auxiliary variables. In Sects. 8.1 and 8.2, after exchanging the auxiliary variables between neighboring agents, the variables were updated until they converge to a constant. Then, the converged variables are converted to the orientation angles of agents. In Sects. 8.3 and 8.4, the orientations of agents have been controlled until they converge to a common value. Since the orientations are controlled, we combined both the orientation estimation and the orientation control. Since the estimation and control schemes are simultaneously updated, the convergence property was also analyzed simultaneously. Additionally, in Sect. 8.4, we showed that the orientation control law can be combined to achieve the desired formation of agents, which means that the orientations and positions of agents can be controlled together for orientation synchronization and for achieving the desired formation. Table 8.1 shows properties of the formation control via the global orientation control in $\mathbb{R}^2$. From the comparison with Table 6.3, which ensures a quasi-global convergence, we can see that the global convergence is achieved with a more cost in communications and computations.

Let us consider five agents in Fig. 6.6 for the verification of the orientation-control-based formation control in $\mathbb{R}^3$. The sensing is undirected while the communications and actuations are directed. Also, the underlying sensing topology and communications/actuations topologies are not identical. For the simulation, the orientation control law (8.75), the auxiliary update law (8.74) with Ω_i given in (8.92), the Gram–Schmidt orthonormalization procedure (8.73), and the position control law (8.100) are used. It is also required to select initial values of $\hat{z}_{i,k}(t_0)$ for $k = 1, 2$. Except the initial variables $\hat{z}_{i,k}(t_0)$, all. Other values are given as the ones in Sect. 6.4. We first choose the initial values $\hat{z}_{i,k}(t_0)$, and then conduct the Gram–Schmidt orthonormalization procedure (8.73). From the orthonormalization procedure, we compute the matrix $B_i(t)$, which is used for the computation of Ω_i. Then, each agent actively controls its orientation according to the orientation control law (8.75). Also, agents communicate with neighboring agents for the update of $\hat{z}_{i,k}(t)$, assuming that the

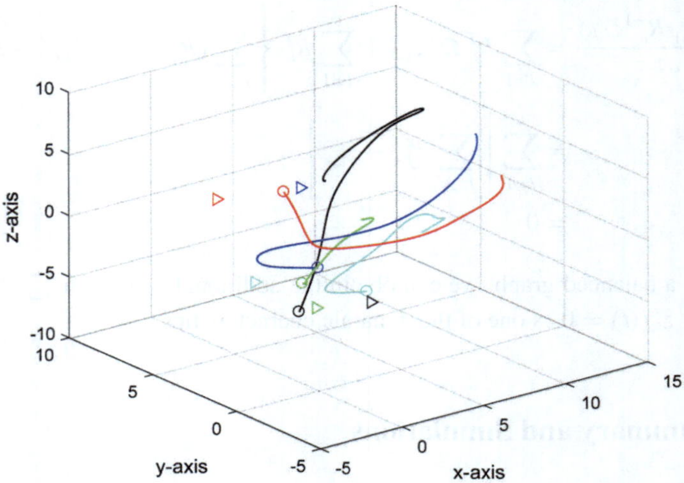

Fig. 8.4 Trajectories of positions of five agents, without compensations of the rotations and translations. The trajectories of agents are distinguished by the colors. The initial positions are marked by ∘ and the desired positions are marked by ▹

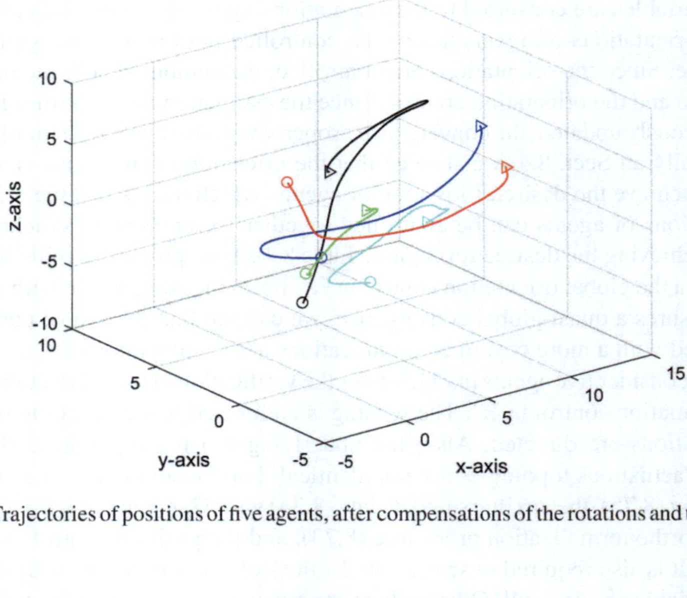

Fig. 8.5 Trajectories of positions of five agents, after compensations of the rotations and translations

required relative orientations R_{ji}^{-1} are provided from the sensing topology. Then, agents update their positions according to the position control law (8.100). These processes are repeated until they converge to the desired relative displacements.

The initial orientations of the agents are same as the simulation in Figs. 7.5 and 7.6. Figures 8.4 and 8.5 depict the trajectories of positions of agents in $\mathbb{R}^3$.

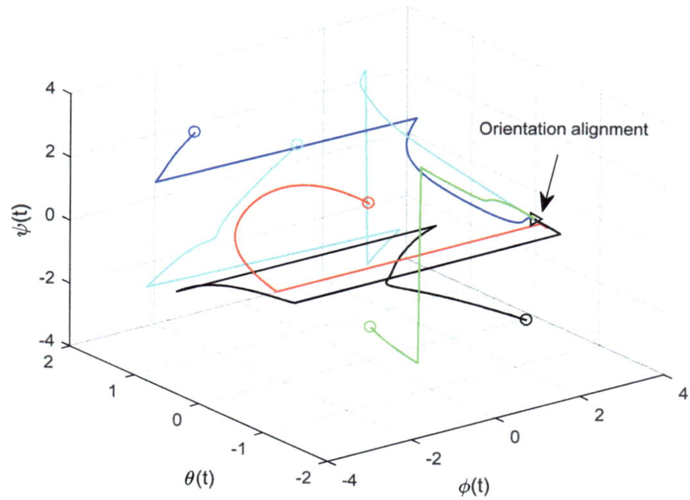

Fig. 8.6 Trajectories of orientations of five agents (Euler angles ϕ_i, θ_i, and ψ_i). The trajectories of agents are distinguished by the colors. The initial angles are marked by ∘ and the final angles are marked by ▹

The formation control law presented in this chapter ensures a convergence to the desired relative displacements up to a common rotation and common translation, i.e, $p_i(t) \to R^c p_i^* + p_\infty$. This is due to the fact that the control law transforms the distance-based setup to the displacement-based setup. The trajectories in Fig. 8.4 are the results without compensation of the offsets (i.e., the common rotations and common translation). After the compensation of the offsets, in Fig. 8.5, we can see that the agents converge to the desired positions exactly. Figure 8.6 depicts the trajectories of Euler angles ϕ_i, θ_i, and ψ_i of agents, i.e., orientations of agents. Since the orientation control law (8.75) aims to achieve a synchronization of the orientations, as expected, the angles have converged to common values. That is, $\phi_i \to \phi^*$, $\theta_i \to \theta^*$, and $\psi_i \to \psi^*$ for all $i \in \mathcal{V}$. When we compare the simulation results of Figs. 8.5 and 8.6 and the results of Figs. 7.5 and 7.6, it is observed that the control laws in Sect. 7.2 require less efforts than the control laws in Sect. 8.4. Figures 8.7 and 8.8 depict the trajectories of positions and orientations of the same agents as Figs. 8.5–8.6, but with a slight different setting. All the initial and desired positions, and initial orientations are the same, with the same control gains. Only the initial values of $\hat{z}_{i,k}(t_0)$, $i \in \mathcal{V}$, $k = 1, 2$ are different. Figures 8.7 and 8.8 show that the trajectories of agents are highly dependent upon the choice of initial values $\hat{z}_{i,k}$. Thus, it may be beneficial to investigate the effects of the initial values $\hat{z}_{i,k}$ to the overall performance of the formation control.

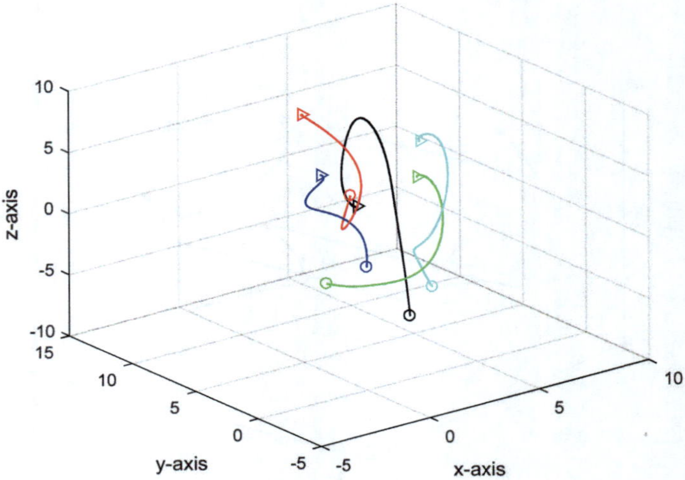

Fig. 8.7 Trajectories of positions with different $\hat{z}_{i,k}(t_0)$, $i \in \mathcal{V}$, $k = 1, 2$ from Fig. 8.5

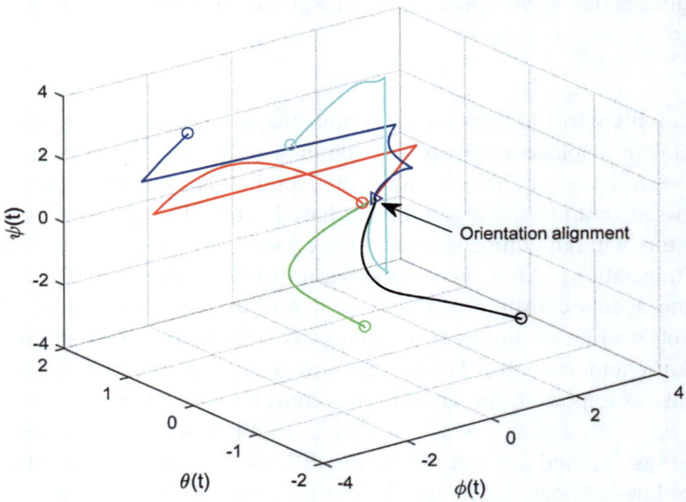

Fig. 8.8 Trajectories of orientations with different $\hat{z}_{i,k}(t_0)$, $i \in \mathcal{V}$, $k = 1, 2$ from Fig. 8.6

8.6 Notes

The formation control problem studied in this chapter is composed of two steps: global orientation estimation, and formation control. To enable a global convergence with local relative measurements, the relative sensings, communications, computation and control processes are connected in a feedback closed loop. So, the overall loop is quite similar to the traditional closed-loop systems except the communications. In the traditional closed-loop systems, there is a fixed global coordinate frame that defines all the states; but in the distributed formation control, there is no fixed frame; instead all the sensings, computations, and control efforts are defined in the local coordinate frames.

This chapter was written on the basis of [4–7, 9]. Section 8.1 was reused from [6], Sect. 8.2 was reproduced from [8, 9], Sect. 8.3 was reused from [7], and Sect. 8.4 was reused from [4]. The following copyright and permission notices are acknowledged.

References

1. Cardoso, J.R., Leite, F.S.: Exponentials of skew-symmetric matrices and logarithms of orthogonal matrices. J. Comput. Appl. Math. **233**(11), 2867–2875 (2010)
2. Gonano, C.A., Zich, R.E.: Cross product in n dimensions – the doublewedge product, pp. 1–6 (2014). arXiv:1408.5799v1 [math.GM]
3. Khalil, H.K.: Nonlinear Systems. Prentice Hall, Upper Saddle River (2002)
4. Lee, B.-H.: Consensus on the Special Orthogonal Group: Theory and Applications to Formation Control. Ph.D. thesis, Gwangju Institute of Science and Technology (2017)
5. Lee, B.-H., Ahn, H.-S.: Distributed estimation for the unknown orientation of the local reference frames in n-dimensional space. In: Proceedings of the International Conference on Control, Automation, Robotics and Vision (ICARCV), pp. 1–6 (2016)
6. Lee, B.-H., Ahn, H.-S.: Distributed formation control via global orientation estimation. Automatica **73**, 125–129 (2016)
7. Lee, B.-H., Ahn, H.-S.: Distributed control for synchronization on the circle. In: Proceedings of the 56th IEEE Conference on Decision and Control, pp. 4169–4174 (2017)
8. Lee, B.-H., Ahn, H.-S.: Formation control and network localization via distributed global orientation estimation in 3-D, pp. 1–8 (2017). arXiv:1708.03591v1 [cs.SY]
9. Lee, B.-H., Kang, S.-M., Ahn, H.-S.: Distributed orientation estimation in SO(d) and applications to formation control and network localization. IEEE Trans. Control Netw. Syst. 1–11 (2019). https://doi.org/10.1109/TCNS.2018.2888999
10. Moon, T.K., Stirling, W.C.: Mathematical Methods and Algorithms for Signal Processing. Prentice Hall, Upper Saddle River (2000)
11. Scardovi, L., Sarlette, A., Sepulchre, R.: Synchronization and balancing on the $n-torus$. Syst. Control Lett. **56**(5), 335–341 (2007)
12. Sepulchre, R., Paley, D.A., Leonard, N.E.: Stabilization of planar collective motion: all-to-all communication. IEEE Trans. Autom. Control **52**(5), 811–824 (2007)

Part IV
Bearing-Based Formation Control

Chapter 9
Formation Control via Bearing Measurements

Abstract This chapter presents bearing-based formation control schemes via orientation alignments. In the bearing-based control schemes, the sensing variables are bearing vectors in the aligned coordinate frames and the control variables are also bearing vectors in the aligned coordinate frames. But, as considered in the previous chapters, the initial coordinate frames of agents are not aligned. In order to align the coordinate frames, we conduct orientation alignments with simultaneous bearing-based formation control. Note that, the bearing vectors are normalized relative displacements; hence as commented in Chap. 1, the bearing vectors include the information of bearing angles. Thus, in bearing-based approaches under orientation alignment setups, the formation configuration that will be achieved is congruent up to translations and dilation.

9.1 Background

In the gradient control laws presented from Chaps. 3 to 5, the control variables are the distances between agents, while the sensing variables are the relative displacements. In the formation control via orientation alignment presented from Chaps. 6 to 8, both the control and sensing variables are the relative displacements. The measurements of relative displacements include angle and pure distance between agents. It is also natural to raise a question whether a formation control can be realized only using angle measurements or pure-distance measurements. The pure-distance-based approaches have been studied in [5, 10]. But, in pure-distance- based approach, it is difficult to realize a control input that usually needs to be a vector in Euclidean space. Thus, in [10], it attempts to estimate displacements only using a set of distance information.

In angle-based approaches, there are two different approaches, bearing-based approach and subtended angle approach, as depicted in Fig. 9.1. The bearing measurements are described when the coordinate frames of agents are aligned. The bearing measurements are given between two neighboring agents. For agent i with two neighboring agents j and k, the subtended angle is the angle given as the difference between two vectors z_{ji} and z_{ki} as shown in Fig. 9.1. In the literature, there are some recent works on subtended angle-based formation control. For example, in [3],

H.-S. Ahn, *Formation Control*, Studies in Systems, Decision and Control 205, https://doi.org/10.1007/978-3-030-15187-4_9

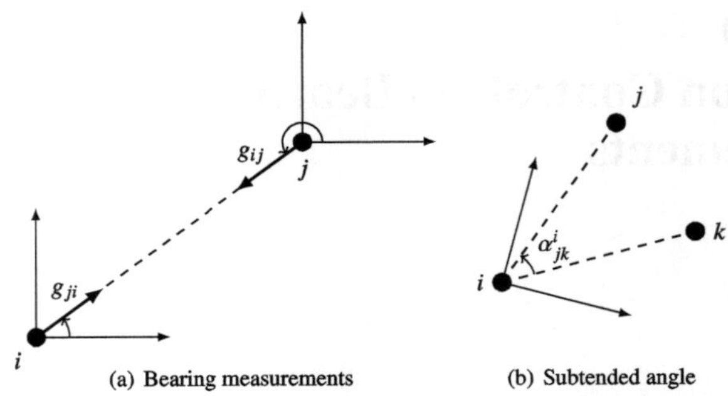

Fig. 9.1 Bearing vectors g_{ji} and g_{ij} versus subtended angle α^i_{jk}: The bearing vectors are defined in orientation aligned coordinate frames; but it can be generalized to orientation misaligned agents. The subtended angles are scalar components independent of the orientation of local coordinate frames

subtended angle-based control law was designed for a triangular formation, and in [7], a triangular formation was controlled to achieve the desired subtended angle constraints with measurements of bearing angles, under misaligned orientation setup. In [11], the concepts of weak rigidity and global weak rigidity were introduced for transforming the distance constraints to subtended angle constraints. It was argued that the rigidity of formations under subtended angle constraints could be checked by distance-based rigidity. The infinitesimally weak rigidity without incident edges and formation control of triangular agents under weak rigidity concept were developed in [9], and formation control in 3-dimensional space was further studied in [8] (refer to Sect. 2.2.4 for weak rigidity theory). The distance constraints and bearing constraints may be also combined to achieve a better performance in formation control [4]. It is beneficial to note that the bearing measurements are actually vector measurements, while the subtended angles are scalar variables.

The subtended angles are independent of the orientations of local coordinate frames. For example, consider a set of agents in a triangular formation shape; then the subtended angles are invariant whatever the orientations of local coordinate frames are. However, the bearing vectors are functions of the orientations of local coordinate frames. Thus, it appears that, if a multi-agent system has only the subtended angles as both sensing variables and control variables, it is not possible to control the formation in a vector space. That is, if control variables are subtended angles, then we may need to have subtended angles as well as some more variables such as displacements for sensing variables. On the other hand, if the subtended angles are only the sensing variables, we may have to estimate some more variables such as displacements or bearing vectors. In [8], one of the control variables is subtended angle for an agent, but for this agent, we need to have relative displacements for sensing variables. In [3], the desired subtended angles have been achieved in triangular graphs; but in the computation of control inputs that are implemented in local coordinate frames, they use a maximum or minimum magnitude of bearing angles of other agents that are measured in an agent's local frame. Thus, it seems that a subtended angle-based control may have to be formulated with other vector variables that may be measured, estimated, or controlled. Also note that, a difference of two bearing vectors can be considered as a subtended angle, i.e., $\alpha^i_{jk} = \angle g^i_{ji} - \angle g^i_{ki}$, where $\angle g^i_{ji}$ and $\angle g^i_{ki}$ are angles determined by the bearing vectors g^i_{ji} and g^i_{ki} defined in the local coordinate frame. Thus, formation control with subtended angles may be formulated by the bearing vectors. Recall that a similar comment was also given at the end of Chap. 1 and more discussions on subtended angles are given in Sect. 13.4. Thus, in this chapter, we would like to focus on bearing vector-based formation control problems. Specifically, we introduce formation control laws for directed graphs only using bearing measurements. A bearing-based formation control law for acyclic persistent graphs when the agents' local coordinate frames are aligned with respect to a common coordinate frame was introduced in [16], and a control law when agents are not aligned was developed in [18]. This chapter summarizes these results in a concise way.

Consider a directed graph $\vec{\mathcal{G}}$. With $x \in \mathbb{R}^n$, the orthogonal projection matrix of a nonzero vector x is an $n \times n$ matrix defined as

$$\mathbb{P}_x \triangleq \mathbb{I}_n - \frac{x}{\|x\|}\frac{x^T}{\|x\|} \tag{9.1}$$

The above projection matrix has several nice properties including symmetry and idempotence, i.e., $\mathbb{P}_x = \mathbb{P}_x^T = \mathbb{P}_x^2 = x^{\perp}(x^{\perp})^T$ in $\mathbb{R}^2$. Letting $\hat{x} = \frac{x}{\|x\|}$, for any nonzero vector v, we can have

$$v^T \mathbb{P}_x v = v^T \left(\mathbb{I}_n - \frac{x}{\|x\|}\frac{x^T}{\|x\|} \right) v = v^T v - v^T \hat{x}\hat{x}^T v$$

Since it is clear that $\|v\| \geq \|\hat{x}^T v\|$, we can have $v^T \mathbb{P}_x v \geq 0$, which means that the matrix $\mathbb{P}_x$ is positive semi-definite. From $(\mathbb{I}_n - \hat{x}\hat{x}^T)(\mathbb{I}_n - \hat{x}\hat{x}^T) = \mathbb{I}_n - \hat{x}\hat{x}^T - \hat{x}\hat{x}^T + \hat{x}\hat{x}^T\hat{x}\hat{x}^T = \mathbb{I}_n - \hat{x}\hat{x}^T$, we can see that it is idempotent. Also from $(\mathbb{I}_n - \hat{x}\hat{x}^T)\hat{x} = \hat{x} - \hat{x} = 0$, we can find that the matrix $\mathbb{P}_x$ has the nullspace of span$\{x\} \subseteq$ *null*$(\mathbb{P}_x)$. From this nullspace property, we can see that zero is an eigenvalue of the matrix $\mathbb{P}_x$, with a corresponding eigenvector $\hat{x}$. Furthermore, for any vector $\hat{y}$, which is orthogonal to the vector $\hat{x}$, we have $(\mathbb{I}_n - \hat{x}\hat{x}^T)\hat{y} = \hat{y}$. Thus, the remaining eigenvalues of $\mathbb{P}_x$ are 1, with multiplicity of $n - 1$. The relative bearing vector $g_{ji} \in \mathbb{R}^d$ between two points p_i and p_j is defined as a unit vector pointing from p_i (tail) to p_j (head) such as

$$g_{ji} \triangleq \frac{p_j - p_i}{\|p_j - p_i\|} = \frac{z_{ji}}{\|z_{ji}\|} \tag{9.2}$$

The *bearing congruency* indicates that the two realized configurations differ only by translations and dilation. Given a desired framework $(\vec{\mathcal{G}}, p^*)$, the desired bearing vectors are given as $\Gamma \triangleq \{g^*_{ji} : \forall (i, j)^{\bar{e}} \in \overrightarrow{\mathcal{E}^s}\}$ where $\overrightarrow{\mathcal{E}^s}$ is the set of edges given by the desired sensing topology and $g^*_{ji} \triangleq \frac{p^*_j - p^*_i}{\|p^*_j - p^*_i\|}$. For more background about bearing rigidity, refer to Sect. 2.2.3.

9.2 Bearing-Based Formation Control

Let us consider a directed graph composed of n agents where the vertex v_1 is the root node. Figure 9.2 shows an example of bearing-based formation control. The agent 2 measures the bearing vector g_{12}, and agent 3 measures the bearing vectors g_{23} and g_{13}. The agent 4 has two bearing measurements as g_{34} and g_{24}. For a convenience of presentation in the bearing-based formation control, the underlying topology is defined as a graph with the same vertex set and the same edge set as the sensing graph; but the directions of edges are reversed. So, if edges are given as $(i_p, i_q)^{\bar{e}} \in \overrightarrow{\mathcal{E}^s}$, $i_p, i_q \in \mathcal{V}$, in the sensing topology, then the underlying graph topology is considered as having edges $(i_q, i_p)^{\bar{e}} \in \overrightarrow{\mathcal{E}}$. It is supposed that the underlying graph has an arborescence, with the root node 1. In Fig. 9.2a, the agents 2 and 3 are directing toward the leader; but in Fig. 9.2c, the directions are opposite from the directions of Fig. 9.2a.

Any graph $\overrightarrow{\mathcal{G}} = (\mathcal{V}, \overrightarrow{\mathcal{E}})$ of n vertices constructed by Henneberg construction is minimally acyclic. The minimally acyclic digraph $\overrightarrow{\mathcal{G}}$ has exactly $2n - 3$ directed edges in 2-dimensional space and has a rooted spanning tree (arborescence). In the bearing-based networks, an edge in the sensing topology represents a bearing vector assignment. Let us consider acyclic minimally persistent formations (AMPF) in $\mathbb{R}^2$ in terms of sensing topology. Note that the terminology *acyclic minimally persistent formations* (AMPF) is rather a topological definition given in Chap. 5 and Sect. 2.2.2. In the literature of rigidity theory or distributed formation control, it is also called *leader-first-follower (LFF) structure* [1, 13]. So, if agent i has an outgoing node j, then agent i senses the bearing vector g_{ji} and it also controls the vector g_{ji} such that it becomes g^*_{ji}. Thus, the sensing and actuation topologies are

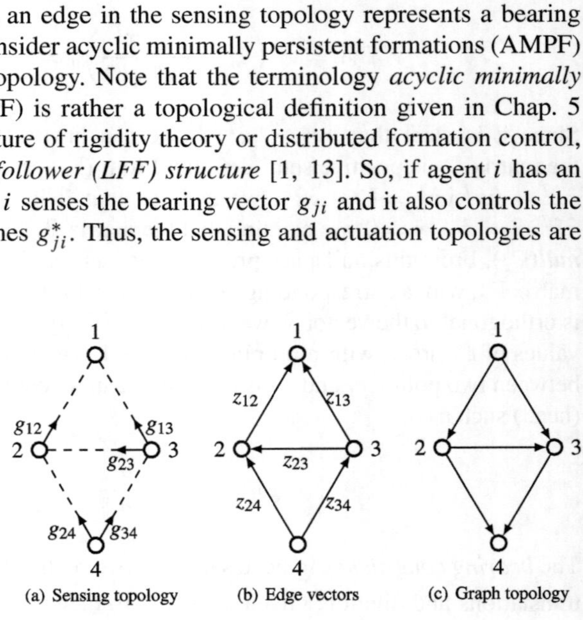

Fig. 9.2 The sensing topology determines the edge vectors. From the edge vectors, the graph topology is characterized. In the graph topology, the directions of edges are opposite to the directions of sensing edges

same, $\vec{\mathcal{G}} = (\mathcal{V}, \vec{\mathcal{E}^s}) = (\mathcal{V}, \vec{\mathcal{E}^a})$. The following lemmas are developed for analyzing the convergence property of AMPF under bearing setups [18].

Lemma 9.1 *Consider an acyclic minimally persistent (AMP) formation with the position of the leader p_1^*. The desired distance between agents 1 and 2 is given as $d_{21}^* = \|p_2^* - p_1^*\|$ and the desired bearing vectors are given as g_{ji}^* for all $(i, j)^{\bar{e}} \in \mathcal{E}^s$. If each agent $i \geq 3$ has two neighbors j, k, $1 \leq j \neq k < i$ with $g_{ji}^* \neq g_{ki}^*$, the location p_i^* is uniquely determined from its neighbors' positions and the desired bearing vectors. Moreover, p_i^*, where $i \geq 3$, is computed iteratively by*

$$p_i^* = \left(\sum_{j \in \mathcal{N}_i} \mathbb{P}_{g_{ji}^*}\right)^{-1} \left(\sum_{j \in \mathcal{N}_i} \mathbb{P}_{g_{ji}^*} p_j^*\right) \tag{9.3}$$

Proof For agent 2, since $g_{12}^* = \frac{p_1^* - p_2^*}{d_{21}^*}$, it is direct to have $p_2^* = p_1^* - d_{21}^* g_{12}^*$. For agent 3, p_3^* satisfies two bearing constraints g_{13}^* and g_{23}^*. Using the property of the projection operator $\mathbb{P}_x$, we can have

$$\mathbb{P}_{g_{13}^*}(p_3^* - p_1^*) = 0$$
$$\mathbb{P}_{g_{23}^*}(p_3^* - p_2^*) = 0$$

from which, we have

$$(\mathbb{P}_{g_{13}^*} + \mathbb{P}_{g_{23}^*})p_3^* = \mathbb{P}_{g_{13}^*} p_1^* + \mathbb{P}_{g_{23}^*} p_2^* \tag{9.4}$$

The nullspace of $\mathbb{P}_{g_{13}^*} + \mathbb{P}_{g_{23}^*}$ is only 0 due to $g_{ji}^* \neq g_{ki}^*$. In this case, p_3^* can be uniquely determined as (9.3). For other agents $i > 3$, the above process is exactly applied, which completes the proof.

Example 9.1 Let the desired positions of agents 1 and 2 be $p_1^* = (0, 0)^T$ and $p_2^* = (1, 0)^T$, and the desired bearing vectors g_{13}^* and g_{23}^* be given as

$$g_{13}^* = \frac{p_1^* - p_3^*}{\|p_1^* - p_3^*\|} = \frac{(1, 1)^T}{\sqrt{2}}$$
$$g_{23}^* = \frac{p_2^* - p_3^*}{\|p_2^* - p_3^*\|} = \frac{(2, 1)^T}{\sqrt{5}}$$

Then, we can have $\mathbb{P}_{g_{13}^*}$ and $\mathbb{P}_{g_{23}^*}$ as

$$\mathbb{P}_{g_{13}^*} = \mathbb{I}_2 - \frac{(1, 1)^T(1, 1)}{2} = \begin{bmatrix} \frac{1}{2} & -\frac{1}{2} \\ -\frac{1}{2} & \frac{1}{2} \end{bmatrix}$$
$$\mathbb{P}_{g_{23}^*} = \mathbb{I}_2 - \frac{(2, 1)^T(2, 1)}{5} = \begin{bmatrix} \frac{1}{5} & -\frac{2}{5} \\ -\frac{2}{5} & \frac{4}{5} \end{bmatrix}$$

Thus, we have

$$\sum_{j\in\mathcal{N}_i}\mathbb{P}_{g^*_{ji}}=\mathbb{P}_{g^*_{13}}+\mathbb{P}_{g^*_{23}}=\begin{bmatrix}0.7 & -0.9\\ -0.9 & 1.3\end{bmatrix} \text{ and } \sum_{j\in\mathcal{N}_i}\mathbb{P}_{g^*_{ji}}p^*_j=\begin{bmatrix}\frac{1}{5} & -\frac{2}{5}\\ -\frac{2}{5} & \frac{4}{5}\end{bmatrix}\begin{bmatrix}1\\ 0\end{bmatrix}=\begin{bmatrix}\frac{1}{5}\\ -\frac{2}{5}\end{bmatrix}$$

Finally, we obtain p^*_3 as $(-1,-1)^T$, which is the desired location of the agent 3.

Lemma 9.2 *Suppose that for a AMPF, desired distances d^*_{21} and desired bearing vectors g^*_{ji} for all $(i,j)^{\bar{e}}\in\mathcal{E}^s$ are given. Then, the offset of the leader's position determines the offset of the entire formation.*

Proof Consider an offset δ for the leader as $q^*_1=p^*_1+\delta$. Then for agent 2, we have $q^*_2=q^*_1-d^*_{21}g^*_{12}=p^*_1+\delta-d^*_{21}g^*_{12}=p^*_2+\delta$. For agent 3, it follows

$$\begin{aligned}q^*_3&=(\mathbb{P}_{g^*_{13}}+\mathbb{P}_{g^*_{23}})^{-1}(\mathbb{P}_{g^*_{13}}q^*_1+\mathbb{P}_{g^*_{23}}q^*_2)\\&=(\mathbb{P}_{g^*_{13}}+\mathbb{P}_{g^*_{23}})^{-1}(\mathbb{P}_{g^*_{13}}p^*_1+\mathbb{P}_{g^*_{23}}p^*_2+(\mathbb{P}_{g^*_{13}}+\mathbb{P}_{g^*_{23}})\delta)\\&=p^*_3+\delta\end{aligned}$$

which completes the proof.

In a bearing-based formation system, the distances are not actively controlled. Thus, the size of overall formation may vary without a boundary. For this, let us define the size of a formation as

Definition 9.1 The formation size is defined as the average of all the interagent distances of the edges:

$$\zeta(\mathcal{G}(p))\triangleq\frac{1}{|\mathcal{E}|}\sum_{(i,j)^e\in\mathcal{E}}\|p_i-p_j\|=\frac{1}{|\mathcal{E}|}\sum_{(i,j)^e\in\mathcal{E}}d_{ij}\tag{9.5}$$

Lemma 9.3 *For the AMPF with p^*_1 and desired bearing angles g^*_{ji} for all $(i,j)^{\bar{e}}\in\mathcal{E}^s$, the distance d^*_{21} uniquely determines the formation size.*

Proof From Lemma 9.1, we obtain

$$\begin{aligned}p^*_3-p^*_1&=(\mathbb{P}_{g^*_{13}}+\mathbb{P}_{g^*_{23}})^{-1}(\mathbb{P}_{g^*_{13}}p^*_1+\mathbb{P}_{g^*_{23}}p^*_2)-p^*_1\\&=(\mathbb{P}_{g^*_{13}}+\mathbb{P}_{g^*_{23}})^{-1}(\mathbb{P}_{g^*_{13}}p^*_1+\mathbb{P}_{g^*_{23}}p^*_2-(\mathbb{P}_{g^*_{13}}+\mathbb{P}_{g^*_{23}})p^*_1)\\&=(\mathbb{P}_{g^*_{13}}+\mathbb{P}_{g^*_{23}})^{-1}\mathbb{P}_{g^*_{23}}(p^*_2-p^*_1)\end{aligned}\tag{9.6}$$

Now, suppose that $p^*_2-p^*_1$ is changed as $\alpha(p^*_2-p^*_1)$, where $\alpha\neq 0$. Then, from (9.6), it follows that $p^*_3-p^*_1$ is changed to $\alpha(p^*_3-p^*_1)$ since $p^*_3-p^*_1$ is a linear mapping of $p^*_2-p^*_1$. For other agents, the same argument is applied. Thus, the formation size is determined by $p^*_2-p^*_1$.

Let us suppose that the set of desired bearing vectors, Γ, is given for the actuation topology. Also, it is assumed that the desired bearing vectors for agent $i>2$ are not

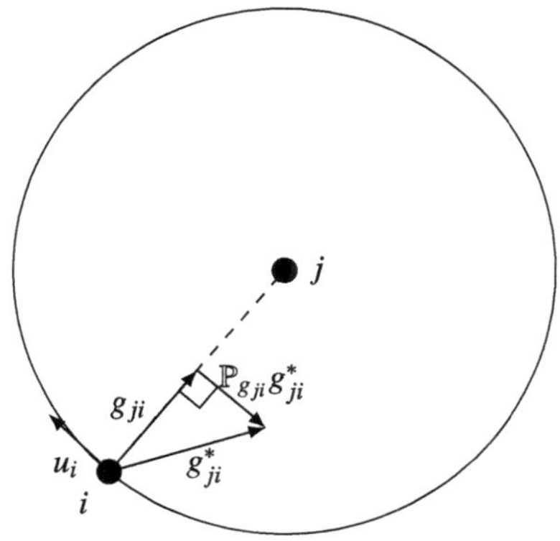

Fig. 9.3 Geographical interpretation of the bearing-based control law

collinear in the sense of $g^*_{ki} \neq \pm g^*_{ji}$, for $j \neq k < i$, and the initial positions of agents are not collocated. The bearing-only control law for agent i is proposed as

$$\dot{p}_i = u_i = -\sum_{j \in \mathcal{N}_i^o} \mathbb{P}_{g_{ji}} g^*_{ji} \tag{9.7}$$

Figure 9.3 depicts the physical meaning of the control law (9.7). The term $\mathbb{P}_{g_{ji}}$ decides the control direction orthogonal to the vector g_{ji} and the term $-g^*_{ji}$ determines the amount of control force. So, the overall term $-\mathbb{P}_{g_{ji}} g^*_{ji}$ determines the orthogonally projected control force corresponding to the agent j.

Since the agents in the acyclic minimally persistent (AMP) formation structure are interacted by an acyclic persistence topology, the analysis can be carried out sequentially. Based on (9.7), the control law for agent 2 is written as

$$\dot{p}_2 = u_2 = -\mathbb{P}_{g_{12}} g^*_{12} \tag{9.8}$$

Example 9.2 Let agent 1 be located at $(0, 0)^T$ and the desired bearing vector be given as $\frac{(1,1)^T}{\sqrt{2}}$. It means that the desired location of agent 2 is on the line $y = x,\ x \leq 0$ in 2-dimensional space. Let the current position of agent 2 be $(-1, 0)^T$. Then, we have

$$\mathbb{P}_{g_{12}} = \begin{bmatrix} 0 & 0 \\ 0 & 1 \end{bmatrix} \text{ and } \mathbb{P}_{g_{12}} g^*_{12} = \begin{bmatrix} 0 \\ \frac{1}{\sqrt{2}} \end{bmatrix}$$

Thus, we have $\dot{p}_2 = u_2 = -\begin{bmatrix} 0 \\ \frac{1}{\sqrt{2}} \end{bmatrix}$ at the point $p_2 = (-1, 0)^T$. This control input matches well to the intuition since it forces the agent 2 to move along the negative y-axis.

Lemma 9.4 *For the AMPF, the control law (9.8) does not change the length of d_{21}, and ensures an exponential convergence to the point $p_2 = p^*_1 + d_{12} g^*_{12}$.*

Proof See the appendix.

Next, let us analyze the dynamics of agent 3, which is governed by the following equation:

$$\dot{p}_3 = u_3 = -\mathbb{P}_{g_{13}} g_{13}^* - \mathbb{P}_{g_{23}} g_{23}^* \tag{9.9}$$

It is possible to consider the control law (9.9) as a cascade system, where the motion of agent 2 is an input to the unforced system. When the agent 2 is at the desired position p_{2a}^*, the control input u_3 is a function of p_3 only, i.e., $u_3(p_3, p_{2a}^*)$. However, if the agent is at an undesired equilibrium point p_{2b}^*, the dynamics of agent 3 will be $\dot{p}_3 = u_3(p_3, p_{2b}^*)$.

Example 9.3 Consider the same setup as *Example* 9.1. But, the agent 3 is located at $(-2, -1)^T$. Since we have $\mathbb{P}_{g_{13}} = \begin{bmatrix} 0.2 & -0.4 \\ -0.4 & 0.8 \end{bmatrix}$, $g_{13}^* = (0.7071, 0.7071)^T$, $\mathbb{P}_{g_{23}} = \begin{bmatrix} 0.1 & -0.3 \\ -0.3 & 0.9 \end{bmatrix}$, and $g_{23}^* = (0.8944, 0.4472)^T$, we compute the control input as $u_3 = (0.1861, -0.4170)^T$. The control direction does not directly match to the intuition since it is expected that the agent 3 would move from $(-2, -1)^T$ to $(-1, -1)^T$.

The counter-intuition in the above example can be explained as follows. With the input term $-\mathbb{P}_{g_{13}} g_{13}^*$, the agent 3 tries to achieve the desired bearing vector g_{13}^* by rotating in a counterclockwise direction around the agent 1, while keeping the distance d_{13}, and with the input term $-\mathbb{P}_{g_{23}} g_{23}^*$, the agent 3 tries to achieve the desired bearing vector g_{23}^* by rotating in a counterclockwise direction around the agent 2, while keeping the distance d_{23}. Thus, by the combination of these terms, the agent 3 is forced to move toward the direction $(0.1861, -0.4170)^T$. Consequently, it can be commented that the bearing-based formation control is not so efficient in terms of the trajectory. To resolve this issue, a new bearing vector-based formation control law was proposed in [17], where the control law is modified as

$$\dot{p}_3 = -\mathbb{P}_{g_{13}^*} g_{13} - \mathbb{P}_{g_{23}^*} g_{23} \tag{9.10}$$

which may send the agent 3 from the current location p_3 to the desired location p_3^* in a straight line.

Lemma 9.5 *The unforced system $\dot{p}_3 = u_3(p_3, p_{2a}^*)$ has a unique equilibrium point $p_{3a}^* = (\mathbb{P}_{g_{13}^*} + \mathbb{P}_{g_{23}^*})^{-1}(\mathbb{P}_{g_{13}^*} p_1^* + \mathbb{P}_{g_{23}^*} p_{2a}^*)$ corresponding to $g_{13} = g_{13}^*$ and $g_{23} = g_{23}^*$, while the unforced system $\dot{p}_3 = u_3(p_3, p_{2b}^*)$ has a unique equilibrium point $p_{3b}^* = (\mathbb{P}_{g_{13}^*} + \mathbb{P}_{g_{23}^*})^{-1}(\mathbb{P}_{g_{13}^*} p_1^* + \mathbb{P}_{g_{23}^*} p_{2b}^*)$ corresponding to $g_{13} = -g_{13}^*$ and $g_{23} = -g_{23}^*$.*

Proof To get the equilibrium set of (9.9), from $\mathbb{P}_{g_{13}} g_{13}^* + \mathbb{P}_{g_{23}} g_{23}^* = 0$, by pre-multiplying g_{13}^T to the both sides, we have $g_{13}^T(\mathbb{P}_{g_{13}} g_{13}^* + \mathbb{P}_{g_{23}} g_{23}^*) = 0$ from which the following is true

$$g_{13}^T \mathbb{P}_{g_{23}} g_{23}^* = 0 \tag{9.11}$$

The equality of (9.11) is true if and only if $g_{13} = \pm g_{23}$ or $g_{23}^* = \pm g_{23}$. The condition $g_{13} = \pm g_{23}$ takes place if and only if agents 1, 2 and 3 are collinear. In this case, we have $\mathbb{P}_{g_{13}} = \mathbb{P}_{g_{23}} = \mathbb{P}_{g_{12}^*}$, which further implies $\mathbb{P}_{g_{12}^*}(g_{13}^* + g_{23}^*) = 0$, or equivalently

$$g_{23}^* + g_{13}^* = k g_{12}^* \tag{9.12}$$

where k is a nonzero constant. On the other hand, since the desired formation is coplanar in 2-dimensional space, there should exist positive constants k_{12}^*, k_{32}^* and k_{31}^* such as

$$k_{12}^* g_{12}^* - k_{23}^* g_{23}^* + k_{13}^* g_{13}^* = 0 \tag{9.13}$$

Now, from (9.12) and (9.13), it follows that

$$(k_{12}^* + k k_{13}^*) g_{13}^* + (k_{12}^* - k k_{23}^*) g_{23}^* = 0 \tag{9.14}$$

which means that g_{13}^* is parallel to g_{23}^*. However, this argument is a contradiction to the assumption of non-collinearity of desired positions of three agents. Thus, the case of $g_{13} = \pm g_{23}$ cannot occur. Let us consider the case $g_{23}^* = \pm g_{23}$. In this case, $\mathbb{P}_{g_{23}} g_{23}^* = 0$; thus from $\mathbb{P}_{g_{13}} g_{13}^* + \mathbb{P}_{g_{23}} g_{23}^* = 0$, it should be true that $g_{13} = \pm g_{13}^*$. Consequently, there are four cases for equilibrium such as $(g_{13}, g_{23}) \in \{(g_{13}^*, g_{23}^*), (g_{13}^*, -g_{23}^*), (-g_{13}^*, g_{23}^*), (-g_{13}^*, -g_{23}^*)\}$. From Lemma 9.1, when $p_2 = p_{2a}^*$, the unique position of agent 3 is only when $(g_{13}, g_{23}) = (g_{13}^*, g_{23}^*)$. Similarly, however when $p_2 = p_{2b}^*$, the unique position of agent 3 from Lemma 9.1 is only when $(g_{13}, g_{23}) = (-g_{13}^*, -g_{23}^*)$.

In the next lemma, the stability of equilibria of the system is investigated.

Lemma 9.6 *The equilibrium point p_{3a}^* of the unforced system $\dot{p}_3 = u_3(p_3, p_{2a}^*)$ is globally asymptotically stable, while the equilibrium point p_{3b}^* of the unforced system $\dot{p}_3 = u_3(p_3, p_{2b}^*)$ is unstable.*

Proof Consider the Lyapunov candidate $V = \frac{1}{2}\|p_3 - p_{3a}^*\|^2$, which is continuously differentiable and radially unbounded. Denoting $M = \left(\frac{\mathbb{P}_{g_{13}}}{\|z_{13}^*\|} + \frac{\mathbb{P}_{g_{23}}}{\|z_{23}^*\|}\right)$, the derivative of V can be obtained as

$$\begin{aligned}
\dot{V} &= -(p_3 - p_{3a}^*)^T (\mathbb{P}_{g_{13}} g_{13}^* + \mathbb{P}_{g_{23}} g_{23}^*) \\
&= -(p_3 - p_{3a}^*)^T \frac{\mathbb{P}_{g_{13}}}{\|z_{13}^*\|} (p_1 - p_3 + p_3 - p_{3a}^*) \\
&\quad - (p_3 - p_{3a}^*)^T \frac{\mathbb{P}_{g_{23}}}{\|z_{23}^*\|} (p_2 - p_3 + p_3 - p_{3a}^*) \\
&= -(p_3 - p_{3a}^*)^T M (p_3 - p_{3a}^*)
\end{aligned} \tag{9.15}$$

Since $\mathbb{P}_{g_{13}}$ and $\mathbb{P}_{g_{23}}$ are positive semi-definite matrices, the matrix M is also positive semi-definite. Thus, $\dot{V} \leq 0$ and $\dot{V} = 0$ if and only if $p_3 - p_{3a}^* \in null(M)$. Consider the following two cases:

- If $g_{13} = \pm g_{23}$, then $null(M) = \text{span}\{g_{13}\}$. Since p_{3a}^* is not collinear with p_1 and p_{2a}^*, we have $p_3 - p_{3a}^* \notin null(M)$. Therefore, following the same process outlined in (A.60) and (A.61), we have

$$\dot{V} \leq -\gamma \sin^2 \alpha \| p_3 - p_{3a}^* \|^2 = -2\gamma \sin^2 \alpha \leq 0$$

  where α is the angle of the three points p_1, p_3 and p_{3a}^*, i.e., $\alpha = \angle(p_1, p_3, p_{3a}^*)$ and $\gamma = \|z_{13}^*\|^{-1} + \|z_{23}^*\|^{-1}$.
- If $g_{13} \neq \pm g_{23}$, the matrix M is positive definite. Thus, it follows

$$\dot{V} \leq -\lambda_{min}(M(t)) \| p_3 - p_{3a}^* \|^2 = -2\lambda_{min}(M(t))V \tag{9.16}$$

  where $\lambda_{min}(M(t)) > 0$ is the smallest eigenvalue of M.

Now, choose $\kappa \triangleq \min\{\inf_t \{2\lambda_{min}(M(t))\}, 2\gamma \sin^2 \alpha\} > 0$. Then, it is obtained as $\dot{V} \leq -\kappa V \leq 0$. Consequently, $\dot{V}$ is negative definite and $\dot{V} = 0$ if and only if $p_3 = p_{3a}^*$. Thus, $p_3 = p_{3a}^*$ is globally asymptotically stable.

Next, to show a repulsiveness of p_{3b}^*, we use the Lyapunov function $V = \frac{1}{2}\| p_3 - p_{3b}^* \|^2$. The derivative of V can be written as

$$\begin{aligned}
\dot{V} &= (p_3 - p_{3b}^*)^T (\mathbb{P}_{g_{13}}(-g_{13}^*) + \mathbb{P}_{g_{23}}(-g_{23}^*)) \\
&= (p_3 - p_{3b}^*)^T \frac{\mathbb{P}_{g_{13}}}{\|z_{13}^*\|}(p_1 - p_3 + p_3 - p_{3b}^*) \\
&\quad + (p_3 - p_{3b}^*)^T \frac{\mathbb{P}_{g_{23}}}{\|z_{23}^*\|}(p_2 - p_3 + p_3 - p_{3b}^*) \\
&= (p_3 - p_{3b}^*)^T M (p_3 - p_{3b}^*)
\end{aligned} \tag{9.17}$$

Therefore, $\dot{V} > 0$ if $p_3 \neq p_{3b}^*$, which means that $p_3 = p_{3b}^*$ is unstable.

Lemma 9.7 *For the unforced dynamics $\dot{p}_3 = u_3(p_3, p_{2a}^*)$, where agent 2 is on the desired position, the agent 3 never collides agents 1 and 2 if*

$$\| p_3(t_0) - p_3^* \| < \min\{\| p_3^* - p_1^* \|, \| p_3^* - p_{2a}^* \|\} \tag{9.18}$$

Proof It is known that $\| p_3 - p_1^* \| = \| p_3 - p_3^* + p_3^* - p_1^* \| \geq \| p_3^* - p_1^* \| - \| p_3 - p_3^* \|$ and $p_3(t) \to p_3^*$ asymptotically. Thus, if $\| p_3(t_0) - p_3^* \| < \| p_3^* - p_1^* \|$, due to $\dot{V} \leq 0$, we have $\| p_3 - p_1^* \| > 0$ for all $t \geq t_0$. Similarly, if $\| p_3(t_0) - p_3^* \| < \| p_3^* - p_{2a}^* \|$, we have $\| p_3 - p_{2a}^* \| > 0$ for all $t \geq t_0$.

Theorem 9.1 *When the agent 3 is governed by (9.9), it has an almost globally asymptotically stable equilibrium $p_3 = p_{3a}^*$ corresponding to $g_{13} = g_{13}^*$ and $g_{23} = g_{23}^*$.*

Proof Consider a Lyapunov candidate $V = \frac{1}{2}\|p_3 - p^*_{3a}\|^2$ which is positive definite, radially unbounded, and continuously differentiable. If $p_2(t_0) \neq p^*_{2b}$, the derivative of V along the trajectory of the system (9.9) is given as

$$\begin{aligned}
\dot{V} &= -(p_3 - p^*_3)^T(\mathbb{P}_{g_{13}} g^*_{13} + \mathbb{P}_{g_{23}} g^*_{23}) \\
&= -(p_3 - p^*_3)^T\left(\frac{\mathbb{P}_{g_{13}}}{\|z^*_{13}\|}(p_1 - p_3 + p_3 - p^*_{3a}) + \frac{\mathbb{P}_{g_{23}}}{\|z^*_{23}\|}(p^*_{2a} - p_2 + p_2 - p_3 + p_3 - p^*_{3a})\right) \\
&= -(p_3 - p^*_3)^T\left(\frac{\mathbb{P}_{g_{13}}}{\|z^*_{13}\|} + \frac{\mathbb{P}_{g_{23}}}{\|z^*_{23}\|}\right)(p_3 - p^*_{3a}) + (p_3 - p^*_3)^T\frac{\mathbb{P}_{g_{23}}}{\|z^*_{23}\|}(p_2 - p^*_{2a}) \\
&\leq -(p_3 - p^*_3)^T\left(\frac{\mathbb{P}_{g_{13}}}{\|z^*_{13}\|} + \frac{\mathbb{P}_{g_{23}}}{\|z^*_{23}\|}\right)(p_3 - p^*_{3a}) + \|(p_3 - p^*_3)\|\frac{\|\mathbb{P}_{g_{23}}\|}{\|z^*_{23}\|}\|(p_2 - p^*_{2a})\|
\end{aligned} \tag{9.19}$$

Notice that as $p_3 \to \pm\infty$, the first term of (9.19) dominates the second term since $\frac{\|\mathbb{P}_{g_{23}}\|}{\|z^*_{23}\|}\|(p_2 - p^*_{2a})\|$ is bounded. Consequently, $\|p_3 - p^*_{3a}\|$ is ultimately bounded and $\|p_3\|$ is bounded. Since, as shown in Lemma 9.6, the unforced system $\dot{p}_3 = u_3(p_3, p^*_{2a})$ is globally asymptotically stable, and it is ultimately bounded, the system (9.9) is input-to-state stable (ISS) with regard to the input p_2 by the Lemma 2.10. Also, Lemma 9.4 shows that p_2 converges to p^*_{2a} globally exponentially fast. Thus, the equilibrium $p_3 = p^*_{3a}$ is almost globally asymptotically stable by Lemma 2.11 [2].

In Theorem 9.1, it was shown that the desired equilibrium $(p_2^T, p_3^T) = ((p^*_{2a})^T, (p^*_{3a})^T)^T$ of the cascade systems

$$\begin{aligned}
\dot{p}_2 &= u_2(p_2) \\
\dot{p}_3 &= u_3(p_3, p_2)
\end{aligned}$$

is almost globally asymptotically stable. That is, all trajectories starting at arbitrary initial point except for those starting at $p_2(t_0) = p^*_{2b}$ will converge to the desired positions. Let us extend Theorem 9.1 to the general ith agent where $i > 3$. Since the ith agent has two neighbors j and k, where $j, k \leq i - 1$, the control law for agent i can be explicitly written as

$$\dot{p}_i = u_i(p_i, p_{i-1}, \ldots, p_2) = -\mathbb{P}_{g_{ji}} g^*_{ji} - \mathbb{P}_{g_{ki}} g^*_{ki} \tag{9.20}$$

From Lemmas 9.5 and 9.6, it is clear that for all $i = 3, 4, \ldots, n$, there is a unique desired position $p_i = p^*_{ia}$, and it is globally asymptotically stable when agents $2, \ldots, i-1$ are at the desired positions such as $p_j = p^*_{ja}$, $j = 2, \ldots, i-1$. That is, for the following unforced subsystem

$$\dot{p}_i = u_i(p_i, p^*_{(i-1)a}, \ldots, p^*_{2a}) \tag{9.21}$$

the desired equilibrium $p_i = p^*_{ia}$ is globally asymptotically stable. Combining from agent 1 to agent n, the dynamics of n agents can be compactly written in the form of a cascade system:

$$\dot{p} = \begin{bmatrix} \dot{p}_1 = 0 \\ \dot{p}_2 = u_2(p_2) \\ \vdots \\ \dot{p}_i = u_i(p_i, \ldots, p_2) \\ \vdots \\ \dot{p}_n = u_n(p_n, \ldots, p_2) \end{bmatrix} \tag{9.22}$$

where $3 \leq i \leq n$. The following theorem is the main result of this section.

Theorem 9.2 *For the acyclic minimally persistent (AMP) formations with realizable bearing constraints, if the initial positions are not collocated, then the cascade system (9.22) almost globally asymptotically converges to desired equilibrium* $p_a^* = ((p_1^*)^T, (p_{2a}^*)^T, \ldots, (p_{na}^*)^T)^T$. *Moreover, if* $p_2(t_0) \neq p_{2b}^*$, *then the equilibrium* $p_b^* = ((p_1^*)^T, (p_{2b}^*)^T, \ldots, (p_{nb}^*)^T)^T$ *is unstable, and the desired equilibrium* p_a^* *is globally asymptotically stable.*

Proof Let us prove this theorem by induction. For the agents $l = 2$ and $l = 3$, the argument is clearly true by Lemma 9.1. Now, suppose that the argument is true for $3 \leq l \leq i-1$. We need to show that the argument is also true for $l = i$. Using the same process as the proof of Theorem 9.6, the equilibrium of the unforced system (9.20) is globally asymptotically stable. To employ the input-to-state stability, we show that $p_i(t)$ is bounded. To this end, suppose that agent i has two neighbors j and k. Then from the Lyapunov function $V = \frac{1}{2}\|p_i - p_{ia}^*\|^2$, the derivative can be obtained as

$$\begin{aligned}
\dot{V} &= -(p_i - p_{ia}^*)^T(\mathbb{P}_{g_{ji}} g_{ji}^* + \mathbb{P}_{g_{ki}} g_{ki}^*) \\
&= -(p_i - p_{ia}^*)^T \left(\frac{\mathbb{P}_{g_{ji}}}{\|z_{ji}^*\|}(p_{ja}^* - p_j + p_j - p_i + p_i - p_{ia}^*) \right. \\
&\quad \left. + \frac{\mathbb{P}_{g_{ki}}}{\|z_{ki}^*\|}(p_{ka}^* - p_k + p_k - p_i + p_i - p_{ia}^*) \right) \\
&= -(p_i - p_{ia}^*)^T \left(\frac{\mathbb{P}_{g_{ji}}}{\|z_{ji}^*\|} + \frac{\mathbb{P}_{g_{ki}}}{\|z_{ki}^*\|} \right)(p_i - p_{ia}^*) \\
&\quad - (p_i - p_{ia}^*)^T \left[\frac{\mathbb{P}_{g_{ji}}}{\|z_{ji}^*\|}(p_{ja}^* - p_j) + \frac{\mathbb{P}_{g_{ki}}}{\|z_{ki}^*\|}(p_{ka}^* - p_k) \right] \\
&\leq -(p_i - p_{ia}^*)^T \left(\frac{\mathbb{P}_{g_{ji}}}{\|z_{ji}^*\|} + \frac{\mathbb{P}_{g_{ki}}}{\|z_{ki}^*\|} \right)(p_i - p_{ia}^*) \\
&\quad + \|p_i - p_{ia}^*\| \left[\frac{\|\mathbb{P}_{g_{ji}}\|}{\|z_{ji}^*\|}\|p_{ja}^* - p_j\| + \frac{\|\mathbb{P}_{g_{ki}}\|}{\|z_{ki}^*\|}\|p_{ka}^* - p_k\| \right]
\end{aligned} \tag{9.23}$$

Thus, since $\frac{\|\mathbb{P}_{g_{ji}}\|}{\|z^*_{ji}\|}\|p^*_{ja} - p_j\| + \frac{\|\mathbb{P}_{g_{ki}}\|}{\|z^*_{ki}\|}\|p^*_{ka} - p_k\|$ is bounded, we can see that $\|p_i\|$ is also bounded. Consequently, the equilibrium p^*_{ia} is globally asymptotically stable, and all trajectories globally converge to the desired positions p^*_{ia} for $l = i$ by Lemma 2.11 if $p_2(t_0) \neq p^*_{2b}$. Since the above argument does hold for all $l \geq 3$, the cascade system (9.22) almost globally asymptotically converges to the desired equilibrium. Note that if $p_2(t_0) = p^*_{2b}$, then there exists an undesired equilibrium where $g_{ij} = -g^*_{ij}$; but this undesired equilibrium is unstable by Lemma 9.6. Therefore, we can conclude that the desired equilibrium p^*_a is globally asymptotically stable if $p_2(t_0) \neq p^*_{2b}$.

In this section, we have considered AMPFs in $\mathbb{R}^2$. For undirected graphs, the control law (9.7) is also used for the formation stabilization to the desired configuration. For undirected graphs, the control law is written as

$$\dot{p}_i = u_i = -\sum_{j\in\mathcal{N}_i} \mathbb{P}_{g_{ji}} g^*_{ji} \tag{9.24}$$

where the set of neighboring agents is changed as the set of agents $\mathcal{N}_i$ in undirected graphs. The stability analysis of (9.24) is given in [19], resulting in the following theorem.

Theorem 9.3 *If the desired bearing vectors given as $\Gamma \triangleq \{g^*_{ji} : \forall (i, j)^e \in \mathcal{E}\}$ are feasible and ensure infinitesimal bearing rigidity, the control law* (9.24) *ensures an almost global exponential convergence to the desired configuration.*

Proof See [19].

Similarly to Sect. 5.4, we can also ensure a finite-time convergence of the bearing-based control law (9.24). For the finite-time convergence, (9.24) is modified as

$$\dot{p}_i = -\sum_{j\in\mathcal{N}_i} \frac{\mathbb{P}_{g_{ji}} g^*_{ji}}{\|\mathbb{P}_{g_{ji}} g^*_{ji}\|^{\alpha}} \tag{9.25}$$

and

$$\dot{p}_i = -\sum_{j\in\mathcal{N}_i} \mathbb{P}_{g_{ji}} \text{sig}(\mathbb{P}_{g_{ji}} g^*_{ji})^{\alpha} \tag{9.26}$$

where $0 < \alpha < 1$. The following analyses are reused from [14], which further can be found in [19]. For a stability analysis, let us denote the stacked vectors of bearing vectors as $g^T = (g_1^T, \ldots, g_m^T)^T \in \mathbb{R}^{dm}$ and $(g^*)^T = ((g^*_1)^T, \ldots, (g^*_m)^T)^T \in \mathbb{R}^{dm}$ where $m = |\mathcal{E}|$. Then, (9.25) and (9.26) can be written in concise forms as

$$\dot{p} = (\mathbb{H} \otimes \mathbb{I}_d)^T \cdot \text{blkdg}\left(\frac{\mathbb{P}_{g_k}}{\|\mathbb{P}_{g_k} g_k^*\|^\alpha}\right) g^*$$
$$= (\mathbb{H} \otimes \mathbb{I}_d)^T \cdot \text{blkdg}(\mathbb{P}_{g_k})\text{blkdg}\left(\frac{1}{\|\mathbb{P}_{g_k} g_k^*\|^\alpha} \otimes \mathbb{I}_d\right) g^* \quad (9.27)$$

and

$$\dot{p} = (\mathbb{H} \otimes \mathbb{I}_d)^T \cdot \text{blkdg}(\mathbb{P}_{g_k}) \cdot \text{sig}(\text{blkdg}(\mathbb{P}_{g_k}) g^*)^\alpha, \ \forall k = 1, \ldots, m \quad (9.28)$$

where $\mathbb{H} \in \mathbb{R}^{m \times n}$ denotes the corresponding incidence matrix. It is a well-known fact that the bearing rigidity matrix and incidence matrix are related as

$$\mathbb{R}_B = \text{blkdg}\left(\frac{\mathbb{P}_{g_k}}{\|z_k\|}\right)(\mathbb{H} \otimes \mathbb{I}_d) = \text{blkdg}\left(\frac{1}{\|z_k\|} \otimes \mathbb{I}_d\right) \text{blkdg}(\mathbb{P}_{g_k})(\mathbb{H} \otimes \mathbb{I}_d) \quad (9.29)$$

Then, (9.27) and (9.28) can be changed as

$$\dot{p} = \mathbb{R}_B^T \cdot \text{blkdg}(\|z_k\| \otimes \mathbb{I}_d) \cdot \text{blkdg}\left(\frac{1}{\|\mathbb{P}_{g_k} g_k^*\|^\alpha} \otimes \mathbb{I}_d\right) g^* \quad (9.30)$$

and

$$\dot{p} = \mathbb{R}_B^T \cdot \text{blkdg}(\|z_k\| \otimes \mathbb{I}_d) \cdot \text{sig}(\text{blkdg}(\mathbb{P}_{g_k}) g^*)^\alpha \quad (9.31)$$

For the main analysis, let us suppose that the desired framework and realized frameworks are infinitesimally bearing rigid. Then, the null space of the rigidity matrix is given as $\text{null}(\mathbb{R}_B) = \text{span}\{\mathbf{1} \otimes \mathbb{I}_d, p\}$. Also since $\text{null}(\mathbb{R}_B) = \text{null}(\mathbb{R}_B P)$, for any nonsingular matrix P, and $\text{range}(\mathbb{R}_B^T) \perp \text{null}(\mathbb{R}_B)$, we can see that $\dot{p} \in \text{range}(\mathbb{R}_B^T)$ and $\dot{p} \perp \text{null}(\mathbb{R}_B)$, and equivalently $\dot{p} \perp \text{span}\{\mathbf{1} \otimes \mathbb{I}_d, p\}$ [19]. Consequently, under the assumption of infinitesimally bearing rigidity of the underlying topology, we can see that $\dot{p} = 0$ implies that $g^* \in \text{null}(\mathbb{R}_B^T)$ for both (9.27) and (9.28). Let the centroid p_c and scale p_s of the formation be defined as

$$p_c = \frac{1}{n}\sum_{i=1}^{n} p_i = \frac{1}{n}(\mathbf{1} \otimes \mathbb{I}_d)^T p$$
$$p_s = \sqrt{\frac{1}{n}\sum_{i=1}^{n} \|p_i - p_c\|^2} = \frac{1}{\sqrt{n}}\|p - \mathbf{1} \otimes p_c\|$$

Now, the following lemmas can be used to characterize the equilibrium points of (9.30) and (9.31).

Lemma 9.8 *Under the dynamics either* (9.30) *or* (9.31), *the centroid* p_c *and scale* p_s *of the formation are invariant.*

Proof From $\dot{p}_c = 1/n \sum_{i=1}^{n} \dot{p}_i = \frac{1}{n}(\mathbf{1} \otimes \mathbb{I}_d)^T \dot{p}$ and $\dot{p} \perp \text{span}\{\mathbf{1} \otimes \mathbb{I}_d, p\}$, for both (9.30) and (9.31), we can see that $\dot{p}_c = 0$. Similarly, for the scaling, we have $\dot{p}_s = 1/\sqrt{n}(p - \mathbf{1} \otimes p_c)^T / \|p - \mathbf{1} \otimes p_c\| \dot{p} = 0$.

Lemma 9.9 (Theorem 10 of [19]) *The finite-time dynamics* (9.30) *and* (9.31) *have two equilibrium points* $p^* = p_a^*$ *corresponding to the vectors* $g_k = g_k^*$, $\forall k = 1, \ldots, m$, *and* $p^* = p_b^*$ *corresponding to the vectors* $g_k = -g_k^*$, $\forall k = 1, \ldots, m$.

For the convergence analysis to a desired configuration, let $\delta_i = p_i - p_i^*$, $\forall i = 1, \ldots, n$, and $\delta = (\delta_1^T, \ldots, \delta_n^T)^T$. Then, we can rewrite (9.30) and (9.31) as

$$\dot{\delta} = \mathbb{R}_B^T \cdot \text{blkdg}(\|z_k\| \otimes \mathbb{I}_d) \cdot \text{blkdg}\left(\frac{1}{\|\mathbb{P}_{g_k} g_k^*\|^{\alpha}} \otimes \mathbb{I}_d\right) g^* \tag{9.32}$$

and

$$\dot{\delta} = \mathbb{R}_B^T \cdot \text{blkdg}(\|z_k\| \otimes \mathbb{I}_d) \cdot \text{sig}(\text{blkdg}(\mathbb{P}_{g_k}) g^*)^{\alpha} \tag{9.33}$$

Letting $r(t) = p(t) - \mathbf{1} \otimes p_c$ and $r^* = p^* - \mathbf{1} \otimes p_c^*$, we can also write $\delta(t) = r(t) - r^*$. Further, from $\|r(t)\| = \|p(t) - \mathbf{1} \otimes p_c\|$, we can have

$$\frac{d\|r(t)\|}{dt} = \frac{(p - \mathbf{1} \otimes p_c)^T}{\|p - \mathbf{1} \otimes p_c\|} \dot{p} = 0, \tag{9.34}$$

which implies that $\|r(t)\| = \|r^*\|$ and $\|r(t)\| = \|\delta(t) + r^*\| = \|r^*\|$. Thus, the dynamics $\delta_i = p_i - p_i^*$ evolves on the following dn-sphere:

$$\mathcal{S} = \{\delta \in \mathbb{R}^{dn} \mid \|\delta(t) + r^*\| = \|r^*\|\} \tag{9.35}$$

Lemma 9.10 *For the dynamics* (9.25) *and* (9.26), *the desired configuration* $p = p^* = p_a^*$ *is asymptotically stable, while the undesired configuration* $p = p_b^*$ *is unstable.*

Proof Let us first consider a Lyapunov candidate $V = 1/2\|\delta\|^2 = 1/2\|p - p^*\|^2$, which is positive definite, radially unbounded, and continuously differentiable. Taking a derivative along (9.25), we have

$$\begin{aligned}
\dot{V} &= (p - p^*)^T (\mathbb{H} \otimes \mathbb{I}_d)^T \cdot \text{blkdg}\left(\frac{\mathbb{P}_{g_k}}{\|\mathbb{P}_{g_k} g_k^*\|^{\alpha}}\right) g^* \\
&= (z - z^*)^T \cdot \text{blkdg}\left(\frac{\mathbb{P}_{g_k}}{\|\mathbb{P}_{g_k} g_k^*\|^{\alpha}}\right) g^* \\
&= -(z^*)^T \text{blkdg}\left(\frac{\mathbb{P}_{g_k}}{\|\mathbb{P}_{g_k} g_k^*\|^{\alpha}}\right) g^*
\end{aligned}$$

$$
\begin{aligned}
&= -(g^*)^T \mathrm{blkdg}\left(\frac{\|z_k^*\|\mathbb{P}_{g_k}}{\|\mathbb{P}_{g_k} g_k^*\|^\alpha}\right) g^* \\
&= -\sum_{k=1}^{m} \|z_k^*\| \frac{(g_k^*)^T \mathbb{P}_{g_k}^T \mathbb{P}_{g_k} g_k^*}{\|\mathbb{P}_{g_k} g_k^*\|^\alpha} \\
&= -\sum_{k=1}^{m} \|z_k^*\| \, \|\mathbb{P}_{g_k} g_k^*\|^{2-\alpha} \le 0
\end{aligned} \tag{9.36}
$$

Consequently, since $0 < \alpha < 1$, we can see that $\dot{V} = 0$ if and only if $p = p^* = p_a^*$ or $p = p_b^*$. However, if we take another Lyapunov candidate as $V = 1/2\|\delta\|^2 = 1/2\|p - p_b^*\|^2$, where p_b^* corresponds to $g = -g^*$, its derivative along (9.25) can be shown as

$$
\begin{aligned}
\dot{V} &= (p - p_b^*)^T (\mathbb{H} \otimes \mathbb{I}_d)^T \cdot \mathrm{blkdg}\left(\frac{\mathbb{P}_{g_k}}{\|\mathbb{P}_{g_k} g_k^*\|^\alpha}\right) g^* \\
&= (z + z^*)^T \cdot \mathrm{blkdg}\left(\frac{\mathbb{P}_{g_k}}{\|\mathbb{P}_{g_k} g_k^*\|^\alpha}\right) g^* \\
&= (z^*)^T \mathrm{blkdg}\left(\frac{\mathbb{P}_{g_k}}{\|\mathbb{P}_{g_k} g_k^*\|^\alpha}\right) g^* \\
&= \sum_{k=1}^{m} \|z_k^*\| \, \|\mathbb{P}_{g_k} g_k^*\|^{2-\alpha} > 0
\end{aligned} \tag{9.37}
$$

for any point in a neighborhood of p_b^*. Thus, from Chetaev theorem, $p = p_b^*$ is unstable. Likewise, along the dynamics (9.26), we can reach the same conclusion.

Now, we can make the following main result.

Theorem 9.4 *When agents are updated by finite-time control laws as described in the dynamics* (9.25) *and* (9.26), *the desired configuration p^* is almost globally asymptotically stable.*

Proof By Lemma 9.10, if agents are not in $p = p_b^*$ initially, they will reach to the desired configuration p_a^* for any other initial states.

Let us further scrutinize the convergence in a finite-time. For this purpose, we use the following lemmas:

Lemma 9.11 (Corollary 2 of [19]) *For the dynamics* (9.25) *and* (9.26), *the lengths of edges are upper bounded as $\|z_k\| \le 2p_s\sqrt{n-1}$, $\forall k = 1, \dots, m$.*

Lemma 9.12 (Theorem 4 of [19]) *A framework $f_p = (\mathcal{V}, \mathcal{E}, p)$ is infinitesimally bearing rigid if and only if $null(\mathbb{R}_B) = span\{range(\mathbf{1} \otimes \mathbb{I}_d, p - \mathbf{1} \otimes p_c)\}$.*

Theorem 9.5 *In Theorem 9.4, the agents converge to the desired configuration p^* in a finite-time.*

Proof Using the fact that $(g_k^*)^T \mathbb{P}_{g_k} g_k^* = (g_k^*)^T (\mathbb{I}_d - g_k g_k^T) g_k^* = \mathbb{I}_d - g_k^T g_k^* (g_k^*)^T g_k = g_k^T (\mathbb{I}_d - g_k^* (g_k^*)^T) g_k = g_k^T \mathbb{P}_{g_k^*} g_k$, we can rewrite (9.36) as

$$\begin{aligned} \dot{V} &\leq -\epsilon \sum_{k=1}^{m} ((g_k^*)^T \mathbb{P}_{g_k} g_k^*)^{\frac{2-\alpha}{2}} \\ &= -\epsilon \sum_{k=1}^{m} (g_k^T \mathbb{P}_{g_k^*} g_k)^{\frac{2-\alpha}{2}} \\ &= -\epsilon \sum_{k=1}^{m} \left(\frac{z_k^T \mathbb{P}_{z_k^*} z_k}{\|z_k\|^2} \right)^{\frac{2-\alpha}{2}} \end{aligned} \tag{9.38}$$

where $\epsilon \triangleq \min_{\forall k} \|z_k^*\|$. With the help of Lemma 9.11, we can further obtain the following inequality:

$$\begin{aligned} \dot{V} &\leq -\frac{\epsilon}{(2p_s\sqrt{n}-1)^{2-\alpha}} \sum_{k=1}^{m} \left(z_k^T \mathbb{P}_{z_k^*} z_k \right)^{\frac{2-\alpha}{2}} \\ &= -\frac{\epsilon}{(2p_s\sqrt{n}-1)^{2-\alpha}} \sum_{k=1}^{m} \left((z_k^T - (z_k^*)^T) \mathbb{P}_{z_k^*} (z_k - z_k^*) \right)^{\frac{2-\alpha}{2}} \\ &= -\frac{\epsilon}{(2p_s\sqrt{n}-1)^{2-\alpha}} \left((p-p^*)^T (\mathbb{H}\otimes \mathbb{I}_d)^T \text{blkdg}(\mathbb{P}_{g_k^*})(\mathbb{H}\otimes \mathbb{I}_d)(p-p^*) \right)^{\frac{2-\alpha}{2}} \\ &= -\frac{\epsilon}{(2p_s\sqrt{n}-1)^{2-\alpha}} (\delta^T (\bar{\mathbb{R}}_B^*)^T \bar{\mathbb{R}}_B^* \delta)^{\frac{2-\alpha}{2}} \end{aligned} \tag{9.39}$$

where $\bar{\mathbb{R}}_B^* = \text{blkdg}(\|z_k^*\| \otimes \mathbb{I}_d)\mathbb{R}_B^*$ and $\mathbb{R}_B^*$ is the bearing rigidity matrix when $p = p^*$. Further, it is obvious that $\delta \notin \text{range}(\mathbf{1} \otimes \mathbb{I}_d)$ since the subspace δ is on the surface determined by $\|\delta(t) + r^*\| = \|r^*\|$, while the subspace range$(\mathbf{1} \otimes \mathbb{I}_d)$ corresponds to the translations; so, they are different subspaces. Also noticing that $\text{null}((\bar{\mathbb{R}}_B^*)^T \bar{\mathbb{R}}_B^*) = \text{null}(\bar{\mathbb{R}}_B^*) = \text{null}(\mathbb{R}_B^*) = \text{span}\{\text{range}(\mathbf{1} \otimes \mathbb{I}_d), r^*\}$, we can see that $\bar{\mathbb{R}}_B^* \delta \neq 0$, if $\delta \notin \text{span}\{\text{range}(\mathbf{1} \otimes \mathbb{I}_d), r^*\}$. Thus, from $\delta = r - r^* = p - \mathbf{1} \otimes p_c - (p^* - \mathbf{1} \otimes p_c^*) = p - p^*$, we can see that if and only if $p \neq p^*$, then $\bar{\mathbb{R}}_B^* \delta \neq 0$. That is, if the vector δ is parallel to the vector r^*, i.e., $\delta \parallel r^*$, then we have $\bar{\mathbb{R}}_B^* \delta = 0$. Otherwise, if the vector δ is orthogonal to the vector r^*, i.e., $\delta \perp r^*$, then we have $\bar{\mathbb{R}}_B^* \delta \neq 0$. Thus, considering only the orthogonal component of δ with respect to r^*, which is defined as $\delta \sin(\xi)$,[1] where ξ is the inter-distance angle between the vectors δ and $-r^*$, as far as $p \neq p^*$, we can have

$$\delta^T (\bar{\mathbb{R}}_B^*)^T \bar{\mathbb{R}}_B^* \delta \geq \lambda_{min} \sin^2(\xi) \|\delta\|^2 \tag{9.40}$$

where λ_{min} is nonzero minimum eigenvalue of $(\bar{\mathbb{R}}_B^*)^T \bar{\mathbb{R}}_B^*$. Thus, we can finally obtain

[1] The orthogonal component is $\delta \sin(\xi)$ and the parallel component is $\delta \cos(\xi)$.

$$\dot{V} \leq -\frac{\epsilon}{(2p_s\sqrt{n-1})^{2-\alpha}}(\lambda_{min}\sin^2(\xi)\|\delta\|^2)^{\frac{2-\alpha}{2}}$$
$$\leq -\frac{\epsilon}{(2p_s\sqrt{n-1})^{2-\alpha}}(\lambda_{min}\sin^2(\xi))^{\frac{2-\alpha}{2}}(2V)^{\frac{2-\alpha}{2}} \quad (9.41)$$

where $\sin(\xi)$ would increase as $r(t) \rightarrow r^*$ since the scaling of the formation does not change by Lemma 9.8. In more detail, from $\frac{d\|\delta\|^2}{dt} \leq 0$, we can see that $\|\delta\|$ decreases as time passes. Thus, $\delta(t)$ will move toward the point p^*, to make $r(t) \rightarrow r^*$, on the dn-sphere $\mathcal{S}$. Hence, we can have

$$\dot{V} \leq -\frac{\epsilon}{(p_s\sqrt{2(n-1)})^{2-\alpha}}(\lambda_{min}\sin^2(\xi(t_0)))^{\frac{2-\alpha}{2}}V^{\frac{2-\alpha}{2}} \quad (9.42)$$

where $\xi(t_0) = \cos^{-1}\left(\frac{\delta(t_0)^T r^*}{\|r^*\|^2}\right)$. Consequently, by Lemma 2.15, it is finite-time stable.

Example 9.4 Consider a triangular network with three agents located at $p_1 = (0, 1)^T$, $p_2 = (1, -2)^T$, and $p_3 = (-1, -2)^T$, with desired locations $p_1^* = (0, 0)^T$, $p_2^* = (0, 1)^T$, and $p_3^* = (-1, -1)^T$. Let the agents be updated by the finite-time control law (9.25) or (9.26). To compute the finite-time, it is required to find the bearing rigidity matrix. For this, we first assign the directions of edges arbitrarily as $2 \rightarrow 1$, $3 \rightarrow 1$, and $3 \rightarrow 2$, which makes the following incidence matrix:

$$\mathbb{H}_+ = \begin{bmatrix} 1 & -1 & 0 \\ 1 & 0 & -1 \\ 0 & 1 & -1 \end{bmatrix}$$

We also compute the orthogonal projection operators as

$$\mathbb{P}_{g_{12}} = \mathbb{I}_2 - \frac{1}{\|z_{12}\|^2}(p_1 - p_2)(p_1 - p_2)^T$$
$$\mathbb{P}_{g_{13}} = \mathbb{I}_2 - \frac{1}{\|z_{13}\|^2}(p_1 - p_3)(p_1 - p_3)^T$$
$$\mathbb{P}_{g_{23}} = \mathbb{I}_2 - \frac{1}{\|z_{23}\|^2}(p_2 - p_3)(p_2 - p_3)^T$$

The bearing rigidity matrix $\mathbb{R}_B$ is computed as

$$\mathbb{R}_B(p) = \text{blkdg}\left(\frac{\mathbb{P}_{g_k}}{\|z_k\|}\right)\mathbb{H}_+ \otimes \mathbb{I}_2 = \begin{bmatrix} \frac{1}{\|z_{12}\|}\mathbb{P}_{g_{12}} & \mathbf{0}_{2\times 2} & \mathbf{0}_{2\times 2} \\ \mathbf{0}_{2\times 2} & \frac{1}{\|z_{13}\|}\mathbb{P}_{g_{13}} & \mathbf{0}_{2\times 2} \\ \mathbf{0}_{2\times 2} & \mathbf{0}_{2\times 2} & \frac{1}{\|z_{23}\|}\mathbb{P}_{g_{23}} \end{bmatrix}\begin{bmatrix} \mathbb{I}_2 & -\mathbb{I}_2 & \mathbf{0}_{2\times 2} \\ \mathbb{I}_2 & \mathbf{0}_{2\times 2} & -\mathbb{I}_2 \\ \mathbf{0}_{2\times 2} & \mathbb{I}_2 & -\mathbb{I}_2 \end{bmatrix}$$

When $p = p^*$, the eigenvalues of $(\bar{\mathbb{R}}_B^*)^T\bar{\mathbb{R}}_B^*$ are computed as 0, 0, 0, 0.1840, 3.3155, 3.9005. Thus, the nonzero minimum eigenvalue is $\lambda_{min} = 0.1840$. We further compute $\epsilon \triangleq \min_{\forall k}\|z_k^*\| = 1$, $p_c = \frac{1}{3}(p_1 + p_2 + p_3) = (0, -1)^T$, and $p_s =$

$\sqrt{\frac{1}{3}\sum_{i=1}^{3}\|p_i - p_c\|^2} = 1.6330$. To compute $\sin(\xi)$, the initial configuration and desired configuration, which are defined by the positions of agents, they need to be matched to the origin. For this, we need to compute r and r^* such as

$$r = p - \mathbf{1} \otimes p_c = ((0, 2), (1, -1), (-1, -1))^T$$
$$r^* = p^* - \mathbf{1} \otimes p_c^* = ((1/3, 0), (1/3, 1), (-2/3, -1))^T$$

Since the scaling does not matter in bearing-based formation control, we further normalize r and r^* as

$$\bar{r} = \frac{1}{\|r\|} r = \frac{1}{2.8284}((0, 2), (1, -1), (-1, -1))^T$$
$$\bar{r}^* = \frac{1}{\|r^*\|} r^* = \frac{1}{1.6330}((1/3, 0), (1/3, 1), (-2/3, -1))^T$$

Thus, we have $\delta(t_0) = \bar{r} - \bar{r}^* = (-0.2041, 0.7071, 0.1494, -0.9659, 0.05457, 0.2588)^T$. Therefore, we compute $\xi(t_0) = 2.0712$ and $\sin^2(\xi(t_0)) = 0.7698$. Now, we can calculate

$$k = \frac{\epsilon}{(p_s\sqrt{2(n-1)})^{2-\alpha}}(\lambda_{min}\sin^2(\xi(t_0)))^{\frac{2-\alpha}{2}}$$
$$= \frac{1}{(2 \times 1.6330)^{2-\alpha}}(0.1840 \times 0.7698)^{\frac{2-\alpha}{2}}$$

Also we compute $V(p(t_0)) = \frac{1}{2}\|\delta(t_0)\|^2 = 0.7834$. Finally, replacing $\alpha' = \frac{2-\alpha}{2}$, we can have

$$T \leq \frac{V(p(t_0))^{1-\alpha'}}{k'(1-\alpha')} = \frac{0.7834^{1-\alpha'}}{k'(1-\alpha')} \tag{9.43}$$

where $k' = \frac{1}{(2\times 1.6330)^{2\alpha'}}(0.1840 \times 0.7698)^{\alpha'}$ and $0.5 < \alpha' < 1$.

It is worth remarking that the right-hand side of (9.43) may be minimized by choosing $0.5 < \alpha' < 1$ to an optimal value.

9.3 Bearing-Based Formation Control via Orientation Alignment

In the previous bearing-based control, it is assumed that the orientations of the local coordinate frames of agents are aligned already to a common reference frame. Under such an assumption, the bearing-based formation control law (9.7) is developed. However, it is natural to suppose that the orientations of agents are not aligned initially. In this section, we would like to implement the control law (9.7) into local coordinate frames even though the orientations of agents are not aligned initially. Further, in this section, we would like to extend the bearing-based formation control of the previous section to 3-dimensional space.

Let us rewrite (9.7) into the local frame as

$$\begin{aligned} R_i \dot{p}_i = u_i^i &= -\sum_{j \in \mathcal{N}_i^o} R_i \mathbb{P}_{g_{ji}} g_{ji}^* \\ &= -\sum_{j \in \mathcal{N}_i^o} \left[R_i g_{ji}^* - \frac{R_i g_{ji}}{\|g_{ji}\|} \frac{g_{ji}^T R_i^T}{\|g_{ji}\|} R_i g_{ji}^* \right] \end{aligned} \tag{9.44}$$

where $R_i g_{ji}^*$ are the desired bearing vectors given in the local coordinate frame of agent i, and $R_i g_{ji}$ are bearing vectors measured in the local coordinate frame of agent i. Denoting $R_i g_{ji}^* \triangleq (g_{ji}^i)^*$ and $R_i g_{ji} = g_{ji}^i$, we can rewrite the above equation as

$$u_i^i = -\sum_{j \in \mathcal{N}_i^o} \mathbb{P}_{g_{ji}^i} (g_{ji}^i)^* \tag{9.45}$$

Thus, the control law (9.7) can be implemented into the local coordinate frames using only local measurements although the coordinate frames are not aligned. But, since the bearing rigidity is defined for a group of agents whose coordinate frames are aligned [19], even though each agent achieves the desired bearing constraints by the control law (9.45), the achieved formation configuration may not be as the desired one when the orientations of agents are not aligned. To overcome this issue, in this section, we design a bearing-based formation control law via orientation alignment. The underlying idea is similar to the orientation alignment-based control laws introduced in Chap. 6.

In $\mathbb{R}^3$, the linear and angular velocities of agent i with respect to ${}^i\Sigma$ are denoted as $u_i^i = (u_x^i, u_y^i, u_z^i)^T$ and $\omega_i = (\omega_{1i}, \omega_{2i}, \omega_{3i})^T$. Let the coordinate transformation from ${}^i\Sigma$ to ${}^g\Sigma$ be denoted as R^i, i.e., $R^i = R_i^{-1}$. Then, the position and orientation dynamics of agent i in a global coordinate frame can be written as

$$\dot{p}_i = R^i u_i^i = R_i^T u_i^i \tag{9.46}$$

$$\dot{R}_i = R_i \Omega_i \tag{9.47}$$

where the skew-symmetric matrix Ω_i is a function of angular velocities (see (8.72) and (8.75))

$$\Omega_i = \begin{bmatrix} 0 & -\omega_{3i} & \omega_{2i} \\ \omega_{3i} & 0 & -\omega_{1i} \\ -\omega_{2i} & \omega_{1i} & 0 \end{bmatrix}$$

The bearing vector g_{ji} is defined with respect to a global coordinate frame. Let us transform it to the local frame as $g_{ji}^i = R_i g_{ji}$. As done in Chaps. 6 and 7, it is assumed that agent i can obtain the relative orientation $R_j R_i^{-1}$ of the outgoing neighboring agents, where $j \in \mathcal{N}_i^o$.

For the orientation alignment, the control law of (7.32) is used as

$$\omega_i = -\sum_{j\in\mathcal{N}_i} (R_j R_i^{-1} - R_i R_j^{-1})^\vee \tag{9.48}$$

In the above equation, the term $R_j R_i^{-1}$ is measured by agent i, while the term $R_i R_j^{-1}$ is measured by agent j. Thus, both agents need to sense the relative orientation in their own coordinate frames, which implies that the sensing topology is bidirectional, i.e., undirected. But, agent i does not need to provide its measurement to agent j if it controls its orientation toward its outgoing nodes only. But, still it needs to receive the measurements $R_i R_j^{-1}$ from the agent j. Thus, the communication and actuation topologies could be a directed one. We assume that the communication and actuation topologies include an arborescence. From (9.47) and (9.48), the orientation of agent i is updated by

$$\dot{R}_i = -\sum_{j\in\mathcal{N}_i} R_i(R_j R_i^{-1} - R_i R_j^{-1}) = \sum_{j\in\mathcal{N}_i} (R_j R_i^{-1} - R_i R_j^{-1}) R_i \tag{9.49}$$

Based on Lemma 7.5 and Theorem 7.2, the following statement can be obtained directly.

Lemma 9.13 *Assume that, given a graph $\mathcal{G} = (\mathcal{V}, \mathcal{E})$, the sensing topology has a spanning tree and the communication and actuation topologies include an arborescence. If $R_j(t_0) R_i^T(t_0)$ for all edges $(i, j)^e \in \mathcal{E}$ are positive definite, then the orientation alignment scheme (9.49) ensures $\lim_{t\to\infty} R_i^T R_j = \lim_{t\to\infty} R_j R_i^T = \mathbb{I}_3$. Furthermore, if the leader is stationary, then the orientations of other agents converge to $R_1(t_0)$ exponentially fast as $t \to \infty$.*

For the formation control, the first follower is updated in its local frame as

$$u_2^2 = -\mathbb{P}_{g_{12}^2}(\mathbb{I}_3 + R_2 R_1^T)(g_{12}^2)^* \tag{9.50}$$

For other agents i, $3 \le i \le n$, the position is updated in ${}^i\Sigma$ as

$$u_i^i = -\sum_{j\in\mathcal{N}_i} \mathbb{P}_{g_{ji}^i}(\mathbb{I}_3 + R_i R_j^T)(g_{ji}^i)^* \tag{9.51}$$

where $\mathbb{P}_{g_{ji}^i} = \mathbb{I}_3 - g_{ji}^i (g_{ji}^i)^T$ is the orthogonal projection matrix of the local bearing vector g_{ji}^i. To implement the above control laws (9.50) and (9.51), the neighboring agent j needs to send the information of $R_i R_j^T$ to the agent i. For an analysis, let us transform the control input (9.51) to a global coordinate frame as

$$
\begin{aligned}
u_i &= R^i u_i^i \\
&= -R^i \mathbb{P}_{g_{ji}^i}(\mathbb{I}_3 + R_i R_j^T)(g_{ji}^i)^* \\
&= -R^i(\mathbb{I}_3 - R_i g_{ji}(g_{ji})^T R_i^T)(\mathbb{I}_3 + R_i R_j^T)(g_{ji}^i)^* \\
&= -R_i^T R_i(\mathbb{I}_3 - g_{ji}(g_{ji})^T) R_i^T(\mathbb{I}_3 + R_i R_j^T)(g_{ji}^i)^* \\
&= -\mathbb{P}_{g_{ji}}(R_i^T + R_j^T)(g_{ji}^i)^*
\end{aligned}
$$

Then, the dynamics of agent 2 and other agents can be written in a global coordinate frame as

$$
\dot{p}_2 = \underbrace{-2\mathbb{P}_{g_{12}} R_1^T (g_{12}^2)^*}_{\triangleq f_2(p,t)} + \underbrace{\mathbb{P}_{g_{12}}(R_1^T - R_2^T)(g_{12}^2)^*}_{\triangleq h_2(p,t)} \tag{9.52}
$$

$$
\dot{p}_i = \underbrace{-2\sum_{j\in\mathcal{N}_i} \mathbb{P}_{g_{ji}} R_1^T (g_{ji}^i)^*}_{\triangleq f_i(p,t)} + \underbrace{\sum_{j\in\mathcal{N}_i} \mathbb{P}_{g_{ji}}(2R_1^T - R_i^T - R_j^T)(g_{ji}^i)^*}_{\triangleq h_i(p,t)} \tag{9.53}
$$

Then, the overall position dynamics can be written as

$$
\dot{p} = f(p) + h(p,t) \tag{9.54}
$$

where $f(p) = (f_1^T, \ldots, f_n^T)^T$, $h(p,t) = (h_1^T, \ldots, h_n^T)^T$, and $f_1 = 0$ and $h_1 = 0$. In (9.54), the term $h(p,t)$ can be considered as an input to the nominal system

$$
\dot{p} = f(p) \tag{9.55}
$$

Lemma 9.14 *Under the same assumption as Lemma 9.13, $h(t)$ is bounded and converges to 0 asymptotically.*

Proof The boundedness of $h_i(t)$ is clear since all the states in bearing-based formation control are bounded. The convergence is also direct from Lemma 9.13.

Based on Lemma 9.4, the following lemmas can be generated:

Lemma 9.15 *The unforced system $\dot{p}_2 = f_2(p_2)$ has two equilibria. The first equilibrium $p_2 = p_{2a}^*$ corresponding to $g_{12} = R_1^T(g_{12}^2)^*$ is almost globally asymptotically stable. The second equilibrium $p_2 = p_{2b}^*$ corresponding to $g_{12} = -R_1^T(g_{12}^2)^*$ is unstable.*

Proof The unforced system is written as

$$
\begin{aligned}
\dot{p}_2 &= -2\mathbb{P}_{g_{12}} R_1^T (g_{12}^2)^* \\
&= -2\mathbb{P}_{g_{12}} R_1^T R_2 g_{12}^* \\
&= -2\mathbb{P}_{g_{12}} g_{12}^* - 2\mathbb{P}_{g_{12}} \Delta R_{12} g_{12}^*
\end{aligned}
$$

where $R_1^T R_2 = \mathbb{I}_3 + \Delta R_{12}$. From Lemma 9.13, we know that the term $2\mathbb{P}_{g_{12}} \Delta R_{12} g_{12}^*$ converges to zero exponentially fast. Also from Lemma 9.4, the term $-2\mathbb{P}_{g_{12}} g_{12}^*$ also converges to zero exponentially fast. Thus, the unforced dynamics for the equilibrium $p_2 = p_{2a}^*$ is almost globally asymptotically stable. However, for the equilibrium $p_2 = p_{2b}^*$, since the dynamics $\dot{p}_2 = -2\mathbb{P}_{g_{12}} g_{12}^*$ is unstable, the overall dynamics will be unstable.

Lemma 9.16 *The system (9.52) has two equilibria. The equilibrium $p_2 = p_{2a}^*$ is almost globally asymptotically stable, while the equilibrium $p_2 = p_{2b}^*$ is exponentially unstable. All the trajectories starting from $p_2(t_0) \neq p_{2b}^*$ asymptotically converge to the stable equilibrium.*

Proof Consider the potential function $V = \frac{1}{2}\|p_2 - p_{2a}^*\|^2$. Then, using the same idea of (A.60)–(A.61), we have

$$\begin{aligned}
\dot{V} &= (p_2 - p_{2a}^*)^T \dot{p}_2 \\
&= -2(p_2 - p_{2a}^*)^T \mathbb{P}_{g_{12}} R_1^T (g_{12}^2)^* + (p_2 - p_{2a}^*)^T h_2 \\
&= -2(p_2 - p_{2a}^*)^T \frac{\mathbb{P}_{g_{21}}}{d_{21}} (p_2 - p_{2a}^*) + (p_2 - p_{2a}^*)^T h_2 \\
&\leq -\frac{2\sin^2\alpha(t_0)}{d_{21}} \|p_2 - p_{2a}^*\|^2 + \|p_2 - p_{2a}^*\| \|h_2\| \\
&\leq -\kappa V + 2d_{21}\|h_2\|
\end{aligned} \tag{9.56}$$

where $\kappa = \frac{4\sin^2\alpha(t_0)}{d_{21}}$, and α is the angle between $p_2 - p_{2a}^*$ and g_{12}^*. Thus, from (9.56), the state p_2 is shown to be ultimately bounded. Thus, since the unforced nominal system is globally asymptotically stable by Lemma 9.15, and h_2 is ultimately bounded, it is almost globally input-to-state stable (ISS). Finally, since $h_2(t)$ converges to zero by Lemma 9.14, the equilibrium $p_2 = p_{2a}^*$ is almost globally asymptotically stable. The instability of the point $p_2 = p_{2b}^*$ can be shown by following the proof of Lemma 9.6.

The dynamics of the second follower is given as

$$\begin{aligned}
\dot{p}_3 &= f_3(p_3, p_2) + h_3(t) \\
&= -2\mathbb{P}_{g_{13}} R_1^T (g_{13}^3)^* - 2\mathbb{P}_{g_{23}} R_1^T (g_{23}^3)^* + h_3(t)
\end{aligned} \tag{9.57}$$

To utilize the concept of ISS, consider the following two unforced systems

$$\dot{p}_3 = f_3(p_3, p_{2a}^*) \tag{9.58}$$

$$\dot{p}_3 = f_3(p_3, p_{2b}^*) \tag{9.59}$$

with, from (9.49), the update of angular velocities as

$$\dot{R}_3 = R_3(R_1 R_3^{-1} - R_3 R_1^{-1}) + R_3(R_2 R_3^{-1} - R_3 R_2^{-1}) \tag{9.60}$$

Lemma 9.17 *The system (9.58) has an almost globally asymptotically stable equilibrium* p_{3a}^* *where* $g_{13} = R_1^T(g_{13}^3)^*$ *and* $g_{23} = R_1^T(g_{23}^3)^*$*, and the system (9.59) has an unstable equilibrium* p_{3b}^* *where* $g_{13} = -R_1^T(g_{13}^3)^*$ *and* $g_{23} = -R_1^T(g_{23}^3)^*$.

Proof The unforced dynamics is written as $\dot{p}_3 = -2\mathbb{P}_{g_{13}} R_1^T R_3 g_{13}^* - 2\mathbb{P}_{g_{23}} R_1^T R_3 g_{23}^* = -2\mathbb{P}_{g_{13}} g_{13}^* - 2\mathbb{P}_{g_{23}} g_{23}^* - 2\mathbb{P}_{g_{13}} \Delta R_{13} g_{13}^* - 2\mathbb{P}_{g_{23}} \Delta R_{13} g_{23}^*$, where $R_1^T R_3 = \mathbb{I}_3 + \Delta R_{13}$. Consequently, with Lemmas 9.5 and 9.6, and by following a similar proof of Lemma 9.15, the proof is completed.

Theorem 9.6 *The cascade system governed by (9.57) and (9.60) has an almost globally asymptotically stable equilibrium* $p_2 = p_{2a}^*$, $p_3 = p_{3a}^*$*, while it has unstable equilibrium* $p_2 = p_{2b}^*$, $p_3 = p_{3b}^*$*. All trajectories starting out of the undesired equilibrium* $p_2 = p_{2b}^*$ *will converge to the stable equilibrium.*

Proof Using the Lyapunov candidate $V = \frac{1}{2}\|p_3 - p_{3a}^*\|^2$, the derivative is obtained as

$$
\begin{aligned}
\dot{V} &= -2(p_3 - p_{3a}^*)^T (\mathbb{P}_{g_{13}} R_1^T (g_{13}^3)^* + \mathbb{P}_{g_{23}} R_1^T (g_{23}^3)^* - h_3(t)) \\
&= -2(p_3 - p_{3a}^*)^T \left(\frac{\mathbb{P}_{g_{13}}}{\|z_{13}^*\|} + \frac{\mathbb{P}_{g_{23}}}{\|z_{23}^*\|} \right) (p_3 - p_{3a}^*) + 2(p_3 - p_{3a}^*)^T \frac{\mathbb{P}_{g_{23}}}{\|z_{23}^*\|} (p_2 - p_{2a}^*) \\
&\quad + 2(p_3 - p_{3a}^*)^T h_3(t) \\
&\leq -2(p_3 - p_{3a}^*)^T M (p_3 - p_{3a}^*) + \|2(p_3 - p_{3a}^*)^T\| \left(\frac{\|\mathbb{P}_{g_{23}}\|}{\|z_{23}^*\|} \|p_2 - p_{2a}^*\| + \|h_3(t)\| \right) \\
&\leq -2\kappa \|p_3 - p_{3a}^*\|^2 + \|2(p_3 - p_{3a}^*)^T\| \left(\frac{\|\mathbb{P}_{g_{23}}\|}{\|z_{32}^*\|} \|p_2 - p_{2a}^*\| + \|h_3(t)\| \right)
\end{aligned}
\tag{9.61}
$$

where κ was defined in Lemma 9.6. Also it is clear that the term $\frac{\|\mathbb{P}_{g_{23}}\|}{\|z_{23}^*\|} \|(p_2 - p_{2a}^*)\| + \|h_3(t)\|$ is bounded. Thus, as $\|p_3 - p_{3a}^*\|$ becomes large, we have $\dot{V} < 0$. Consequently, $\|p_3 - p_{3a}^*\|$ is bounded. Denoting $\xi_{max} \triangleq \frac{\|\mathbb{P}_{g_{23}}\|}{\|z_{23}^*\|} \|(p_2 - p_{2a}^*)\|$ and taking a φ such that $\varphi > \max_{t \geq t_0} \|p_3 - p_{3a}^*\|$, we can have

$$
\dot{V} \leq -4\kappa V + 2\xi_{max}\varphi + 2\varphi\|h_3(t)\| \tag{9.62}
$$

Thus, the system is ultimately bounded. Finally, since the unforced system (9.58) is almost globally asymptotically stable, the dynamics of angular velocity (9.60) is globally asymptotically stable under the conditions of Lemma 9.13, and $h_3(t)$ is bounded (but it converges to zero as $t \to \infty$), the overall cascade system governed by (9.58) and (9.60) is almost globally asymptotically stable. The instability of the cascade system governed by (9.59) and (9.60) is direct from Lemmas 9.16 and 9.17.

Now, consider the n-agent system (9.54), where $h(p, t)$ can be considered as an input. With the arguments thus far, the following lemma is obtained.

Lemma 9.18 *The unforced system (9.55) has two equilibria. The first equilibrium $p = p_a^*$ corresponding to $g_{ji} = R_1^T(g_{ji}^i)^*$ is almost globally asymptotically stable, while the second equilibrium $p = p_b^*$ corresponding to $g_{ji} = -R_1^T(g_{ji}^i)^*$ is unstable.*

Then, with the above lemma, we can obtain the main result of this section as follows:

Theorem 9.7 *Consider the system (9.46)–(9.47) under conditions of Lemma 9.13 with the position control law (9.51) and orientation alignment law (9.49). If the graph is connected and the leader agent is stationary, we then have $R_i(t) \to R_1$ and $p(t) \to p_a^*$ for all $i = 2, \ldots, n$ as $t \to \infty$ if $R_2(t_0) \neq R_1$ and $p_2(t_0) \neq p_{2b}^*$.*

Proof From Lemma 9.13, we have $R_i(t) \to R_1$. Then, using the results of Lemmas 9.14, 9.18, and Theorem 9.6, we can get the desired conclusion by taking mathematical induction as done in the proof of Theorem 9.2.

So far, in this chapter, we have presented the bearing-based formation control schemes with and without orientation alignment. When the orientations of agents are aligned, it is a kind of displacement-based scheme as shown in Sect. 9.2; but when the orientations are not aligned, it is a fully distributed formation control scheme as outlined in Sect. 9.3. In terms of implementation cost, the control scheme of Sect. 9.2 is much simple than the scheme presented in Sect. 9.3. But, the control scheme of Sect. 9.3 does not use any global information, which makes the algorithm of Sect. 9.3 more advanced and more favorable. As another advantage of the algorithm of Sect. 9.3, it may enable the agent to escape the saddle point. The following example provides a comparison of the algorithms of Sects. 9.2 and 9.3.

Example 9.5 Let the agent 2 be located at $p_2 = p_{2b}^*$ initially. For the case of $R_2 = R_1$, we have $\dot{p}_2 = -2\mathbb{P}_{g_{12}} g_{12}^* = 0$. Thus, if the agent 2 was at the point p_{2b}^* initially, it will be staying there forever. But, for the case of $R_2 \neq R_1$, we have

$$\begin{aligned}
\dot{p}_2 &= -2\mathbb{P}_{g_{12}} g_{12}^* - 2\mathbb{P}_{g_{12}} \Delta R_{12} g_{12}^* \\
&= -2(\mathbb{I}_2 - g_{12} g_{12}^T) \Delta R_{12} g_{12}^* \\
&= -2\left(\mathbb{I}_2 - \frac{(p_1 - p_{2b}^*)(p_1 - p_{2b}^*)^T}{\|p_1 - p_{2b}^*\|^2}\right) \Delta R_{12} \frac{p_1 - p_{2a}^*}{\|p_1 - p_{2a}^*\|} \neq 0
\end{aligned} \tag{9.63}$$

For example, let $p_1 = (0, 0)^T$, and $p_{2a}^* = (-1, 0)^T$ and $p_{2b}^* = (1, 0)^T$. Also suppose that $\Delta R_{12} = \begin{bmatrix} \frac{1}{2} & -\frac{\sqrt{3}}{2} \\ \frac{\sqrt{3}}{2} & \frac{1}{2} \end{bmatrix}$. Then, when $p_2(t_0) = p_{2b}^*$, we have $\dot{p}_2 = -\begin{bmatrix} 0 \\ \sqrt{3} \end{bmatrix}$. Thus, the agent 2 is forced to move along the negative direction of the y-axis. It is also observed that when $p_2(t_0) = p_{2a}^*$, with the same orientation, the relative orientation angles in ΔR_{12} will have opposite signs. Then, we would have $\Delta R_{12} = \begin{bmatrix} \frac{1}{2} & \frac{\sqrt{3}}{2} \\ -\frac{\sqrt{3}}{2} & \frac{1}{2} \end{bmatrix}$; thus, the agent 2 would be forced to move along the positive direction of the y-axis. Consequently, with the case of either $p_2(t_0) = p_{2b}^*$ or $p_2(t_0) = p_{2a}^*$, the agent 2 is forced to move. This is due to the fact that the orientation alignment dynamics (9.49) and position dynamics (9.51) are coupled. In the orientation alignment-based formation control, the misalignment in the orientations acts as a disturbance so that the agents may escape from the saddle points in certain cases.

9.4 Bearing-Based Control of a Cyclic Triangular Formation

Thus far, in this chapter, we have studied acyclic minimally persistent formations under directed graphs. It is also natural to postulate that agents sense each other in a cycle. However, it is usually considered difficult to examine a convergence property of formations that have cycles. In this section, we introduce a result of bearing-based formation control for a cyclic triangular graph that was developed in [15]. Actually, the result of this section can be considered as a counterpart, reformulated in the bearing-based setup, of the result of Sect. 3.3. Let us use the control law given in (9.7) for the pairs of neighboring agents $(1, 2)^{\bar{e}}$, $(2, 3)^{\bar{e}}$, and $(3, 1)^{\bar{e}}$:

$$\dot{p}_i = u_i = -\sum_{j \in \mathcal{N}_i^o} \mathbb{P}_{g_{ji}} g_{ji}^* \tag{9.64}$$

where the bearing vector g_{ji}^* provides the desired directional vector from agent i to j, and the matrix $\mathbb{P}_{g_{ji}}$ is determined from the current direction. For a simplicity, in (9.64), the orientations of agents are assumed to be aligned. For the three agents, it is obvious that the desired bearing vectors g_{21}^*, g_{32}^* and g_{13}^* will be realizable if and only if $g_{13}^* \neq \pm g_{32}^*$, $g_{13}^* \neq \pm g_{21}^*$, and $g_{21}^* \neq \pm g_{32}^*$. This condition is equivalent to an existence of positive constants m_1, m_2,, and m_3 such that

$$m_1 g_{21}^* + m_2 g_{32}^* + m_3 g_{13}^* = 0 \tag{9.65}$$

From (9.64), the equilibrium set can be determined as

$$E_{p^*} \triangleq \{p \in \mathbb{R}^6 : g_{ji} = \pm g_{ji}^*, (i, j)^{\bar{e}} \in \overrightarrow{\mathcal{E}}\} \tag{9.66}$$

It can be shown that the set E_{p^*} can be divided explicitly into two disjoint sets as

Lemma 9.19 *The equilibrium set E_{p^*} is divided into two sets as $E_{p^*} = E_{p^*}^+ \cup E_{p^*}^-$ where $E_{p^*}^+ \triangleq \{p \in \mathbb{R}^6 : g_{ji} = g_{ji}^*, (i, j)^{\bar{e}} \in \overrightarrow{\mathcal{E}}\}$ and $E_{p^*}^- \triangleq \{p \in \mathbb{R}^6 : g_{ji} = -g_{ji}^*, (i, j)^{\bar{e}} \in \overrightarrow{\mathcal{E}}\}$.*

Proof Without loss of generality, let us assume that $g_{21} = -g_{21}^*$, $g_{32} = g_{32}^*$, and $g_{13} = g_{13}^*$. By inserting these values into (9.65), we can have

$$-m_1 g_{21} + m_2 g_{32} + m_3 g_{13} = 0 \tag{9.67}$$

But, it is easy to see that the vector $m_2 g_{32} + m_3 g_{13}$ is pointing the same direction as the vector $-m_1 g_{21}$, which cannot make $-m_1 g_{21} + m_2 g_{32} + m_3 g_{13}$ be equal to zero. For the cases of $g_{ji} = g_{ji}^*$ and $g_{ji} = -g_{ji}^*$, $\forall (i, j)^{\bar{e}} \in \overrightarrow{\mathcal{E}}$, they can satisfy the condition of (9.65), which completes the proof.

Example 9.6 Let us suppose that agents are located as $p_1 = (0, 0)^T$, $p_2 = (2, 1)^T$, $p_3 = (3, -1)^T$. Then, we have

$$g_{21} = \frac{1}{\sqrt{5}} \begin{bmatrix} 2 \\ 1 \end{bmatrix};\ g_{32} = \frac{1}{\sqrt{5}} \begin{bmatrix} 1 \\ -2 \end{bmatrix};\ g_{13} = \frac{1}{\sqrt{10}} \begin{bmatrix} -3 \\ 1 \end{bmatrix}$$

Thus, when $g_{21} = -g_{21}^*$, $g_{32} = g_{32}^*$, and $g_{13} = g_{13}^*$, there must exist positive constants m_1, m_2, m_3 such that the following holds:

$$-m_1 \begin{bmatrix} 2 \\ 1 \end{bmatrix} + m_2 \begin{bmatrix} 1 \\ -2 \end{bmatrix} + m_3 \begin{bmatrix} -3 \\ 1 \end{bmatrix} = 0$$

The above equation would be satisfied if and only if $m_2 = m_3 = -m_1$. Thus, the signs of m_1, m_2, m_3 are not same. On the other hand, if $g_{21} = g_{21}^*$, $g_{32} = g_{32}^*$, and $g_{13} = g_{13}^*$, we have

$$m_1 \begin{bmatrix} 2 \\ 1 \end{bmatrix} + m_2 \begin{bmatrix} 1 \\ -2 \end{bmatrix} + m_3 \begin{bmatrix} -3 \\ 1 \end{bmatrix} = 0$$

which will be satisfied if and only if $m_1 = m_2 = m_3$.

For the stability analysis of (9.64), we use the subtended angle between g_{ji} and g_{ji}^*. Let these angles be denoted as α_i, which are computed as $\cos(\alpha_i) = (g_{ji}^*)^T g_{ji}$. Due to Lemma 9.19, it is then easy to see that if the subtended angles between g_{ji} and g_{ji}^* are given as $0 \leq \alpha_i \leq \pi$, $i = 1, 2, 3$, then the subtended angles between $-g_{ji}$ and g_{ji}^* are $-\pi \leq \alpha_i \leq 0$, $i = 1, 2, 3$. It is clear that if $\alpha_i, i = 1, 2, 3$, converge to zero, then it corresponds to the equilibrium set $E_{p^*}^+$, which is the desired one, or $\alpha_i, i = 1, 2, 3$, converge to $-\pi$, then it corresponds to the equilibrium set $E_{p^*}^-$, which is the undesired one. Thus, by analyzing the dynamics of α_i, we can evaluate the stability of the system. From $\cos(\alpha_i) = (g_{ji}^*)^T g_{ji}$, by taking a derivative along the time, we can obtain the dynamics for α_i as

$$\dot{\alpha}_i \sin \alpha_i = -(g_{ji}^*)^T \frac{\mathbb{P}_{g_{ji}}}{d_{ji}} (\dot{p}_j - \dot{p}_i) \tag{9.68}$$

where $d_{ji} = \|p_j - p_i\|$. Now, using (9.64) and the idempotent property, and substituting $\mathbb{P}_{g_{ij}} = g_{ji}^{\perp} (g_{ji}^{\perp})^T$ into the above equation, we can have

$$\begin{aligned} d_{ji} \dot{\alpha}_i \sin \alpha_i &= (g_{ji}^*)^T g_{ji}^{\perp} (g_{ji}^{\perp})^T (g_{kj}^{\perp}) (g_{kj}^{\perp})^T g_{kj}^* - (g_{ji}^*)^T g_{ji}^{\perp} (g_{ji}^{\perp})^T g_{ji}^* \\ &= \pm \sin \alpha_i \cos \beta_i (\pm \sin \alpha_j) - \sin^2 \alpha_i \end{aligned} \tag{9.69}$$

where $j = i + 1(\text{modulo}3)$ and $k = i + 2(\text{modulo}3)$, and β_i is the subtended angle between two adjacent vectors g_{ji} and g_{kj}. Thus, we can have the dynamics for angles α_i as

$$\dot{\alpha}_1 = -\frac{\sin\alpha_1}{d_{21}} \pm \frac{\sin\alpha_2 \cos\beta_1}{d_{21}} \tag{9.70}$$

$$\dot{\alpha}_2 = -\frac{\sin\alpha_2}{d_{32}} \pm \frac{\sin\alpha_3 \cos\beta_2}{d_{32}} \tag{9.71}$$

$$\dot{\alpha}_3 = -\frac{\sin\alpha_3}{d_{13}} \pm \frac{\sin\alpha_1 \cos\beta_3}{d_{13}} \tag{9.72}$$

For the stability analysis of the system (9.70)–(9.72), we will use a linearization. Denoting $\alpha = (\alpha_1, \alpha_2, \alpha_3)^T$ and $\beta = (\beta_1, \beta_2, \beta_3)^T$, we rewrite (9.70)–(9.72) in a concise form as

$$\dot{\alpha} = f(\alpha, \beta)$$

The Jacobian of the above dynamics can be obtained as

$$\frac{\partial f(\alpha, \beta)}{\partial \alpha} = \left[\frac{\partial f_p}{\partial \alpha_q}\right]_{p=\{1,2,3\},\ q=\{1,2,3\}} \tag{9.73}$$

where $\frac{\partial f_i}{\partial \alpha_i}$, $\frac{\partial f_i}{\partial \alpha_j}$, and $\frac{\partial f_i}{\partial \alpha_k}$, $j = i + 1(\text{modulo}3)$ and $k = i + 2(\text{modulo}3)$ are computed as

$$\frac{\partial f_i}{\partial \alpha_i} = -\frac{\cos\alpha_i}{d_{ji}} + \frac{\sin\alpha_i \pm \sin\alpha_j \cos\beta_i}{d_{ji}^2}\frac{\partial d_{ji}}{\partial \alpha_i} \pm \frac{\sin\alpha_j \sin\beta_i}{d_{ji}}\frac{\partial \beta_i}{\partial \alpha_i} \tag{9.74}$$

$$\frac{\partial f_i}{\partial \alpha_j} = \pm\frac{\cos\alpha_j \cos\beta_i}{d_{ji}} + \frac{\sin\alpha_i \pm \sin\alpha_j \cos\beta_i}{d_{ji}^2}\frac{\partial d_{ji}}{\partial \alpha_j} \pm \frac{\sin\alpha_j \sin\beta_i}{d_{ji}}\frac{\partial \beta_i}{\partial \alpha_j} \tag{9.75}$$

$$\frac{\partial f_i}{\partial \alpha_k} = \frac{\sin\alpha_i \pm \sin\alpha_j \cos\beta_i}{d_{ji}^2}\frac{\partial d_{ji}}{\partial \alpha_k} \tag{9.76}$$

Using the fact that the equilibrium set of (9.64) is divided as E_{p*}^+ and E_{p*}^-, we check the stability of these sets.

Lemma 9.20 *The equilibrium set* E_{p*}^+ *is stable while the equilibrium set* E_{p*}^- *is unstable.*

Proof It is clear that in the set E_{p*}^+, $\alpha_1 = \alpha_2 = \alpha_3 = 0$, while in E_{p*}^-, $\alpha_1 = \alpha_2 = \alpha_3 = -\pi$. Thus, inserting these cases into (9.73), we have

$$\left[\frac{\partial f_p}{\partial \alpha_q}\right]_{\alpha_1=\alpha_2=\alpha_3=0} = \begin{bmatrix} -\frac{1}{d_{21}^*} & \pm\frac{\cos\beta_1^*}{d_{21}^*} & 0 \\ 0 & -\frac{1}{d_{32}^*} & \pm\frac{\cos\beta_2^*}{d_{32}^*} \\ \pm\frac{\cos\beta_3^*}{d_{13}^*} & 0 & -\frac{1}{d_{13}^*} \end{bmatrix} \tag{9.77}$$

$$\left[\frac{\partial f_p}{\partial \alpha_q}\right]_{\alpha_1=\alpha_2=\alpha_3=-\pi} = \begin{bmatrix} \frac{1}{d_{21}^*} & \pm\frac{\cos\beta_1^*}{d_{21}^*} & 0 \\ 0 & \frac{1}{d_{32}^*} & \pm\frac{\cos\beta_2^*}{d_{32}^*} \\ \pm\frac{\cos\beta_3^*}{d_{13}^*} & 0 & \frac{1}{d_{13}^*} \end{bmatrix} \tag{9.78}$$

where * is used to denote the final state when reaching to the equilibrium. Thus, by Gershgorin's theorem [6], it is clear that (9.77) is Hurwitz if $\beta_i^* \neq 0$. Since the desired configuration is not a line, it is always Hurwitz stable. On the other hand, (9.78) is unstable.

Now, with the above results, we can make the following main conclusion [15]:

Theorem 9.8 *Assume that initial angles satisfy* $0 \leq \alpha_i \leq \frac{\pi}{2}$, *for* $i = 1, 2, 3$. *Then,* $\alpha_i \to 0$ *asymptotically.*

Proof For the proof, let us choose a Lyapunov candidate $V = \alpha_{max} \triangleq \max_{i=1,2,3} \alpha_i$. Let us suppose, without loss of generality, that at interval $t \in [T_1, T_2)$, $\alpha_1 > \alpha_2 \geq \alpha_3 \geq 0$ holds. Then, from (9.70), it is clear that $\dot{V} = \dot{\alpha}_1 = -\frac{\sin\alpha_1}{d_{21}} \pm \frac{\sin\alpha_2\cos\beta_1}{d_{21}} < 0$ due to $\sin\alpha_1 > |\sin\alpha_2\cos\beta_1| > 0$. Next, at $t = T_2$, we suppose that α_{max} becomes $\alpha_{max} = \alpha_2 = \alpha_1$. Immediately after T_2, we further suppose that α_{max} is α_2 until reaching $t = T_3$, i.e., $\alpha_{max} = \alpha_2$ at $t \in (T_2, T_3)$. Then, at $t = T_2$, based on Clark's generalized gradient [12], we can have

$$\partial V(\alpha) = \overline{\text{co}}\{(1,0,0)^T, (0,1,0)^T\} \triangleq \{(\eta_1, \eta_2, 0)^T : \eta_i \in [0,1], \eta_1 + \eta_2 = 1\}$$

where $\overline{\text{co}}$ is the convex closure. Then, it is said that $\dot{V}$ exists almost everywhere and at $t = T_2$, we have $\dot{V} \in \mathcal{V}(T_2)$ where the set $\mathcal{V}(T_2)$ is given as

$$\mathcal{V}(T_2) \triangleq \bigcup_{\eta \in \partial V(\alpha)} \eta^T \begin{bmatrix} \dot{\alpha}_1 \\ \dot{\alpha}_2 \\ \dot{\alpha}_3 \end{bmatrix} = \{\eta_1\dot{\alpha}_1 + \eta_1\dot{\alpha}_2 : \eta_i \in [0,1], \eta_1 + \eta_2 = 1\} \tag{9.79}$$

By inserting $\dot{\alpha}_1$ of (9.70) and $\dot{\alpha}_2$ of (9.71) into the above equation, we can see that $\eta_1\dot{\alpha}_1 + \eta_1\dot{\alpha}_2 < 0$ for all η_1 and η_2. On the other hands, we suppose that α_{max} becomes $\alpha_{max} = \alpha_3 = \alpha_1$ at $t = T_2$. Also, immediately after T_2, assume that α_{max} is α_3 until reaching $t = T_3$, i.e., $\alpha_{max} = \alpha_3$ at $t \in (T_2, T_3)$. Then, at $t = T_2$, also using the Clark's generalized gradient, we have $\dot{V} < 0$ for any η_1 and η_3. Lastly, it is also required to consider the case of $\alpha_1 = \alpha_2 = \alpha_3$. Let us suppose that this case occurs at $t = T_4$. Then, the Clark's generalized gradient becomes

$$\begin{aligned} \partial V(\alpha) &= \overline{\text{co}}\{(1,0,0)^T, (0,1,0)^T, (0,0,1)^T\} \\ &\triangleq \{(\eta_1, \eta_2, \eta_3)^T : \eta_i \in [0,1], \eta_1 + \eta_2 + \eta_3 = 1\} \end{aligned}$$

In this case, the set $\mathcal{V}$ at T_4 becomes as

$$\mathcal{V}(T_2) \triangleq \bigcup_{\eta \in \partial V(\alpha)} \eta^T \begin{bmatrix} \dot{\alpha}_1 \\ \dot{\alpha}_2 \\ \dot{\alpha}_3 \end{bmatrix} = \{\eta_1 \dot{\alpha}_1 + \eta_1 \dot{\alpha}_2 + \eta_3 \dot{\alpha}_3 : \eta_i \in [0, 1], \eta_1 + \eta_2 + \eta_3 = 1\} \tag{9.80}$$

Then, by using $\dot{\alpha}_1$ of (9.70), $\dot{\alpha}_2$ of (9.71), and $\dot{\alpha}_3$ of (9.72), we also can see that $\dot{V} < 0$ for any η_1, η_2, and η_3. Consequently, we can see that V is a nonincreasing function and $0 \leq \alpha_i \leq \alpha_{max} \leq \alpha_{max}(t_0) \leq \frac{\pi}{2}$. Moreover, since β_i cannot be 0 or π, $\dot{V} = 0$ if and only if $\alpha_1 = \alpha_2 = \alpha_3 = 0$. Thus, $E_{p^*}^{+}$ is asymptotically stable if $0 \leq \alpha_i \leq \frac{\pi}{2}$, for $i = 1, 2, 3$.

The fact that the equilibrium $\alpha_i = 0$ is locally exponentially stable (from Lemma 9.20 and Theorem 9.8) implies p converges to a desired configuration when the initial conditions $0 \leq \alpha_i \leq \frac{\pi}{2}$ are satisfied.

9.5 Summary and Simulations

In bearing-based formation control, there could be undesired equilibrium points since the controller tries to align the current bearing vector to the desired bearing vector. So, there could be undesired equilibrium points that may correspond to an opposite direction of the current and desired bearing vectors. Thus, the key issue in the convergence analysis is to show the instability of undesired equilibrium sets. In Sect. 9.2, the orientations of axes of coordinate frames of agents are assumed aligned. Thus, the agents do not need to exchange the sensing variables with neighboring agents. However, in Sect. 9.3, the orientations of agents' local coordinate frames are not aligned. Thus, to make the alignment, the agents need to exchange the relative orientation measurements $R_j R_i^{-1}$ and $R_i R_j^{-1}$. However, since the actuation topology is directed, the communications also just need to be directed, although the directions of actuations and communications could be opposite. It is noticeable that in the bearing-based formation control under misaligned orientations, the agents only need to sense the bearing vectors on their local coordinate frames. Compared to the distance-based formation control, it uses less sensing variables. Compare Tables 8.1 and 9.1. Although the bearing-based formation control focuses on acyclic minimally persistent formations, it clearly requires less sensing and control information. Note that, the relative displacements p_{ji}^i in Table 8.1 include both the bearing and distance information. From this comparison, we can see that the distance-based formation control utilizes more information than the bearing-based formation control. Although the bearing-based approaches ignore the dilations of formation, it can be still controlled if a distance of an edge can be controlled, as studied in Sect. 12.2.2. Thus, if the underlying topology is rigid (bearing rigid or distance rigid), a desired formation configuration can be achieved with less sensing information by the bearing-based

Table 9.1 Bearing-based formation via global orientation alignment

	Variables	Topology	Edge direction
Sensing	δ_{ji}	Acyclic minimally persistent	Undirected
Control	δ_{ji}, θ_i	Acyclic minimally persistent	Directed
Communications	$\delta_{ij}, \hat{z}_i$	Acyclic minimally persistent	Directed
Computation	$u_i^i, \dot{\hat{z}}_i$	None	N/A

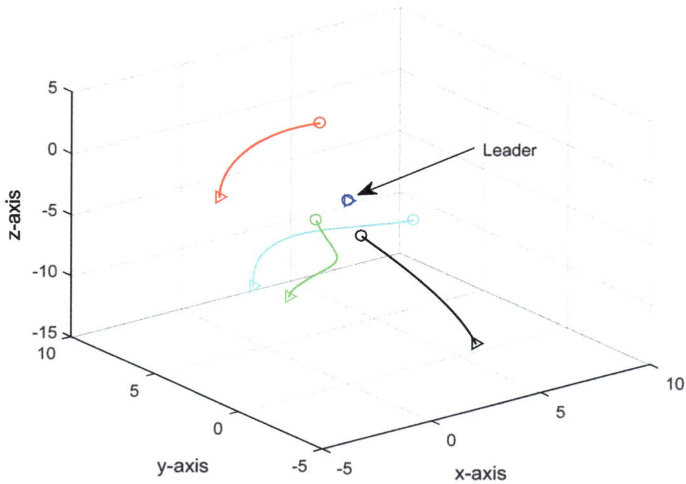

Fig. 9.4 Trajectories of positions of agents. The initial positions are marked by ∘ and the desired positions are marked by ▷

approaches. But, it is also important to know that the distance-based formation control under orientation alignment setups can achieve a desired formation even though the underlying topology has only a spanning tree. The reason for this is quite clear from the fact that the displacement constraints under a tree can define a unique formation.

For the simulation, let us consider five agents with sensing, communication, and actuation topologies as depicted in Fig. 5.4. But, the sensings in this chapter are assumed undirected. Like the distance-based approaches, the sensing is undirected, while the communications and actuations could be directed. Also note that the directions of communications and actuations are opposite. To have acyclic minimally persistent formation in $\mathbb{R}^3$, the agents 4 and 5 need to have three neighboring agents. The initial orientation angles of agents are the same as those in the simulation in Figs. 7.3 and 7.4. The desired positions of agents are also the same to Figs. 7.3 and 7.4. However, in acyclic minimally persistent formation, the leader agent does not control itself. We use the desired configuration given in the simulations of Chap. 7 only to compute the desired reference bearing vectors. After reaching a steady state, we compensate the rotation and scaling, and translations of the final formation.

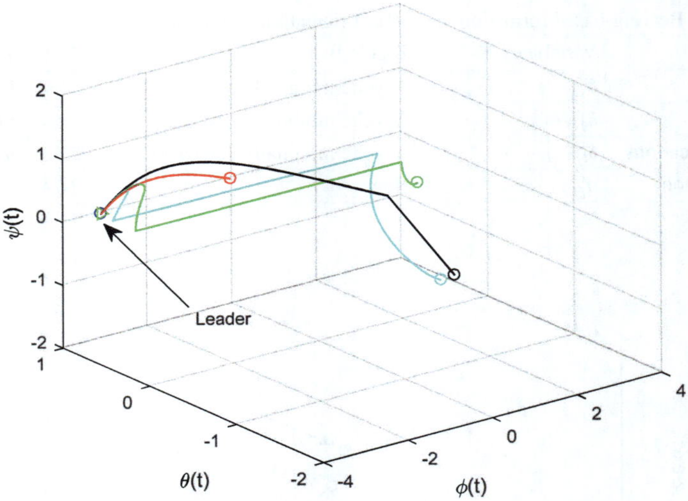

Fig. 9.5 Trajectories of orientations of agents. The initial Euler angles are marked by ∘ and the final Euler angles are marked by ▷

Figures 9.4 and 9.5 show the trajectories of positions and orientation angles of the agents. Since the leader agent, i.e., agent 1, is stationary, the initial and final positions of agent 1 are the same. Also, since it does not control its orientations, other agents have converged to the orientation angles of the leader.

9.6 Notes

In this chapter, we have presented bearing-based formation control of acyclic minimally persistent formations and cyclic triangular formations. It seems quite difficult to generalize the results of this chapter to general directed graphs. But, it may be possible to extend the results to formations with *leader-remote-follower* and *three coleaders* [13]. The results of Sects. 9.2 and 9.3 were mainly reused from [14, 16, 18], and the results of Sect. 9.4 were reused from [15]. Thus, the following copyright and permission notices are acknowledged.

- The text of Sect. 9.2, the statements of Lemmas 9.1–9.7, Theorems 9.1, 9.2, Lemmas 9.15–9.17, Theorems 9.6, and 9.7 with the proofs are © [2019] IEEE. Reprinted, with permission, from [18].
- The text of Sect. 9.2 and a part of the proofs of Sect. 9.2 are © [2016] IFAC. Reprinted, with permission, from [16].
- The statements of Lemmas 9.8, 9.10, and Theorem 9.5 with the proofs are © [2017] IEEE. Reprinted, with permission, from [14].

- The text of Sect. 9.4, and the statements of Lemmas 9.19, 9.20 and Theorem 9.8 with the proofs are © Reproduced from [15] by permission of John Wiley and Sons.

References

1. Anderson, B.D.O., Dasgupta, S., Yu, C.: Control of directed formations with a leader-first follower structure. In: Proceedings of the 46th IEEE Conference on Decision and Control, pp. 2882–2887 (2007)
2. Angeli, D.: An almost global notion of input to state stability. IEEE Trans. Autom. Control **49**(6), 866–874 (2004)
3. Basiri, M., Bishop, A.N., Jensfelt, P.: Distributed control of triangular formations with angle-only constraints. Syst. Control Lett. **59**(2), 147–154 (2010)
4. Bishop, A.N., Deghat, M., Anderson, B.D.O., Hong, Y.: Distributed formation control with relaxed motion requirements. Int. J. Robust Nonlinear Control **25**(17), 3210–3230 (2015)
5. Cao, M., Yu, C., Anderson, B.D.O.: Formation control using range-only measurements. Automatica **47**(4), 776–781 (2011)
6. Feingold, D.G., Varga, R.S.: Block diagonally dominant matrices and generalizations of the Gershgorin circle theorem. Pac. J. Math. **12**(4), 1241–1250 (1962)
7. Guo, J., Zhang, C., Lin, M.: Angle-based cooperation control of triangle formation. In: Proceedings of the 34th Chinese Control Conference, pp. 7386–7391 (2015)
8. Kwon, S.-H., Trinh, M.H., Oh, K.-H., Zhao, S., Ahn, H.-S.: Infinitesimal weak rigidity, formation control of three agents, and extension to 3-dimensional space, pp. 1–12. arXiv:1803.09545
9. Kwon, S.-H., Trinh, M.H., Oh, K.-H., Zhao, S., Ahn, H.-S.: Infinitesimal weak rigidity and stability analysis on three-agent formations. In: Proceedings of the SICE Annual Conference, pp. 266–271 (2018)
10. Park, M.-C., Ahn, H.-S.: Displacement estimation by range measurements and application to formation control. In: 2015 IEEE International Symposium on Intelligent Control, pp. 888–893 (2015)
11. Park, M.-C., Kim, H.-K., Ahn, H.-S.: Rigidity of distance-based formations with additional subtended-angle constraints. In: Proceedings of the International Conference on Control, Automation and Systems, pp. 111–116 (2017)
12. Shevitz, D., Paden, B.: Lyapunov stability theory of nonsmooth systems. IEEE Trans. Autom. Control **39**(9), 1910–1914 (1994)
13. Summers, T.H., Yu, C., Dasgupta, S., Anderson, B.D.O.: Control of minimally persistent leader-remote-follower and coleader formations in the plane. IEEE Trans. Autom. Control **56**(12), 2778–2792 (2011)
14. Trinh, M.-H., Mukherjee, D., Zelazo, D., Ahn, H.-S.: Finite-time bearing-only formation control. In: Proceedings of the 56th IEEE Conference on Decision and Control, pp. 1578–1583 (2017)
15. Trinh, M.H., Mukherjee, D., Zelazo, D., Ahn, H.-S.: Formations on directed cycles with bearing-only measurements. Int. J. Robust Nonlinear Control **28**(3), 1074–1096 (2018)
16. Trinh, M.H., Oh, K.-K., Jeong, K., Ahn, H.-S.: Bearing-only control of leader first follower formations. In: Proceedings of the 14th IFAC Symposium on Large Scale Complex Systems: Theory and Applications, pp. 7–12 (2016)
17. Trinh, M.-H., Pham, V.H., Hoang, P.H., Son, J.-H., Ahn, H.-S.: A new bearing-only navigation law. In: Proceedings of the 2017 International Conference on Control, Automation and Systems, pp. 117–122 (2017)

18. Trinh, M.H., Zhao, S., Sun, Z., Zelazo, D., Anderson, B.D.O., Ahn, H.-S.: Bearing-based formation control of a group of agents with leader-first follower structure. IEEE Trans. Autom. Control **64**(2), 598–613 (2019). https://doi.org/10.1109/TAC.2018.2836022
19. Zhao, S., Zelazo, D.: Bearing rigidity and almost global bearing-only formation stabilization. IEEE Trans. Autom. Control **61**(5), 1255–1268 (2016)

Chapter 10
Bearing-Based Formation Control via Global Orientation Alignment

Abstract This chapter studies bearing-based formation control while globally aligning the orientations of agents' local coordinate frames. Two topics are addressed. The first topic is to transform the formation systems with misaligned orientations to a formation system with aligned orientations. For this, while aligning the orientations of agents, we apply the typical bearing-based formation control laws to the distributed agents. The second topic is to control the positions of agents. But, since the positions of agents are not measured in local coordinate frames, they need to be estimated with the measurements of bearing vectors. The second topic is the counterpart of Chap. 8 in bearing setups.

10.1 Displacement-Transformed Formation Control

The bearing-based formation control via global orientation alignment is called displacement-transformed formation control, which was studied in [3]. Since the sensing variables are fully distributed ones, we need to first transform these variables into displacement-based ones that are represented in an aligned common coordinate frame. Let us consider the following multi-agent systems

$$\dot{p}_i^i = u_i^i, \; i = 1, \ldots, n \tag{10.1}$$

which are described in a local coordinate frame ${}^i\Sigma$. The agent i can sense the bearing vector of neighboring agent j in the following manner:

$$g_{ji}^i = \frac{p_j^i - p_i^i}{\|p_j^i - p_i^i\|} = R_i \frac{p_j - p_i}{\|p_j - p_i\|} = R_i g_{ji} \tag{10.2}$$

in the local coordinate frame ${}^i\Sigma$. It is noticeable that $g_{ji} = -g_{ij}$ while $g_{ji}^i \neq -g_{ij}^j$. The projection operator in the local coordinate frame is defined as

$$\mathbb{P}_{g_{ji}^i} = \mathbb{I}_d - g_{ji}^i (g_{ji}^i)^T \tag{10.3}$$

H.-S. Ahn, *Formation Control*, Studies in Systems, Decision and Control 205, https://doi.org/10.1007/978-3-030-15187-4_10

which has the following relationships

$$\mathbb{P}_{g^i_{ji}} = \mathbb{I}_d - g^i_{ji}(g^i_{ji})^T = \mathbb{I}_d - R_i g_{ji} g^T_{ji} R^T_i$$
$$= R_i(\mathbb{I}_d - g_{ji} g^T_{ji}) R^T_i$$
$$= R_i \mathbb{P}_{g_{ji}} R^T_i \tag{10.4}$$

Let agent i measure the relative orientations $R_{ji} = R_j R_i^{-1}$ of agents j in its own coordinate frame. The orientation estimation problem is to estimate the orientation R_i up to a common coordinate rotation X. Following the same procedure as in Chap. 8, we use the subsidiary variables $z_{i,k} \in \mathbb{R}^d$, $k \in \{1, \ldots, d-1\}$ for the computation of orientations of agents. Using the update law (8.23), it was shown that the matrix $B_i(t)$ converges to $R_i X$ exponentially fast with almost all initial conditions, which implies that $\hat{R}_i(t) = B_i(t) \in \mathrm{SO}(3)$ in Theorem 8.3. As studied in Chap. 9, let us consider an acyclic minimally persistent (AMP) graph that can be constructed by a bearing-based Henneberg sequence. For the given AMP structure, let the topology of the AMP graph be denoted by a directed graph $\vec{\mathcal{G}} = (\mathcal{V}, \vec{\mathcal{E}})$. Also let $\mathbb{L}$ be the graph Laplacian of $\vec{\mathcal{G}}$. Let us write (8.23) as follows:

$$\dot{z}_{i,k} = \sum_{j \in \mathcal{N}_i} a_{ij}(R_{ij} z_{j,k} - z_{i,k}), \ \forall i \in \mathcal{V}, \ \forall k \in \{1, \ldots, d-1\} \tag{10.5}$$

Note that the relative orientation R_{ij} can be changed as $R_{ij} = R_i R_j^T = (R_j R_i^T)^T = R_{ji}^{-1}$. With (10.5), we can obtain the matrix $B_i(t)$ using the Gram–Schmidt process. So, let us suppose that we have obtained $B_i(t)$ from (10.5). Also, since $B_i(t)$ is a rotation matrix, we can see that there exists $R_{\Delta i} \in \mathrm{SO}(d)$ such that $B_i(t) = R_i R_{\Delta i} X$ for a common offset rotation matrix X before converging to a steady-state value. In the case of the AMP formations, a follower agent senses the relative bearing angles of leaders in its own coordinate frame. For a notational purpose, similarly to Sect. 9.2, the underlying topology of the graph is defined with edges whose directions are opposite of the sensing directions. Thus, the underlying topology is assumed to have an arborescence. To design a formation control law for the AMP graph under global orientation estimation, we first make the following result:

Corollary 10.1 *Suppose that the AMP structure has an arborescence as its underlying topology, and the root agent (the leader) has information about the global coordinate frame and it does not update its orientation, i.e.,* $\dot{z}_{1,k} = 0$*, while other agents update their orientations by (10.5). Then, the estimated orientations of other agents converge to* $B_i(t) = R_i R_1$ *as* $t \to \infty$.

Proof With the replacement $z_{i,k} = R_i y_{i,k}$, the dynamics (10.5) can be rewritten in terms of $y_{i,k}$ as follows:

$$\dot{y}_k(t) = -(\mathbb{L} \otimes \mathbb{I}_d) y_k(t), \ \forall k \in \{1, \ldots, d-1\}. \tag{10.6}$$

Since the first agent does not have neighboring agents and it does not update, i.e., $\dot{z}_{1,k} = 0$, the Laplacian matrix $\mathbb{L}$ has the following form:

$$\mathbb{L} = \left[\begin{array}{c|c} 0 & 0 \cdots 0 \\ \hline -a_{21} & \\ \vdots & \mathbb{L}_r \\ -a_{n1} & \end{array}\right], \tag{10.7}$$

where $\mathbb{L}_r$ is a reduced Laplacian matrix. Due to the property of triangular block matrix, the eigenvalues of $\mathbb{L}_r$ are the same as the eigenvalues of $\mathbb{L}$ except for one additional zero eigenvalue. Thus, $\mathbb{L}_r$ is a Hurwitz matrix. Let us denote $\tilde{y}_{i,k} = y_{i,k} - y_{1,k}$. Defining $\tilde{y}_k = (\tilde{y}_{2,k}, \ldots, \tilde{y}_{n,k})$, we can get the derivatives of $\tilde{y}_k$ as

$$\dot{\tilde{y}}_k(t) = -(\mathbb{L}_r \otimes \mathbb{I}_d)\tilde{y}_k(t), \;\forall k \in \{1, \ldots, n\}. \tag{10.8}$$

Obviously, $\tilde{y}_k(t)$ exponentially converges to 0, which means that $y_{i,k} \rightarrow y_{1,k}$ for all $i \in \mathcal{V}$. Thus, following the same procedure of the proof of Theorem 8.3, we can have

$$\lim_{t\rightarrow\infty} B_i(t) = R_i R_1, \tag{10.9}$$

for all $i = 2, \ldots, n$.

Denoting $g_{ji}^{\infty} \triangleq X g_{ji}^*$, and $h_i(t) \triangleq \sum_{j\in\mathcal{N}_i} \mathbb{P}_{g_{ji}}(\mathbb{I}_d - R_{\Delta i})X g_{ji}^*$, we propose the following distributed formation controller for agent i to achieve the desired formation defined by g_{ji}^*:

$$u_i^i = -\sum_{i\in\mathcal{N}_i} \mathbb{P}_{g_{ji}^i} B_i g_{ji}^* \tag{10.10}$$

Then, the dynamics of each agent i is written in the global reference frame as

$$\begin{aligned}
\dot{p}_i &= -R_i^T \sum_{j\in\mathcal{N}_i} \mathbb{P}_{g_{ji}^i} B_i g_{ji}^* \\
&= -R_i^T \sum_{j\in\mathcal{N}_i} R_i \mathbb{P}_{g_{ji}} R_i^T R_i R_{\Delta i} X g_{ji}^* \\
&= -\sum_{j\in\mathcal{N}_i} \mathbb{P}_{g_{ji}} R_{\Delta i} X g_{ji}^* \\
&= -\sum_{j\in\mathcal{N}_i} \mathbb{P}_{g_{ji}} X g_{ji}^* + \sum_{j\in\mathcal{N}_i} \mathbb{P}_{g_{ji}}(\mathbb{I}_d - R_{\Delta i})X g_{ji}^* \\
&= -\sum_{j\in\mathcal{N}_i} \mathbb{P}_{g_{ji}} g_{ji}^{\infty} + h_i(t)
\end{aligned} \tag{10.11}$$

Let $p = (p_1^T, \dots, p_n^T)^T \in \mathbb{R}^{dn}$, $h(t) = (h_1^T, \dots, h_n^T)^T \in \mathbb{R}^{dn}$, and $g^\infty = ((g_1^\infty)^T, \dots, (g_m^\infty)^T)^T \in \mathbb{R}^{dm}$. Then, the above Eq. (10.11) can be compactly written as

$$\dot{p} = f(p) + h(t) \tag{10.12}$$

From the above equation, it is shown that the term $h(t)$ may be considered as an input to the unforced system:

$$\dot{p} = f(p) = (\mathbb{H}_+ \otimes \mathbb{I}_d)^T \text{diag}\left(\mathbb{P}_{g_k}\right) g^\infty, \tag{10.13}$$

where $\mathbb{H}_+ \in \mathbb{R}^{m \times n}$ is an incidence matrix corresponding to $\overrightarrow{\mathcal{E}}$ of the directed graph.

Lemma 10.1 *The input $h(t)$ is bounded and $h(t) \to 0$ exponentially fast.*

Proof Since $\|h_i(t)\| \le \sum_{j \in \mathcal{N}_i} \|\mathbb{P}_{g_{ji}}\| \|\mathbb{I}_d - R_{\Delta i}\| \|X\| \|g_{ji}^*\|$, it is bounded. From the Corollary 10.1, $R_{\Delta i} \to \mathbb{I}_d$ exponentially as $t \to \infty$. Thus $\|h_i(t)\| \to 0$ exponentially fast.

Next, based on the results of Chap. 9, we can obtain the following conclusion about the stability of the system (10.13) at a desired equilibrium point $p = p_d^\infty$ [3].

Corollary 10.2 *The unforced system* (10.13) *has two equilibrium points. The equilibrium $p = p_d^\infty$ corresponding to $g_{ji} = g_{ji}^\infty$, $\forall (i, j)^{\bar{e}} \in \overrightarrow{\mathcal{E}}$ is almost globally exponentially stable, while the equilibrium $p = p_u^\infty$ corresponding to $g_{ji} = -g_{ji}^\infty$, $\forall (i, j)^{\bar{e}} \in \overrightarrow{\mathcal{E}}$ is exponentially unstable.*

In the above corollary and in the results of Chap. 9, the terminology *almost* can be also understood geographically. For example, Fig. 10.1a shows the desired configuration, where the agent 2 stays at the position 2 satisfying $g_{12} = g_{12}^*$ and agent 3 stays at the position 3 satisfying $g_{13} = g_{13}^*$ and $g_{23} = g_{23}^*$. If agent 2 stays at a position, i.e., $2'$ as shown in Fig. 10.1b, satisfying $g_{12} = -g_{12}^*$, the agent 3 cannot stay at the desired position 3 since $g_{23} \neq \pm g_{23}^*$. Similarly, when the agent 2 stays at the desired position 2, the agent 3 cannot stay at an undesired position $3'$ since still $g_{23} \neq \pm g_{23}^*$. But, both agents 2 and 3 can stay at undesired equilibrium positions $2'$ and $3'$ since we have $g_{12} = -g_{12}^*$, and $g_{13} = -g_{13}^*$ and $g_{23} = -g_{23}^*$, as depicted in Fig. 10.1d. Consequently, only when all the follower agents are at undesired equilibrium points, i.e., $g_{ji} = -g_{ji}^*$, the desired configuration would not be achieved and there will be no more nonzero update in the control efforts. In a more general case, as shown in Fig. 10.1b, when one of the follower agents is at an undesired equilibrium point, all the remaining follower agents would not converge to the desired locations. Figure 10.1c also shows an example of interesting interpretation. When the agent 2 stays at the desired position, the agent 3 cannot stay at the undesired equilibrium point $3'$ due to the fact that $g_{23} \neq \pm g_{23}^*$. Thus, as far as agent 2 does not satisfy $g_{12} = -g_{12}^*$ initially, the desired configuration would be achieved.

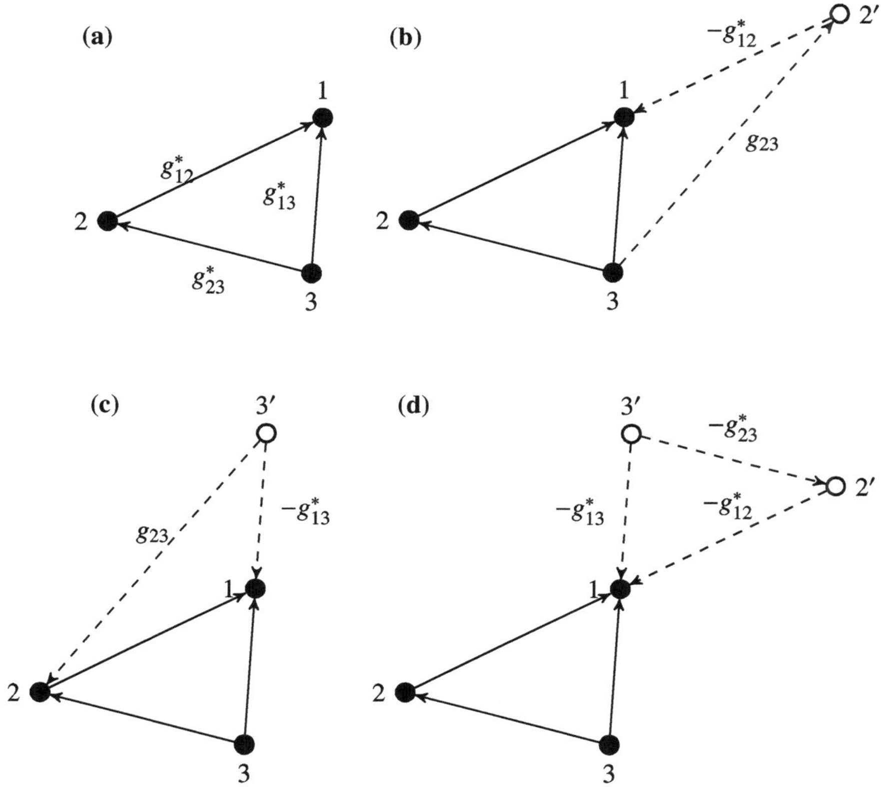

Fig. 10.1 Geographical interpretation of *almost*

Example 10.1 Let $g_{12}^* = \frac{1}{\sqrt{5}}(2, 1)^T$, $g_{13}^* = \frac{1}{2}(0, 2)^T$, and $g_{23}^* = \frac{1}{\sqrt{5}}(-2, 1)^T$. Let the leader agent be located at the origin. Then, for example, if agents 2 and 3 stay at $(-2, -1)^T$ and $(0, -2)^T$, respectively, the desired configuration can be considered as achieved. Also, due to $\dot{p}_2 = -\mathbb{P}_{g_{12}} g_{12}^* = 0$, and $\dot{p}_3 = -\mathbb{P}_{g_{13}} g_{13}^* - \mathbb{P}_{g_{23}} g_{23}^* = 0$, we can see that the agents 2 and 3 would be stationary (i.e., no more update in control efforts). But, when the agent 2 is at $(2, 1)^T$ initially, we also have $\dot{p}_2 = \mathbb{P}_{g_{12}} g_{12}^* = 0$; so the agent 2 is stationary. But, for the agent 3, when it is at $(0, -2)^T$, we have

$$\begin{aligned}
\dot{p}_3 &= -\mathbb{P}_{g_{13}} g_{13}^* - \mathbb{P}_{g_{23}} g_{23}^* \\
&= -\left(\begin{bmatrix} 1 & 0 \\ 0 & 1 \end{bmatrix} - \begin{bmatrix} 0 & 0 \\ 0 & 1 \end{bmatrix}\right)\begin{bmatrix} 0 \\ 1 \end{bmatrix} - \left(\begin{bmatrix} 1 & 0 \\ 0 & 1 \end{bmatrix} - \begin{bmatrix} \frac{4}{13} & \frac{6}{13} \\ \frac{6}{13} & \frac{9}{13} \end{bmatrix}\right)\begin{bmatrix} -\frac{2}{\sqrt{5}} \\ \frac{1}{\sqrt{5}} \end{bmatrix} \\
&= \begin{bmatrix} 0.8256 \\ -0.5504 \end{bmatrix}
\end{aligned} \tag{10.14}$$

However, when $p_3 = (0, 2)^T$, which is the undesired equilibrium point, we have

$$\begin{aligned}\dot{p}_3 &= -\mathbb{P}_{g_{13}} g_{13}^* - \mathbb{P}_{g_{23}} g_{23}^* \\ &= -\left(\begin{bmatrix} 1 & 0 \\ 0 & 1 \end{bmatrix} - \begin{bmatrix} 0 & 0 \\ 0 & 1 \end{bmatrix}\right)\begin{bmatrix} 0 \\ 1 \end{bmatrix} - \left(\begin{bmatrix} 1 & 0 \\ 0 & 1 \end{bmatrix} - \begin{bmatrix} \frac{4}{5} & -\frac{2}{5} \\ -\frac{2}{5} & \frac{1}{5} \end{bmatrix}\right)\begin{bmatrix} -\frac{2}{\sqrt{5}} \\ \frac{1}{\sqrt{5}} \end{bmatrix} \\ &= \begin{bmatrix} 0 \\ 0 \end{bmatrix}\end{aligned} \tag{10.15}$$

Thus, when both agents 2 and 3 are on the undesired equilibrium points initially, there will be no control update; thus, these points may be considered as saddle points. As far as the agents are not on undesired equilibrium points initially, the desired configuration would be achieved.

Theorem 10.1 *Let us suppose that orientations of agents are estimated by way of the update of auxiliary variables* (10.5). *Then, the equilibrium $p = p_d^\infty$ corresponding to $g_{ji} = g_{ji}^\infty$, $\forall (i, j)^{\bar{e}} \in \vec{\mathcal{E}}$ of the system* (10.1) *is almost globally asymptotically stable.*

Proof Following the proof of Theorem 9.6, and using a Lyapunov candidate $V = \frac{1}{2}\|p - p_d^\infty\|^2$, the following inequality can be obtained:

$$\dot{V} \leq -2\kappa V + 4\kappa\gamma + 2\gamma\|h(t)\|$$

where $\kappa > 0$ is defined in Lemma 9.6 and γ satisfies $\max_{t \geq t_0}\{\|p - p_d^\infty\|\} \leq \gamma < \infty$. Thus, the system is ultimately bounded. Now using Corollary 10.2 and ultimately boundedness of the states, we can ensure that if the initial points of agents are not in $p = p_u^\infty$, the trajectory will reach the desired equilibrium point by a mathematical induction.

This section has used orientation estimation-based formation control under the bearing setup. If we use (8.74) for the update of the auxiliary variables and (8.73) for Gram–Schmidt normalization, and (8.92) for the orientation control of agents, we can actively achieve a synchronization in the orientations of agents' coordinate frames. Then, repeating the same procedure of this section, we can guarantee a convergence to the desired configuration, which is summarized in the following theorem:

Theorem 10.2 *Let us suppose that orientations of agents are actively controlled by way of the update of auxiliary variables* (8.74), *and the orientation control* (8.92). *Then, for the AMP structure, the equilibrium $p = p_d^\infty$ corresponding to $g_{ji} = g_{ji}^\infty$, $\forall (i, j)^{\bar{e}} \in \vec{\mathcal{E}}$ of the system* (10.1) *is almost globally asymptotically stable, while all agents reach a synchronization in their orientations.*

10.2 Position-Transformed Formation Control

In the previous section, the distributed sensing variables have been transformed into the orientation-aligned displacement setup. Now, we further transform the orientation-aligned displacement setup into a position-based setup [2]. In the bearing-based formation control laws, each agent measures the bearing vectors of neighboring agents $g_{ji} = \frac{p_j - p_i}{\|p_j - p_i\|}$, $j \in \mathcal{N}_i$, under the assumption that the agents have an aligned directional information or aligned directions. The bearing vector is a normalized displacement information. So, there may exist an overshoot during the convergence in the displacement-based control; or it may take time until the convergence. In this section, we would like to estimate the position of agents using the bearing measurements. Then, with the estimated position information, we would like to control the positions of agents directly. For the development of these results, we use a concept of matrix-weighted consensus [4]. Assigning a positive semi-definite weighted matrix $\mathbf{A}_{ij} \in \mathbb{R}^{d \times d}$ to each edge, from

$$\dot{x}_i = -\sum_{j \in \mathcal{N}_i} \mathbf{A}_{ij}(x_i - x_j) \tag{10.16}$$

we obtain the *matrix-weighted Laplacian*, denoted as $\mathbb{L}(\mathcal{G})$. The property of the matrix-weighted Laplacian is given as follows:

Theorem 10.3 ([4]) *The matrix-weighted Laplacian* $\mathbb{L}$ *is symmetric, positive semi-definite, and has null space as* $null(\mathbb{L}) = span\{\mathbf{1}_n \otimes \mathbb{I}_d, \{x = (x_1^T, \ldots, x_n^T)^T \in \mathbb{R}^{dn} | (x_i - x_j) \in null(\mathbf{A}_{ij}), \forall (i, j)^e \in \mathcal{E}\}\}$.

The bearing Laplacian matrix of the framework $\mathcal{G}(p)$, denoted as $\mathbb{L}^b(\mathcal{G}(p)) \in \mathbb{R}^{dn \times dn}$, is defined as [6, 7]:

$$[\mathbb{L}^b(\mathcal{G}(p))]_{ji} = \begin{cases} \mathbf{0}_{d \times d}, & i \neq j, (i, j)^e \notin \mathcal{E} \\ -\mathbb{P}_{g_{ji}}, & i \neq j, (i, j)^e \in \mathcal{E} \\ \sum_{k \in \mathcal{N}_i} \mathbb{P}_{g_{ki}}, & i = j, i \in \mathcal{V}. \end{cases} \tag{10.17}$$

Since each element of the Laplacian matrix is a matrix, the bearing Laplacian can be considered as a matrix-weighted Laplacian [4]. In this section, we also consider leader–follower type formation systems. Let us suppose that two leaders know their positions, and denote these leaders as 1 and 2. Let $\mathcal{V}_{\{1,2\}}$ denote the set of leader agents, and $p_{\{1,2\}} = (p_1^T, p_2^T)^T$ denote the positions of the leaders. For the remaining agents, let $\mathcal{V}_{\{3:n\}}$ denote the set of follower agents, and the stacked vector of positions is $p_{\{3:n\}} = (p_3^T, \ldots, p_n^T)^T \in \mathbb{R}^{d(n-2)}$. Then, the bearing Laplacian can be partitioned into the following form:

$$\mathbb{L}^b(\mathcal{G}(p)) = \left[\begin{array}{c|c} \mathbb{L}_{ll}^b & \mathbb{L}_{lf}^b \\ \hline (\mathbb{L}_{lf}^b)^T & \mathbb{L}_{ff}^b \end{array}\right], \tag{10.18}$$

where $\mathbb{L}^b_{ll} \in \mathbb{R}^{2d\times 2d}$ and $\mathbb{L}^b_{ff} \in \mathbb{R}^{(n-2)d\times(n-2)d}$ correspond to the set of leaders and set of followers, respectively, and $\mathbb{L}^b_{lf} \in \mathbb{R}^{2d\times(n-2)d}$ corresponds to the cross-terms. The framework is *infinitesimally bearing rigid* (IBR) if all the infinitesimal motions are the infinitesimal translations and dilations spanned by $\{\mathbf{1}_n \otimes \mathbb{I}_d, p\}$. If the framework $\mathcal{G}(p)$ is IBR, then we have $\text{null}(\mathbb{L}^b(\mathcal{G}(p))) = \text{span}\{\mathbf{1}_n \otimes \mathbb{I}_d, p\}$. Note that the IBR property can be checked by Theorem 2.12 from a given bearing rigidity matrix $\mathbb{R}_B$. The following lemma is implied from [7].

Lemma 10.2 *For a framework $\mathcal{G}(p)$ with two leaders, the submatrix $\mathbb{L}^b_{ff}$ is positive definite if and only if the augmented framework $\bar{\mathcal{G}}(p)$ obtained from $\mathcal{G}(p)$ by connecting two leaders is IBR.*

Consider a system of n agents in the bearing-based formation control setup in $\mathbb{R}^d$, $(d \geq 2,\ n \geq 2)$, whose interaction graph is IBR.

Definition 10.1 (*Configuration Trajectory*) A configuration trajectory is the trajectory of a framework $p(t)$ over a time interval $\mathcal{D} \subseteq \mathbb{R}_+$, denoted as $\mathcal{T}(p(t), \mathcal{D}) \triangleq \{\{p(t)\}|p(t) = (p_{\{1,2\}}(t)^T, p_{\{3:n\}}(t)^T)^T,\ t \in \mathcal{D} \subseteq \mathbb{R}_+\}$.

It is supposed that given the desired positions $p^* = (p_1^*, \ldots, p_n^*)$, the equivalent desired bearings $b^* = \{g^*_{ji} = \frac{p^*_{ji}}{\|p^*_{ji}\|}, (i, j)^e \in \mathcal{E}\}$ can be given to each agent. Then, we would like to study the problem of estimating agents' positions based on relative bearing information.

Problem 10.1 Given a graph $\mathcal{G} = (\mathcal{V}, \mathcal{E})$, for each agent $i \in \mathcal{V}$, design an estimation law to achieve an exponential convergence of $\hat{p}_i(t)$ to $p_i(t) + \tilde{p}^\infty$ for a certain $\tilde{p}^\infty \in \mathbb{R}^d$, by using inter-neighbor bearings $\{g_{ji}(t)\}_{j\in\mathcal{N}_i}$.

Once the estimated position $\hat{p}_i$ is available to each agent, the second problem is to drive agents to the desired formation up to a certain offset. This problem is defined as follows:

Problem 10.2 Design a position control law to achieve

$$\lim_{t\to\infty} p_i(t) = p_i^* + e_p^\infty \tag{10.19}$$

for a certain $e_p^\infty \in \mathbb{R}^d$, by using the estimated position $\hat{p}_i(t)$ in Problem 10.1, for all $i \in \mathcal{V}$.

Assume that there exist at least two neighboring agents that know the distance between them. For simplicity of presentation, let us suppose that agents 1 and 2 know the distance between them as $d_{12} = \|p_1 - p_2\|$. Since each agent also senses the bearing direction to the other agent under the IBR condition, it can be considered that agents 1 and 2 could measure the relative positions p_{12} and p_{21}, respectively. That is, agents 1 and 2 have bearing measurements as well as distance between them; the combination of these measurements is a displacement. We propose distributed position estimation laws as follows:

$$\dot{\hat{p}}_1 = u_1 - k_O \sum_{j \in \mathcal{N}_1/\{2\}} \mathbb{P}_{g_{j1}}(\hat{p}_1 - \hat{p}_j) - k_O[\hat{p}_1 - \hat{p}_2 - (p_1 - p_2)] \tag{10.20}$$

$$\dot{\hat{p}}_2 = u_2 - k_O \sum_{j \in \mathcal{N}_2/\{1\}} \mathbb{P}_{g_{j2}}(\hat{p}_2 - \hat{p}_j) - k_O[\hat{p}_2 - \hat{p}_1 - (p_2 - p_1)] \tag{10.21}$$

$$\dot{\hat{p}}_i = u_i - k_O \sum_{j \in \mathcal{N}_i} \mathbb{P}_{g_{ji}}(\hat{p}_i - \hat{p}_j), \ \forall i \in \{3, \ldots, n\} \tag{10.22}$$

In the above estimation laws, the perpendicular projection term $\mathbb{P}_{g_{ji}}(\hat{p}_i - \hat{p}_j)$ drives the estimated position $\hat{p}_i$ such that it achieves the bearing constraint $\hat{g}_{ji} = g_{ji}$ and the control input u_i is designed by the estimated position. It can be understood that the position estimation $\hat{p}_i$ is updated until the estimated bearing vectors become equivalent to the measured true bearing vectors.

For any nonzero bearing vector g_{ji}, $\forall (i, j)^{\bar{e}} \in \overrightarrow{\mathcal{E}}$, we have $\mathbb{P}_{g_{ji}} p_{ji} = \mathbb{P}_{g_{ji}} p_{ij} = 0$. Thus, from $\mathbb{P}_{g_{ji}}(\hat{p}_{ji} - p_{ij}) = \mathbb{P}_{g_{ji}} \hat{p}_{ji}$, the projections of $\hat{p}_{ji}$ and $\hat{p}_{ji} - p_{ji}$ onto the orthogonal supplement of g_{ji} are identical. Now, using $\dot{p}_i = u_i$ and denoting $\tilde{p} \triangleq \hat{p}_i - p_i$, (10.20) can be rewritten as

$$\begin{aligned} \dot{\tilde{p}}_1 &= -k_O \sum_{j \in \mathcal{N}_1/\{2\}} \mathbb{P}_{g_{j1}}(\hat{p}_1 - \hat{p}_j - p_{1j}) - k_O[\hat{p}_1 - p_1 - (\hat{p}_2 - p_2)] \\ &= -k_O \sum_{j \in \mathcal{N}_1/\{2\}} \mathbb{P}_{g_{j1}}(\tilde{p}_1 - \tilde{p}_j) - k_O(\tilde{p}_1 - \tilde{p}_2) \end{aligned} \tag{10.23}$$

Similarly, we can have

$$\dot{\tilde{p}}_2 = -k_O \sum_{j \in \mathcal{N}_2/\{1\}} \mathbb{P}_{g_{j2}}(\tilde{p}_2 - \tilde{p}_j) - k_O(\tilde{p}_2 - \tilde{p}_1) \tag{10.24}$$

$$\dot{\tilde{p}}_i = -k_O \sum_{j \in \mathcal{N}_i} \mathbb{P}_{g_{ji}}(\tilde{p}_i - \tilde{p}_j), \ \forall i \in \{3, \ldots, n\} \tag{10.25}$$

Then, in a vector form, the estimation error dynamics can be given as

$$\dot{\tilde{p}} = -k_O \mathbb{L}_o^b(p) \tilde{p} \tag{10.26}$$

where $[\mathbb{L}_o^b]_{12} = [\mathbb{L}_o^b]_{21} = -\mathbb{I}_d$, $[\mathbb{L}_o^b]_{ij} = -\mathbb{P}_{g_{ji}}$, for $i \neq j, \forall (i, j)^{\bar{e}} \in \overrightarrow{\mathcal{E}} \setminus \{(1, 2)^e, (2, 1)^e\}$, and $[\mathbb{L}_o^b]_{ii} = \sum_{j \in \mathcal{N}_i} [\mathbb{L}_o^b]_{ij}$, $\forall i \in \mathcal{V}$.

Lemma 10.3 *Let us suppose that the framework $\mathcal{G}(p(t))$ is infinitesimally bearing rigid along the realized configuration trajectory. Then, the bearing Laplacian $\mathbb{L}_o^b(p)$ is positive semi-definite and satisfies null$(\mathbb{L}_o^b) = $ span$\{\mathbf{1}_n \otimes \mathbb{I}_d\}$.*

Proof From the assumption that the framework $\mathcal{G}(p(t))$ is infinitesimally bearing rigid, it is clear that null$(\mathbb{L}^b(p)) = $ span$\{\mathbf{1}_n \otimes \mathbb{I}_d, p\}$, which implies that $\mathbb{L}_o^b$ is positive semi-definite. Further, the (i, j)th weighted matrix elements of $\mathbb{L}_o^b$ are the same as those of $\mathbb{L}^b$, except the two elements $\{(1, 2)^e, (2, 1)^e\}$. Note that $\text{null}(\mathbb{I}_d) = \emptyset$; so from the Theorem 10.3, it follows that $\text{null}(\mathbb{L}_o^b) = \text{span}\{\mathbf{1}_n \otimes \mathbb{I}_d\}$.

Theorem 10.4 *Let us suppose that the framework $\mathcal{G}(p(t))$ is infinitesimally bearing rigid along the realized configuration trajectory, with the underlying graph being undirected (or balanced directed). Then, the estimation errors $\tilde{p}(t)$ given in (10.23)–(10.25) globally exponentially converge to the average consensus $\tilde{p}^* = \boldsymbol{1}_n \otimes \bar{\tilde{p}}(t_0)$, where $\bar{\tilde{p}}(t_0)$ is the average of initial position estimation errors.*

Proof From Lemma 10.3, we know that $\text{null}(\mathbb{L}_o^b) = \text{span}\{\mathbf{1}_n \otimes \mathbb{I}_d\}$; thus, from the property of consensus dynamics, the proof can be completed directly.

It is now clear from Theorem 10.4 that the positions of all agents are globally exponentially estimated up to a constant vector $\tilde{p}^\infty := \bar{\tilde{p}}(t_0)$. Next, assume that there are two leader agents that know their absolute positions. Then, the estimation law (10.22) can be rewritten as

$$\dot{\tilde{p}}_i = -k_0 \sum_{j \in \mathcal{N}_i} \mathbb{P}_{g_{ji}}(\tilde{p}_i - \tilde{p}_j), \ \forall i \in \{3, \ldots, n\} \tag{10.27}$$

which can be further concisely expressed as

$$\begin{aligned} \dot{\tilde{p}}_{\{3:n\}} &= -k_0 \mathbb{L}_{ff}^b(p) \tilde{p}_{\{3:n\}} - k_0 \mathbb{L}_{fl}^b(p) \tilde{p}_{\{1,2\}} \\ &= -k_0 \mathbb{L}_{ff}^b(p) \tilde{p}_{\{3:n\}} \end{aligned} \tag{10.28}$$

due to the fact $\tilde{p}_{\{1,2\}} = 0$.

Theorem 10.5 *Let us suppose that the framework $\mathcal{G}(p(t))$ is infinitesimally bearing rigid along the realized configuration trajectory. Then, under the estimation dynamics (10.22), the state $\hat{p}(t)$ globally exponentially converges to the true value $p(t)$ as $t \to \infty$.*

Proof Consider a Lyapounov candidate $V(\tilde{p}) = \frac{1}{2}\tilde{p}_{\{3:n\}}^T \tilde{p}_{\{3:n\}}$, which is radically unbounded and continuous differentiable for all $t > t_0$. The time derivative of $V(\tilde{p})$ is given by

$$\begin{aligned} \frac{d}{dt} V(\tilde{p}) &= \tilde{p}_{\{3:n\}}^T \dot{\tilde{p}}_{\{3:n\}} \\ &= -k_0 \tilde{p}_{\{3:n\}}^T \mathbb{L}_{ff}^b(p) \tilde{p}_{\{3:n\}} \end{aligned} \tag{10.29}$$

Under the IBR condition, $\mathbb{L}_{ff}^b(p(t))$ is positive definite by Lemma 10.2. Thus, $\dot{V}(\tilde{p}) = 0$ if and only if $\tilde{p}_{\{3:n\}} = 0$, which implies that $\tilde{p}_{\{3:n\}} = 0$ is globally exponentially stable for the error dynamics. Consequently, the proof is completed.

Next, to solve Problem 10.2, we propose the following estimation-based feedback control law:

$$\dot{p}_i = u_i = k_p(p_i^* - \hat{p}_i), \ \forall i \in \mathcal{V}$$
$$= k_p[p_i^* - p_i - (\hat{p}_i - p_i)], \ \forall i \in \mathcal{V} \quad (10.30)$$

which can be rewritten as

$$\dot{e}_p = -k_p e_p - k_p \tilde{p} \quad (10.31)$$

where $e_p \triangleq p - p^* \in \mathbb{R}^{dn}$ is the position error. Then, the position estimation error dynamics (10.26) and position control error dynamics (10.31) can be combined as

$$\dot{\tilde{p}} = -k_0 \mathbb{L}_o^b(p)\tilde{p} \quad (10.32a)$$
$$\dot{e}_p = -k_p e_p - k_p \tilde{p} \quad (10.32b)$$

Now, we can obtain the following main conclusion of this section [2]:

Theorem 10.6 *Assume that the underlying graph is infinitesimally bearing rigid, without fixed (stationary) leaders. Then, the position estimation law* (10.22) *and position feedback control law* (10.30) *ensure that $p_i(t)$ globally exponentially converges to $p_i^* - \tilde{p}^\infty$, $i \in \mathcal{V}$.*

Proof It is clear that $\tilde{p} \to \bar{\tilde{p}}(t_0)$ exponentially fast by the Theorem 10.4, if the framework $\mathcal{G}(p(t))$ is IBR. Thus, the dynamics (10.32a) is globally exponentially stable to $\bar{\tilde{p}}(t_0)$, which implies that there exist constants $k_{\tilde{p}}$ and $\lambda_{\tilde{p}}$ such that

$$\|\tilde{p}(t) - \mathbf{1}_n \otimes \tilde{p}^\infty\| \leq k_{\tilde{p}} \|\tilde{p}(t_0) - \mathbf{1}_n \otimes \tilde{p}^\infty\| e^{-\lambda_{\tilde{p}}(t-t_0)} \quad (10.33)$$

The unforced system of (10.32b), i.e., $\dot{e}_p = -k_p e_p$, is also globally exponentially stable and the input of (10.32b), i.e., $-k_p\tilde{p}$, is bounded and globally exponentially converges to $k_p \mathbf{1}_n \otimes \tilde{p}^\infty$. Therefore, (10.32b) is globally input-to-state stable. Hence, by Lemmas 2.12 and 2.13, the overall system (10.32a)–(10.32b) can be considered as globally asymptotically stable. Furthermore, from (10.32b), we have

$$\begin{aligned}
\|e_p(t) - \mathbf{1}_n \otimes \tilde{p}^\infty\| &\leq \|e_p(t_0) - \mathbf{1}_n \otimes \tilde{p}^\infty\| e^{-k_p(t-t_0)} \\
&\quad + \int_{t_0}^{t} e^{-k_p(t-\tau)} k_p \|\tilde{p} + \mathbf{1}_n \otimes \tilde{p}^\infty\| d\tau \\
&\leq \|e_p(t_0) - \mathbf{1}_n \otimes \tilde{p}^\infty\| e^{-k_p(t-t_0)} + \frac{k_p}{k_p} \sup_{t_0 \leq \tau \leq t} \|\tilde{p} + \mathbf{1}_n \otimes \tilde{p}^\infty\| \\
&\leq \|e_p(t_0) - \mathbf{1}_n \otimes \tilde{p}^\infty\| e^{-k_p(t-t_0)} + k_{\tilde{p}} \|\tilde{p}(t_0) + \mathbf{1}_n \otimes \tilde{p}^\infty\| e^{-\lambda_{\tilde{p}}(t-t_0)}
\end{aligned}$$

which completes the proof.

If we consider a formation control law with two leaders that know their positions, we may confirm that the agents move in straight lines. In such a circumstance, let us assume that the underlying graph is infinitesimally bearing rigid with two leaders.

Let us also assume that the estimation dynamics is much faster than the position control loop. Then, due to the exponential convergence of the estimation dynamics, there may exist a finite-time t_s such that the dynamics (10.32a) reaches a steady-state, i.e., $\tilde{p}_i(t) = \tilde{p}^\infty, \forall i \in \mathcal{V}, \forall t \geq t_s$.[1] Then, $p_i^\dagger \triangleq p_i^* + \tilde{p}^\infty,\ \forall i \in \mathcal{V}$ can be considered as the modified desired positions. Hence, from the feedback control law (10.30), we can obtain

$$p_i(t) = p_i^\dagger + (p_i(t_s) - p_i^\dagger)e^{-k_p(t-t_s)},\ \forall i \in \mathcal{V},\ t \geq t_s \tag{10.34}$$

Since $e^{-k_p(t-t_s)}$ is monotonically decreasing from 1 to 0 after t_s, the trajectory $p_i(t) = p_i^\dagger + (p_i(t_s) - p_i^\dagger)e^{-k_p(t-t_s)}$ is a straight line connecting two points $p_i(t_s)$ and $p_i^\dagger$, for $t \geq t_s$.

We remark that the position estimation dynamics (10.20)–(10.22) are not dependent upon the desired formation configurations, and it uses the control inputs u_i to cancel the time dependency of the formation control law (10.30). Thus, when the positions of agents are estimated once by (10.20)–(10.22). Then, even though agents are moving by (10.30), the positions of agents would be estimated up to a common offset $\tilde{p}^\infty$. Thus, the agents would move in straight lines. That is, if agents are controlled after estimating the positions of agents, the agents will move along straight lines from the initial t_o to onwards.

10.3 Summary and Simulations

In Sect. 10.1, we have presented a global stabilization of formations via orientation alignment for any initial conditions. Since bearing and relative orientation measurements are used for the alignment and formation control, it is a fully distributed control law. Since the orientations and positions of agents are controlled, the overall problem is formulated in SE(3). Note that a gradient-style formation control law, under bearing-based setups, was designed for a group of agents in SE(2) [5]. Related to the SE(3) formation, a bearing rigidity theory in SE(3) was presented in [1]. It may be possible to design a distributed control law for formation control of SE(3) groups with the theory of [1]. In Sect. 10.2, the bearing-based displacement setup under aligned orientations has been transformed into the position-based setup. For this purpose, we used position estimation laws. The position estimation laws for leader agents and other follower agents are designed differently. For the position estimations of the leaders, we also used the relative displacement measurements so as to satisfy the formation size. For the other agents, only the bearing measurements and exchanges of the estimated positions are used.

Let us verify Theorem 10.2 through numerical simulations. Consider the underlying topology as depicted in Fig. 5.4. The initial orientations of agents are the same as the simulations in Figs. 7.5 and 7.6. These initial orientations do not satisfy the

[1] The estimation dynamics may be designed in a finite-time convergence law.

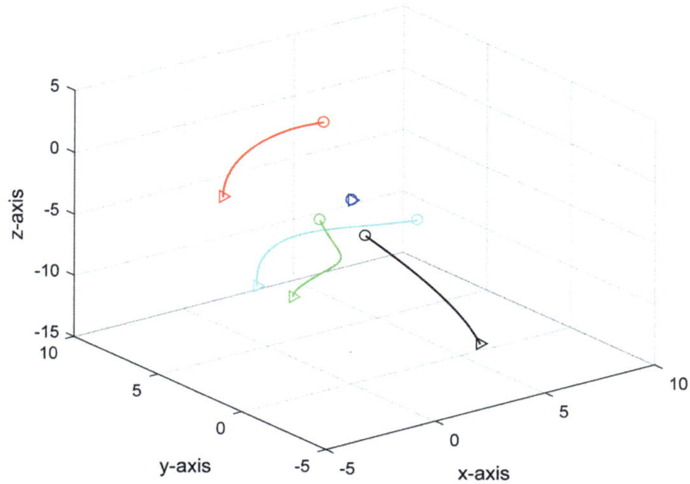

Fig. 10.2 Evolutions of positions of agents

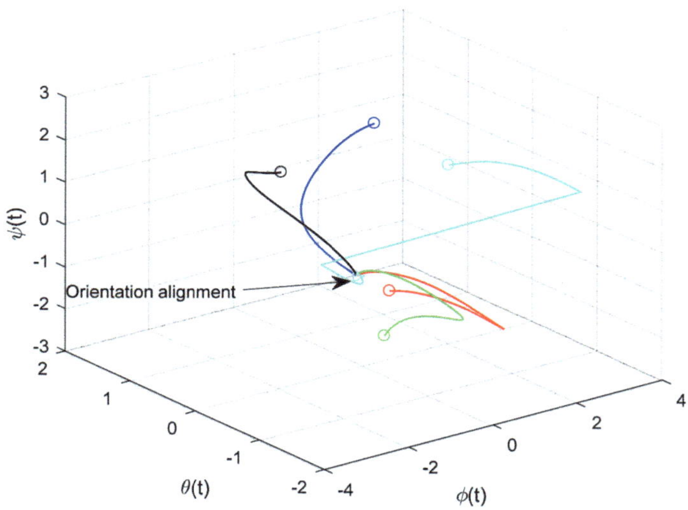

Fig. 10.3 Evolutions of orientations of agents

condition of Lemma 9.13. Figure 10.2 shows the trajectories of agents. It is observed that the position of the leader, i.e., agent 1, does not change since the position of the leader is not updated. Figure 10.3 shows the trajectories of Euler angles of agents. Unlike Fig. 10.2, however, the orientation of the leader has been also updated. This is due to the updates of auxiliary variables (8.74) and the orientation control law (8.92). In (8.92), the Lie algebra $\Omega_i(t)$ is a function of $B_i(t)$. Thus, given initial auxiliary variables of agent 1 as $\hat{z}_{1,k}$, $k = 1, 2$, the updating Eq. (8.74) can be updated although there is no neighboring agent in the communication topology, i.e., $a_{ij} = 0$. Therefore, the orientation of the leader will be updated according to (8.75), until the orientations of agents reach a consensus.

10.4 Notes

The results of this chapter were reused from [2, 3]. It is remarkable that the results of Sects. 10.1 and 10.2 can be combined to achieve a formation control law using estimated positions under the setup of misaligned bearing measurement setup. That is, the sensing variables are bearing vectors under misaligned orientations. From these measurements, the positions of agents are measured up to translations and rotations. Then, these estimated positions can be directly used for the formation control, as done in Chap. 7. The following copyright and permission notices are acknowledged.

- The text of Sect. 10.1 and the statements of Lemma 10.1 and Theorem 10.1 with the proofs are © [2018] IEEE. Reprinted, with permission, from [3].
- The text of Sect. 10.2 and the statements of Lemma 10.3, Theorems 10.4, 10.5 and 10.6 with the proofs are © [2018] IEEE. Reprinted, with permission, from [2].

References

1. Michieletto, G., Cenedese, A., Franchi, A.:. Bearing rigidity theory in SE(3). In: Proceedings of the 55th IEEE Conference on Decision and Control, pp. 5950–5955 (2016)
2. Tran, Q.V., Ahn, H.-S.: Bearing-based formation control via distributed position estimation. In: Proceedings of the IEEE Conference on Control Technology and Applications, pp. 658–663 (2018)
3. Trinh, M.-H., Lee, B.-H., Ye, M., Ahn, H.-S.: Bearing-based formation control and network localization via global orientation estimation. In: Proceedings of the IEEE Conference on Control Technology and Applications, pp. 1084–1089 (2018)
4. Trinh, M.H., Nguyen, C.V., Lim, Y.-H., Ahn, H.-S.: Matrix-weighted consensus and its applications. Automatica **89**(3), 415–419 (2018)
5. Zelazo, D., Giordano, P.R., Franchi, A.: Bearing-only formation control using an SE(2) rigidity theory. In: Proceedings of the 54h IEEE Conference on Decision and Control, pp. 6121–6126 (2015)
6. Zhao, S., Zelazo, D.: Bearing rigidity and almost global bearing-only formation stabilization. IEEE Trans. Autom. Control **61**(5), 1255–1268 (2016)
7. Zhao, S., Zelazo, D.: Localizability and distributed protocols for bearing-based network localization in arbitrary dimensions. Automatica **69**, 334–341 (2016)

Part V
Advanced Topics

Chapter 11
Moving Formation

Abstract In the previous chapters, we have studied formation control of multi-agents ignoring the ambiguity of translations. That is, if the goal of the formation control is to achieve a desired formation configuration that can be determined by desired distances $\| p_i^* - p_j^* \|$ or desired bearing vectors $g_{ji}^* = \frac{p_j^* - p_i^*}{\| p_j^* - p_i^* \|}$, we can ignore a translation motion of a group of agents as far as the formation shape does not diverge. However, depending upon applications, it may be important to put a constraint on the motions or on the movements of the agents. For example, in a surveillance application, it will be important to consider not only the formation configuration but also the movement of the formation of the agents. First, this chapter considers the movement of agents under distance-based formation control setup. Each agent measures relative displacements in its own coordinate frame. Then, using the measured displacements, each agent estimates the velocity of leader agents. Then, using this estimated velocity, the agent controls the relative displacements as well as the velocity. It is supposed that there is one leader agent and the leader agent is moving with a constant velocity; so, the follower agents need to match their velocities to the velocity of leader agents. For a simplicity, we consider the AMP structure in 2-D. Sections 11.1 and 11.2 are dedicated to this problem. Second, we also consider double integrator dynamics for formulating the velocity matching among agents. Then, under the velocity consensus, agents can achieve the desired formation configuration under the moving formation. Section 11.3 is dedicated to the double integrator dynamics.

11.1 Moving Formation of 3 Agents

This section studies a formation control of a group of agents that is moving with a constant velocity [4, 6]. Consider the single integrator dynamics

$$\dot{p}_i = u_i, \ \forall i \in \mathcal{V} \tag{11.1}$$

The agent 1, which is the leader agent, is moving with the constant velocity v_0 such as

H.-S. Ahn, *Formation Control*, Studies in Systems, Decision and Control 205, https://doi.org/10.1007/978-3-030-15187-4_11

$$\dot{p}_1 = v_0 \tag{11.2}$$

The first follower, i.e., agent 2, needs to estimate the velocity of the leader with the measurement p_1^2. For agent 2, the neighbor agent in the sensing topology is 1, i.e., $\mathcal{N}_2^O = \{1\}$. Since only the relative displacement p_j^i is available, we need to estimate the velocity from the displacement. Let the estimated velocity be denoted as $\hat{v}_i$ at agent i. The control law and adaptation law for the first follower are proposed as

$$u_2 = \hat{v}_2 + z_{12}\bar{e}_{12} \tag{11.3}$$

$$\dot{\hat{v}}_2 = k_v z_{12}\bar{e}_{12} \tag{11.4}$$

where $k_v > 0$ is the gain for the velocity estimation. The above laws can be implemented in the local frame as $u_2^2 = \hat{v}_2^2 + p_1^2\bar{e}_{12}$ and $\dot{\hat{v}}_2^2 = k_v p_1^2 \bar{e}_{12}$, respectively. For the stability analysis, we employ the Lyapunov candidate $V_1 = \frac{1}{2}\bar{e}_{12}^2 + \frac{1}{k_v}\|v_0 - \hat{v}_2\|^2$, which is continuously differentiable and radially unbounded. The derivative of it along the trajectory of (11.3)–(11.4) can be expressed as

$$\begin{aligned}
\dot{V}_1 &= \bar{e}_{12}\dot{\bar{e}}_{12} - \frac{2}{k_v}(v_0 - \hat{v}_2)^T\dot{\hat{v}}_2 \\
&= 2\bar{e}_{12}z_{12}^T(\dot{p}_1 - \dot{p}_2) - 2(v_0 - \hat{v}_2)^T\bar{e}_{12}z_{12} \\
&= 2\bar{e}_{12}z_{12}^T(v_0 - \hat{v}_2 - \bar{e}_{12}z_{12}) - 2(v_0 - \hat{v}_2)^T\bar{e}_{12}z_{12} \\
&= -2\bar{e}_{12}^2\|z_{12}\|^2 \le 0
\end{aligned} \tag{11.5}$$

which is negative semi-definite. The above inequality (11.5) implies that the signals $\bar{e}_{12}$ and $v_0 - \hat{v}_2$ are bounded.

Lemma 11.1 *For the first follower, one of the following two cases holds:*

1. *The distance error $\bar{e}_{12}$ converges to zero.*
2. *The position of the first follower becomes coincident with the position of leader (i.e., $z_{12} \to 0$).*

Proof Since $\ddot{V}_1$ is bounded, it is clear that $\dot{V}_1$ is continuously differentiable. Thus, by Barbalat's lemma (see Lemma 2.8) and Theorem 2.23, $\dot{V}_1 \to 0$ as $t \to \infty$, which means that either $\bar{e}_{12} = 0$ or $z_{12} = 0$.

Now, with the above lemma, we can make the following result.

Theorem 11.1 *For the first follower, the estimated velocity $\hat{v}_2$ by (11.4) converges to the reference velocity v_0 as $t \to \infty$.*

Proof First, for the case of $z_{12} = 0$ in Lemma 11.1, we have $\frac{d(z_{12})}{dt} = \dot{p}_1 - \dot{p}_2 = 0$. Thus, since $z_{12}\bar{e}_{12} = 0$, it is clear that $\hat{v}_2 = u_2 = \dot{p}_2 = \dot{p}_1 = v_0$ from (11.3). Second, for the case of $\bar{e}_{12} = 0$, to show the convergence, let us use the potential function $V_2 = \frac{1}{2}\|z_{12}\|^2$. The derivative of V_2 is given as $\dot{V}_2 = z_{12}^T(v_0 - \hat{v}_2 - \bar{e}_{12}z_{12})$. Since the states $\bar{e}_{12}$, z_{12}, and $\hat{v}_2 - v_0$ are bounded, it is shown that $\ddot{V}_2$ is also bounded and $\dot{V}_2$ is uniformly continuous. Since it was assumed that $\bar{e}_{12}$ converges to zero, it can be concluded that $\|z_{12}\|^2 \to (d_{12}^*)^2$. Thus, by the Barbalat's lemma, we have $\dot{V}_2 = z_{12}^T(v_0 - \hat{v}_2 - \bar{e}_{12}z_{12}) \to 0$. Let $V_3 = z_{12}^T(v_0 - \hat{v}_2 - \bar{e}_{12}z_{12})$, which yields $\dot{V}_3 = \|v_0 - \hat{v}_2 - \bar{e}_{12}z_{12}\|^2 - \bar{e}_{12}\|z_{12}\|^2 - 2\|z_{12}\|^2 z_{12}^T(v_0 - \hat{v}_2 - \bar{e}_{12}z_{12}) - \bar{e}_{12}z_{12}^T(v_0 - \hat{v}_2 - \bar{e}_{12}z_{12})$. Moreover, due to the fact that $\ddot{V}_3$ is bounded, it is clear that $\dot{V}_3$ is uniformly continuous. Finally, since V_3 exists and finite, by Barbalat's lemma, $\dot{V}_3$ converges to zero. Now it was shown that $z_{12}^T(v_0 - \hat{v}_2 - \bar{e}_{12}z_{12}) \to 0$ and $\bar{e}_{12} \to 0$. Thus, to have $\dot{V}_3 = 0$, it is required $\hat{v}_2 = v_0$, which completes the proof.

In the above theorem, two cases, i.e., (i) $\bar{e}_{12} = 0$ and $\hat{v}_2 = v_0$ and (ii) $z_{12} = 0$ and $\hat{v}_2 = v_0$, were analyzed. The following theorem shows that the second case cannot occur.

Theorem 11.2 *The control law (11.3) and velocity estimation law (11.4) ensure the first follower to achieve the desired configuration such as* $\|z_{12}\| \to d_{12}^*$ *and* $\dot{p}_2 \to v_0$.

Proof Denote $f_1 \triangleq v_0 - \hat{v}_2 - \bar{e}_{12}z_{12}$ and $f_2 \triangleq -k_v\bar{e}_{12}z_{12}$, and $z \triangleq (z_{12}^T, (v_0 - \hat{v}_2)^T)^T$ and $f \triangleq (f_1^T, f_2^T)^T$. Then, we have a dynamics $\dot{z} = f$. Linearizing it around $z = 0$ yields

$$\begin{aligned} \left.\frac{\partial f}{\partial z}\right|_{z=0} &= \left.\left[\begin{array}{c|c} -\bar{e}_{12}\mathbb{I}_2 - 2z_{12}z_{12}^T & \mathbb{I}_2 \\ \hline -k_v\bar{e}_{12}\mathbb{I}_2 - 2k_v z_{12}z_{12}^T & \mathbf{0}_{2\times 2} \end{array}\right]\right|_{z=0} \\ &= \left[\begin{array}{c|c} (d_{12}^*)^2\mathbb{I}_2 & \mathbb{I}_2 \\ \hline k_v(d_{12}^*)^2\mathbb{I}_2 & \mathbf{0}_{2\times 2} \end{array}\right] \end{aligned} \tag{11.6}$$

From the following characteristic polynomial of the above matrix

$$\lambda^4 - 2(d_{12}^*)^2\lambda^3 + ((d_{12}^*)^4 - 2k_v(d_{12}^*)^2)\lambda^2 + 2k_v(d_{12}^*)^4\lambda + k_v^2(d_{12}^*)^4 = 0$$

it is easy to see that the linearized system (11.6) is unstable according to the Routh–Hurwitz stability criterion. Hence, since the case $z_{12} = 0$ is unstable, from Lemma 11.1 and from Theorem 11.1, it is clear that $\bar{e}_{12} \to 0$ and $v_2 = \dot{p}_2 \to \hat{v}_2 \to v_0$.

For the agent 3, the following control and adaptation laws are used:

$$u_3 = \hat{v}_3 + z_{13}\bar{e}_{13} + z_{23}\bar{e}_{23} \tag{11.7}$$

$$\dot{\hat{v}}_3 = k_v(z_{13}\bar{e}_{13} + z_{23}\bar{e}_{23}) \tag{11.8}$$

where $k_v > 0$ is the gain for the adaptation law. Note that the above laws also can be implemented in local frame. The convergence property is evaluated by the Lyapunov candidate $V_4 = \frac{1}{4}\bar{e}_{13}^2 + \frac{1}{4}\bar{e}_{23}^2 + \frac{1}{2k_v}\|v_0 - \hat{v}_3\|^2$. The derivative of V_4 is obtained as

$$\dot{V}_4 = \bar{e}_{13}z_{13}(v_0 - \dot{p}_3) + \bar{e}_{23}z_{23}(\dot{p}_2 - \dot{p}_3) - (v_0 - \hat{v}_3)(z_{13}\bar{e}_{13} + z_{23}\bar{e}_{23})$$
$$= -\|\bar{e}_{13}z_{13} + \bar{e}_{23}z_{23}\|^2 + \bar{e}_{23}z_{23}^T(\dot{p}_2 - v_0)$$
$$\leq -\|\bar{e}_{13}z_{13} + \bar{e}_{23}z_{23}\|^2 + \|\bar{e}_{23}z_{23}\|\|\dot{p}_2 - v_0\| \quad (11.9)$$

For a simplicity of convergence analysis, denote $\alpha(z) \triangleq \|\bar{e}_{13}z_{13} + \bar{e}_{23}z_{23}\|^2$ and $\beta(z) \triangleq \|\bar{e}_{23}z_{23}\|\|\dot{p}_2 - v_0\|$.

Lemma 11.2 *For the agent* 3, *with the control law (11.7) and adaptation law (11.8), the states* $\bar{e}_{13}$, $\bar{e}_{23}$, *and* $v_0 - \hat{v}_3$ *are bounded.*

Proof Suppose that $\|z_{13}\|$ and $\|z_{23}\|$ are sufficiently large. Then, it can be approximated as $\bar{e}_{13} \simeq \bar{e}_{23} \simeq \|z_{13}\|^2 \simeq \|z_{23}\|^2$. Thus, the right-hand side of (11.9) is dominated by $-\|\bar{e}_{13}z_{13} + \bar{e}_{23}z_{23}\|^2$ when errors become large. Thus, when the errors become large, $\dot{V}_4 < 0$, which implies V_4 is bounded and $\bar{e}_{13}$, $\bar{e}_{23}$, and $v_0 - \hat{v}_3$ are bounded.

Lemma 11.3 *The term* $\|\bar{e}_{13}z_{13} + \bar{e}_{23}z_{23}\|$ *in (11.9) converges to zero.*

Proof Since V_4 is bounded, there exists a finite M such that $\|\bar{e}_{23}z_{23}\| \leq M$. Also since $\dot{p}_2 - v_0 \to 0$, for any ϵ, there exists T such that $\|\dot{p}_2 - v_0\| < \epsilon$ for all $t > T$. Then, it follows for $t > T$ that

$$-\|\bar{e}_{13}z_{13} + \bar{e}_{23}z_{23}\|^2 - \epsilon' < \dot{V}_4(t) < -\|\bar{e}_{13}z_{13} + \bar{e}_{23}z_{23}\|^2 + \epsilon'$$

where $\epsilon' = \epsilon M$. Since ϵ' can be taken arbitrarily small, by Barbalat's lemma, $\dot{V}_4 \to 0$, which means that $\|\bar{e}_{13}z_{13} + \bar{e}_{23}z_{23}\| \to 0$.

In Lemma 11.3, it was shown that $\|\bar{e}_{13}z_{13} + \bar{e}_{23}z_{23}\| \to 0$. Note that $\|\bar{e}_{13}z_{13} + \bar{e}_{23}z_{23}\| = 0$ can be true if one of the following conditions is satisfied:

- Case-1: $\bar{e}_{13} = 0$ and $\bar{e}_{23} = 0$.
- Case-2: $z_{13} = -\frac{\bar{e}_{23}}{\bar{e}_{13}}z_{23}$ and $\bar{e}_{13} \neq 0$.
- Case-3: $z_{23} = -\frac{\bar{e}_{13}}{\bar{e}_{23}}z_{13}$ and $\bar{e}_{23} \neq 0$.

It is easy to see that the case-2 and case-3 may not occur in most practical cases, with the help of the following example.

Example 11.1 Let the desired inter-agent distances be given as $\bar{d}_{12}^* = 2$, $\bar{d}_{13}^* = 5$, and $\bar{d}_{23}^* = 5$. Let us also suppose that the current inter-agent distances are $\bar{d}_{12} = 2$ and $\bar{d}_{23} = 5$; but $\bar{d}_{13} \neq 5$. Then, we have $\bar{e}_{23} = 0$ and $\bar{e}_{13} \neq 0$. Thus, to make $\|\bar{e}_{13}z_{13} + \bar{e}_{23}z_{23}\| = 0$, it is required to have $z_{13} = 0$. But this is the contradiction to $\bar{e}_{23} = 0$.

Now, we provide detailed analyses for the case-2 and case-3 through the following lemmas.

Lemma 11.4 *For the case-*2 *and case-*3, *it is true that* $z_{13}^x z_{23}^y - z_{13}^y z_{23}^x = 0$, *where* $z_{13} = (z_{13}^x, z_{13}^y)^T$ *and* $z_{23} = (z_{23}^x, z_{23}^y)^T$.

Proof In the case-2, $z_{13}^x z_{23}^y - z_{13}^y z_{23}^x = -\frac{\bar{e}_{23}}{\bar{e}_{13}} z_{23}^x z_{23}^y + \frac{\bar{e}_{23}}{\bar{e}_{13}} z_{23}^y z_{23}^x = 0$. In the case-3, we have $z_{13}^x z_{23}^y - z_{13}^y z_{23}^x = -z_{13}^x \frac{\bar{e}_{13}}{\bar{e}_{23}} z_{13}^y + z_{13}^y \frac{\bar{e}_{13}}{\bar{e}_{23}} z_{13}^x = 0$.

Lemma 11.5 *For the case-2 and case-3, the system is unstable. Thus, case-1 is the only case that may occur.*

Proof For the stability analysis, we use the following dynamics:

$$\dot{z}_{13} = \dot{p}_1 - \hat{v}_3 - z_{13}\bar{e}_{13} - z_{23}\bar{e}_{23} \triangleq f_1$$
$$\dot{z}_{23} = \dot{p}_2 - \hat{v}_3 - z_{13}\bar{e}_{13} - z_{23}\bar{e}_{23} \triangleq f_2$$

which can be compactly written as $\dot{z} = f$, where $z = (z_{13}^T, z_{23}^T)^T$ and $f = (f_1^T, f_2^T)^T$. Since $f_i = \dot{p}_i - \hat{v}_3 - z_{13}\bar{e}_{13} - z_{23}\bar{e}_{23} = \dot{p}_i - \hat{v}_3 - z_{13}(\|z_{13}\|^2 - (d_{13}^*)^2) - z_{23}(\|z_{23}\|^2 - (d_{23}^*)^2)$, the Jacobian matrix F_z of f can be obtained as

$$F_z \triangleq \begin{bmatrix} \frac{\partial f_1}{\partial z_{13}} & \frac{\partial f_1}{\partial z_{23}} \\ \frac{\partial f_2}{\partial z_{13}} & \frac{\partial f_2}{\partial z_{23}} \end{bmatrix} = \left[\begin{array}{c|c} -\bar{e}_{13}\mathbb{I}_2 - 2z_{13}z_{13}^T & -\bar{e}_{23}\mathbb{I}_2 - 2z_{23}z_{23}^T \\ -\bar{e}_{13}\mathbb{I}_2 - 2z_{13}z_{13}^T & -\bar{e}_{23}\mathbb{I}_2 - 2z_{23}z_{23}^T \end{array}\right] \tag{11.10}$$

Let us investigate the eigenvalues of F_z. Let

$$\lambda\mathbb{I}_4 - F_z = \left[\begin{array}{c|c} \lambda\mathbb{I}_2 + \bar{e}_{13}\mathbb{I}_2 + 2z_{13}z_{13}^T & \bar{e}_{23}\mathbb{I}_2 + 2z_{23}z_{23}^T \\ \hline \bar{e}_{13}\mathbb{I}_2 + 2z_{13}z_{13}^T & \lambda\mathbb{I}_2 + \bar{e}_{23}\mathbb{I}_2 + 2z_{23}z_{23}^T \end{array}\right]$$

By eliminations, the above matrix can be changed as

$$\lambda\mathbb{I}_4 - F_z = \left[\begin{array}{c|c} \lambda\mathbb{I}_2 + \bar{e}_{13}\mathbb{I}_2 + 2z_{13}z_{13}^T + \bar{e}_{23}\mathbb{I}_2 + 2z_{23}z_{23}^T & \bar{e}_{23}\mathbb{I}_2 + 2z_{23}z_{23}^T \\ \hline \mathbf{0}_{2\times 2} & \lambda\mathbb{I}_2 \end{array}\right]$$

Thus, it is clear that two eigenvalues of F_z are 0 and the other two are eigenvalues of $L_z \triangleq -(\bar{e}_{13}\mathbb{I}_2 + 2z_{13}z_{13}^T + \bar{e}_{23}\mathbb{I}_2 + 2z_{23}z_{23}^T)$. For the case-3, using the relationship $z_{23}z_{23}^T = (\bar{e}_{13}/\bar{e}_{23})^2 z_{13}z_{13}^T$, L_z can be further changed as

$$L_z = -2\left[\begin{array}{c|c} (z_{13}^x)^2\left(\frac{\bar{e}_{13}^2+\bar{e}_{23}^2}{\bar{e}_{23}^2}\right) + \frac{\bar{e}_{13}+\bar{e}_{23}}{2} & z_{13}^x z_{13}^y\left(\frac{\bar{e}_{13}^2+\bar{e}_{23}^2}{\bar{e}_{23}^2}\right) \\ \hline z_{13}^x z_{13}^y\left(\frac{\bar{e}_{13}^2+\bar{e}_{23}^2}{\bar{e}_{23}^2}\right) & (z_{13}^y)^2\left(\frac{\bar{e}_{13}^2+\bar{e}_{23}^2}{\bar{e}_{23}^2}\right) + \frac{\bar{e}_{13}+\bar{e}_{23}}{2} \end{array}\right] \in \mathbb{R}^{2\times 2}$$

Thus, using the fact that $\bar{e}_{13}^2\|z_{13}\|^2 = \bar{e}_{23}^2\|z_{23}\|^2$, the determinant of L_z is obtained as

$$\begin{aligned}
\det L_z &= (\bar{e}_{13} + \bar{e}_{23})\left(\frac{2\bar{e}_{13}^2\|z_{13}\|^2 + 2\bar{e}_{23}^2\|z_{13}\|^2}{\bar{e}_{23}^2} + \bar{e}_{13} + \bar{e}_{23}\right) \\
&= (\bar{e}_{13} + \bar{e}_{23})(2\|z_{13}\|^2 + 2\|z_{23}\|^2 + \bar{e}_{13} + \bar{e}_{23}) \\
&= (\bar{e}_{13} + \bar{e}_{23})(-\bar{d}_{13}^* - \bar{d}_{23}^* + 3\|z_{13} - z_{23}\|^2 + 6z_{13}^T z_{23}) \\
&= (\bar{e}_{13} + \bar{e}_{23})(-\bar{d}_{13}^* - \bar{d}_{23}^* + 3d_{12}^2 + 6z_{13}^T z_{23})
\end{aligned}$$

Next, let us take the trace of L_z as

$$\begin{aligned}
\text{trace}L_z &= -2(z_{13}^x)^2\left(\frac{\bar{e}_{13}^2+\bar{e}_{23}^2}{\bar{e}_{23}^2}\right)-(\bar{e}_{13}+\bar{e}_{23})-2(z_{13}^y)^2\left(\frac{\bar{e}_{13}^2+\bar{e}_{23}^2}{\bar{e}_{23}^2}\right)-(\bar{e}_{13}+\bar{e}_{23})\\
&= 2\left(-(\bar{e}_{13}+\bar{e}_{23})-\frac{\|z_{13}\|^2(\bar{e}_{13}^2+\bar{e}_{23}^2)}{\bar{e}_{23}^2}\right)\\
&= 2(-(\bar{e}_{13}+\bar{e}_{23})-\|z_{13}\|^2-\|z_{23}\|^2)\\
&= 2(\bar{d}_{13}^*+\bar{d}_{23}^*-2\|z_{13}-z_{23}\|^2-4z_{13}^Tz_{23})\\
&= 2(\bar{d}_{13}^*+\bar{d}_{23}^*-2d_{12}^2-4z_{13}^Tz_{23})
\end{aligned}$$

Now, using the fact that $\bar{e}_{13}+\bar{e}_{23}<0$ (see Lemmas 5.2 or 5.7), we can consider two cases as (i) $3d_{12}^2+6z_{13}^Tz_{23}-\bar{d}_{13}^*-\bar{d}_{23}^*>0$ and (ii) $3d_{12}^2+6z_{13}^Tz_{23}-\bar{d}_{13}^*-\bar{d}_{23}^*\leq 0$. If (i) $3d_{12}^2+6z_{13}^Tz_{23}-\bar{d}_{13}^*-\bar{d}_{23}^*>0$, then $\det L_z<0$, which implies that the eigenvalues of L_z have opposite signs. On the other hand if (ii) $2d_{12}^2+4z_{13}^Tz_{23}\leq \frac{2}{3}(\bar{d}_{13}^*+\bar{d}_{23}^*)$, then $\text{trace}L_z\geq \frac{2}{3}(\bar{d}_{13}^*+\bar{d}_{23}^*)$, which means that one of the eigenvalues of L_z has a positive real part. Similarly to the case 3, we can show that one of the eigenvalues of L_z for the case 2 is positive. Consequently, the equilibrium points under the case-2 and case-3 are unstable.

Theorem 11.3 *The follower agent* 3 *converges to the desired configuration with respect to agents* 1 *and* 2 *by the control law (11.7) and adaptation law (11.8).*

Proof From Lemmas 11.3 and 11.5, it is clear that $\bar{e}_{13}\to 0$ and $\bar{e}_{23}\to 0$ as $t\to\infty$. Since the underlying topology is persistent, it must be true that $\dot{p}_3=v_0$ when $\bar{e}_{13}=0$ and $\bar{e}_{23}=0$.

The result of the 3-agent case thus far can be extended to n-agent systems. For the general ith agent, similarly to the agent 3, the following control and adaptation laws are used:

$$u_i=\hat{v}_i+\sum_{j\in\mathcal{N}_i}\bar{e}_{ji}z_{ji} \tag{11.11}$$

$$\dot{\hat{v}}_i=k_v\sum_{j\in\mathcal{N}_i}\bar{e}_{ji}z_{ji} \tag{11.12}$$

where $j\in\mathcal{N}_i^O$ are the outgoing neighboring nodes of agent i. Since it is an AMP formation, the outgoing agents are leader agents of agent i. All the processes for the convergence analysis to the desired configuration are exactly the same to the processes in the previous 3-agent case. Assuming that agents $1,\ldots,n$ have converged to the desired configuration, the convergence of agent $n+1$ to the desired configuration can be analyzed. By induction, the general n agents will converge to the desired configuration. However, the convergence cannot be global unlike the 3-agent case. For the minimally acyclic persistent formations, there are two positions that satisfy the distance constraints. In the 3-agent case, when the agent 3 satisfies two constraints, the overall configuration is still congruent. But, from agent 4, the achieved formation will not be congruent even though agents satisfy the two desired constraints. Thus, if agent 4 has converged to an undesired position in terms of desired formation configuration, then agent 5 may not reach an equilibrium point. Thus, given a desired formation configuration, in the case of general n-agent group, only the local asymptotic convergence can be ensured. To achieve a global convergence, the technique used in Sect. 5.3 can be employed that uses a virtual variable to impose another constraint to the follower agents. In such a case, the follower agents after agent 4 will have three constraints.

11.2 Shape and Direction Angle Control

In the previous section, the follower agents need to match their velocity to the velocity of leaders. But, in such a case, the positions of follower agents are not unique. As shown in Fig. 11.1, when agent 2 is supposed to satisfy the distance d_{12}^* from the agent 1 and velocity v_0, the position of agent 2, for example, could be ambiguous. In this figure, both the positions $2'$ and $2''$ satisfy the distance and velocity matching conditions. In real implementation, the agents may rotate around the leader 1 with radius d_{12}^*. To overcome this drawback, we would like to impose one more constraint to the follower agents to remove the possibility of rotations. Figure 11.1 shows the angle θ between the vector z_{12} and vector v_2, which is called *directional angle*. The formation configuration with an additional angle constraint, i.e., formation with constraints of d_{ij}^*, v_0, and θ, is called *formation shape*. The shape and direction angle control under the distance-based formation control setup was developed in [5].

The leader (agent 1) has its own local coordinate frame ${}^1\Sigma$. It is considered that the leader is moving with a constant speed v_0 along the y-axis of ${}^1\Sigma$. The follower agents also have their local coordinate frames ${}^i\Sigma$, and the y-axis of ${}^i\Sigma$ is along the

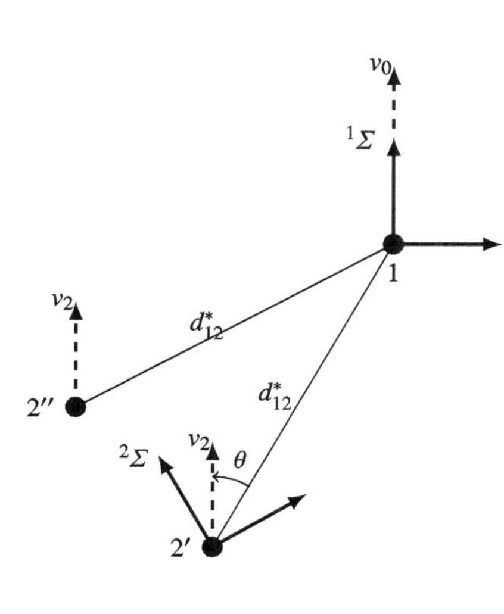

Fig. 11.1 Ambiguity of rotations: at two different points $2'$ and $2''$, agent 2 can satisfy the desired inter-agent distance and velocity matching with respect to the leader 1

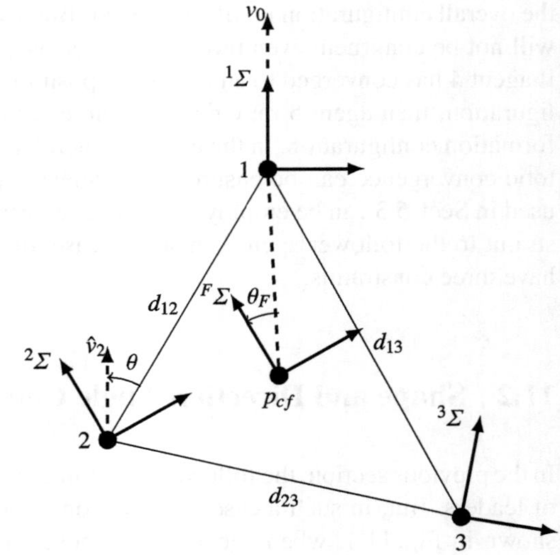

Fig. 11.2 A formation configuration with a unique shape and unique directional angle in a moving formation. © Reprinted from [5], with permission from Elsevier

direction of the estimated velocity $\hat{v}_i$. The local coordinate frames of agents are not aligned each other. The angle between the line connecting p_1 and p_2 and $\hat{v}_2$ is denoted as θ. Denote the center of formation as $p_{cf} = \frac{\sum p_i}{|\mathcal{V}|}$. Further, we define a coordinate frame $^F\Sigma$ that has the origin at p_{cf} and aligned to $^2\Sigma$ as shown in Fig. 11.2.

Then, the direction of formation is defined as follows:

Definition 11.1 (*Direction of formation*) The direction of formation is represented by an angle θ_F that is the angle difference between the vector $z_{1,cf} = p_1 - p_{cf}$ and the y-axis of the coordinate frame $^F\Sigma$.

Remark that if the directional angle θ_F is given as a constraint, then the formation shape is unique. Let us consider a triangular formation. Without the directional constraint, as already mentioned, the formation may rotate around the leader agent 1. The direction of the vector $z_{1,cf} = p_1 - p_{cf}$ will be also varying with respect to a global coordinate frame. But, when the angle θ_F between the coordinate frame $^F\Sigma$ and the vector $z_{1,cf} = p_1 - p_{cf}$ is fixed with the velocity $\hat{v}_2 = v_0$, the overall formation shape would be unique. But it is observed that when the angle θ_F is given without angular direction, i.e., counterclockwise or clockwise, there will be two points that satisfy the angular constraints. Thus, we also assign a sign to the angle θ_F. Thus, as the main goal of this section, we would like to ensure the direction angle control such as $\theta_F \to \theta_{F_d}$ in addition to the formation control such as $\dot{p}_i \to v_0, \forall i \in \mathcal{V}$ and $\|z_{i,j}\| \to d^*_{i,j}, \ \forall (i,j)^{\bar{e}} \in \overrightarrow{\mathcal{E}}$. The angle θ_F can be stabilized by controlling the angle between the reference velocity vector v_0 and the vector z_{12}. Thus, the angle θ_F can be controlled by manipulating the motion of agent 2 around the leader agent. But, in fact, this is not true. From $p_{cf} = \frac{\sum p_i}{|\mathcal{V}|}$, if $p_i, i = 1, \ldots, n$ are determined, then the p_{cf} is uniquely determined. However, given p_{cf}, the positions of agents are not unique. Thus, we rely upon the following assumption.

Assumption 11.2.1 The center of the agents, i.e., p_{cf}, changes according to the position of agent 2. That is, p_{cf} is uniquely determined by the position of p_2.

Example 11.2 Let us consider the AMP graphs. Let us suppose that all the agents are moving with the same velocity (in the reference velocity v_0) as the leader p_1, under the formation of a desired configuration. Thus, the position of the leader agent 1 can be considered as the origin in a common moving frame. Let the center of agents p_{cf} be given as p_{cf}^1 with respect to ${}^1\Sigma$. Then, the center of agents can be calculated as $p_{cf}^1 + p_1$ in the global reference coordinate frame ${}^g\Sigma$. Let agent 2 be located at $(-\sqrt{3}, -1)^T$ with respect to ${}^1\Sigma$, at t_0. Then, suppose that, at t_1, the agent 2 has moved to $(-1, -\sqrt{3})^T$, also with respect to ${}^1\Sigma$. Then, it can be considered that the agent 2 has rotated around p_1 by an amount of $\frac{\pi}{6}$ angle. Since the graph is an AMP graph, all the agents can be considered as rotated by an amount of $\frac{\pi}{6}$ angle around the p_1. Thus, the positions of agents can be expressed as

$$p_i^1(t_1) = \begin{bmatrix} \cos\frac{\pi}{6} & -\sin\frac{\pi}{6} \\ \sin\frac{\pi}{6} & \cos\frac{\pi}{6} \end{bmatrix} p_i^1(t_0) \triangleq R(\pi/6)p_i^1(t_0), \ \forall i \in \mathcal{V} \tag{11.13}$$

Thus, the center position p_{cf}^1 is updated as $p_{cf}^1 = \frac{1}{n}R(\pi/6)\sum_{i=2}^n p_i^1(t_0)$, where $n = |\mathcal{V}|$. Finally, the center position in ${}^g\Sigma$ is computed as $\frac{1}{n}R(\pi/6)\sum_{i=2}^n p_i^1(t_0) + p_1 = R(\pi/6)p_{cf}^1 + p_1$.

Now, with the result of the above example, we can see that, having equalized the center positions as $R(\pi/6)p_{cf}^1 + p_1 = p_{cf}^1 + p_1$, the equality needs to hold $(R(\pi/6) + \mathbb{I}_2)p_{cf}^1 = 0$. Consequently, the exceptions of the Assumption 11.2.1 are when $R(\pi/6) = -\mathbb{I}_2$ or $p_{cf}^1 = 0$.

For the control laws, by slightly modifying (11.11)–(11.12), let us propose the following position control and velocity adaptation laws for agents $i \in \{3, 4, \ldots, n\}$:

$$u_i = [\mathbb{I}_2, \mathbf{0}_{2\times 2(|\mathcal{N}_i|-1)}]\hat{v}_i + k\sum_{j\in\mathcal{N}_i} \bar{e}_{j,i}z_{j,i} \tag{11.14}$$

$$\dot{\hat{z}}_i = A_i\hat{z}_i + (K_i + \Gamma_i\Lambda_i\Gamma_i^T)(z_i - \hat{z}_i) - \mathbf{1}_{|\mathcal{N}_i|\times 1} \otimes [\mathbb{I}_2, \mathbf{0}_{2\times 2(|\mathcal{N}_i|-1)}]\hat{v}_i + \hat{v}_i \tag{11.15}$$

$$\dot{\hat{v}}_i = \Lambda_i\Gamma_i^T(z_i - \hat{z}_i) \tag{11.16}$$

where $z_i = [\ldots, z_{j,i}^T, \ldots]^T \in \mathbb{R}^{2|\mathcal{N}_i|}$, $A_i = \mathbf{1}_{|\mathcal{N}_i|\times 1} \otimes [\ldots, -k\bar{e}_{j,i}\mathbb{I}_2, \ldots] \in \mathbb{R}^{2|\mathcal{N}_i|\times 2|\mathcal{N}_i|}$ and $K_i = \mathbf{1}_{|\mathcal{N}_i|\times 1} \otimes [\ldots, -k\bar{e}_{j,i}\mathbb{I}_2, \ldots] + \mathbb{I}_{2|\mathcal{N}_i|} \in \mathbb{R}^{2|\mathcal{N}_i|\times 2|\mathcal{N}_i|}$ and $\Gamma_i \in \mathbb{R}^{2|\mathcal{N}_i|\times 2|\mathcal{N}_i|}$ is the matrix generated by

$$\dot{\Gamma}_i = [A_i - K_i]\Gamma_i + \mathbb{I}_{2|\mathcal{N}_i|} \tag{11.17}$$

and $\Lambda_i \in \mathbb{R}^{2|\mathcal{N}_i|\times 2|\mathcal{N}_i|}$ is any symmetric positive definite matrix, and $\hat{z}_i$ and $\hat{v}_i$ are the estimations of z_i and the reference velocity, respectively. The above control and adaptation laws (11.14)–(11.17) can be implemented via local coordinate frame since all the states and signals can be measured and updated in a local frame [5]. Since the input $\mathbb{I}_{2|\mathcal{N}_i|}$ in (11.17) can be considered as a persistent excitation [8], it is globally input-to-state stable since $A_i - K_i = -\mathbb{I}_{2|\mathcal{N}_i|}$ is Hurwitz. For the analysis, we need the following lemma.

Lemma 11.6 ([13]) *Let $\Gamma_i(t) \in \mathbb{R}^{p\times q}$ be a bounded and piecewise continuous matrix and $\Lambda \in \mathbb{R}^{p\times p}$ be any symmetric positive definite matrix. If there exist $T, \alpha, \beta > 0$ such that*

$$\alpha\mathbb{I}_p \leq \int_t^{t+T} \Gamma_i(\tau)^T \Gamma_i(\tau) d\tau \leq \beta\mathbb{I}_p$$

then the system $\dot{x}(t) = -\Lambda\Gamma_i(t)^T\Gamma_i(t)x(t)$ is globally exponentially stable.

To proceed, let us substitute (11.16) into (11.15) to get

$$\dot{\hat{z}}_i = A_i\hat{z}_i + K_i(z_i - \hat{z}_i) + \Gamma_i\dot{\hat{v}}_i - \mathbf{1}_{|\mathcal{N}_i|\times 1} \otimes [\mathbb{I}_2, \mathbf{0}_{2\times 2(|\mathcal{N}_i|-1)}]\hat{v}_i + \hat{v}_i \tag{11.18}$$

Define the estimation errors as $\tilde{z}_i = \hat{z}_i - z_i$ and $\tilde{v}_i = \hat{v}_i - \dot{p}_{\mathcal{N}_i}$ where $\dot{p}_{\mathcal{N}_i} = (\dot{p}^T_{\mathcal{N}_i(1)}, \dot{p}^T_{\mathcal{N}_i(2)}, \ldots, \dot{p}^T_{\mathcal{N}_i(|\mathcal{N}_i|)})^T$. Then, we have

$$\dot{\tilde{z}}_i = (A_i - K_i)\tilde{z}_i + \tilde{v}_i + \Gamma_i\dot{\hat{v}}_i \tag{11.19}$$

$$\dot{\tilde{v}}_i = -\Lambda_i\Gamma_i^T\tilde{z}_i - \ddot{p}_{\mathcal{N}_i} \tag{11.20}$$

By defining $\eta_i \triangleq \tilde{z}_i - \Gamma_i\tilde{v}_i$, the derivative of it can be obtained as

$$\dot{\eta}_i = (A_i - K_i)\eta_i + \Gamma_i\ddot{p}_{\mathcal{N}_i} \tag{11.21}$$

Also the derivative of the velocity estimation error $\tilde{v}_i$ can be represented using η_i as

$$\dot{\tilde{v}}_i = -\Lambda_i\Gamma_i^T\Gamma_i\tilde{v}_i - \Lambda_i\Gamma_i^T\eta_i - \ddot{p}_{\mathcal{N}_i} \tag{11.22}$$

Denoting $f_{\eta_i}(\eta_i, \ddot{p}_{\mathcal{N}_i}) \triangleq (A_i - K_i)\eta_i + \Gamma_i\ddot{p}_{\mathcal{N}_i}$ and $f_{\tilde{v}_i}(\tilde{v}_i, \eta_i, \ddot{p}_{\mathcal{N}_i}) \triangleq -\Lambda_i\Gamma_i^T\Gamma_i\tilde{v}_i - \Lambda_i\Gamma_i^T\eta_i - \ddot{p}_{\mathcal{N}_i}$, we can make the following theorem.

Theorem 11.4 *Consider the following cascade system*

$$\dot{\eta}_i = f_{\eta_i}(\eta_i, \ddot{p}_{\mathcal{N}_i})$$

$$\dot{\tilde{v}}_i = f_{\tilde{v}_i}(\tilde{v}_i, \eta_i, \ddot{p}_{\mathcal{N}_i})$$

where $\ddot{p}_{\mathcal{N}_i}$ is an input. If $\ddot{p}_{\mathcal{N}_i}$ is bounded and asymptotically converges to zero, then the velocity estimation error $\tilde{v}_i$ is bounded and asymptotically converges to zero.

Proof When $\ddot{p}_{\mathcal{N}_i}$ is zero, the system (11.21) becomes $\dot{\eta}_i = (A_i - K_i)\eta_i$, which is globally exponentially stable. Thus, if $\ddot{p}_{\mathcal{N}_i}$ is bounded, then it is input-to-state stable. Moreover, if η_i and $\ddot{p}_{\mathcal{N}_i}$ are zero, then the system (11.22) becomes $\dot{\tilde{v}}_i = -\Lambda_i \Gamma_i^T \Gamma_i \tilde{v}_i$. Since (11.17) satisfies the persistent excitation condition, by Lemma 11.6, we can see that $\tilde{v}_i$ converges to zero. Thus, if η_i and $\ddot{p}_{\mathcal{N}_i}$ are bounded, then the system (11.22) is also input-to-state stable. Consequently, if $\ddot{p}_{\mathcal{N}_i}$ converges to zero, then η_i converges to zero due to the property of the input-to-state stability. Since η_i and $\ddot{p}_{\mathcal{N}_i}$ converge to zero, $\tilde{v}_i$ also converges to zero, again due to the input-to-state stability.

The dynamics of distance error with respect to the global coordinate system ${}^g\Sigma$ is expressed as

$$\dot{\bar{e}}_{\mathcal{N}_i(1),i} = 2z_{\mathcal{N}_i(1),i}^T \left(\dot{p}_{\mathcal{N}_i(1)} - [\mathbb{I}_2, \mathbf{0}_{2\times 2(|\mathcal{N}_i|-1)}]\hat{v}_i - k \sum_{j\in\mathcal{N}_i} \bar{e}_{j,i} z_{j,i} \right)$$

$$\dot{\bar{e}}_{\mathcal{N}_i(2),i} = 2z_{\mathcal{N}_i(2),i}^T \left(\dot{p}_{\mathcal{N}_i(2)} - [\mathbb{I}_2, \mathbf{0}_{2\times 2(|\mathcal{N}_i|-1)}]\hat{v}_i - k \sum_{j\in\mathcal{N}_i} \bar{e}_{j,i} z_{j,i} \right)$$

$$\vdots$$

$$\dot{\bar{e}}_{\mathcal{N}_i(|\mathcal{N}_i|),i} = 2z_{\mathcal{N}_i(|\mathcal{N}_i|),i}^T \left(\dot{p}_{\mathcal{N}_i(|\mathcal{N}_i|)} - [\mathbb{I}_2, \mathbf{0}_{2\times 2(|\mathcal{N}_i|-1)}]\hat{v}_i - k \sum_{j\in\mathcal{N}_i} \bar{e}_{j,i} z_{j,i} \right)$$

Let us denote $e_i = (e_{\mathcal{N}_i(1),i}, \ldots, e_{\mathcal{N}_i(|\mathcal{N}_i|),i})^T$. Then the above error dynamics can be compactly written as

$$\dot{e}_i = f(e_i, z_{j,i}, \dot{p}_{\mathcal{N}_i} - \mathbf{1}_{|\mathcal{N}_i|\times 1} \otimes [\mathbb{I}_2, \mathbf{0}_{2\times 2(|\mathcal{N}_i|-1)}]\hat{v}_i) \tag{11.23}$$

Theorem 11.5 *The distance error e_i is bounded and locally asymptotically converges to zero if the velocity estimation error $\dot{p}_{\mathcal{N}_i} - \mathbf{1}_{|\mathcal{N}_i|\times 1} \otimes [\mathbb{I}_2, \mathbf{0}_{2\times 2(|\mathcal{N}_i|-1)}]\hat{v}_i$ is bounded and asymptotically converges to zero.*

Proof The convergence of the distance error to zero can be proved using the concept of input-to-state stability. To this aim, the velocity estimation error $\dot{p}_{\mathcal{N}_i} - \mathbf{1}_{|\mathcal{N}_i|\times 1} \otimes [\mathbb{I}_2, \mathbf{0}_{2\times 2(|\mathcal{N}_i|-1)}]\hat{v}_i$ is considered as an input. To prove the input-to-state stability, consider the unforced system where the velocity estimation error is zero. Then, in the unforced system, it is true that $\dot{z}_{ji} = \dot{p}_j - \dot{p}_i = v_j - u_i = v_j - [\mathbb{I}_2, \mathbf{0}_{2\times 2(|\mathcal{N}_i|-1)}]\hat{v}_i - k\sum_{j\in\mathcal{N}_i} \bar{e}_{j,i} z_{j,i} = -k\sum_{j\in\mathcal{N}_i} \bar{e}_{j,i} z_{j,i}$ where $[\mathbb{I}_2, \mathbf{0}_{2\times 2(|\mathcal{N}_i|-1)}]\hat{v}_i - v_j = 0$ since the velocity estimation error is zero.

Now, for the unforced system, consider the potential function V_i as $V_i(p_i) = \frac{1}{4}\sum_{j\in\mathcal{N}_i} \bar{e}_{j,i}^2$. Then, the derivative is obtained as

$$
\begin{aligned}
\dot{V}_i(p_i) &= -k\left(\sum_{j\in\mathcal{N}_i}\bar{e}_{j,i}z_{j,i}\right)^T\left(\sum_{j\in\mathcal{N}_i}\bar{e}_{j,i}z_{j,i}\right)\\
&= -k\left\|\sum_{j\in\mathcal{N}_i}\bar{e}_{j,i}z_{j,i}\right\|^2\\
&= -k\|\nabla_{p_i}V_i\|^2 \le 0
\end{aligned}
\tag{11.24}
$$

Given the set of desired inter-agent distances d_{ij}^*, let us suppose that the desired configuration with the positions $p_i^*, \forall i \in \mathcal{V}$ is given. Then, it is clear that $V_i(p_i^*) = 0$, and there exists a neighborhood $U_{p_i^*}$ of p_i^* such that $V(p_i) > 0$ for all $p_i \neq p_i^*$ and $p_i \in U_{p_i^*}$. From the Lojasiewicz's inequality [1] (see Lemma 4.2), there also exists a neighborhood $U'_{p_i^*}$ of p_i^* and constants $c_i > 0$ and $\rho_i \in [0, 1)$ such that $\|\nabla V_i\| \ge c_i\|V_i\|^{\rho_i}$ for any $p_i \in U'_{p_i^*}$. Thus since, at any $p_i \in U'_{p_i^*}$, V_i is positive and $\dot{V}_i$ is negative definite, $e_i = 0$ is locally exponentially stable. Thus, the distance error dynamics described by (11.23) can be considered locally input-to-state stable with the input $\dot{p}_{\mathcal{N}_i} - \mathbf{1}_{|\mathcal{N}_i|\times 1} \otimes [\mathbb{I}_2, \mathbf{0}_{2\times 2(|\mathcal{N}_i|-1)}]\hat{v}_i$, which completes the proof.

Since the underlying topology of the formation control problem is a persistent graph, the desired formation up to translations and rotations can be achieved if the desired inter-agent distances are satisfied. However, as aforementioned, the directional angle of formation, i.e., θ, may be arbitrary. Assuming that the desired inter-agent distances are satisfied while the agents move with the constant reference velocity, it is seen that $\theta_F \to \theta_{F_d}$ if and only if $\theta \to \theta_d$, where θ_{F_d} is the desired directional angle of formation and θ_d is the subtended angle between the two vectors $\hat{v}_2$ and z_{12}. Hence, given θ_d, if $\theta \to \theta_d$, the desired directional angle of the formation can be achieved. It means that the directional angle can be achieved if and only if we control the angle θ. To this aim, the control and velocity adaptation laws for agent 2 are designed as follows [5]:

$$u_2 = \hat{v}_2 + k\bar{e}_{1,2}z_{1,2} + u_\theta \tag{11.25}$$

$$u_\theta = -\text{sgn}\left(\frac{z_{1,2}^\perp}{\|z_{1,2}^\perp\|}\cdot\frac{z_d}{\|z_d\|}\right)\left\|\frac{z_d}{\|z_d\|} - \frac{z_{1,2}}{\|z_{1,2}\|}\right\|\frac{z_{1,2}^\perp}{\|z_{1,2}^\perp\|} \tag{11.26}$$

$$\dot{\hat{z}}_{1,2} = A_2\hat{z}_{1,2} + (K_2 + \Gamma_2\Lambda_2\Gamma_2^T)(z_{1,2} - \hat{z}_{1,2}) - u_\theta \tag{11.27}$$

$$\dot{\hat{v}}_2 = \Lambda_2\Gamma_2^T(z_{1,2} - \hat{z}_{1,2}) \tag{11.28}$$

where

$$z_d \triangleq \begin{bmatrix}\cos\theta_d & -\sin\theta_d\\ \sin\theta_d & \cos\theta_d\end{bmatrix}\hat{v}_2;\; z_{1,2}^\perp \triangleq \begin{bmatrix}\cos\pi/2 & -\sin\pi/2\\ \sin\pi/2 & \cos\pi/2\end{bmatrix}z_{1,2}$$

For the convergence analysis of agent 2, substituting (11.28) into (11.27) yields $\dot{\hat{z}}_{1,2} = A_2\hat{z}_{1,2} + K_2(z_{1,2} - \hat{z}_{1,2}) + \Gamma_2\dot{\hat{v}}_2 - u_\theta$. By defining the adaptation errors as $\tilde{z}_{1,2} = \hat{z}_{1,2} - z_{1,2}$ and $\tilde{v}_2 = \hat{v}_2 - v_0$, it follows that

$$
\begin{aligned}
\dot{\tilde{z}}_{1,2} &= A_2\hat{z}_{1,2} + K_2(z_{1,2} - \hat{z}_{1,2}) + \Gamma_2\dot{\hat{v}}_2 - u_\theta - (\dot{p}_1 - \dot{p}_2) \\
&= A_2\hat{z}_{1,2} + K_2(z_{1,2} - \hat{z}_{1,2}) + \Gamma_2\dot{\hat{v}}_2 - u_\theta - v_0 + \hat{v}_2 + k\bar{e}_{1,2}z_{1,2} + u_\theta \\
&= (A_2 - K_2)\tilde{z}_{1,2} + \tilde{v}_2 + \Gamma_2\dot{\hat{v}}_2 \qquad (11.29) \\
\dot{\tilde{v}}_2 &= -\Lambda_2\Gamma_2^T\tilde{z}_{1,2} - \dot{v}_0 \qquad (11.30)
\end{aligned}
$$

We remark that Eqs. (11.29) and (11.30) are exactly the same as Eqs. (11.19) and (11.20). Thus, the agent 2 will reach a point that is away from the agent 1 by an amount of the desired distance, and it will move with the reference velocity v_0. This argument is clear from the following equation:

$$
\begin{aligned}
\dot{\bar{e}}_{1,2} &= 2(p_1 - p_2)^T(\dot{p}_1 - \dot{p}_2) \\
&= 2z_{1,2}^T(v_0 - \hat{v}_2 - k\bar{e}_{1,2}z_{1,2} - u_\theta) \qquad (11.31)
\end{aligned}
$$

which is obtained from (11.25). Since $z_{1,2}^T z_{1,2}^\perp = 0$, the above equation can be changed as

$$
\dot{\bar{e}}_{1,2} = 2z_{1,2}^T(v_0 - \hat{v}_2 - k\bar{e}_{1,2}z_{1,2}) \qquad (11.32)
$$

Thus, similar to Theorem 11.5, we can make the following result.

Theorem 11.6 *The angle error $\theta - \theta_d$ is bounded and asymptotically converges to zero if the velocity estimation error $v_0 - \hat{v}_2$ in (11.31), which can be considered as an input, is bounded and asymptotically converges to zero.*

Proof For the analysis of stability in angle, let us use the potential function $V_\theta = \left\| \frac{z_d}{\|z_d\|} - \frac{z_{1,2}}{\|z_{1,2}\|} \right\|^2$. Assuming that the term $v_0 - \hat{v}_2$ in (11.31) is equal to 0, the derivative of V_θ is obtained as

$$
\begin{aligned}
\dot{V}_\theta &= -\left(\frac{z_d}{\|z_d\|} - \frac{z_{1,2}}{\|z_{1,2}\|}\right)^T \left(\frac{\dot{z}_{1,2}}{\|z_{1,2}\|} - \frac{z_{1,2}}{\|z_{1,2}\|}\frac{z_{1,2}^T\dot{z}_{1,2}}{\|z_{1,2}\|^2}\right) \\
&= -\left(\frac{z_d}{\|z_d\|}\right)^T \left(\frac{\dot{z}_{1,2}}{\|z_{1,2}\|} - \frac{z_{1,2}}{\|z_{1,2}\|}\frac{z_{1,2}^T\dot{z}_{1,2}}{\|z_{1,2}\|^2}\right) + \underbrace{\left(\frac{z_{1,2}}{\|z_{1,2}\|}\right)^T \left(\frac{\dot{z}_{1,2}}{\|z_{1,2}\|} - \frac{z_{1,2}}{\|z_{1,2}\|}\frac{z_{1,2}^T\dot{z}_{1,2}}{\|z_{1,2}\|^2}\right)}_{=0} \\
&= -\left(\frac{z_d}{\|z_d\|}\right)^T \left(\frac{\dot{z}_{1,2}}{\|z_{1,2}\|} - \frac{z_{1,2}}{\|z_{1,2}\|}\frac{z_{1,2}^T\dot{z}_{1,2}}{\|z_{1,2}\|^2}\right) \\
&= -\left(\frac{z_d}{\|z_d\|}\right)^T \frac{1}{\|z_{1,2}\|}\left(1 - \frac{z_{1,2}z_{1,2}^T}{\|z_{1,2}\|^2}\right)\dot{z}_{1,2}
\end{aligned}
$$

$$
\begin{aligned}
&= -\left(\frac{z_d}{\|z_d\|}\right)^T \frac{1}{\|z_{1,2}\|}\left(1-\frac{z_{1,2}z_{1,2}^T}{\|z_{1,2}\|^2}\right)(v_0-\hat{v}_2-k\bar{e}_{1,2}z_{1,2}-u_\theta) \\
&= -\left(\frac{z_d}{\|z_d\|}\right)^T \frac{1}{\|z_{1,2}\|}\left(1-\frac{z_{1,2}z_{1,2}^T}{\|z_{1,2}\|^2}\right)(\underbrace{v_0-\hat{v}_2}_{=0}-k\bar{e}_{1,2}z_{1,2}-u_\theta) \\
&= \underbrace{k\bar{e}_{1,2}\left(\frac{z_d}{\|z_d\|}\right)^T \frac{1}{\|z_{1,2}\|}\left(1-\frac{z_{1,2}z_{1,2}^T}{\|z_{1,2}\|^2}\right)z_{1,2}}_{=0}+\left(\frac{z_d}{\|z_d\|}\right)^T \frac{1}{\|z_{1,2}\|}\left(1-\frac{z_{1,2}z_{1,2}^T}{\|z_{1,2}\|^2}\right)u_\theta \\
&= \left(\frac{z_d}{\|z_d\|}\right)^T \frac{1}{\|z_{1,2}\|}u_\theta - \underbrace{\left(\frac{z_d}{\|z_d\|}\right)^T \frac{1}{\|z_{1,2}\|}\left(\frac{z_{1,2}z_{1,2}^T}{\|z_{1,2}\|^2}\right)u_\theta}_{=0} \\
&= \frac{z_d^T u_\theta}{\|z_d\|\|z_{1,2}\|} \\
&= -\frac{1}{\|z_{1,2}\|}\left|\frac{z_d^T z_{1,2}^\perp}{\|z_d\|\|z_{1,2}^\perp\|}\right|\left\|\frac{z_d}{\|z_d\|}-\frac{z_{1,2}}{\|z_{1,2}\|}\right\| \\
&\le -\frac{1}{\|z_{1,2}\|}\left|\frac{z_d^T z_{1,2}^\perp}{\|z_d\|\|z_{1,2}^\perp\|}\right| V_\theta^{1/2}
\end{aligned}
$$

Thus, the unforced system exponentially converges to zero. Note that $V_\theta = 0$ means that the vector $\hat{v}_2$ is aligned from the vector $z_{1,2}$ by the angle θ_d. Consequently, from the concept of input-to-state stability, if $v_0 - \hat{v}_2 = 0$, then the angle error asymptotically converges to zero.

Taking account of the results thus far, the following theorem can be directly obtained:

Theorem 11.7 *In AMP formation, the agents locally asymptotically converge to the desired formation shape when the reference velocity v_0 is constant. That is, the tasks of $\dot{p}_i \to v_0$, $\|z_{j,i}\| \to d_{j,i}^*$, and $\theta_F \to \theta_{F_d}$ can be achieved by the control and adaptation laws (11.14), (11.15), and (11.16) for agent $i \geq 3$, and by (11.25), (11.26), (11.27), and (11.28) for the agent 2.*

Proof Suppose that the leader moves with the velocity of v_0. Then, for the agent 2, by Theorem 11.4, $\tilde{v}_2$ converges to zero, which further implies that the error $e_{1,2}$ also converges to zero by Theorem 11.5. Consequently, the angle error $\theta - \theta_d$ also converges to zero by Theorem 11.6. Next, let us consider agent 3. Since the leader agents converge to a desired formation shape, we can see that $\ddot{p}_{\mathcal{N}_i} \to 0$ locally asymptotically. Thus, by Theorems 11.4 and 11.5, the velocity error $\tilde{v}_3$ and the distance error e_3 converge to zero. Hence, the agent 3 reaches the desired formation shape with respect to the leader agents. For the remaining agents, we can use the same argument as agent 3 based on mathematical induction.

11.3 Double Integrator Dynamics

When we consider velocity control of agents, it may be natural to consider double integrator dynamics in agents' model. Under the assumption that the agents are connected under the undirected graph topology, let us consider the following double integrators [9, 11]:

$$\dot{p}_i = v_i \tag{11.33}$$

$$\dot{v}_i = u_i \tag{11.34}$$

The goal of the moving formation is to make agents move in a same velocity while keeping the desired configuration. To achieve this aim, the following control law can be proposed [11]:

$$u_i^i = -k_v \sum_{j \in \mathcal{N}_i} (v_i - v_j)^i + k_p \sum_{j \in \mathcal{N}_i} \bar{e}_{ij} z_{ji}^i \tag{11.35}$$

where k_v and k_p are positive control gains associated with the velocity and shape term, and $(v_i - v_j)^i$ is the relative velocity of agent i with respect to agent j which is assumed available. The above local control input can be transformed into a global coordinate frame as

$$u_i = -k_v \sum_{j \in \mathcal{N}_i} (v_i - v_j) + k_p \sum_{j \in \mathcal{N}_i} \bar{e}_{ij} (p_j - p_i) \tag{11.36}$$

Then, the position and velocity dynamics of all agents can be concisely expressed as

$$\dot{p} = v \tag{11.37}$$

$$\dot{v} = -k_v \mathbb{L} v - k_p \mathbb{R}_{\mathcal{G}}^T \bar{e} \tag{11.38}$$

It is noticed that the vector $(\mathbf{1}_n \otimes \mathbb{I}_2)$ is in the null space of Laplacian $\mathbb{L}$ and the rigidity matrix $\mathbb{R}_{\mathcal{G}}$. So, we have $(\mathbf{1}_n \otimes \mathbb{I}_2)^T \dot{v} = -k_v (\mathbf{1}_n \otimes \mathbb{I}_2)^T \mathbb{L} v - k_p (\mathbf{1}_n \otimes \mathbb{I}_2)^T \mathbb{R}_{\mathcal{G}}^T \bar{e} = 0$ since the matrix $\mathbb{L}$ is symmetric, which implies that $\sum_{i=1}^n v_i(t) = const.$ for all $t \geq t_0$. Thus, the average of velocities does not change, and it is fixed as $v_c = \frac{1}{n} \sum_{i=1}^n v_i(t)$. Let us consider the formation of agents over the moving frame with the velocity v_c. For this, we define the following states:

$$p'(t) = p(t) - (\mathbf{1}_n \otimes \mathbb{I}_2) v_c \cdot t$$

$$v'(t) = v(t) - (\mathbf{1}_n \otimes \mathbb{I}_2) v_c$$

Then, we can have

$$\dot{p}' = v' \tag{11.39}$$

$$\dot{v}' = \dot{v} = -k_v \mathbb{L} v' - k_p \mathbb{R}_{\mathcal{G}}(p')^T \bar{e} \tag{11.40}$$

where the elements of the rigidity matrix are functions of p' due to $p_i' - p_j' = p_i - v_c \cdot t - p_j + v_c \cdot t = p_i - p_j$. Also from $\sum_{i=1}^n \dot{p}_i' = 0$, we can have $p_c' = \frac{1}{n}\sum_{i=1}^n p_i' = const$. Then, with the transformed dynamics (11.39) and (11.40), we can analyze the convergence of states to the set $E_{z,v} \triangleq \{z \in \mathbb{R}^n, v \in \mathbb{R}^{2n} : \|z_{ji}\| = d_{ji}^*, (i, j)^e \in \mathcal{E}, v_1 = \cdots = v_n = v_c\}$.

Theorem 11.8 *Let us suppose that the underlying graph topology is infinitesimally minimally rigid. Then, with the formation control law* (11.35), *the velocity matching, i.e.,* $v_1 = \cdots, v_n = v_c$, *and desired formation are achieved if the initial formation is close enough to the desired formation, i.e., a local asymptotic convergence.*

Proof Let us select a Lyapunov candidate as $V = \frac{1}{2}(v')^T v' + \frac{1}{4}\sum_{(i,j)\in\mathcal{E}} \bar{e}_{ij}^2$. The derivative along the trajectory is obtained as

$$\begin{aligned}\dot{V} &= (v')^T \dot{v}' - \bar{e}^T \mathbb{R}_{\mathcal{G}} \mathbb{R}_{\mathcal{G}}^T \bar{e} \\ &= -k_v (v')^T \mathbb{L} v' - \bar{e}^T \mathbb{R}_{\mathcal{G}} \mathbb{R}_{\mathcal{G}}^T \bar{e} \\ &\leq -k_v (v')^T \mathbb{L} v' \leq 0\end{aligned} \tag{11.41}$$

Thus, since v' and $\bar{e}$ are bounded, by LaSalle's invariance principle, we have $v_i' = v_j', \forall i \neq j$, which implies that $v_i = v_j$. Thus, we can have $\sum_{i=1}^n v_i' = 0$, which leads to the conclusion that $v_i' \to 0$ for all $i \in \mathcal{V}$. Consequently, we can see that $v_1 = \cdots, v_n = v_c$. Also, from the proof of Lemma 4.4, we can see that there exists ρ such that $\mathbb{R}_{\mathcal{G}} \mathbb{R}_{\mathcal{G}}^T$ is positive definite at $\bar{e} \in \mathcal{D}$, where $\mathcal{D} = \{\bar{e} \in \mathbb{R}^{|\mathcal{E}|} | \|\bar{e}\| < \rho, \underline{\mathbf{h}}_{\mathcal{G}}(p) \in \mathcal{U}_{edge}\}$. Consequently, from $\dot{V} \leq -\bar{e}^T \mathbb{R}_{\mathcal{G}} \mathbb{R}_{\mathcal{G}}^T \bar{e} \leq 0$, we can conclude that the set $E_{z,v}$ is locally asymptotically stable.

The control law (11.35) is based on the gradient control law, when orientations of agents are not aligned. If we use the orientation alignment law, then even though the coordinate frames of agents are not aligned, a global convergence could be achieved. The formation stabilization problems of double integrator dynamics and exponential convergence property have been examined in [9, 11]. On the other hand, when the orientations are aligned, we may use the bearing-based approach that requires less information than the displacement-based approach. For the bearing-based moving formation, we change the control law (11.36) as [12]

$$u_i = -k_v \sum_{j \in \mathcal{N}_i} (v_i - v_j) - k_p \sum_{j \in \mathcal{N}_i} \frac{\mathbb{P}_{g_{ji}}}{\|z_{ji}\|} g_{ji}^* \tag{11.42}$$

For a simplicity, let the bearing edge function $\mathbf{h}_{\mathcal{G}_B}$ be denoted as g and the desired bearing edge function as g^*. Then, the overall system can be written as

$$\dot{p} = v \tag{11.43}$$
$$\dot{v} = -k_v \mathbb{L} v + k_p \mathbb{R}_B^T g^* \tag{11.44}$$

where $\mathbb{R}_B$ is the bearing rigidity matrix, which can be further transformed as

$$\dot{p}' = v' \tag{11.45}$$
$$\dot{v}' = -k_v \mathbb{L} v' + k_p \mathbb{R}_B (p')^T g^* \tag{11.46}$$

The trivial vector $(\mathbf{1}_n \otimes \mathbb{I}_2)$ is also in the null space of the bearing rigidity matrix $\mathbb{R}_B$. Thus, the average of velocities is fixed. For the convergence analysis, let us define an equilibrium set $E_{p',v'} \triangleq \{(p', v') : v' = 0, g = \pm g^*\}$. Then, we can make the following lemma.

Lemma 11.7 *When the double integrator dynamic agents are updated by the bearing-based formation control law* (11.42)*, the states of agents will converge to the set* $E_{p',v'}$ *as* $t \rightarrow \infty$.

Proof Let us consider the Lyapunov candidate $V = \frac{1}{2}(v')^T v' + \frac{1}{2} k_p \|g - g^*\|^2$. Then, with the fact that $\|g - g^*\|^2 = (g - g^*)^T (g - g^*) = g^T g - g^T g^* - (g^*)^T g + (g^*)^T g^* = 2 - 2 g^T g^*$, the derivative of V is given as

$$\begin{aligned} \dot{V} &= -k_v (v')^T \mathbb{L} v' + k_p (v')^T \mathbb{R}_B^T g^* - k_p \left[\frac{\partial g}{\partial p'} \dot{p}' \right] g^* \\ &= -k_v (v')^T \mathbb{L} v' + k_p (v')^T \mathbb{R}_B^T g^* - k_p (v')^T \mathbb{R}_B^T g^* \\ &= -k_v (v')^T \mathbb{L} v' \end{aligned} \tag{11.47}$$

Due to the same reason as the proof of Theorem 11.8, we can see that $v_i' \rightarrow 0$ for all $i \in \mathcal{V}$. Since $\mathbb{R}_B = \text{blkdg}\left(\frac{\mathbb{P}_{g_{ji}}}{\|z_{ji}\|}\right)(\mathbb{H} \otimes \mathbb{I}_2)$, from (11.46), equalizing $v' = \dot{v}' = 0$, we obtain $(\mathbb{H} \otimes \mathbb{I}_2)^T \text{blkdg}\left(\frac{\mathbb{P}_{g_{ji}}}{\|z_{ji}\|}\right) g^* = 0$. It was observed in [14] that the condition $(\mathbb{H} \otimes \mathbb{I}_2)^T \text{blkdg}\left(\frac{\mathbb{P}_{g_{ji}}}{\|z_{ji}\|}\right) g^* = 0$ implies both $(g_{ji}^*)^T \mathbb{P}_{g_{ji}} g_{ji}^* = 0$ and $g_{ji}^T \mathbb{P}_{g_{ji}^*} g_{ji} = 0$. Thus, denoting $z_{ji}' = p_j' - p_i'$, we can see that $(z_{ji}')^T \mathbb{P}_{g_{ji}^*} z_{ji}' = 0$ for all edges. Therefore, we can have the following relationship:

$$(z')^T \text{blkdg}\left(\mathbb{P}_{g_{ji}^*}\right) z' = (p')^T (\mathbb{H} \otimes \mathbb{I}_2)^T \text{blkdg}\left(\mathbb{P}_{g_{ji}^*}\right) \text{blkdg}\left(\mathbb{P}_{g_{ji}^*}\right) (\mathbb{H} \otimes \mathbb{I}_2) p' = 0 \tag{11.48}$$

which implies $\text{blkdg}\left(\mathbb{P}_{g_{ji}^*}\right)(\mathbb{H} \otimes \mathbb{I}_2) p' = \mathbb{R}_B((p')^*) p' = 0$, where $(p')^*$ is p' when $p = p^*$. Note that the bearing-based framework f_p is infinitesimally bearing rigid if and only if the null space of the rigidity matrix includes only the trivial motions (see Definition 2.11), i.e., when $p = p^*$, $\text{null}(\mathbb{R}_B(p^*)) = \text{span}(\text{range}(\mathbf{1}_n \otimes \mathbb{I}_2), p^* - \mathbf{1}_n \otimes p_c^*)$ where p_c is the center of the realization. Thus, since the underlying topology is infinitesimally bearing rigid, we have $\text{null}(\mathbb{R}_B((p')^*)) = \text{span}(\text{range}(\mathbf{1}_n \otimes$

$\mathbb{I}_2), (p')^* - \mathbf{1}_n \otimes (p'_c)^*)$. Moreover, the condition $\mathbb{R}_B((p')^*)p' = 0$ implies that $\mathbb{R}_B((p')^*)(p' - \mathbf{1}_n \otimes p'_c) = 0$. Therefore, we can see that

$$p' - \mathbf{1}_n \otimes p'_c = \text{span}(\text{range}(\mathbf{1}_n \otimes \mathbb{I}_2), (p')^* - \mathbf{1}_n \otimes (p'_c)^*) \tag{11.49}$$

It is clear that the subspace $p' - \mathbf{1}_n \otimes p'_c$ is for scaling while the subspace $\mathbf{1}_n \otimes \mathbb{I}_2)$ is for translations. Thus, they are orthogonal to each other. Therefore, it is clear that $p' - \mathbf{1}_n \otimes p'_c = k_{scale}((p')^* - \mathbf{1}_n \otimes (p'_c)^*)$ where k_{scale} is a constant. Consequently, $p'_j - p'_i = k_{scale}((p'_j)^* - (p'_i)^*)$, which implies $g_{ji} = g^*_{ji}$ or $g_{ji} = -g^*_{ji}$ for all $(i, j)^e \in \mathcal{E}$.

In what follows, we will show that the equilibrium point with $g = -g^*$ is unstable, while the one with $g = g^*$ is stable. For evaluating the stability, let us obtain the Jacobian of the right-hand side of (11.45)–(11.46) as

$$J_{p',v'} = \begin{bmatrix} \mathbf{0}_{2n\times 2n} & \mathbb{I}_{2n} \\ J_{p'} & -k_v\mathbb{L} \end{bmatrix} \tag{11.50}$$

where $J_{p'}$ is the Jacobian of (11.46) with respect to p'. From

$$J_{p'} = k_p \frac{\partial \mathbb{R}_B^T}{\partial p'} = k_p(\mathbb{H} \otimes \mathbb{I}_2)^T \frac{\partial \text{blkdg}\left(\mathbb{P}_{g^*_{ji}}\right) g^*}{\partial p'}$$

we can see that each row element of the vector $\text{blkdg}\left(\mathbb{P}_{g^*_{ji}}\right) g^*$ is taken by a derivative with each p'_i. Let the kth row of $\text{blkdg}\left(\mathbb{P}_{g^*_{ji}}\right) g^*$ be denoted as ψ_k and write $\frac{\partial \psi_k}{\partial p'_i}$ as $\bar{J}_{k,i}$. Then, we can calculate $\bar{J}_{k,i}$ as

$$\begin{aligned} \bar{J}_{k,i} &= \frac{1}{\|z'_{ji}\|^2} \mathbb{P}_{g_{ji}} g^*_{ji} (p'_i)^T + \frac{1}{\|z'_{ji}\|} \frac{\partial (g^*_{ji} - g_{ji} g^T_{ji} g^*_{ji})}{\partial g_{ji}} \frac{\partial g_{ji}}{\partial p'_i} \\ &= \frac{1}{\|z'_{ji}\|^2} \mathbb{P}_{g_{ji}} g^*_{ji} (p'_i)^T + \frac{1}{\|z'_{ji}\|^2} (g^T_{ji} g^*_{ji} \mathbb{I}_2 + g_{ji} (g^*_{ji})^T) \mathbb{P}_{g_{ji}} \end{aligned}$$

So, at $p' = (p')^*$, with the fact that $(g^*_{ji})^T g^*_{ji} = 1$, we have

$$\begin{aligned} \bar{J}_{k,i}((p')^*) &= \frac{1}{\|z'_{ji}\|^2} (\mathbb{I}_2 + g^*_{ji} (g^*_{ji})^T)(\mathbb{I}_2 - g^*_{ji} (g^*_{ji})^T) \\ &= \frac{1}{\|z'_{ji}\|^2} \mathbb{P}_{g^*_{ji}} \end{aligned} \tag{11.51}$$

By following the same procedure, we can calculate $\frac{\partial \psi_k}{\partial p'_j}$, which is denoted as $\bar{J}_{k,j}$, as

$$\bar{J}_{k,j}((p')^*) = -\frac{1}{\|z'_{ji}\|^2}\mathbb{P}_{g^*_{ji}} \tag{11.52}$$

It is certain that $\bar{J}_{k,q}((p')^*) = 0$ when $q \neq \{i, j\}$. Therefore, denoting $\bar{J}_{p'} = [\bar{J}_{k,i}]$, we can obtain

$$\bar{J}_{p'=(p')^*} = -\text{blkdg}\left(\frac{1}{\|z'_{ji}\|^2}\mathbb{P}_{g^*_{ji}}\right)(\mathbb{H} \otimes \mathbb{I}_2) \tag{11.53}$$

Therefore, $J_{p'}$ is obtained as

$$\begin{aligned}
J_{p'}(p' = (p')^*) &= k_p \frac{\partial \mathbb{R}_B^T}{\partial p'} = k_p(\mathbb{H} \otimes \mathbb{I}_2)^T \bar{J}_{p'} \\
&= -k_p(\mathbb{H} \otimes \mathbb{I}_2)^T \text{blkdg}\left(\frac{1}{\|z'_{ji}\|^2}\mathbb{P}_{g^*_{ji}}\right)(\mathbb{H} \otimes \mathbb{I}_2) \\
&= -k_p(\mathbb{H} \otimes \mathbb{I}_2)^T \text{blkdg}\left(\frac{1}{\|z'_{ji}\|}\mathbb{P}_{g^*_{ji}}\right)\text{blkdg}\left(\frac{1}{\|z'_{ji}\|}\mathbb{P}_{g^*_{ji}}\right)(\mathbb{H} \otimes \mathbb{I}_2) \\
&= -k_p \mathbb{R}_B((p')^*)^T \mathbb{R}_B((p')^*) \leq 0
\end{aligned} \tag{11.54}$$

Let us repeat the above procedure with p' corresponding to $g = -g^*$. In this case, from the fact that $g_{ji} = \frac{p_j - p_i}{\|p_j - p_i\|} = -g^*_{ji} = 0\frac{p^*_j - p^*_i}{\|p^*_j - p^*_i\|}$, we can equalize as $p'_i = -(p'_i)^*$ for all $i \in \mathcal{V}$. Thus, we can have $J_{p'}(p' = -(p')^*) = k_p \mathbb{R}_B((p')^*)^T \mathbb{R}_B((p')^*) \geq 0$. Consequently, when $g = -g^*$, the Jacobian matrix $J_{p'}$ has positive eigenvalues. With the results so far, let us compute the eigenvalues of $J_{p',v'}$. For this purpose, with an eigenvalue λ corresponding to the eigenvector $(v_p^T, v_v^T)^T$, we can write

$$\begin{bmatrix} \lambda \mathbb{I}_{2n} & -\mathbb{I}_{2n} \\ -J_{p'} & \lambda \mathbb{I}_{2n} + k_v \mathbb{L} \end{bmatrix} \begin{bmatrix} v_p \\ v_v \end{bmatrix} = 0 \tag{11.55}$$

In the above equation, there are two equalities. Thus, we can replace v_v by v_p and obtain the following equality:

$$\|v_p\|^2 \lambda^2 + k_v v_p^T \mathbb{L} v_p \lambda - v_p^T J_{p'} v_p = 0 \tag{11.56}$$

Notice that when $\lambda = 0$, we have $v_v = 0$, which means that the eigenvector corresponding to the zero eigenvalue is $(v_p^T, 0)^T$. Thus, reminding that $\text{null}(J_{p'}) = \text{span}\{\mathbf{1}_n \otimes \mathbb{I}_2, p'\}$, we should have $v_p \in \text{span}\{\mathbf{1}_n \otimes \mathbb{I}_2, p'\}$ when $\lambda = 0$. Further observing that there are 3 independent bases in $\text{span}\{\mathbf{1}_n \otimes \mathbb{I}_2, p'\}$, we can see that there are three zero eigenvalues in 2-dimensional space. Consequently, the matrix $J_{p',v'}$ has three zero eigenvalues. Moreover, if v_p is not in the null space of $J_{p'}$ and $\mathbb{L}$, it is true that $k_v v_p^T \mathbb{L} v_p > 0$ and $v_p^T J_{p'} v_p > 0$. Thus, since $\|v_p\|^2 > 0$, $k_v v_p^T \mathbb{L} v_p > 0$, and $v_p^T J_{p'} v_p > 0$ in (11.56), there should exist a positive λ to satisfy the equality in (11.56). That is, denoting $a_1 = \|v_p\|^2$, $a_2 = k_v v_p^T \mathbb{L} v_p$, and $a_3 = v_p^T J_{p'} v_p$, the solutions of $a_1\lambda^2 + a_2\lambda - a_3 = 0$ are computed as $\lambda = \frac{-a_2 \pm \sqrt{a_2^2 + 4a_1a_3}}{2a_1}$. Thus, it is clear that $\lambda = \frac{-a_2 + \sqrt{a_2^2 + 4a_1a_3}}{2a_1} > 0$. The above result is summarized as follows.

Lemma 11.8 *For the double integrator dynamics, the equilibrium set corresponding to the case $g = -g^*$ is unstable.*

Although it was shown that the undesired equilibrium point is unstable, it is also necessary to elaborate on the desired case $g = g^*$ since there are multiple zero eigenvalues in the Jacobian matrix $J_{p'}$. The convergence property of a linear system with multiple zero real-part eigenvalues and stable eigenvalues can be analyzed by center manifold theory, which is briefly summarized in what follows [7]. Let us consider an autonomous system $\dot{x} = f(x),\ x \in \mathbb{R}^d$. Let the Jacobian matrix after linearization at the origin $x = 0$ be given as $A = \frac{\partial f(x)}{\partial x}$, which has ξ zero real-part eigenvalues and $d - \xi$ eigenvalues with negative real parts. After a similarity transformation, let the linearized system be decomposed as

$$\dot{y} = A_1 y + g_1(y, z) \tag{11.57}$$
$$\dot{z} = A_2 z + g_2(y, z) \tag{11.58}$$

where $x = (y^T, z^T)^T$, $y \in \mathbb{R}^\xi$, and $z \in \mathbb{R}^{d-\xi}$, and the matrix A_1 has zero real-part eigenvalues and A_2 has eigenvalues with negative real parts. In the above equations, assuming that g_1 and g_2 are twice continuously differentiable, it is required to satisfy $g_i(0, 0) = 0$, $\frac{\partial g_i}{\partial y}(0, 0) = 0$, and $\frac{\partial g_i}{\partial z}(0, 0) = 0$. Let the states y and z be related as $z = h(y)$. Let $\mathcal{M} \triangleq \{(y, z) \in \mathbb{R}^\xi \times \mathbb{R}^{d-\xi} : z = h(y)\}$. The set $\mathcal{M}$ is called center manifold set if it is invariant for (11.57)–(11.58) and $h(0) = 0$ and $\frac{\partial h}{\partial y}(0) = 0$.

Theorem 11.9 ([7]) *Let us consider an autonomous system $\dot{x} = f(x)$ with f being twice continuously differentiable, which is decomposed as* (11.57)–(11.58). *Then, there exists a constant $\delta > 0$, which defines $\|y\| < \delta$, such that $z = h(y)$ is a center manifold for* (11.57)–(11.58).

Let the equilibrium set $\mathcal{M}_1 \triangleq \{x \in \mathbb{R}^d : f(x) = 0\}$ contain the origin. Then, when there is a manifold of equilibria, we can state the center manifold theory as follows.

Theorem 11.10 ([10]) *Under the same assumptions as in Theorem 11.9, suppose that there is a smooth ξ-dimensional manifold of equilibrium set $\mathcal{M}_1$ that contains the origin. Then, $\mathcal{M}_1$ is a center manifold for $\dot{x} = f(x)$ and there are compact neighborhoods Ω_1 and Ω_2 of the origin such that $\mathcal{M}_2 = \mathcal{M}_1 \cap \Omega_2$ is locally exponentially stable. Furthermore, for each $x(t_0) \in \Omega_1$, $x(t)$ converges to a point in $\mathcal{M}_2$.*

From the discussions given in the above, it is clear that the matrix $J_{p',v'}$ has three zero eigenvalues and other eigenvalues with negative real parts, when $g = g^*$. When $g = g^*$, we can write $p' = (p')^*$ and $v' = 0$. Let us conduct a similarity transformation of $J_{p',v'}$ to have $T J_{p',v'} T^{-1} = \text{blkdg}(P_1, P_2)$ where P_1 is the matrix with three zero eigenvalues and P_2 is the matrix with eigenvalues with negative real parts. To use the center manifold theory, denoting $x' = ((p')^T, (v')^T)^T$, we write (11.45)–(11.46) as $\dot{x}' = f(x')$, which can be expressed as

$$\dot{x}' = J_0 x' + \underbrace{(f(x') - J_0 x')}_{=f'(x')} \tag{11.59}$$

where $J_0 = J_{p'=p^*, v'=0}$. Then, using $(y^T, z^T)^T = Tx'$, we can rewrite the above equation as

$$\dot{y} = P_1 y + g_1(y, z) \tag{11.60}$$

$$\dot{z} = P_2 z + g_2(y, z) \tag{11.61}$$

where g_1 and g_2 are obtained from $f'(x')$. By a change of variables as $x' \triangleq x' - ((p')^{*T}, 0)^T$, we can ensure that $g_i(0, 0) = 0$ and $\frac{\partial g_i}{\partial y}(0, 0) = 0$, and $\frac{\partial g_i}{\partial z}(0, 0) = 0$. Let the invariant set for (11.60)–(11.61) be given as

$$\mathcal{M}' \triangleq \{(y^T, z^T)^T : (y^T, z^T)^T = Tx', \ x' = ((p')^{*T}, 0)^T\} \tag{11.62}$$

Lemma 11.9 *The set $\mathcal{M}'$ is a center manifold set for* (11.60)–(11.61).

Proof It is clear that the set $\mathcal{M}'$ is an invariant set since it is an equilibrium set for (11.60)–(11.61). To show it is a center manifold, it is required to show the $\mathcal{M}'$ is tangent to the center subspaces of (11.60)–(11.61), which is equivalent to showing that $\mathcal{S}^* \triangleq \{(p'^T, v'^T) : p' = (p')^*, \ v' = 0\}$ is tangent to the eigenvectors $(v_p^T, 0)^T$ corresponding to the zero eigenvalues, where $v_p \in \text{null}(J_{p'})$, of $J_{p',v'}$. It is obvious that $\mathcal{S}^* \subset \bar{\mathcal{S}} \triangleq \{(p'^T, v'^T)^T : \bar{F}(p', v') = (g^T - g^{*T}, v'^T)^T$. The Jacobian of $\bar{F}(p', v')$ is given as

$$J_F = \begin{bmatrix} \mathbb{R}((p')^T & 0 \\ 0 & \mathbb{I}_{2n} \end{bmatrix}$$

Thus, we have $J_F(v_p^T, 0)^T = 0$. Consequently, since $\mathcal{S}^*$ is a subset of $\bar{\mathcal{S}}$, we can conclude that $\mathcal{S}^*$ is tangent to $(v_p^T, 0)^T$ at the origin.

Now, summarizing the results thus far, we can make the following theorem.

Theorem 11.11 *Let us consider double integrator dynamic systems described by* (11.43)–(11.46). *Then, there exist compact neighborhoods Ω_1 and Ω_2 of the set $\mathcal{M}'$ such that any trajectory starting within Ω_1 would exponentially converge to a point in $\mathcal{M}^\dagger = \mathcal{M}' \cap \Omega_2$.*

11.4 Summary and Simulations

In Sect. 11.1, when a leader moves with a constant velocity, the follower agents estimate the velocity of the leader by way of adaptation. Since the inputs to the adaptation law are relative displacements, it is a distance-based formation control law. By matching the velocity and controlling the distances, the desired formation configuration is achieved. In Sect. 11.2, to avoid a possibility of rotations of the follower agents with respect to the leader, we additionally considered the directional angle constraints of the formation. Since the directional angle control law and formation shape control law are designed in an orthogonal way, the two control laws are independently implemented into local coordinate frames. Related with this chapter, it is remarkable that an adaptive control setup was also presented to achieve a desired configuration for general mobile robot dynamics in [2]. In this chapter, the reference velocity of the leader is assumed to be constant. Otherwise, the proposed adaptation law cannot estimate it since the reference velocity is time-variant. From a practical sense, if the velocity is changing slowly, we can use a large gain k_v in (11.4), (11.8), and (11.12) for a rapid estimation of the velocity in a moving formation. Also, we can use a big gain matrix Λ_i for the velocity estimation in (11.16) for the shape and directional angle control. With big gains in the velocity estimation, the slowly varying reference velocity may be estimated in a fast dynamics, while the position is controlled in a slow dynamics.

Figure 11.3 shows the simulation results of five agents with a moving leader under acyclic persistence formation setup as shown in Fig. 5.4. But, in this simulation, the agents are moving in 2-dimensional space. The control law (11.11) and estimation law (11.12) are used for tracking the motion of the moving leader. The leader is assumed to move with the constant velocity $v_0 = (0.5, 0.5)^T$ in x- and $y-$axes of $\mathbb{R}^2$. From a number of numerical simulations, it was observed that the performance of the moving formation is dependent upon the gain k_v of (11.12) and the magnitude of the velocity of the leader. Figure 11.3 shows the results with the gain $k_v = 10$. With the selected gain, the velocity of the leader, i.e., $v_0 = (0.5, 0.5)^T$, is estimated by the followers as shown in the left-bottom and right-bottom plots. For a comparison purpose, we also conducted simulations without a velocity matching term, i.e., the follower agents are purely updated by (11.11) without velocity estimation. Figure 11.4 shows the simulation results without velocity estimation. The overall formation patterns of Figs. 11.3 and 11.4 look similar; but as shown in the plots of the errors, the formation system with velocity estimation outperforms the formation without the velocity estimation. Figure 11.5 shows a moving formation with a leader that is moving with slowly varying velocities. The time-variant velocity is modeled as $v_0 = (0.5, 0.5)^T + \frac{1}{4}\sin(2\pi f_v t)$ where $f_v = \frac{1}{2}$. For the slowly varying velocity, the gains are selected as $k_v = 50$ to improve the convergence speed in the estimation dynamics. As shown in the plots of Fig. 11.5, even with the time-variant leader, the overall errors of the formation system are smaller than those of the case of Fig. 11.4.

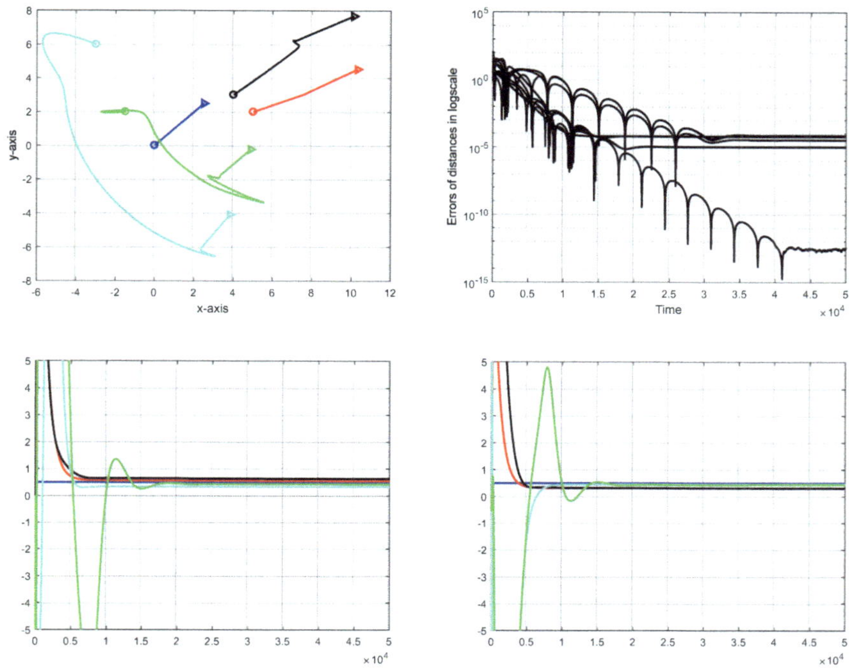

Fig. 11.3 Moving formation with velocity estimation. Left-top: Trajectories of five agents. Right-top: Errors of the distances. Left-bottom: Estimated speeds in x-axis (y-axis of this figure represents the speed, and x-axis of this figure represents the sampling instants). Right-bottom: Estimated speeds in y-axis (y-axis of this figure represents the speed, and x-axis of this figure represents the sampling instants)

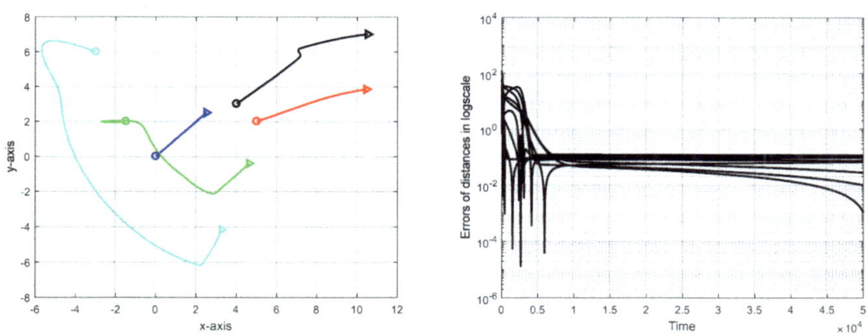

Fig. 11.4 Moving formation without velocity estimation. Left: Trajectories of five agents. Right: Errors of the distances

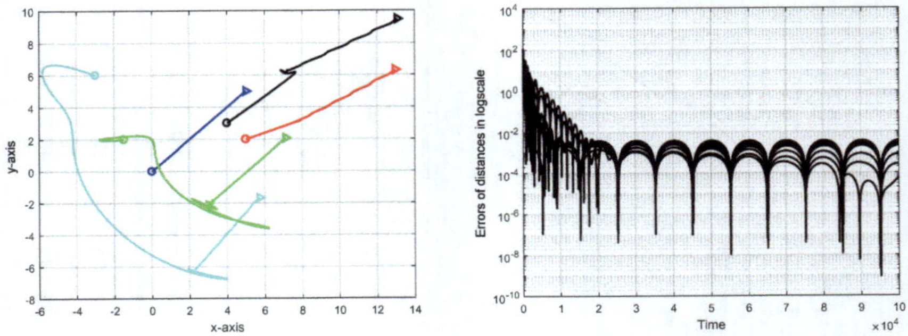

Fig. 11.5 Moving formation with time-variant leader with velocity estimation. Left: Trajectories of five agents. Right: Errors of the distances

11.5 Notes

The flocking behavior combined with formation shape control was introduced [3]. It is recommended to refer to [3] for a moving formation of double integrator dynamics under a displacement-based setup. The results of Sect. 11.1 have been reused from [4, 6] and results of Sect. 11.2 were reused from [5]. Section 11.3 is based on the results of [12]. The following copyright and permission notices are acknowledged.

- The text of Sect. 11.1 is © [2014] IEEE. Reprinted, with permission, from [6].
- The statements of Theorems 11.1, 11.2, Lemmas 11.3, and 11.5 with the proofs are © [2016] IEEE. Reprinted, with permission, from [4].
- The text of Sect. 11.2, and the statements of Theorems 11.4, 11.5, 11.6, and 11.7 with the proofs are © Reprinted from [5], with permission from Elsevier.
- The text of Sect. 11.3, and the statements of Lemmas 11.7 and 11.9 with the proofs are © [2018] The Institute of Control, Robotics and Systems (ICROS) of Korea. Reuse and reproduced by *Returned Rights*, with royalty-free permission, from [12]

References

1. Absil, P.-A., Kurdyka, K.: On the stable equilibrium points of gradient systems. Syst. Control Lett. **55**(7), 573–577 (2006)
2. Cai, X., de Queiroz, M.: Adaptive rigidity-based formation control for multirobotic vehicles with dynamics. IEEE Trans. Control Syst. Technol. **23**, 389–396 (2015)
3. Deghat, M., Anderson, B.D.O., Lin, Z.: Combined flocking and distance-based shape control of multi-agent formations. IEEE Trans. Autom. Control **61**(7), 1824–1837 (2016)
4. Kang, S.-M., Ahn, H.-S.: Design and realization of distributed adaptive formation control law for multi-agent systems with moving leader. IEEE Trans. Ind. Electron. **63**(2), 1268–1279 (2016)
5. Kang, S.-M., Ahn, H.-S.: Shape and orientation control of moving formation in multi-agent systems without global reference frame. Automatica **92**(6), 210–216 (2018)

6. Kang, S.-M., Park, M.-C., Lee, B.-H., Ahn, H.-S.: Distance-based formation control with a single moving leader. In: Proceedings of the American Control Conference, pp. 305–310 (2014)
7. Khalil, H.K.: Nonlinear Systems. Prentice Hall, New Jersey (2002)
8. Narendra, K.S., Annaswamy, A.M.: Persistent excitation in dynamical system. In: Proceedings of the American Control Conference, pp. 336–338 (1984)
9. Oh, K.-K., Ahn, H.-S.: Distance-based undirected formations of single- and double-integrator modeled agents in n-dimensional space. Int. J. Robust Nonlinear Control **24**(12), 1809–1820 (2014)
10. Summers, T.H., Yu, C., Dasgupta, S., Anderson, B.D.O.: Control of minimally persistent leader-remote-follower and coleader formations in the plane. IEEE Trans. Autom. Control **56**(12), 2778–2792 (2011)
11. Sun, Z., Anderson, B.D.O., Deghat, M., Ahn, H.-S.: Rigid formation control of double-integrator systems. Int. J. Control **90**(7), 1403–1419 (2017)
12. Tran, Q.V., Ahn, H.-S.: Flocking control and bearing-based formation control. In: Proceedings of the International Conference on Control, Automation and Systems, pp. 124–129 (2018)
13. Zhang, Q.: Adaptive observer for multiple-inputmultiple-output (MIMO) linear time-varying systems. IEEE Trans. Autom. Control **47**(3), 525–529 (2002)
14. Zhao, S., Zelazo, D.: Bearing rigidity and almost global bearing-only formation stabilization. IEEE Trans. Autom. Control **61**(5), 1255–1268 (2016)

Chapter 12
$K(n)$ Formation and Resizing

Abstract In Chaps. 3, 4 and 5, we provided global stabilization, local stabilization, and stabilization of directed persistent formations under gradient-based control laws. However, as shown in Fig. 4.5, it is not possible to stabilize a general formation to the desired configuration under the gradient control laws. There have been some efforts to improve the performance of a formation system by adding a perturbation term into the gradient control law. But, as commented in Trinh et al. (Automatica 77:393–396, 2018 [10]), there is no clear clue for an improvement of a formation system even with perturbation terms. Hence, it is generally considered that the gradient control laws may be effective for local stabilizations. However, the gradient control laws still can be a good solution for certain circumstances. This chapter provides some advanced results in gradient control laws. As the first result, in Sect. 12.1 we provide a gradient control law enhanced by virtual variables for ensuring a global stabilization of general complete graphs in general dimensional space. However, note that since the control law proposed in Sect. 12.1 uses virtual variables that should be communicated between neighboring agents, it has a drawback in terms of computations and communications over the traditional gradient control laws. As the second results, in Sect. 12.2 we present a gradient control law for resizing the formation in the distance-based setup, and also solve a scaling problem under the bearing-based setup. Since a resizing of formation can be achieved by controlling the distance of an edge, it is convenient for reorganizing the positions of agents for some applications. For example, when a group of unmanned aerial vehicles needs to rescale the size of formation for surveillance or for gas detection, it can be done by simply changing the distance between the leader and the first follower (see Lemma 9.3).

12.1 $K(n)$ Formation in d-Dimensional Space

In Chap. 3, global stabilizations of formations were addressed for a specific set of agents under gradient control laws. The triangular formation, polygon formations, $K(3) + 1$ edge formations, and $K(4) - 1$ edge formations in 2-dimensional space were stabilized to desired configurations. Also, $K(4)$ formation in 3-dimensional space was shown to be achieved by gradient control law. For general graphs, local

H.-S. Ahn, *Formation Control*, Studies in Systems,
Decision and Control 205, https://doi.org/10.1007/978-3-030-15187-4_12

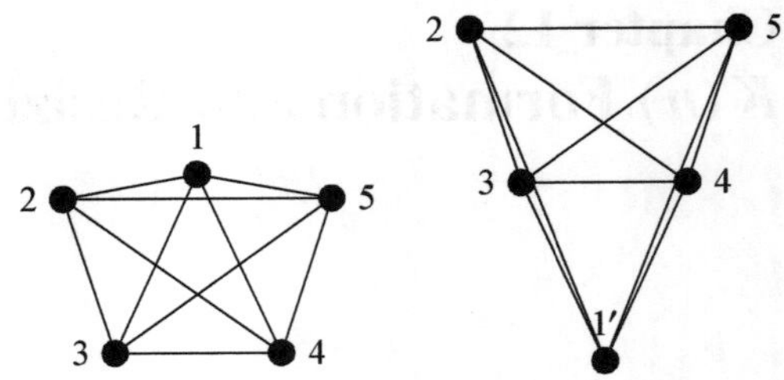

Fig. 12.1 $K(5)$ graph composed of five agents: **a** Initial configuration. **b** Desired configuration

stabilizations were shown to be ensured in Chap. 4. In this chapter, we would like to develop a control law that can further stabilize more generalized formations. For example, let us consider $K(5)$ formation in 2-dimensional space as depicted in Fig. 12.1. The initial formation is the $K(5)$ formation composed of vertices 1, 2, 3, 4, 5, and the desired formation is the $K(5)$ formation composed of vertices $1'$, 2, 3, 4, 5. All the initial and desired edge lengths are the same except $d_{12} \neq d_{1'2}$ and $d_{15} \neq d_{1'5}$. In this situation, it looks difficult to send the agent 1 from the initial position to the desired position, by using a gradient control law. From numerous numerical simulations, however, as shown in Fig. 12.2, we can observe that wherever the agent 1 is located initially under the $K(5)$ formation configuration of Fig. 12.1, the desired formation can be stabilized. Even though the agent 1 is away from the desired point $1'$ initially, we could observe that the formation is stabilized to the desired configuration easily. However, although the desired configuration has been achieved easily for the $K(5)$ formation from numerical simulations, it is still not clear how the formation could be stabilized to the desired configuration. It is noticeable that as shown in Fig. 4.5, when there is no edge between nodes 2 and 3, the desired configuration has not been achieved. But, when the edge $(2, 3)^e$ is added, the desired configuration has been readily achieved. Thus, it seems that a complete graph tends to be more stabilizable to a desired configuration than other (non-complete) rigid graphs. The reason is not quite clear from an intuition. This chapter attempts to provide an answer to this issue. Specifically, we try to design a gradient-based control law to answer this problem using virtual variables under the similar setup as Sect. 3.6. Although the gradient control law using virtual variables is not a typical gradient control law, it is still a form of gradient laws. Thus, with this result, we can infer that a complete graph may tend to be stabilized to a desired configuration easily under the gradient control law.

12.1.1 Degenerate Formation

Suppose that a desired formation configuration is given in d-dimensional space; but if an initial formation is in $d-1$ dimensional space, the desired formation cannot be achieved under the gradient control law. If all the agents are in $d-1$ dimensional space, the dimension of the subspace spanned by the control vector fields of gradient

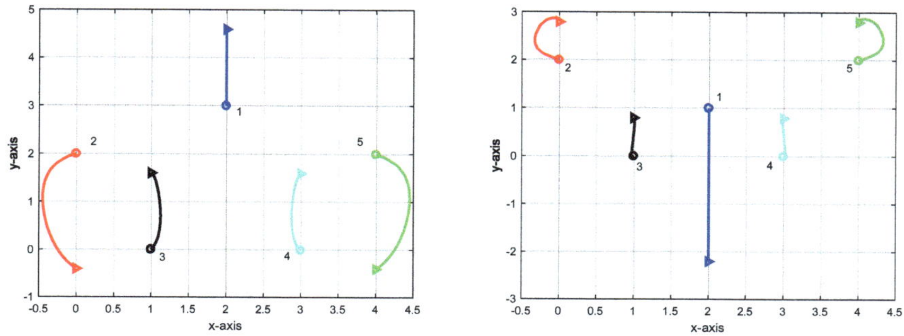

Fig. 12.2 Stabilization of $K(5)$ formation under gradient control law (the initial positions are marked by $\circ$ and the desired positions are marked by $\triangleright$)

control laws ((3.2) or (3.3)) will be also given in the same $d-1$ dimensional space. That is, the control efforts of the gradient control law will be constrained to the $d-1$ dimensional space because there is no control effort toward the dth dimensional axis. Thus, the group of agents will be staying in the $d-1$ dimensional space forever. We remark that the dimensional properties for formation control systems, and for general networked control systems, are discussed in [9]. The degenerate formation can be defined for this circumstance. The desired configuration is given in d-dimensional space; but the initial configuration or during the motions of agents, if the set of agents stays in a lower dimensional space, it will be staying there, and such circumstance is called *degenerate formation*. This section provides some basic properties of the degenerate space, which will be used in the design of a $K(n)$ formation in d-dimensional space.

Consider a set $\mathcal{S} = \{s_1, \ldots, s_k\}$ where $s_i \in \mathbb{R}^d$. The affine hull of $\mathcal{S}$ is defined by

$$\text{aff hull}\mathcal{S} \triangleq \{w \in \mathbb{R}^d | w = \sum_{i=1}^{k} a_i s_i, s_i \in \mathcal{S}, a_i \in \mathbb{R}, \sum_{i=1}^{k} a_i = 1\}$$

Given a formation configuration realized as $p = (p_1, \ldots, p_n) \in \mathbb{R}^{dn}$, the dimension of the formation is defined as the dimension of aff hull$\mathcal{S}$. The formation is said to be *degenerate* if the dimension of the formation is less than $\min\{d, n-1\}$. The following lemma was developed to characterize a property of infinitesimal rigidity using the concept of the affine hull [3]:

Lemma 12.1 *Given a framework* $(\mathcal{G}, p)$ *in* $\mathbb{R}^d$*, if it is infinitesimally rigid in* $\mathbb{R}^d$*, then the dimension of aff hull*$\mathcal{S}$ *is equal to* $\min\{d, n-1\}$.

It is important to note that in the above lemma, and in the definition of degenerate space, the dimension d is determined from the dimensions of the affine hull spanned by the points p_i. For example, given a d-dimensional space, where $d > 3$, if we consider three agents, then the set of agents organizes a 2-dimensional space in the

d-dimension space. In such a case, the triangular formation may be considered as a degenerate formation, which is actually not right. Thus, we need to interpret the dimension of triangular formation as 2, because the points of the triangular are only defined in 2-dimensional space.

It is an unsolved problem to design a distributed formation control law for general graphs under gradient control laws, although $K(3)$ formation in 2-D, and $K(4)$ formation in 3-D have been completely solved. Extending the approaches of $K(3)$ formation in 2-D, and $K(4)$ formation in 3-D to general d-dimensional space, it may be possible to solve $K(n)$ formation in $n-1$ dimensional space.

Let a realization of formation in $\mathbb{R}^{n-1}$ be denoted as $p = (p_1, \ldots, p_n)$, where $p_i \in \mathbb{R}^{n-1}$. The potential function for $K(n)$ graphs can be defined as

$$V(p) = \frac{1}{4} \sum_{1 \leq i \leq j \leq n} (\|p_i - p_j\|^2 - \bar{d}_{ij}^*)^2 \tag{12.1}$$

The gradient control law is obtained as

$$\dot{p} = -\left(\frac{\partial V}{\partial p}\right)^T = -\mathbb{R}_{\mathcal{G}}^T \bar{e} \tag{12.2}$$

Lemma 12.2 *Suppose that the target formation is infinitesimally rigid in $\mathbb{R}^{n-1}$. For the $K(n)$ graph driven by the gradient control law (12.2), any formation corresponding to an incorrect equilibrium point is degenerate.*

Proof At any incorrect equilibrium point $p^\dagger$ of (12.2), we have $\sum_{j \in \mathcal{V}\setminus\{i\}} (p_i^\dagger - p_j^\dagger)e_{ij} = 0$ although $V \neq 0$. Since it is a complete graph, $n-1$ vectors in $\{z_{ij}^\dagger = p_i^\dagger - p_j^\dagger : i \in \mathcal{V}, \forall j \in \mathcal{V} \setminus \{i\}\}$ are linearly dependent. Thus for each n, the maximum number of the linearly independent vectors is at most $n-2$. Therefore, the dimension of aff hull$\{p_1^\dagger, \ldots, p_n^\dagger\}$ is at most $n-2$, which implies that the formation corresponding to $p^\dagger$ is degenerate.

The stability of the formation at incorrect equilibrium point plays a key role in analyzing a convergence of the overall formation. In [8], there are discussions on the stability of formation in general d-dimensional space. The following lemma is the summary of the results of [8].

Lemma 12.3 *Let the target formation be infinitesimally rigid and the potential function is given as (12.1) with gradient control law (12.2). Then, the Hessian of V at any incorrect equilibrium point has at least one negative eigenvalue.*

Based on Lemma 12.1, we can see that the result of Lemma 12.2 gives rise to a contradiction to the assumption that the target formation is infinitesimally rigid. From this argument, we can now generate the following lemma.

Lemma 12.4 *Let the target formation be infinitesimally rigid. Then, the Hessian of $V(p) = \frac{1}{4}\sum_{1 \leq i \leq j \leq n}(\|p_i - p_j\|^2 - \bar{d}_{ij}^*)^2$ has at least one negative eigenvalue at any degenerate equilibrium point. Thus, any degenerate formations are unstable.*

Proof Based on Lemma 12.2, the formation corresponding to any incorrect equilibrium point is degenerate. Thus, by combining it with Lemma 12.3, the Hessian of $V(p)$ of any degenerate formation has at least one negative eigenvalue, which completes the proof.

From the above results, we can see that the target $K(n)$ formation in $\mathbb{R}^{n-1}$ would be achieved almost globally if the initial formation is not in degenerate spaces. In the following subsection, we will design a control law for $K(n)$ formation in a general d-dimensional space, using virtual variables. This problem was solved in [7] on the basis of the tetrahedron formation control that was studied in $\mathbb{R}^3$ [6].

12.1.2 $K(n)$ Formation with Virtual Variables

It is meaningful to consider n agents in a space with dimension less than or equal to $n-1$ dimensional space, and it is less meaningful to consider n agents in a space with dimension higher than or equal to n. If we consider n agents in $n+d$ dimension, where $d \geq 0$, then, there are extra dimensions as many as $1+d$, since the overall position of agents stays in $n-1$ dimensional space at most.

Let us first study $K(n)$ formation in general d-dimensional space, where $d \leq n-2$. Then a realization p is in $\mathbb{R}^{dn}$, i.e., $p = (p_1^T, \ldots, p_n^T)^T \in \mathbb{R}^{dn}$. Also, the desired formation is given in $\mathbb{R}^{dn}$ as $p^* = (p_1^{*T}, \ldots, p_n^{*T})^T \in \mathbb{R}^{dn}$. As studied in the previous chapters, there are complete solutions for $K(3)$ formation in 2-D and $K(4)$ formation in 3-D, and in general $K(n)$ formation in $n-1$ dimensional space. In this section, we would like to use these existing solutions to solve $K(n)$ formation in general d dimensional space using virtual variables. Let $\eta = n-1-d$ and $\omega = \frac{(n-d)(n-1-d)}{2}$, and define the set of scalar variables as:

$$\mathcal{W}_v \triangleq \{w_j^i, 1 \leq i \leq j \leq \eta\}$$

whose cardinality is ω. Then, using virtual variables $w_j^i \in \mathbb{R}$ defined from the set $\mathcal{W}_v$, we denote the locations of agents in the higher dimensional space as

$$\begin{aligned}
q_1 &= (p_1, w_1^1) \in \mathbb{R}^{d+1} \\
q_2 &= (p_2, w_2^1, w_2^2) \in \mathbb{R}^{d+2} \\
&\vdots \\
q_\eta &= (p_\eta, w_\eta^1, w_\eta^2, \ldots, w_\eta^\eta) \in \mathbb{R}^{d+\eta} \\
q_{\eta+1} &= p_{\eta+1} \in \mathbb{R}^d \\
&\vdots \\
q_n &= p_n \in \mathbb{R}^d
\end{aligned}$$

Note that the variables $q_{\eta+1}, \ldots, q_n$ are in $\mathbb{R}^d$, while other variables live in virtual higher dimensional space, i.e., $q_i \in \mathbb{R}^{d+i}$; thus, $q = (q_1^T, \ldots, q_n^T)^T \in \mathbb{R}^{dn+\omega}$. The variables $q_{\eta+1}, \ldots, q_n$ have desired locations in $\mathbb{R}^d$; but the desired locations of other variables are virtually defined in $\mathbb{R}^{d+i}$. Let us define desired augmented locations in $\mathbb{R}^{n-1}$ as $\breve{q}_i^* = ((p_i^*)^T, \alpha(\mathbf{1}_i)^T, (\mathbf{0}_{\eta-i})^T)^T$ for all $i = 1, \ldots, \eta$ where $\alpha > 0$, and $\breve{q}_i^* = ((p_i^*)^T, (\mathbf{0}_\eta)^T)^T$ for all $i = \eta+1, \ldots, n$. Then, following the traditional gradient control law, denoting $d_{ij}^a = \|\breve{q}_i^* - \breve{q}_j^*\|$, we can write $d_{ij}^a = \sqrt{\|p_i^* - p_j^*\|^2 + |i-j|\alpha^2}$ in an augmented space. To make the dimensions of variables $q_1, \ldots, q_n$ be agreed, we also write $\breve{q}_i = (q_i^T, (\mathbf{0}_{\eta-i})^T)^T$ for all $i = 1, \ldots, \eta$ and $\breve{q}_i = (q_i^T, (\mathbf{0}_\eta)^T)^T$ for all $i = \eta+1, \ldots, n$. These variables are called augmented virtual vectors.

Example 12.1 Let us consider five agent p_1, p_2, p_3, p_4, p_5 in 2-dimensional space under the complete graph, i.e., $p_i \in \mathbb{R}^2, \forall i = 1, \ldots, 5$. Then, we have $q_1 = (p_1^T, w_1^1)^T \in \mathbb{R}^3, q_2 = (p_1^T, w_2^1, w_2^2)^T \in \mathbb{R}^4, q_3 = p_3 \in \mathbb{R}^2, q_4 = p_4 \in \mathbb{R}^2, q_5 = p_5 \in \mathbb{R}^2$. Thus, we can write $\breve{q}_1 = (q_1^T, 0)^T \in \mathbb{R}^4, \breve{q}_2 = q_2 \in \mathbb{R}^4, \breve{q}_3 = (q_3^T, 0, 0)^T \in \mathbb{R}^4, \breve{q}_4 = (q_4^T, 0, 0)^T \in \mathbb{R}^4, \breve{q}_5 = (q_5^T, 0, 0)^T \in \mathbb{R}^4$. Consequently, the realization of the augmented vectors $\breve{q} = (\breve{q}_1^T, \ldots, \breve{q}_5^T)^T$ stays in 4-dimensional space under a complete graph. Thus, it is $K(5)$ graph in 4-D.

Denoting $\bar{d}_{ij}^a = {d_{ij}^a}^2$, the potential function is now selected as $V^a = \frac{1}{4}\sum_{1\le i\le j\le n} (\|\breve{q}_i - \breve{q}_j\|^2 - \bar{d}_{ij}^a)^2$. Then, since the components corresponding to zero are not updated, we can have the following control law:

$$\dot{q} = -\left[\frac{\partial V^a}{\partial q}\right]^T \tag{12.3}$$

in which the virtual variables w_i^k, $i = 1, \ldots, \eta$ and $k = 1, \ldots, \eta$, are arbitrarily updated as

$$\dot{w}_i^k = s_i^k \tag{12.4}$$

where s_i^k are constants. Note that the virtual variables w_i^k are used to transform the formation in d-dimensional space into $(n-1)$ dimensions. For a complete solution in $\mathbb{R}^{n(n-1)}$, we can also map p_i from $\mathbb{R}^d$ to $\mathbb{R}^{n-1}$ so that we could get $K(n)$ formation realization in $\mathbb{R}^{n-1}$ as the augmented virtual vector $\breve{p} = (\breve{p}_1^T, \ldots, \breve{p}_n^T)^T \in \mathbb{R}^{n(n-1)}$.

For the following results, with an abuse of notation, we use the terminology *congruence* for two realizations under different dimensions. Two realizations $\breve{p}$ and q in different dimensions are said to be congruent if $\|\breve{p}_i - \breve{p}_j\| = \|q_i - q_j\|$ for any pair of nodes $i, j \in \mathcal{V}$. Now, for the main result, we need the following lemma.

Lemma 12.5 *Consider two realizations $\breve{p} \in \mathbb{R}^{n(n-1)}$ and $q \in \mathbb{R}^{dn+\omega}$ which are congruent. Assume that $\breve{p}$ and q are critical points of the potential functions $\breve{V}(\breve{p})$ and $V^a(q)$ of the form (12.1), and $\breve{V}$ and V^a are generated by the same desired inter-agent distances. Then $\breve{p}$ is not a local minimizer of $\breve{V}$ if and only if q is not a local minimizer of V^a.*

Proof Suppose that $\breve{p}$ is not a local minimizer of $\breve{V}$. Then, for any $\delta > 0$, there exists $\breve{p}' \in \{x \in \mathbb{R}^{n(n-1)} | \|\breve{p} - x\| < \delta\}$ such that $\breve{V}(\breve{p}') < \breve{V}(\breve{p})$. Considering arbitrarily small $\bar{\delta} > 0$, then there also always exists $q' \in \{x \in \mathbb{R}^{dn+\omega} | \|q - x\| < \bar{\delta}\}$ such that $V^a(q') < V^a(q)$ because we can choose q' so that q' and $\breve{p}'$ are congruent and $\breve{V}(\breve{p}') < \breve{V}(\breve{p})$ with arbitrarily small $\delta > 0$. Consequently, q is not a local minimizer of V^a if $\breve{p}$ is not a local minimizer of $\breve{V}$. The converse can be proved similarly. Let us suppose that q is not a local minimizer of V^a. Then, for any $\bar{\delta} > 0$, there exists $q' \in \{x \in \mathbb{R}^{dn+\omega} | \|q - x\| < \bar{\delta}\}$ such that $V^a(q') < V^a(q)$. For arbitrarily small $\delta > 0$, we can always find $\breve{p}' \in \{x \in \mathbb{R}^{n(n-1)} | \|\breve{p} - x\| < \delta\}$ such that $\breve{V}(\breve{p}') < \breve{V}(\breve{p})$ from the fact that we can take $\breve{p}'$ so that $\breve{p}'$ and q' are congruent and $V^a(q') < V^a(q)$. Thus, $\breve{p}$ is not a local minimizer of $\breve{V}$ if and only if q is not a local minimizer of V^a.

The following lemmas are used in the proof of Theorem 12.1.

Lemma 12.6 (Theorem 2.2 in [2]) *Let $\phi : \mathbb{R}^n \to \mathbb{R}$ be a real analytic function and let $x(t)$ be a continuously differentiable curve in $\mathbb{R}^n$. Assume that there exist a $\delta > 0$ and a real τ such that for $t > \tau$, $x(t)$ satisfies the angle condition $\dot{\phi} = \frac{\partial \phi}{\partial x}\dot{x} \leq -\delta \|\frac{\partial \phi}{\partial x}\| \|\dot{x}\|$, and a weak decrease condition $\dot{\phi} = 0 \Rightarrow \dot{x} = 0$. Then, either $\lim_{t\to+\infty} \|x(t)\| = \infty$ or there exists $x^* \in \mathbb{R}^n$ such that $\lim_{t\to+\infty} x(t) = x^*$.*

Lemma 12.7 *The variables q are bounded when updated by (12.3).*

Proof Let $\bar{e}_{ij} = \|\breve{q}_i - \breve{q}_j\|^2 - (d^a_{ij})^2$. Since $\frac{\partial \bar{e}_{ij}}{\partial q_i} = -\frac{\partial \bar{e}_{ij}}{\partial q_j}$ for each edge, we know that $\sum_{i=1}^n \dot{q}_i = -\sum_{i=1}^n \left[\frac{\partial V^a}{\partial q_i}\right]^T = 0$. Therefore, the centroid of $\breve{q}_1, \ldots, \breve{q}_n$ is stationary under (12.3). Moreover, from the real analytic function V^a, we have

$$\dot{V}^a = \frac{\partial V^a}{\partial q}\dot{q} = -\left\|\frac{\partial V^a}{\partial q}\right\|^2 \leq 0. \tag{12.5}$$

Thus, $\|\breve{q}_i - \breve{q}_j\|$ cannot diverge for any $i, j \in \mathcal{V}$. Consequently, $\breve{q}$ is bounded.

Theorem 12.1 *The solution of (12.3) converges to a point.*

Proof Since, from (12.5), $\dot{V}^a = 0$ implies that $\frac{\partial V^a}{\partial q} = 0$, we have $\dot{q} = 0$. Thus, by Lemma 12.6, we conclude that either $\lim_{t\to+\infty} \|q(t)\| = \infty$ or there exists q^∞ such that $\lim_{t\to+\infty} q(t) = q^\infty$. Also, from Lemma 12.7, we know that q is bounded. Thus, by Lemma 12.6, q converges to a limit point.

By Barbalat's lemma (see Lemma 2.8) and by Theorem 2.23, now we can confirm that $\dot{V}^a \to 0$. Thus, $q(t)$ converges to an equilibrium point of (12.3). To show a convergence to a desired equilibrium point, we use the desired augmented configuration $\breve{q}^* = ((\breve{q}_1^*)^T, \ldots, (\breve{q}_n^*)^T)^T \in \mathbb{R}^{n(n-1)}$ in $\mathbb{R}^{n-1}$. Now, we make the main result of this section:

Theorem 12.2 *Suppose that the target $K(n)$ formation represented by $\breve{q}^*$ is nondegenerate and infinitesimally rigid in $\mathbb{R}^{n-1}$. Then, any incorrect equilibrium point of (12.3) is unstable.*

Proof Let $q^\dagger$ be an incorrect equilibrium point of (12.3) and we suppose that $q^\dagger$ is stable. Since V^a is real analytic and stable, $q^\dagger$ is considered as a local minimizer of V^a [1]. For the variables $\breve{p}$ in $\mathbb{R}^{n-1}$, let us consider a target realization $\breve{p}^*$ such that $\breve{p}^*$ and $q^\dagger$ are congruent. Then, by Lemma 12.5, $\breve{p}^*$ is a local minimizer of $\breve{V}$. But, from Lemma 12.4, since $\breve{p}^*$ is an incorrect equilibrium point, it must be unstable. Consequently, it cannot be a local minimizer of $\breve{V}$, which leads to a contradiction. Thus, we can conclude that any incorrect equilibrium point of (12.3) is unstable.

From the above theorem, we can now conclude that $q(t)$ converges to the desired equilibrium set for almost all initial conditions. Consequently, since $q(t)$ converges to the desired virtual configuration, we can see that p also converges to the desired configuration. To implement the control law (12.3), denoting $w_i = (w_i^1, \ldots, w_i^i)^T$ and $s_i = (s_i^1, \ldots, s_i^i)^T$, let us rewrite it as

$$\dot{q} = \begin{bmatrix} \vdots \\ \dot{q}_i \\ \vdots \end{bmatrix} = \begin{bmatrix} \vdots \\ \begin{bmatrix} \dot{p}_i \\ \dot{w}_i \end{bmatrix} \\ \vdots \end{bmatrix} = \begin{bmatrix} \vdots \\ \begin{bmatrix} -\left[\frac{\partial V^a}{\partial p_i}\right]^T \\ s_i \end{bmatrix} \\ \vdots \end{bmatrix} \tag{12.6}$$

Since $\bar{e}_{ij} = \|\breve{q}_i - \breve{q}_j\|^2 - \bar{d}_{ij}^a = \|p_i - p_j\|^2 + \|w_i - w_j\|^2 - (\|p_i^* - p_j^*\|^2 + |i-j|\alpha^2)$, we write $-\left[\frac{\partial V^a}{\partial p_i}\right]^T$ as

$$-\left[\frac{\partial V^a}{\partial p_i}\right]^T = -\sum_{j=1}^{n}(p_i - p_j)\bar{e}_{ij} \tag{12.7}$$

Thus, at the agent i, the control input is implemented in the local coordinate frame as

$$\dot{p}_i = \sum_{j=1}^{n} p_j^i \bar{e}_{ij} \tag{12.8}$$

In the above control law, the terms p_j^i are measured by distributed sensings and the term $\bar{e}_{ij}$ is computed after exchanging w_i and w_j with neighboring agents via distributed communications. Also note that the term w_i is updated by a distributed computation at each node. Consequently, we can see that a formation modeled by a complete graph can be stabilized to the desired configuration by way of the distributed sensings, distributed communications, and distributed computation. But, when compared to the traditional gradient control laws, since it requires extra distributed communications and distributed computation, it is disadvantageous. Also to avoid a divergence of w_i, it may be needed to select the initial values of $s_i(t_0)$ as $s_i(t_0) < 0$. However, it is still important to notice that the control law (12.8) is in the form of the gradient control law (3.2); only the gains multiplied to the error terms are different. Hence, the simulation result, under the traditional gradient control law, given in Fig. 12.2, can be understood in the sense that gradient control laws may be able to stabilize a complete graph to the desired configuration if the scalar error terms (i.e., $\bar{e}_{ij}$ in (12.8) and $(\|z_{ij}\|^2 - \bar{d}_{ij}^*)$ in (3.3)) are properly modified.

12.2 Formation Resizing and Scaling

One of the topics studied in formation control is the formation scaling. In the formation scaling, we would like to vary the size of formation without changing the formation shape. Let us consider a framework $(\mathcal{G}, p)$ where the realization is in a desired configuration at $t = t_0$. Then, all the desired distances are satisfied, which means that agents are all stationary without putting any control efforts. That is, since agents are all in equilibrium, they do not generate control signals anymore. Let the underlying topology of the graph be rigid or persistent. Then, when all the inter-agent distances need to be changed by a same ratio, we can change the scale of the overall graph by changing one of the inter-agent distances. On the other hand, we may want to change the inter-agent distances by some fixed ratios with respect to the distance of a reference edge. In this case, the distances of edges are not scaled-up or scaled-down in the same ratio. This problem is called resizing. This section is dedicated to these problems. In Sect. 12.2.1, we consider the resizing problem in directed graphs under the distance-based setups and in Sect. 12.2.2, we consider the scaling problem in bearing-based setups. These results are extracted from [5, 11], respectively.

Even though two frameworks are bearing congruent, their edge distances may be different. Thus, while keeping the bearing congruence, we can scale-up or scale-down the formation by changing some distances of certain edges. The resizing and scaling of a formation are achieved by changing the distances of edges. The scaling problem can be defined as the changing of the scale of formation as $d_{ij} \to sd_{ij}$ where $s > 0$ is a scaling constant and $(i, j)^e \in \mathcal{E}$. The agents are in a desired configuration initially, and the desired distances are changed as $\|z_{ij}\| = d_{ij}^* \to sd_{ij}^*$. Here, the ratio s is same to all the edges. However, the resizing problem is to achieve $\|z_{ij}\| = d_{ij}^* \to s_{ij}d_{ij}^*$ where the ratios s_{ij} may be different for edges. Thus, in the scaling problem, the configuration of formation does not vary, while in the resizing problem, the shape of the formation may vary. It is thus remarkable that the scaling problem is a special case of the resizing problem. Figure 12.3 explains the concepts of resizing and scaling in directed graphs.

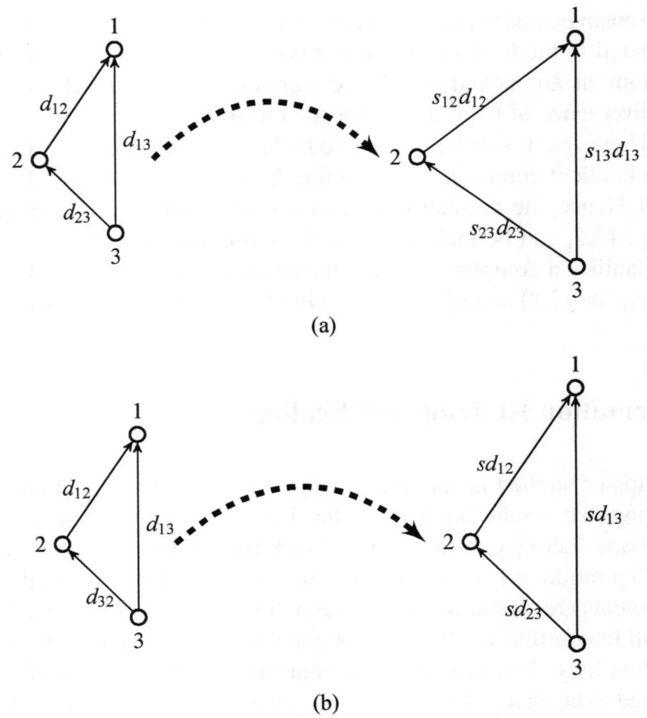

Fig. 12.3 **a** Resizing versus **b** scaling of formation: In **a**, $s_{12} \neq s_{13} \neq s_{23}$, but in **b**, the edges are scaled in the same ratio s

12.2.1 Distance-Based Formation Resizing

As commented, in the scaling problem, we would like to achieve $d_{ij} \to sd_{ij}$ with a common ratio for all edges. Thus, under the distance-based setups, the results of Chaps. 3, 4, and 5 are applicable exactly. That is, when the new desired distances are given, which are all scaled-up, or scaled-down by a same ratio from the former distances, the realizability of formation does not change. Thus, the results of global stabilization of Chap. 3, local stabilization of Chap. 4, and stabilization of persistent graphs of Chap. 5 are still valid. However, the resizing problem is not directly obtained from the former results since the realizability of a formation after the resizing is not clear. This subsection introduces a resizing problem of acyclic minimally persistent formation investigated in [5].

Let us first study an acyclic triangular formation as a base of general acyclic minimally persistent formations. The goal of control task is to achieve $\lim_{t\to\infty} \|z_{12}\| = d_{12}^*$, $\lim_{t\to\infty} \|z_{13}\| = d_{13}^*$, and $\lim_{t\to\infty} \|z_{23}\| = d_{23}^*$. Let us define the ratios between distances as $s_{13} = d_{13}^*/d_{12}^*$ and $s_{23} = d_{23}^*/d_{12}^*$. If the formation is determined by the ratios s_{13} and s_{23}, since they are related by d_{12}^*, it should be formulated carefully taking into account of the realizability of the formation. For this, we use the following assumption:

Assumption 12.2.1 To ensure a realization of the formations defined by s_{13} and s_{23} in 2-dimensional space, the following constraints on s_{13} and s_{23} are supposed to hold:

$$1 < s_{13} + s_{23},\ s_{13} < 1 + s_{23},\ s_{23} < 1 + s_{13} \tag{12.9}$$

It is remarkable that the condition (12.9) comes from the triangular inequalities. When inserting $s_{13} = d_{13}^*/d_{12}^*$ and $s_{23} = d_{23}^*/d_{12}^*$ into (12.9), we have $d_{12}^* < d_{13}^* + d_{23}^*$, $d_{13}^* < d_{12}^* + d_{23}^*$, and $d_{23}^* < d_{13}^* + d_{12}^*$. The squared distance errors are calculated as functions of ratios as

$$\bar{e}_{12} = \|z_{12}\|^2 - d_{12}^{*\,2},\ \bar{e}_{13} = \|z_{13}\|^2 - (s_{13}d_{12}^*)^2,\ \bar{e}_{23} = \|z_{23}\|^2 - (s_{23}d_{12}^*)^2 \tag{12.10}$$

For the followers, the potential function is given as $\phi_i(p) = \frac{1}{4}\sum_{j\in\mathcal{N}_i} \bar{e}_{ji}^2$. By denoting $\hat{e}_{13} = \|z_{13}\|^2 - (s_{13}d_{12})^2$ and $\hat{e}_{23} = \|z_{23}\|^2 - (s_{23}d_{12})^2$, from the following gradient control law,

$$u_i = -\left[\frac{\partial \phi_i}{\partial p_i}\right]^T \tag{12.11}$$

the control inputs for individual agents are given as

$$u_1 = 0 \tag{12.12}$$

$$u_2 = (\|z_{12}\|^2 - \bar{d}_{12}^*)z_{12} = \bar{e}_{12}z_{12} \tag{12.13}$$

$$\begin{aligned} u_3 &= (\|z_{13}\|^2 - (s_{13}d_{12})^2)z_{13} + (\|z_{23}\|^2 - (s_{23}d_{12})^2)z_{23} \\ &= \hat{e}_{13}z_{13} + \hat{e}_{23}z_{23} \end{aligned} \tag{12.14}$$

With the above control inputs, the dynamics in displacements can be expressed as

$$\dot{z}_{12} = -\bar{e}_{12}z_{12} \tag{12.15}$$

$$\dot{z}_{13} = -(\hat{e}_{13}z_{13} + \hat{e}_{23}z_{23}) \tag{12.16}$$

$$\dot{z}_{23} = \bar{e}_{12}z_{12} - (\hat{e}_{13}z_{13} + \hat{e}_{23}z_{23}) \tag{12.17}$$

which can be expressed in a concise form as

$$\dot{z} = f(z),\ z(t_0) = (\mathbb{H}_+ \otimes \mathbb{I}_2)p(t_0) \tag{12.18}$$

where $\mathbb{H}_+$ is the incidence matrix corresponding to the directed edge $\vec{\mathcal{E}}$. The equilibrium sets can be obtained by letting $f(z) = 0$. Notice that if $z_{12} = 0$, then the agents 1 and 2 are collocated, which means that the vectors z_{13} and z_{23} are in the same direction. But, since $\hat{e}_{13} = \hat{e}_{23} > 0$ when $z_{12} = 0$, the only solution for

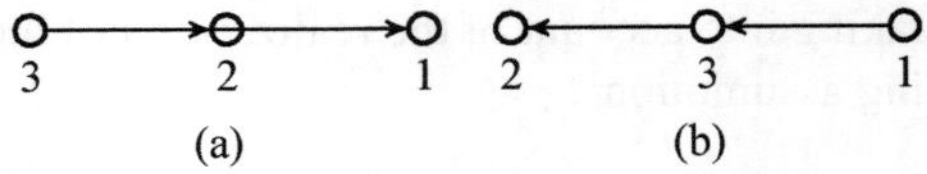

Fig. 12.4 **a** Collinear case being in $\mathcal{C}$. **b** Collinear case being not in $\mathcal{C}$

$\hat{e}_{13}z_{13} + \hat{e}_{23}z_{23} = 0$ is $z_{13} = z_{23} = 0$. Denoting the column space of $(\mathbb{H}_+ \otimes \mathbb{I}_2)$ by $C(\mathbb{H}_+ \otimes \mathbb{I}_2)$, the equilibrium sets can be decomposed as

$$\mathcal{U}_1 \triangleq \{z \in C(\mathbb{H}_+ \otimes \mathbb{I}_2) : z_{12} = z_{13} = z_{23} = 0\}$$
$$\mathcal{U}_2 \triangleq \{z \in C(\mathbb{H}_+ \otimes \mathbb{I}_2) : \bar{e}_{12} = 0, \hat{e}_{13}z_{13} + \hat{e}_{23}z_{23} = 0, \hat{e}_{13} \neq 0 \text{ or } \hat{e}_{23} \neq 0\}$$
$$\mathcal{D}_z \triangleq \{z \in C(\mathbb{H}_+ \otimes \mathbb{I}_2) : \bar{e}_{12} = \hat{e}_{13} = \hat{e}_{23} = 0\}$$

The sets $\mathcal{U}_1$ and $\mathcal{U}_2$ are undesired equilibrium sets and the set $\mathcal{D}_z$ is the desired equilibrium set. Since $\hat{e}_{13} = \bar{e}_{13}$ and $\hat{e}_{23} = \bar{e}_{23}$ when $\bar{e}_{12} = 0$, the desired equilibrium set $\mathcal{D}_z$ is equivalent to $\mathcal{D}_z = \{z \in C(\mathbb{H}_+ \otimes \mathbb{I}_2) : \bar{e}_{12} = \bar{e}_{13} = \bar{e}_{23} = 0\}$. For the stability analysis, the following set is also defined:

$$\mathcal{C} \triangleq \{z \in C(\mathbb{H}_+ \otimes \mathbb{I}_2) : z_{13} \text{ and } z_{23} \text{ are linearly dependent}\} \tag{12.19}$$

Example 12.2 In the set $\mathcal{C}$, the meaning of the linear dependence z_{23} and z_{13} is that the three points are on the same line, i.e., collinear. The incidence matrix $\mathbb{H}_+$ is given as

$$\mathbb{H}_+ = \begin{bmatrix} +1 & -1 & 0 \\ +1 & 0 & -1 \\ 0 & +1 & -1 \end{bmatrix}$$

However, since $C(\mathbb{H}_+ \otimes \mathbb{I}_2)$ is the range of the incidence matrix, Fig. 12.4b cannot occur. Only the case of Fig. 12.4b is considered as an element of the set $\mathcal{C}$.

By taking a derivative of $\bar{e}_{12}$, it follows

$$\begin{aligned}\dot{\bar{e}}_{12} &= 2(p_1 - p_2)^T(\dot{p}_1 - \bar{e}_{12}z_{12}) \\ &= -2\bar{e}_{12}(\bar{e}_{12} + (d_{12}^*)^2)\end{aligned} \tag{12.20}$$

whose solution is uniquely given by

$$\bar{e}_{12}(t) = \frac{\gamma(d_{12}^*)^2 e^{-2(d_{12}^*)^2 t}}{1 - \gamma e^{-2(d_{12}^*)^2 t}} \tag{12.21}$$

where $\gamma = \bar{e}_{12}(t_0)/(\bar{e}_{12}(t_0) + (d_{12}^*)^2)$. Thus, it is obvious that the agent 2 approaches agent 1 exponentially fast. For the analysis of the convergence of $\|z_{13}\|$ and $\|z_{23}\|$, let

$\tilde{e}_{13} = \bar{e}_{13} - \hat{e}_{13} = s_{13}^2 \bar{e}_{12}$, $\tilde{e}_{23} = \bar{e}_{23} - \hat{e}_{23} = s_{23}^2 \bar{e}_{12}$, and $V_3(z(t)) = 1/4(\tilde{e}_{13}^2 + \tilde{e}_{23}^2)$. By taking the derivative of V_3, it follows

$$
\begin{aligned}
\dot{V}_3 &= \frac{1}{2}(\tilde{e}_{13}\dot{\tilde{e}}_{13} + \tilde{e}_{23}\dot{\tilde{e}}_{23}) \\
&= -\|\bar{e}_{13} z_{13} + \bar{e}_{23} z_{23}\|^2 + \bar{e}_{13}\tilde{e}_{13}\|z_{13}\|^2 + \bar{e}_{13}\tilde{e}_{23} z_{13}^T z_{23} + \tilde{e}_{13}\bar{e}_{23} z_{13}^T z_{23} \\
&\quad + \bar{e}_{23}\tilde{e}_{23}\|z_{23}\|^2 + \bar{e}_{23}\bar{e}_{12} z_{23}^T z_{12}
\end{aligned} \tag{12.22}
$$

For a simplicity of analysis, denote $g(z(t)) = -\|\bar{e}_{13} z_{13} + \bar{e}_{23} z_{23}\|^2$ and $h(z(t)) = \bar{e}_{13}\tilde{e}_{13}\|z_{13}\|^2 + \bar{e}_{13}\tilde{e}_{23} z_{13}^T z_{23} + \tilde{e}_{13}\bar{e}_{23} z_{13}^T z_{23} + \bar{e}_{23}\tilde{e}_{23}\|z_{23}\|^2 + \bar{e}_{23}\bar{e}_{12} z_{23}^T z_{12}$.

Lemma 12.8 *The function $V_3(t)$ is bounded along the solution trajectory (12.18). Thus, the signals $\bar{e}_{13}$, $\bar{e}_{23}$, z_{13} and z_{23} are bounded also.*

Proof Since $\bar{e}_{12}$ and z_{12} are bounded, it is clear that $\tilde{e}_{13}$ and $\tilde{e}_{23}$ are bounded. Also, by comparing $g(z(t))$ and $h(z(t))$, it is shown that $|g(z(t))| \gg |h(z(t))|$ as $t \to \infty$ and agent 3 gets away from the other agents. Thus, as agent 3 is away from agents 2 and 1, $\dot{V}_3$ becomes negative. Thus, it is clear that $V_3(t)$ is also bounded, which implies $\bar{e}_{13}$, $\bar{e}_{23}$, z_{13} and z_{23} are bounded.

It is remarkable that since all the signals in $h(z(t))$ are bounded, the signals $\bar{e}_{12}$, $\tilde{e}_{13}$ and $\tilde{e}_{23}$ are decaying to zero exponentially. Thus, $h(z(t))$ is an exponentially decaying function. Using $g(z(t))$ and $h(z(t))$, the integral of $\dot{V}_3$ can be written as

$$
V_3(z(t)) - V_3(z(t_0)) = \int_{t_0}^{t} h(z(\tau))d\tau + \int_{t_0}^{t} g(z(\tau))d\tau \tag{12.23}
$$

In the above equation, the term $\int_{t_0}^{t} h(z(\tau))d\tau$ converges to a constant because $h(z(t))$ is exponentially decaying and the last term satisfies $\int_{t_0}^{t} g(z(\tau))d\tau \leq 0$. Thus, as $t \to \infty$, it can be considered that $V_3(z(t_0)) + \int_{t_0}^{t} h(z(\tau))d\tau$ is a constant. But, since $\int_{t_0}^{t} g(z(\tau))d\tau$ cannot be positive and $V_3(z(t))$ is bounded below, $\int_{t_0}^{t} g(z(\tau))d\tau$ should converge to a negative constant.

Theorem 12.3 *The trajectory of (12.18) converges to $\mathcal{D}_z$ or $\mathcal{U}_2$ if $z_{12}(t_0) \neq 0$. If the solution converges to $\mathcal{D}_z$, the convergence is exponentially fast.*

Proof Since $\int_{t_0}^{t} g(z(\tau))d\tau$ converges to a constant, it is shown that $g(z(t)) \to 0$ as $t \to \infty$ by Barbalat's lemma. Also since $\bar{e}_{12}$ converges to 0 if $z_{12}(t_0) \neq 0$, the trajectory converges to either $\mathcal{D}_z$ or $\mathcal{U}_2$. Let us suppose that the trajectory converges to $\mathcal{D}_z$. Define a subset of $C(\mathbb{H}_+ \otimes \mathbb{I}_2)$ as

$$
\Omega(r) \triangleq \{z \in C(\mathbb{H}_+ \otimes \mathbb{I}_2) : \bar{e}_{12}^2 + \bar{e}_{13}^2 + \bar{e}_{23}^2 \leq r^2\} \tag{12.24}
$$

Since z_{13} and z_{23} are linearly independent for all $z \in \mathcal{D}_z$ and $\mathcal{D}_z \in \Omega(r)$ for a positive constant $r > 0$, there exist t_∞ and r_ϵ such that z_{13} and z_{23} are linearly independent for all $z \in \Omega(r_\epsilon)$ and $z(t) \in \Omega(r_\epsilon)$ for all $t \geq t_\infty$. From the Cauchy–Schwarz inequality, it holds that $z_{13}^T z_{23} < \|z_{13}\|\|z_{23}\|$ for all $z \in \Omega(r_\epsilon)$ because $z_{13} \neq 0$ and $z_{23} \neq 0$ for all $z \in \Omega(r_\epsilon)$. Next, consider $\rho : [t_\infty, \infty) \rightarrow [0, 1)$ for all $t \geq t_\infty$ such that

$$\begin{aligned} -g &= \bar{e}_{13}^2\|z_{13}\|^2 + 2\bar{e}_{13}\bar{e}_{23}z_{13}^T z_{23} + \bar{e}_{23}^2\|z_{23}\|^2 \\ &\geq \bar{e}_{13}^2\|z_{13}\|^2 - 2\rho^2(t)|\bar{e}_{13}||\bar{e}_{23}|\|z_{13}\|\|z_{23}\| + \bar{e}_{23}^2\|z_{23}\|^2 \\ &= [1-\rho^2(t)]\bar{e}_{13}^2\|z_{13}\|^2 + [1-\rho^2(t)]\bar{e}_{23}^2\|z_{23}\|^2 \\ &\quad + \underbrace{\rho^2(t)\bar{e}_{13}^2\|z_{13}\|^2 - 2\rho^2(t)|\bar{e}_{13}||\bar{e}_{23}|\|z_{13}\|\|z_{23}\| + \rho^2(t)\bar{e}_{23}^2\|z_{23}\|^2}_{\geq 0} \\ &\geq \min\left\{\inf_{z\in\Omega(r_\epsilon)}\|z_{13}\|^2, \inf_{z\in\Omega(r_\epsilon)}\|z_{23}\|^2\right\}\left[\inf_{t\in[t_\infty,\infty)}(1-\rho^2(t))\right](\bar{e}_{13}^2+\bar{e}_{23}^2) \\ &= k_g V_3 \end{aligned} \tag{12.25}$$

Since $\rho^2(t) < 1$, it follows $k_g > 0$. Thus, (12.22) can be changed as

$$\dot{V}_3 \leq h - k_g V_3 \tag{12.26}$$

Finally, since h is bounded by an exponentially decaying function, from (12.26), it is concluded that $V_3(t)$ is also an exponentially decaying function from the comparison lemma.

Lemma 12.9 *The trajectory of (12.18) does not converge to $\mathcal{U}_2$ if $z(t_0) \notin \mathcal{C}$.*

Proof Suppose that z converges to $\mathcal{U}_2$ with $z(t_0) \notin \mathcal{C}$. From the definition of $\mathcal{U}_2$, it follows that $\mathcal{U}_2 \in \mathcal{C}$. Because z_{13} and z_{23} are linearly dependent for all $z \in \mathcal{C}$, it holds $\det[z_{13}, z_{23}] = 0$ for all $z \in \mathcal{C}$. Let $\Delta(z(t)) \triangleq \det[z_{13}, z_{23}]$. Taking the derivative of Δ, it follows

$$\begin{aligned} \dot{\Delta} &= \det[\dot{z}_{13}, z_{23}] + \det[z_{13}, \dot{z}_{23}] \\ &= \det[-(\hat{e}_{13}z_{13} + \hat{e}_{23}z_{23}), z_{23}] + \det[z_{13}, \bar{e}_{12}z_{12} - (\hat{e}_{13}z_{13} + \hat{e}_{23}z_{23})] \\ &= -(\hat{e}_{13} + \hat{e}_{23} + \bar{e}_{12})\Delta \end{aligned} \tag{12.27}$$

From the above equation, it can be obtained as

$$\Delta(z(t)) = e^{-\int_\tau^t \sigma(z(s))ds}\Delta(z(\tau)), \ t \geq \tau \geq t_0 \tag{12.28}$$

where $\sigma(z(t)) = \hat{e}_{13} + \hat{e}_{23} + \bar{e}_{12}$. It is shown in [4] that $\sigma < 0$ for all $z \in \mathcal{U}_2$. Since $\mathcal{U}_2$ is a closed set, it is claimed that there always exists an open set Σ such that

$$\Sigma = \{z \in C(\mathbb{H}_+ \otimes \mathbb{I}_2) : \sigma < 0\}, \ \mathcal{U}_2 \subset \Sigma \text{ and } \Sigma \cap \mathcal{D} = \emptyset \tag{12.29}$$

By assumption, since z converges to $\mathcal{U}_2$, there exists a finite-time t_f such that $z(t_f) \notin \mathcal{U}_2$, and $z(t) \in \Sigma$ for all $t \in [t_f, \infty)$. In such a case, Δ could be expressed as

$$\Delta(z(t)) = e^{-\int_{t_0}^{t_f} \sigma(z(s))ds} e^{-\int_{t_f}^{t} \sigma(z(s))ds} \Delta(z(t_0)) \tag{12.30}$$

From the fact that $\sigma < 0$ for all $t \in [t_f, \infty)$, it follows that $e^{-\int_{t_f}^{t} \sigma(z(s))ds} \geq 1$. Also since t_f is finite, $e^{-\int_{t_0}^{t_f} \sigma(z(s))ds}$ is a finite constant. Consequently, Δ does not converge to 0 if $z(t_0) \notin \mathcal{U}_2$, which contradicts to the assumption that z converges to $\mathcal{U}_2$.

With the results thus far, the following results are immediate.

Corollary 12.1 *If the initial triangular formation is not collinear, then z globally converges to $\mathcal{D}$, and the convergence rate is locally exponentially fast.*

Proof It is certain that the trajectory of (12.18) converges to one of the equilibrium sets $\mathcal{U}_1, \mathcal{U}_2$ and $\mathcal{D}$. If the initial triangular formation is not collinear, then it does not converge to $\mathcal{U}$ from Lemma 12.9, which implies the trajectory converges to $\mathcal{D}$. The exponential convergence in a neighborhood of $p \in \mathcal{D}$ was proved in Theorem 12.3.

Corollary 12.2 *The set $\mathcal{C}$ forms a positively invariant set with respect to (12.18).*

Proof From (12.30), if $\Delta(t_0) = 0$, then $\Delta(t) = 0$ for all $t \geq t_0$.

The above corollary is too intuitive. The following example shows the intuition.

Example 12.3 Let the positions of three agents be given as $p_1(t) = (0, 0)^T$, $p_2(t) = (-1, 0)^T$, and $p_3(t) = (-2, 0)^T$ at a specific time t. Then, for example, the instantaneous updates are given as

$$\begin{aligned}
\dot{z}_{12} &= -\bar{e}_{12} \begin{pmatrix} 1 \\ 0 \end{pmatrix} \\
\dot{z}_{13} &= -\hat{e}_{13} \begin{pmatrix} 2 \\ 0 \end{pmatrix} - \hat{e}_{23} \begin{pmatrix} 1 \\ 0 \end{pmatrix} \\
\dot{z}_{23} &= \bar{e}_{12} \begin{pmatrix} 1 \\ 0 \end{pmatrix} - \hat{e}_{13} \begin{pmatrix} 2 \\ 0 \end{pmatrix} - \hat{e}_{23} \begin{pmatrix} 1 \\ 0 \end{pmatrix}
\end{aligned}$$

Thus, the agents are forced to move along the x-axis only, which implies that the agents will be staying on the same line forever. Further, the error values $\bar{e}_{12}$, $\hat{e}_{13}$, and $\hat{e}_{23}$, and the vectors z_{12}, z_{13}, and z_{23} are bounded since the agents are forced back and forward according to the signs of the errors, as far as the leader agent, i.e., agent 1, does not move. Therefore, the agents will be staying in a specific interval range on the line.

Corollary 12.3 *If z converges to $\mathcal{D}$, then p also converges to a stationary point.*

Proof Since $\bar{e}_{12}$, $\hat{e}_{13}$ and $\hat{e}_{23}$ converge to zero exponentially fast, $\dot{p}(t)$ also converges to zero exponentially fast from (12.14), which implies that p converges to a stationary point.

Next, let us consider n agents on the plane. For the agent i, suppose that there are two neighbors i_1 and i_2 for all $i \in \mathcal{V} \setminus \{1, 2\}$. The control law (12.14) is adopted as

$$u_i = \hat{e}_{i_1 i} z_{i_1 i} + \hat{e}_{i_2 i} z_{i_2 i} \tag{12.31}$$

where $\hat{e}_{i_j i} = \|z_{i_j i}\|^2 - (s_{i_j i}\|p_{i_1} - p_{i_2}\|)^2$ for $j = 1, 2$, with the assumptions of $1 < s_{i_1 i} + s_{i_2 i}$ and $s_{i_1 i} < 1 + s_{i_2 i}$, $s_{i_2 i} < 1 + s_{i_1 i}$. Let $\bar{e}_{i_j i} = \|z_{i_j i}\|^2 - (s_{i_j i}\|p_{i_1}(\infty) - p_{i_2}(\infty)\|)^2$, $\tilde{e}_{i_j i} = \bar{e}_{i_j i} - \hat{e}_{i_j i} = s_{i_j i}^2 \bar{e}_{i_1 i_2}$, and $\bar{e}_{i_1 i_2} = \|p_{i_1} - p_{i_2}\|^2 - \|p_{i_1}(\infty) - p_{i_2}(\infty)\|^2$. From the Lyapunov candidate $V_i(z(t)) = 1/4(\bar{e}_{i_1 i}^2 + \bar{e}_{i_2 i}^2)$, it follows

$$\begin{aligned}
\dot{V}_i &= \frac{1}{2}(\bar{e}_{i_1 i}\dot{e}_{i_1 i} + \bar{e}_{i_2 i}\dot{e}_{i_2 i}) \\
&= -\|\bar{e}_{i_1 i} z_{i_1 i} + \bar{e}_{i_2 i} z_{i_2 i}\|^2 + \bar{e}_{i_1 i}\tilde{e}_{i_1 i}\|z_{i_1 i}\|^2 + \bar{e}_{i_1 i}\tilde{e}_{i_2 i} z_{i_1 i}^T z_{i_2 i} \\
&\quad + \tilde{e}_{i_1 i}\bar{e}_{i_2 i} z_{i_1 i}^T z_{i_2 i} + \bar{e}_{i_2 i}\tilde{e}_{i_2 i}\|z_{i_2 i}\|^2 + \bar{e}_{i_1 i} z_{i_1 i}^T \dot{p}_{i_1} + \bar{e}_{i_2 i} z_{i_2 i}^T \dot{p}_{i_2}
\end{aligned} \tag{12.32}$$

Similarly to the agent 3, for an analysis, let $g_i(z(t)) = -\|\bar{e}_{i_1 i} z_{i_1 i} + \bar{e}_{i_2 i} z_{i_2 i}\|^2$ and $h_i(z(t)) = \bar{e}_{i_1 i}\tilde{e}_{i_1 i}\|z_{i_1 i}\|^2 + \bar{e}_{i_1 i}\tilde{e}_{i_2 i} z_{i_1 i}^T z_{i_2 i} + \tilde{e}_{i_1 i}\bar{e}_{i_2 i} z_{i_1 i}^T z_{i_2 i} + \bar{e}_{i_2 i}\tilde{e}_{i_2 i}\|z_{i_2 i}\|^2 + \bar{e}_{i_1 i} z_{i_1 i}^T \dot{p}_{i_1} + \bar{e}_{i_2 i} z_{i_2 i}^T \dot{p}_{i_2}$. Also, let us define the desired set as

$$\mathcal{D}_z \triangleq \{z \in C(\mathbb{H}_+ \otimes \mathbb{I}_2) : \bar{e}_{12} = 0, \hat{e}_{ij} = 0, \forall (i, j)^{\bar{e}} \in \vec{\mathcal{E}} \setminus (1, 2)^{\bar{e}}\} \tag{12.33}$$

Theorem 12.4 *For acyclic minimally persistent graphs, if agents are controlled by (12.12), (12.13) and (12.31), if $z_{12} \neq 0$, then the n-agent system converges to $\mathcal{D}_z$ or a subset in which at least three agents are collinear. Moreover, if the system converges to $\mathcal{D}_z$, it will be locally exponentially fast.*

Proof Let $\mathcal{V}_i \triangleq \{1, \ldots, i\}$, $i \leq |\mathcal{V}|$. Suppose that $\|\dot{p}_j\|$ is bounded by exponentially decaying functions for all $j \in \mathcal{V}_{i-1}$. Then p_{i_1} and p_{i_2} converge to constant as $t \to \infty$. Thus, with the same argument as Lemma 12.8, V_i, $\bar{e}_{i_1 i}$, $\bar{e}_{i_2 i}$, $z_{i_1 i}$, and $z_{i_2 i}$ are bounded. Thus, $h_i(z(t))$ is bounded by an exponentially decaying function. Consequently, with the same argument as done in (5.41)–(5.42) (see also the input-to-state stability in Lemma 2.11), $\dot{V}_i \to -\|\bar{e}_{i_1 i} z_{i_1 i} + \bar{e}_{i_2 i} z_{i_2 i}\|^2$, which implies $\bar{e}_{i_1 i} z_{i_1 i} + \bar{e}_{i_2 i} z_{i_2 i} \to 0$ as $t \to \infty$. Thus, to satisfy the condition $\bar{e}_{i_1 i} z_{i_1 i} + \bar{e}_{i_2 i} z_{i_2 i} = 0$, either the collinearity among three triangular agents or $\bar{e}_{i_1 i} = \bar{e}_{i_2 i} = 0$ is required. The proof for local exponential convergence is exactly the same to the argument of Theorem 12.3.

The repulsiveness of the collinear configuration is not ensured any more. Also, the overall configuration may be different from the desired configuration since the underlying graph topology is only rigid.

12.2.2 Bearing-Based Formation Scaling

Let us suppose that an initial formation is in a desired configuration under the bearing-based acyclic minimally persistent formation (AMPF) setup. That is, at $t = t_0$, $g_{ji} = g_{ji}^*$, $\forall (i, j)^{\bar{e}} \in \overrightarrow{\mathcal{E}}$. Let all the agents except the leader agent, i.e., agent 1, use the same bearing-based control laws as (9.50) and (9.51). Then, the leader agent 1 is responsible for scaling the formation, as per Lemma 9.3. It attempts to change the distance d_{12} to a new one as $d_{12}^* = s d_{12}(t_0)$, with the scale constant $0 < s < \infty$. We propose the following scaling control law for the agent 1:

$$\dot{p}_1(t) = -k(\bar{d}_{12}(t) - \bar{d}_{12}^*)(p_1 - p_2) = -k(\bar{d}_{12}(t) - s^2 \bar{d}_{12}(t_0)) z_{12} \tag{12.34}$$

where $k > 0$ is the control gain. Note that the above control law is implemented into the ith agent's local coordinate frame.

Combining (12.34), (9.52), and (9.53), we can have

$$\dot{p}_1 = \underbrace{-k(\bar{d}_{12}(t) - s^2 \bar{d}_{12}(t_0)) z_{12}}_{\triangleq f_1(p_1, p_2, t)} \tag{12.35}$$

$$\dot{p}_2 = \underbrace{-2 \mathbb{P}_{g_{12}} R_1^T g_{12}^*}_{\triangleq f_2(p,t)} + \underbrace{\mathbb{P}_{g_{12}} (R_1^T - R_2^T) g_{12}^*}_{\triangleq h_2(p,t)} \tag{12.36}$$

$$\dot{p}_i = \underbrace{-2 \sum_{j \in \mathcal{N}_i} \mathbb{P}_{g_{ji}} R_1^T g_{ji}^*}_{\triangleq f_i(p,t)} + \underbrace{\sum_{j \in \mathcal{N}_i} \mathbb{P}_{g_{ji}} (2R_1^T - R_i^T - R_j^T) g_{ji}^*}_{\triangleq h_i(p,t)} \tag{12.37}$$

which can be concisely written as

$$\dot{p} = f(p) + h(p, t) \tag{12.38}$$

where $f(p) = (f_1^T, \ldots, f_n^T)^T$, $h(p, t) = (h_1^T, \ldots, h_n^T)^T$, and $h_1 = 0$. To use the input-to-state stability, let us treat the term $h(p, t)$ as an input to the nominal system

$$\dot{p} = f(p) \tag{12.39}$$

Lemma 12.10 *If $z_{12} \neq 0$ initially, then for the dynamics (12.35), the agent 1 converges to the desired equilibrium p_1^* exponentially fast.*

Proof Let us consider the distance $\bar{d}_{12}(t)$. Using the fact that $\dot{p}_2 \perp z_{12} = 0$ and taking a derivative $\frac{d\bar{d}_{12}}{dt}$, we can have the distance dynamics:

$$\begin{aligned} \dot{\bar{d}}_{12}(t) &= 2(\dot{p}_1 - \dot{p}_2)^T (p_1 - p_2) \\ &= -2k(\bar{d}_{12} - \bar{d}_{12}^*) z_{12}^T (p_1 - p_2) \\ &= -2k \bar{d}_{12} (\bar{d}_{12} - \bar{d}_{12}^*) \end{aligned} \tag{12.40}$$

Taking a Lyapunov candidate $V = (\bar{d}_{12} - \bar{d}_{12}^*)^2$, it is easy to show that $\dot{V} \leq -\kappa V \leq 0$, $\kappa > 0$, which implies that $\bar{d}_{12}$ converges to $\bar{d}_{12}^*$ exponentially fast. Thus, $z_{12} = 0$ does not occur if $z_{12} \neq 0$ initially.

Lemma 12.11 *Under the same assumptions of Lemma 9.13, $h(t)$ in (12.38) is bounded and converges to 0 asymptotically.*

Proof Since all the states are bounded, clearly $h_i(t)$ are bounded for all $i = 2, \ldots, n$. The convergence can be shown by combining several lemmas. From Lemma 9.13, we know that $\lim_{t\to\infty} R_i^T R_j = \mathbb{I}_3$ when the leader agent is stationary. But, in the scaling problem, the agent 1 is not stationary due to the control (12.35). Thus, the alignment dynamics (7.33) and scaling dynamics (12.35) can be considered as a cascade system. Since the alignment dynamics is affected by the motion of agent 1, we can see that (7.33) has an input from the motion of agent 1. From Lemma 7.5 and Theorem 7.2, since the alignment dynamics is exponentially stable, the dynamics (7.33) is input-to-state stable from the Lemma 2.12. From Lemma 12.10, since (12.35) is globally exponentially convergent, the overall system combining (7.33) and (12.35) is globally asymptotically convergent by Lemma 2.13. Thus, $h(t)$ in (12.38) converges to 0.

Based on Lemma 12.11, the following lemma, which is the counterpart of Lemma 9.15, is generated:

Lemma 12.12 *The unforced system $\dot{p}_2 = f_2(p_2)$ of (12.36) has two equilibria. The first equilibrium $p_2 = p_{2a}^*$ corresponding to $g_{21} = R_1 g_{21}^*$ is almost globally asymptotically stable, while the second equilibrium $p_2 = p_{2b}^*$ corresponding to $g_{21} = -R_1 g_{21}^*$ is unstable.*

Proof Similarly to the proof of Lemma 12.11, we can consider the motion of p_1 as an input to the dynamics $f_2(p, t)$ of (12.36). Since it is almost globally asymptotically stable, it can be considered globally input-to-state stable as per the Definition 2.32.[1] Thus, by Lemmas 2.11 and 9.4, the desired equilibrium point $g_{21} = R_1 g_{21}^*$ is almost globally asymptotically stable.

The following lemma is the counterpart of Lemma 9.16:

Lemma 12.13 *The system (12.36) has two equilibria. The equilibrium $p_2 = p_{2a}^*$ is almost globally asymptotically stable, while the equilibrium $p_2 = p_{2b}^*$ is unstable.*

Proof Using the same equation as (9.56), it can be shown that the state p_2 in (12.36) is ultimately bounded. Thus, by combining global input-to-state stability of Lemma 2.11, the convergence of $h_2(t)$ to zero in Lemma 12.11, and global asymptotic stability of unforced dynamics from Lemma 12.12, the proof can be completed.

---

[1]In fact, $f_2(p, t)$ of (12.36) is not affected by the motion of agent 1 since the motion of agent 1 and the motion of agent 2 due to $f_2(p, t)$ in (12.36) are orthogonal. Thus, the proof of Lemma 9.15 can be applicable for the proof of Lemma 12.12.

The stability of agent 3 can be similarly analyzed as the Theorem 9.6. The only difference is that the agent 3 in (12.37) has two inputs. That is, for the same unforced dynamics as (9.58) and (9.59), it has two inputs from angular velocity and scaling motion of agent 1 such as

$$\dot{R}_3 = R_3(R_1 R_3^T - R_3 R_1^T) + R_3(R_2 R_3^T - R_3 R_2^T) \tag{12.41}$$

$$\dot{p}_1 = -k(\bar{d}_{12}(t) - s^2 \bar{d}_{12}(t_0))z_{12} \tag{12.42}$$

Note that $P_{g_{23}}$ is affected by the motion of p_1. However, (12.41) is globally asymptotically stable, and (12.42) is globally exponentially stable; thus, all the analyses conducted in Theorem 9.6 can be applied exactly the same. Consequently, the following theorem can be obtained without proof.

Theorem 12.5 *The cascade system (9.58), (12.41), and (12.42) has an almost globally asymptotically stable equilibrium $p_2 = p_{2a}^*$, $p_3 = p_{3a}^*$, while the cascade system (9.59), (12.41) and (12.42) has an unstable equilibrium $p_2 = p_{2b}^*$, $p_3 = p_{3b}^*$. All trajectories starting out of the undesired equilibrium $p_2 = p_{2b}^*$ will converge to the stable equilibrium.*

Note that the desired point of agent 3, i.e., $p_3 = p_{3a}^*$, is always uniquely determined since it is acyclic minimally persistent graph under bearing-based setup. For a general n-agent system, both $h(p, t)$ and $\dot{p}_1$ can be considered as inputs. Thus, as the counterpart of Theorem 9.7, we can also make the following lemma:

Lemma 12.14 *Consider the system (12.35)–(12.37) under the same conditions as Lemma 9.13, with position control law* (9.51) and orientation alignment law (9.51). Then, we have $p_1(t) \to p_1^*$, $R_i(t) \to R_1$ and $p(t) \to p_a^*$ for all $i = 2, \ldots, n$ as $t \to \infty$ if $R_2(t_0) \neq R_1$ and $p_2(t_0) \neq p_{2b}^*$.

Finally, with the above lemma, we can make the following result for the bearing-based formation scaling.

Theorem 12.6 *Under the same setup as Sect. 10.1, suppose that agent* 1 *can measure $\bar{d}_{12}(t)$ and $z_{21}^1 = p_2^1$. Then, when agent* 1 *is controlled by (12.34), agent* 2 *is controlled by* (9.50), and all other agents are controlled by (9.51), the formation is scaled-up or scaled-down by s in all distances, i.e., $d_{ij} \to s d_{ij}$ and $g_{ij} \to g_{ij}^*$, $\forall (i, j)^{\bar{e}} \in \vec{\mathcal{E}}$.

Proof The proof is direct from Lemmas 9.3 and 12.14.

The results thus far in this section have been developed using input-to-state stability with the fact that the dynamics (12.35) is globally exponentially stable. If we can guarantee a convergence of p_1 to p_1^* in a finite-time, the analysis will be much simpler, without mathematical induction in a cascade form. Let us control the scale of d_{12} with the following modified control law:

$$u_1 = -k \frac{\bar{e}_{12}}{\|\bar{e}_{12}\|} \frac{p_1 - p_2}{\|p_1 - p_2\|^2} \tag{12.43}$$

Then, the inter-agent dynamics between agents 1 and 2 can be obtained as

$$\begin{aligned}\dot{\bar{d}}_{12}(t) &= -2k\frac{\bar{e}_{12}}{\|\bar{e}_{12}\|}\frac{p_1 - p_2}{\|p_1 - p_2\|^2}(p_1 - p_2)\\ &= -2k\text{sign}(\bar{e}_{12}) \end{aligned} \tag{12.44}$$

Let us choose a Lyapunov candidate as $V = (\bar{d}_{12} - \bar{d}_{12}^*)^2 = \bar{e}_{12}^2$. The derivative of V is obtained as

$$\dot{V} = \frac{\partial V}{\partial \bar{d}_{12}}\frac{\partial \bar{d}_{12}}{\partial t} = 2\bar{e}_{12}\dot{\bar{d}}_{12} = -4k\bar{e}_{12}\text{sign}(\bar{e}_{12}) \tag{12.45}$$

Thus, from (12.45), we can see that $\dot{V} < 0$ for all $\bar{e}_{12} \neq 0$, and $\dot{V} = 0$ if and only if $\bar{e}_{12} = 0$. Consequently, we can have $\dot{V} + V^{0.5} = -4k\bar{e}_{12}\text{sign}(\bar{e}_{12}) + |\bar{e}_{12}| = -4k|\bar{e}_{12}| + |\bar{e}_{12}| \leq 0$ if $k \geq 1$. Hence, by Lemma 2.15, $|\bar{e}_{12}|$ converges to zero in a finite-time t_f, where $0 < t_f < \infty$. Since all the states are bounded, if we reset t_f to the initial time point as $t_0 = t_f$, then all the results of Sect. 10.2 can be exactly applied to the bearing-based formation control while conducting the scaling problem. The following theorem is the summary of this argument:

Theorem 12.7 *Under the same conditions of Theorem 12.6, when agent 1 is controlled by (12.43), agent 2 is controlled by (9.50), and all other agents are controlled by (9.51), the formation is scaled-up or scaled-down by s in all distances while achieving the desired formation, i.e.,* $d_{ij} \rightarrow sd_{ij}$ *and* $g_{ij} \rightarrow g_{ij}^*$, $\forall (i,j)^{\bar{e}} \in \vec{\mathcal{E}}$.

12.3 Summary and Simulations

In this chapter, it was shown that any $K(n)$ formations can be stabilized to a desired configuration in general d-dimensional space by way of using augmented virtual vectors. Each agent needs to have distributed communications with other agents and needs to update the virtual variables w_i. Thus, there is a clear weakness of the proposed method. However, since it does not need to align agents' orientations, it is advantageous over the formation control schemes presented in Chap. 6. Also, since it could stabilize any formations modeled by complete graphs with an almost global convergence in general dimensional spaces, it is advantageous over the traditional gradient-based control laws presented in Chap. 3. The resizing problem defined in Sect. 12.2.1 and the scaling problem defined in Sect. 12.2.2 should be distinguished. The resizing problem is a formulation to change the shape of formation by controlling the distances between neighboring agents under a distance-based setup, while the scaling problem is a formulation to change the size of formation by controlling a limited number of inter-agent distances under bearing-based setup. In the resizing problem, the ratios between the desired inter-agent distances and a reference distance of an edge are given. Thus, all the desired ratios need to be assigned to all the

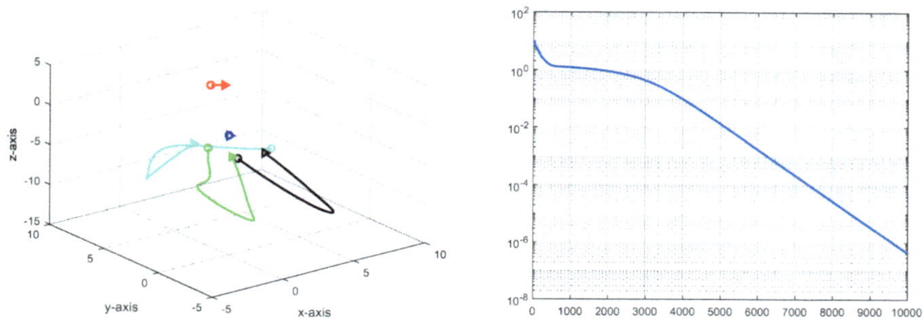

Fig. 12.5 Without scaling, i.e., $s = 1$. Left: Trajectories of positions of agents. Right: Errors of bearing vectors

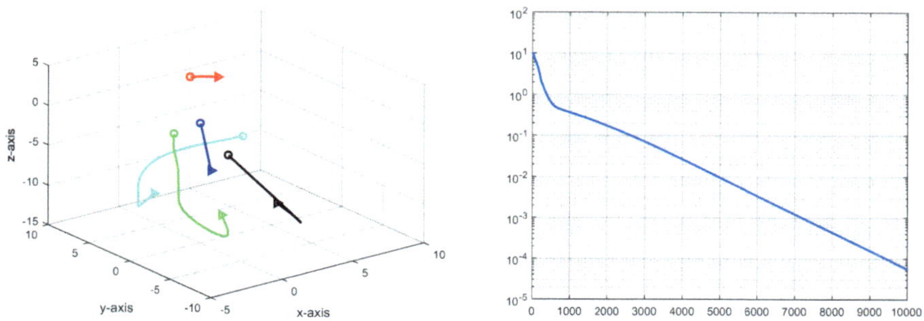

Fig. 12.6 With scaling $s = 2$. Left: Trajectories of positions of agents. Right: Errors of bearing vectors

edges. However, in the scaling problem, which is a special case of the resizing problem, agents attempt to keep their bearing angles to the desired ones continuously. Although the scaling problem can be formulated in a distance-based setup, it is a trivial extension of the schemes presented in Chaps. 3 and 4. It is also clearly preferable to formulate the scaling problem under the bearing-based setup since the scaling of formations can be achieved just by changing the distance of a single edge.

For the simulation, let us consider the bearing-based scaling problem of the graph shown in Fig. 5.4. We conduct two numerical tests. Without scaling, which means $s = 1$, as expected, the agents converge to positions satisfying the desired configuration. The left plots of Fig. 12.5 show the trajectories of agents. Since there is no control in the edge vector z_{12}, the agent 1 is stationary. To check the convergence of the bearing vectors to the desired bearing vectors, we compute $g_{error}(t) = \sum_{(i,j)^{\bar{e}}} \|g_{ji}(t) - g^*_{ji}\|$ for all $(i, j)^{\bar{e}} \in \overrightarrow{\mathcal{E}}$. The right plots of Fig. 12.5 depict $g_{error}(t)$ as time passes in the log scale. Clearly, as the time passes, the error converges to zero. The left plots of Fig. 12.6 show the trajectories of agents with scale $s = 2$, and the right plots of Fig. 12.6 show $g_{error}(t)$. When comparing Figs. 12.5 and 12.6, the formation system without scaling control converges to zero faster than the formation system

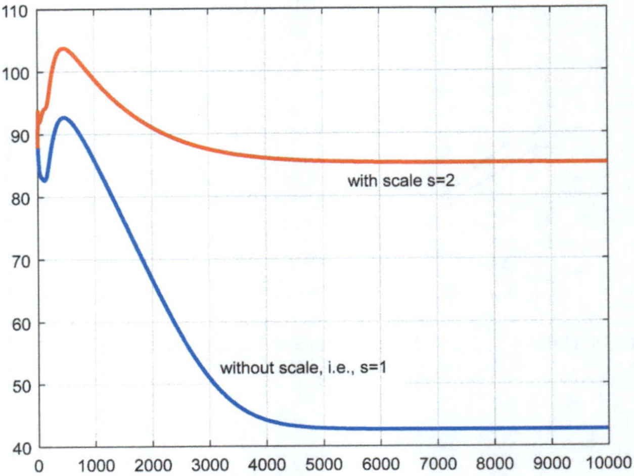

Fig. 12.7 Sum of edge distances: $s = 1$ versus $s = 2$

with the scaling control. But, when comparing Figs. 12.5 and 12.6, the scaling of two configurations is not clear. To verify the scaling control, we compute the sum of edge distances as $z_{sum}(t) = \sum_{(i,j)^{\bar{e}}} \|z_{ji}(t)\|$ for all $(i, j)^{\bar{e}} \in \overrightarrow{\mathcal{E}}$. Figure 12.7 shows the magnitudes of $z_{sum}(t)$ for $s = 1$ and for $s = 2$. From the plots, we can see that during the transient period, the scale is not 2; but as the agents converge to the desired configurations, the ratio of $z_{sum}(t)$ becomes 2, which is the desired scale.

12.4 Notes

This chapter has presented two topics, formation control of $K(n)$ graphs under a gradient-based setup and formation resizing under the distance-based and bearing-based setups. In the scaling control of formation systems, there is a recent work [12] which proposes an estimation-based update law of scaling parameter. The results of Sect. 12.1 and results of Sect. 12.2 have been reused from [5–7, 11], respectively. The following copyright and permission notices are acknowledged.

- The text of Sect. 12.1 is © [2014] IEEE. Reprinted, with permission, from [6].
- The statements of Lemmas 12.2, 12.4, 12.5, 12.7, Theorems 12.1, and 12.2 with the proofs are © [2018] IEEE. Reprinted, with permission, from [7].
- The text of Sect. 12.2.2 is © [2019] IEEE. Reprinted, with permission, from [11].
- The statements of Lemma 12.8, Theorem 12.3, Lemma 12.9, and Theorem 12.4 with the proofs are © Reproduced from [5] by permission of John Wiley and Sons.

References

1. Absil, P.-A., Kurdyka, K.: On the stable equilibrium points of gradient systems. Syst. Control Lett. **55**(7), 573–577 (2006)
2. Absil, P.-A., Mahony, R., Andrews, B.: Convergence of the iterates of descent methods for analytic cost functions. SIAM J. Optim. **16**(2), 531–547 (2005)
3. Asimow, L., Roth, B.: The rigidity of graphs II. J. Math. Anal. Appl. **68**(1), 171–190 (1979)
4. Cao, M., Anderson, B.D.O., Morse, A.S., Yu, C.: Control of acyclic formations of mobile autonomous agents. In: Proceedings of the 47th IEEE Conference on Decision and Control, pp. 1187–1192 (2008)
5. Park, M.-C., Jeong, K., Ahn, H.-S.: Formation stabilization and resizing based on the control of interagent distances. Int. J. Robust Nonlinear Control **25**(14), 2532–2546 (2015)
6. Park, M.-C., Sun, Z., Anderson, B.D.O., Ahn, H.-S.: Stability analysis on four agent tetrahedral formations. In: Proceedings of the 53rd IEEE Conference on Decision and Control, pp. 631–636 (2014)
7. Park, M.-C., Sun, Z., Anderson, B.D.O., Ahn, H.-S.: Distance-based control of $K(n)$ formations in general space with almost global convergence. IEEE Trans. Autom. Control **63**(8), 2678–2685 (2018)
8. Sun, Z., Helmke, U., Anderson, B.D.O.: Rigid formation shape control in general dimensions: an invariance principle and open problems. In: Proceedings of the 54th IEEE Conference on Decision and Control, pp. 6095–6100 (2015)
9. Sun, Z., Yu, C.: Dimensional-invariance principles in coupled dynamical systems: a unified analysis and applications. IEEE Trans. Autom. Control 1–15 (2018). arXiv:1703.07955
10. Trinh, M.H., Pham, V.H., Park, M.-C., Sun, Z., Anderson, B.D.O., Ahn, H.-S.: Comments on global stabilization of rigid formations in the plane. Automatica **77**, 393–396 (2018)
11. Trinh, M.H., Zhao, S., Sun, Z., Zelazo, D., Anderson, B.D.O., Ahn, H.-S.: Bearing-based formation control of a group of agents with leader-first follower structure. IEEE Trans. Autom. Control **64**(2), 598–613 (2019). https://doi.org/10.1109/TAC.2018.2836022
12. Yang, Q., Sun, Z., Cao, M., Fang, H., Chen, J.: Stress-matrix-based formation scaling control. Automatica **101**, 120–127 (2019)

Chapter 13
Network Localization

Abstract The network localization in distributed agent systems may be called as a dual problem of formation control. The mathematical formulation between the network localization and formation control is equivalent. This chapter discusses several network localization problems as dual formulations of formation control problems studied in the previous chapters.

13.1 Background

In linear systems theory, it is well-known that the controllability and observability have a duality relationship. In multi-agent systems theory, a similar duality exists between formation control and network localization, under the gradient control laws and orientation alignment setups. For the n-agent systems $\dot{p}_i = u_i$, let us suppose that the agents measure the relative displacements p^i_j for all $(i, j)^e \in \mathcal{E}$. In formation control problems, we would like to design control laws such that $p_i(t) \to p^*_i$ up to rotations, translations, or scaling. In gradient control laws, we achieve a desired formation configuration by satisfying the desired inter-agent distances, while in orientation alignment laws, we achieve the desired formation by way of estimating the coordinate orientation θ_i. However, in network localization, we would like to design $\hat{p}_i$ such that $\|\hat{p}_j - \hat{p}_i\| = \|p_j - p_i\|$ in gradient descent approaches, or like to design an orientation alignment law for θ_i and estimation law for $\hat{p}_i$ such that $\hat{p}_i(t) \to p_i$ up to a common translation and rotation. Thus, as a dual problem, in network localization, we need to replace the state update dynamics by a state estimation dynamics, under the same structure. As an instance, in the right-hand side of (6.10), the term p^i_j is measured, and the term $p^*_j - p^*_i$ is given. But, in localization, it is required to update the estimated position of agent; so the measurements p^i_j do not change. But, the estimated position is a state, which needs to be updated. So, we would like to ensure the estimated relative positions to be equal to the measured relative positions. In this sense, the term p^i_j can be replaced by $\hat{p}_{ji}$, which is defined as $\hat{p}_{ji} = \hat{p}_j - \hat{p}_i$, and the term $p^*_j - p^*_i$ can be replaced by p^i_j. Figure 13.1 shows the duality concept in multi-agent systems theory, which is analogous to the duality in linear systems theory. In linear systems theory, the dynamics for controllability is determined by

H.-S. Ahn, *Formation Control*, Studies in Systems, Decision and Control 205, https://doi.org/10.1007/978-3-030-15187-4_13

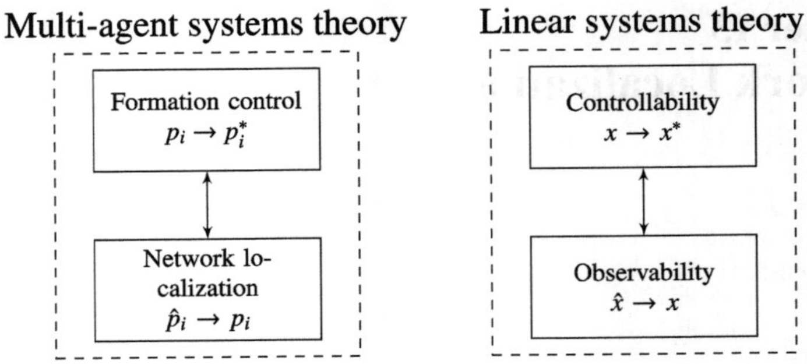

Fig. 13.1 The concept of duality in multi-agent systems theory

$$\dot{x} = Ax(t) + Bu(t) \tag{13.1}$$

while the dynamics for observability is decided by

$$\dot{x} = Ax(t); \;\; y = Cx \tag{13.2}$$

In principle, in the case of observability analysis, given y, we would like to decide a solution x that satisfies the Eq. (13.2), while in the controllability, we would like to send the state x to the desired one x^* using the input $u(t)$. Thus, the observability and controllability have the same dynamics. Similarly, the dynamics for formation control and dynamics for network localization are equivalent (for example, compare (4.1) and (13.4)).

The network localization problems can be formulated under distance-based setup, displacement-based setup, and bearing-based setup. In distance-based setup, we require the graph to be rigid or globally rigid as the underlying graph topology, but in displacement-based setup, we require the graph to be connected. In bearing-based setup, the underlying graph needs to be bearing rigid.

13.2 Network Localization Under Gradient Descent Laws

The gradient control laws for formation control can be readily modified for network localization. Let us denote the estimated position of agent i as $\hat{p}_i$, and its dynamics be represented as:

$$\dot{\hat{p}}_i(t) = \hat{u}_i(t) \tag{13.3}$$

Let agents measure the relative displacements as $p^i_j = p^i_j - p^i_i$. The measurements can be considered as reference displacements, while the estimated relative displacements $\hat{p}_j - \hat{p}_i$ can be considered as inputs in the network localization. Let us denote

$\tilde{d}_{ij}^* \triangleq \|p_j^i - p_i^i\|^2$ and $\tilde{d}_{ij} \triangleq \|\hat{p}_j - \hat{p}_i\|^2$. Then, the errors of the estimated edge distances can be defined as $\tilde{e}_{ij} = \tilde{d}_{ij} - \tilde{d}_{ij}^*$. Let us also suppose that the neighboring agents i and j can communicate with each other to exchange the estimated positions $\hat{p}_i$ and $\hat{p}_j$. When using the inter-agent dynamics laws of Sect. 4.1 in $\mathbb{R}^2$, similarly to (4.1), we can update the estimation input $\hat{u}_i(t)$ as:

$$\underbrace{\begin{bmatrix} \vdots \\ (\hat{p}_i - \hat{p}_j)^T \\ \vdots \end{bmatrix}}_{\triangleq \tilde{A}_i} \hat{u}_i = -\frac{\tilde{k}_{ij}}{4} \underbrace{\begin{bmatrix} \vdots \\ \tilde{e}_{ij} \\ \vdots \end{bmatrix}}_{\triangleq \tilde{b}_i}, \; j \in \mathcal{N}_i, \; i \in \mathcal{V} \tag{13.4}$$

where $\tilde{k}_{ij}$ are constant gains for the localization. Let us define the desired equilibrium set for (13.3) as:

$$E_{\hat{p}^*} \triangleq \{p \in \mathbb{R}^{2n} : \|\hat{p}_i - \hat{p}_j\| = \|p_i - p_j\|, i, j \in \mathcal{V}\} \tag{13.5}$$

Let us use the following estimation law

$$\hat{u}_i = -\frac{\tilde{k}_{ij}}{4}(\tilde{A}_i^T \tilde{A}_i)^{-1} \tilde{A}_i^T \tilde{b}_i \tag{13.6}$$

which can be implemented in a distributed way. Then, as the counterpart of Theorem 4.2, we can make the following result:

Theorem 13.1 *Suppose that the given framework $(\mathcal{G}, p)$ is infinitesimally rigid. Then the estimation law (13.6) ensures a local convergence of $\hat{p}_i$ to a point in E_{p^*}.*

Note that in gradient control laws given in Chaps. 3 and 4, the desired references are given, while, in gradient localization law (13.6), the measured relative displacements play the reference role. Also, in gradient-based formation control, the relative displacements are updated, while in gradient-based network localization, the estimated relative displacements are updated. Thus, in gradient-based formation control, there are no communication variables; but in gradient-based network localization, the estimated position information needs to be exchanged between neighboring agents. Thus, although the mathematical setups are exactly the same, implementation features of gradient control laws and gradient estimation laws should be differentiated. Also in general d-dimensional space, we can modify the control law (4.26) as

$$\hat{u}_i = \hat{k}_p \sum_{j \in \mathcal{N}_i} \frac{\partial \gamma(\tilde{e}_{ij})}{\partial \tilde{e}_{ij}} (\hat{p}_j - \hat{p}_i) \tag{13.7}$$

which will ensure a local asymptotic convergence of $\hat{p}_i$ to a point in E_{p^*}. For the network localization, the underlying topology needs to be infinitesimally rigid. It could be minimally infinitesimally rigid, redundantly infinitesimally rigid, or infinitesimally globally rigid. In any case, in Theorem 13.1, we only ensured a local convergence. Thus, the initial estimations of the positions should be close to the true positions. In existing network localization problems, for a unique solution, it is typically required to have a generically globally rigid graph as an underlying topology [3].

However, when we employ the orientation alignment technique, an existence of spanning tree is enough for a unique solution, as studied in Sect. 13.3. Also, when we consider a particular class of persistent graphs, the unique localization solution can be obtained in a distributed way. Actually, the results of Chap. 5 can be directly modified for network localization of a group of persistent graphs. For example, as a counterpart of Sect. 5.2, let us consider acyclic minimally persistent graphs in $\mathbb{R}^3$, for network localization. Let us use a potential function $\phi(\tilde{e}_{ij}, \tilde{d}_{ij})$, with its gradient $g(\tilde{e}_{ij}, \tilde{d}_{ij})$, that satisfies the conditions of Assumption 5.2.1. Let us define the desired equilibrium set for the localization of acyclic minimally persistent graphs:

$$\hat{\mathcal{U}}_{i,eq}^{C} = \{\hat{p}_i \in \mathbb{R}^3 : g(\tilde{e}_{ii_1}, \tilde{d}_{ii_1}) = g(\tilde{e}_{ii_2}, \tilde{d}_{ii_2}) = g(\tilde{e}_{ii_3}, \tilde{d}_{ii_3}) = 0\} \tag{13.8}$$

where $i_j \in \mathcal{N}_i^O,\ j = 1, 2, 3$. Then, the following gradient estimation law can be proposed for the network localization of acyclic minimally persistent graphs:

$$\dot{\hat{p}}_i = \hat{u}_i = \sum_{j \in \mathcal{N}_i} g^f(\tilde{e}_{ij}, \tilde{d}_{ij})(\hat{p}_j - \hat{p}_i),\ (i, j)^{\bar{e}} \in \overrightarrow{\mathcal{E}} \tag{13.9}$$

where $g^f(\tilde{e}_{ij}, \tilde{d}_{ij})$ is computed either by (5.45) or by (5.46). Then, we can make the following result:

Theorem 13.2 *Consider acyclic minimally persistent graphs in $\mathbb{R}^3$. Suppose that agents are not in $\mathcal{C}_i$, which is given in (5.25), i.e., $p_i \notin \mathcal{C}_i$. Also, let us suppose that the initial assignments of estimated positions, i.e., $\hat{p}_i(t_0)$, are not coplanar with respect to the outgoing leader agents, i.e., rank$[\hat{z}_{ii_1}(t_0), \hat{z}_{ii_2}(t_0), \hat{z}_{ii_3}(t_0)] = 3$, where $\hat{z}_{ii_j} = \hat{p}_i(t_0) - \hat{p}_{i_j}$. Then, the estimated positions $\hat{p}_i$ by the localization law* (13.9) *will asymptotically converge to the set $\hat{\mathcal{U}}_{i,eq}^{C}$.*

In the acyclic minimally persistent graphs in $\mathbb{R}^3$ of the above theorem, the follower agents have only three constraints to the outgoing agents. Thus, Theorem 13.2 will be valid when the initial assignments of the estimated positions, i.e., $\hat{p}_i(t_0)$, are close to the true positions p_i. For a localization with any initial estimations, the follower needs to have four distance constraints to the leaders. For this, we can use the idea of Sect. 5.3 in $\mathbb{R}^3$. For the follower agents $i \geq 5$, we introduce the virtual vertex i^v that is virtually defined in $\mathbb{R}^4$. Let us update the localization of agents as follows:

$$\dot{\hat{p}}_1 = 0 \tag{13.10}$$

$$\dot{\hat{p}}_2 = g_{21}^f \hat{z}_{12} \tag{13.11}$$

$$\dot{\hat{p}}_3 = g_{31}^f \hat{z}_{13} + g_{32}^f \hat{z}_{23} \tag{13.12}$$

$$\dot{\hat{p}}_4 = g^f_{41}\hat{z}_{14} + g^f_{42}\hat{z}_{24} + g^f_{43}\hat{z}_{34} \tag{13.13}$$

$$\dot{\hat{p}}_i = \begin{bmatrix} 1\,0\,0\,0 \\ 0\,1\,0\,0 \\ 0\,0\,1\,0 \end{bmatrix} \dot{\hat{p}}_{i^v}, \; i \geq 5 \tag{13.14}$$

$$\dot{\hat{p}}_{i^v} = g^f_{i^v i_1}\hat{z}_{i_1 i^v} + g^f_{i^v i_2}\hat{z}_{i_2 i^v} + g^f_{i^v i_3}\hat{z}_{i_3 i^v} + g^f_{i^v i_4}\hat{z}_{i_4 i^v} \tag{13.15}$$

where $\hat{z}_{ij} = \hat{p}_i - \hat{p}_j$ and $\hat{p}_{i^v} \in \mathbb{R}^4$. Then, Theorem 13.2 can be changed as

Theorem 13.3 *Consider acyclic persistent graphs in* $\mathbb{R}^3$. *Suppose that agents are not in* $\mathcal{C}_i$, *which is given in (5.25), i.e.,* $p_i \notin \mathcal{C}_i$. *Also, let us suppose that the initial assignments of estimated positions, i.e.,* $\hat{p}_i(t_0)$, *are not coplanar with respect to the outgoing leader agents. Then, the estimated positions* $\hat{p}_i$ *by the localization law* (13.10)–(13.15) *will globally asymptotically converge to the true positions up to a rotation.*

Note that the localization method given in Theorem 13.1 can be implemented simultaneously for all agents. Also in the cases of Theorems 13.2 and 13.3, the localization algorithms may be implemented either in a simultaneous way or in sequential way. As discussed in Sect. 5.4, with the similar argument as Theorem 5.3, the estimated position information by distributed localization does not diverge. Thus, since the positions of the leader agents are estimated in a finite-time, as far as the estimated positions of the follower agents do not diverge, they will be also estimated in a finite-time.

When a follower agent has more than three outgoing agents in Theorem 13.2 and more than four outgoing agents in Theorem 13.3, it needs to select only three or four outgoing agents for the localization. If a node has been localized, then it can be used as a leader agent for the reference of incoming agents. So, the overall procedure should be conducted in a sequentially coordinated way. The computational amount of Theorem 13.3 is heavier than the computational amount of Theorem 13.2 since it needs to compute the update of virtual variables in (13.15). It is also remarkable that sequential localization algorithms have been proposed in [4].

13.3 Network Localization via Orientation Alignment

The network localization under gradient descent laws only can ensure local convergence for general rigid graphs, and almost global convergence for acyclic minimally persistent graphs. When we use the orientation alignment schemes, we can ensure a (quasi-) global convergence with expenses of communications and computational costs.

13.3.1 Distance-Based Network Localization

The network localization via orientation estimation can be similarly designed using the results of Sect. 7.2. Let us first consider 2-dimensional case and let us suppose that $\max_{i\in\mathcal{V}}\theta_i(t_0) - \min_{i\in\mathcal{V}}\theta_i(t_0) < \pi$. For the localization of agents in a distributed way, we need to have the following orientation estimation law first:

$$\dot{\hat{\theta}}_i(t) = k_{\hat{\theta}} \sum_{j\in\mathcal{N}_i} w_{ij}(t)(\hat{\theta}_{ji}(t) - \theta_{ji}(t)) \tag{13.16}$$

Then, using the fact that $\hat{\theta}_{ji} - \theta_{ji} = \hat{\theta}_j - \theta_j - (\hat{\theta}_i - \theta_i)$, the dynamics of the errors in the orientation estimation can be obtained as

$$\dot{\tilde{\theta}} = -k_{\hat{\theta}}\mathbb{L}(t)\tilde{\theta} \tag{13.17}$$

where $\mathbb{L}$ is the Laplacian matrix and $\tilde{\theta} = (\tilde{\theta}_1, \ldots, \tilde{\theta}_n)^T$. It is clear that $\tilde{\theta}$ will converge to a vector $\tilde{\theta}_\infty \mathbf{1}_n$ exponentially fast, where $\tilde{\theta}_\infty$ is a constant. The distributed localization rule for the agents is designed, up to a common coordinate frame ${}^c\Sigma$, as

$$\begin{aligned}
\dot{\hat{p}}_i(t) &= k_{\hat{p}} \sum_{j\in\mathcal{N}_i(t)} w_{ij}(t)\left((\hat{p}_j(t) - \hat{p}_i(t)) - R^{-1}_{\hat{\theta}_i(t)} p^i_{ji}(t)\right), \\
&= k_{\hat{p}} \sum_{j\in\mathcal{N}_i(t)} w_{ij}(t)\left((\hat{p}_j(t) - \hat{p}_i(t)) - R^{-1}_{\hat{\theta}_i(t)} R_{\theta_i(t)+\tilde{\theta}_\infty} p_{ji}\right) \\
&= k_{\hat{p}} \sum_{j\in\mathcal{N}_i(t)} w_{ij}(t)\left((\hat{p}_j(t) - \hat{p}_i(t)) - R_{\tilde{\theta}_\infty - \tilde{\theta}_i(t)} p_{ji}\right).
\end{aligned} \tag{13.18}$$

Defining $e_{\hat{p}_i}(t)$ as $e_{\hat{p}_i}(t) \triangleq \hat{p}_i(t) - p_i$, the error dynamics of the estimated positions can be obtained as follows:

$$\dot{e}_{\hat{p}_i}(t) = k_{\hat{p}} \sum_{j\in\mathcal{N}_i(t)} w_{ij}(t)\left(e_{\hat{p}_j}(t) - e_{\hat{p}_i}(t)\right) + \underbrace{k_{\hat{p}} \sum_{j\in\mathcal{N}_i(t)} w_{ij}(t)\left(\mathbb{I}_2 - R_{\tilde{\theta}_\infty - \tilde{\theta}_i(t)}\right)(p_j - p_i)}_{\triangleq \psi_i}.$$

Then, the overall error dynamics for position estimation can be written as

$$\dot{e}_{\hat{p}}(t) = -k_{\hat{p}}(\mathbb{L}(t) \otimes \mathbb{I}_2)e_{\hat{p}}(t) + \Psi(t), \tag{13.19}$$

where $e_{\hat{p}}(t) = (e^T_{\hat{p}_1}, \ldots, e^T_{\hat{p}_n})^T$ and $\Psi(t) = (\psi_1^T, \ldots, \psi_n^T)^T$. Thus, by analyzing the convergence of (13.17) and (13.19) together, and using the same approach as in Chap. 7.2, we can obtain the following result:

Theorem 13.4 *Suppose that the network localization algorithm is performed as* (13.17) *and* (13.19). *Then,* $\hat{p}$ *exponentially converges to* $\hat{p}^\infty = ((\hat{p}_1^\infty)^T, \ldots, (\hat{p}_n^\infty)^T)^T$

such that $\|\hat{p}_j^\infty - \hat{p}_i^\infty\| = \|p_j - p_i\|$ *for all* $(i, j)^e \in \mathcal{E}$*, for any* $\hat{p}(t_0)$ *and* $\hat{\theta}(t_0)$*, if* $\max_{i\in\mathcal{V}} \theta_i(t_0) - \min_{i\in\mathcal{V}} \theta_i(t_0) < \pi$*,* $\max_{i\in\mathcal{V}} \hat{\theta}_i(t_0) - \min_{i\in\mathcal{V}} \hat{\theta}_i(t_0) < \pi$*, and the graph* $\mathcal{G}(t)$ *is uniformly connected and* $\mathbb{L}(t)$ *is bounded and piecewise continuous.*

Remark that the pair of the position estimation law (13.18) and the orientation estimation law (13.16) can be called SE(2) localization since the orientation and position are estimated simultaneously. Next, let us consider network localization via global orientation estimation in 3-dimensional space. Let us suppose that the orientation of agents could be estimated as per the Theorem 8.3, i.e., after obtaining B_i, we can compute the orientation of agents as $\hat{R}_i(t) = B_i(t) \in \mathrm{SO}(3)$ up to a common rotation. Then, we can design the network localization law as follows:

$$\dot{\hat{p}}_i = k_{\hat{p}} \sum_{j\in\mathcal{N}_i} w_{ij}(\hat{p}_j - \hat{p}_i - \hat{R}_i^T p_{ji}^i) \tag{13.20}$$

where $p_{ji}^i = p_j^i - p_i^i$ is the measured relative displacement. Define the error of position estimation as $e_{\hat{p}_i}(t) \triangleq \hat{p}_i(t) - p_i$. Then, using $p_{ji}^i = p_j^i - p_i^i = R_i(p_j - p_i)$, we can have the error dynamics as follows:

$$\begin{aligned}
\dot{e}_{\hat{p}_i} &= k_{\hat{p}} \sum_{j\in\mathcal{N}_i} w_{ij}(\hat{p}_j - \hat{p}_i - \hat{R}_i^T R_i(p_j - p_i)) \\
&= k_{\hat{p}} \sum_{j\in\mathcal{N}_i} w_{ij}\left(e_{\hat{p}_j} - e_{\hat{p}_i}\right) + \underbrace{k_{\hat{p}} \sum_{j\in\mathcal{N}_i} w_{ij}(R^* - \hat{R}_i^T R_i)(p_j - p_i)}_{\triangleq \psi_i}
\end{aligned} \tag{13.21}$$

where R^* is the converged one of $\hat{R}_i^T R_i$ as the orientation estimation law propagates, i.e., $\hat{R}_i^T R_i \to R^*$. Now, the overall error dynamics can be expressed as

$$\dot{e}_{\hat{p}} = -k_{\hat{p}}(\mathbb{L} \otimes \mathbb{I}_3)e_{\hat{p}} + \Psi \tag{13.22}$$

Then, under the same initial and graph conditions as Theorem 8.4, similarly to the Theorem 13.4, we can make the following theorem:

Theorem 13.5 *With the estimation method outlined in Theorem 8.3, and the position estimation law (13.20), there exist a finite point* $\hat{p}_i^\infty$*,* $\forall i \in \mathcal{V}$ *and a common rotation matrix* $R^c \in SO(3)$ *such that* $\hat{p}_j^\infty - \hat{p}_i^\infty = R^c(p_j - p_i)$*, if* $\mathcal{G}$ *has an arborescence, and* $\hat{\mathbf{q}}_k(t_0)$*,* $\forall k \in \{1, 2\}$ *is not in* $C(\mathbb{L} \otimes \mathbb{I}_3)$*.*

The above theorem is in fact SE(3) localization since the rotation matrices and positions of agents are estimated in $\mathbb{R}^3$. Finally, summarizing the above results, we can formally state the duality of formation control and network localization.

Theorem 13.6 *Given a set of multi-agent systems (6.1), the desired formation can be achieved by the alignment law (6.8) and control law (6.10) if and only if the positions of agents are localized by the alignment law (6.8) and estimation law* (13.18).

Proof Since the alignment law is same to both formation control and network localization, we do not consider the alignment law (6.8). Let us assume that the desired formation has been achieved by the control law (6.10). Then, by replacing p_j^i by $\hat{p}_j(t) - \hat{p}_i(t)$ and replacing $p_j^* - p_i^*$ by p_{ji}^i, we can obtain the convergence of (13.18). The reverse is same, which completes the proof.

13.3.2 Bearing-Based Network Localization

Let the measurements be bearing vectors between agents, as studied in Sect. 10.1. We use the following estimation law for each agent:

$$\dot{\hat{p}}_i(t) = -\sum_{j\in\mathcal{N}_i} \mathbb{P}_{\hat{g}_{ji}} B_i^T g_{ji}^i. \tag{13.23}$$

where $\mathbb{P}_{\hat{g}_{ji}} = \mathbb{I}_d - \hat{g}_{ji}\hat{g}_{ji}^T$ is calculated from $\hat{g}_{ji} = \frac{\hat{p}_j - \hat{p}_i}{\|\hat{p}_j - \hat{p}_i\|}$. Following the same procedure as done in Sect. 10.1, by substituting $B_i(t) = R_i R_{\Delta i} X$ and $g_{ji}^i = R_i g_{ji}$ into (13.23), we have

$$\begin{aligned}
\dot{\hat{p}}_i(t) &= -\sum_{j\in\mathcal{N}_i} \mathbb{P}_{\hat{g}_{ji}} X^T R_{\Delta i}^T g_{ji}. \\
&= -\sum_{j\in\mathcal{N}_i} \mathbb{P}_{\hat{g}_{ji}} X^T g_{ji} + \underbrace{\sum_{j\in\mathcal{N}_i} \mathbb{P}_{\hat{g}_{ji}} X^T (\mathbb{I}_d - R_{\Delta i}^T) g_{ji}}_{\triangleq h_i(t)} \\
&= -\sum_{j\in\mathcal{N}_i} \mathbb{P}_{\hat{g}_{ji}} g_{ji}^\infty + h_i(t)
\end{aligned} \tag{13.24}$$

where $g_{ji}^\infty = X^T g_{ji}$. Letting $h(t) = (h_1^T, \ldots, h_n^T)^T$ and $g^\infty = ((g_1^\infty)^T, \ldots, (g_m^\infty)^T)^T$, we can write the overall dynamics in a concise form:

$$\dot{\hat{p}} = f(\hat{p}) + h(t) \tag{13.25}$$

Note that the above dynamics (13.25) has the same form as (10.12). Since $h(t)$ is bounded and $h(t) \to 0$ as $t \to \infty$, we can have the following result:

Theorem 13.7 *Under the same setup as the Theorem 10.1, with the orientation estimation law (10.5) and the position estimation law* (13.23)*, if* $\hat{p}_i(t_0) \neq \hat{p}_j(t_0)$, $\forall i \neq j$, $i, j = 1, \ldots, n$, *then we can have* $\frac{\hat{p}_j - \hat{p}_i}{\|\hat{p}_j - \hat{p}_i\|} \to X^T g_{ij}$ *almost globally asymptotically, for all* $(i, j)^e \in \mathcal{E}$.

Proof The proof can be completed by combining the proofs of Corollary 10.1 and Theorem 10.1.

Theorem 13.7 means that the positions of agents would be estimated as far as the given measurements g_{ji} constitute a bearing-rigid graph, up to a common rotation X. Thus, the estimated positions of the network may have a different scale from the true network.

13.4 Finite-Time Orientation Localization

In Chaps. 6–10, distance- or bearing-based formation control laws via orientation alignment have been presented. In these chapters, it was shown that the orientations of agents can be estimated up to a common angle offset in 2-D or up to a common rotation in 3-D. For the orientation estimation, the bearing angles between neighboring agents are exchanged. Thus, the orientation estimation or orientation localization schemes developed in Chaps. 6–10 can be called *bearing-based orientation localization*. Also as shown in Sect. 13.3, the orientation localization scheme has been combined to the position estimation. Thus, the localization schemes developed in Sect. 13.3 can be called SE(2) localization for the 2-D cases, and SE(3) localization for the 3-D cases. Since the formation control problem via orientation alignment has a cascade structure (see Fig. 6.3), it would be desirable to achieve the orientation alignment with a faster convergence or in a finite-time. This section presents a generalized orientation localization in a finite-time in general d-dimensional space. This result is reused from [9].

Let the orientation of agent j measured by the agent i be denoted as $R_{ji} = R_j R_i^{-1}$ in $\mathbb{R}^d$. We will modify the update rule of (8.23) in a vector form for general d-dimensional dynamics of auxiliary variables. For this, let the auxiliary variables be represented in a matrix $Z_i \in \mathbb{R}^{d\times d}$. Then, we rewrite (8.23) as

$$\dot{Z}_i(t) = \sum_{j\in\mathcal{N}_i} \frac{1}{\|R_{ji}^{-1}Z_j - Z_i\|_F^{\alpha}}(R_{ji}^{-1}Z_j - Z_i) \tag{13.26}$$

where $0 < \alpha < 1$ and $\|\cdot\|_F$ denotes the Frobenius norm defined as $\|\cdot\|_F = \sqrt{\text{trace}[(\cdot)^T(\cdot)]}$. For a simplicity of analysis, using the following relationship

$$\|R_{ji}^{-1}Z_j - Z_i\|_F^{\alpha} = \|R_j^{-1}Z_j - R_i^{-1}Z_i\|_F^{\alpha}$$

we further rewrite (13.26) as

$$R_i^{-1}\dot{Z}_i(t) = \sum_{j\in\mathcal{N}_i} \frac{1}{\|R_j^{-1}Z_j - R_i^{-1}Z_i\|_F^{\alpha}}(R_j^{-1}Z_j - R_i^{-1}Z_i) \tag{13.27}$$

Denoting $S_i = R_i^{-1}Z_i(t)$, since the orientations of agents are fixed, the above equation can be changed as

$$\dot{S}_i(t) = \sum_{j \in \mathcal{N}_i} \frac{1}{\|S_j - S_i\|_F^\alpha}(S_j - S_i) \tag{13.28}$$

Considering the orientations of all the agents in a vector form, we can express (13.28) as

$$\dot{S}(t) = -(\bar{\mathbb{L}} \otimes \mathbb{I}_d)S(t), \tag{13.29}$$

which is the counterpart of (8.24), where $S(t) = [S_1^T, \ldots, S_n^T]^T$, and the Laplacian matrix $\bar{\mathbb{L}} = [\bar{l}_{ij}]$ is computed as

$$\bar{l}_{ij} = \begin{cases} 0 & \text{if } S_i = S_j,\ (i,j)^e \in \mathcal{E},\ i \neq j,\ \text{or } (i,j)^e \notin \mathcal{E},\ i \neq j \\ -\frac{1}{\|S_j - S_i\|_F^\alpha} & \text{if } (i,j)^e \in \mathcal{E},\ i \neq j,\ S_i \neq S_j \\ \sum_{k \in \mathcal{N}_i} \bar{l}_{ik} & \text{if } i = j \end{cases} \tag{13.30}$$

Denoting the center of matrices S_i, $\forall i \in \mathcal{V}$, as $S^c = \frac{1}{n}(\mathbf{1}_n \otimes \mathbb{I}_d)^T S(t) \in \mathbb{R}^{d \times d}$, from

$$(\mathbf{1}_n \otimes \mathbb{I}_d)^T \dot{S}(t) = -(\mathbf{1}_n \otimes \mathbb{I}_d)^T (\bar{\mathbb{L}} \otimes \mathbb{I}_d)S(t) = 0$$

we can see that S^c is invariant. Let $\delta_i(t) \triangleq S_i(t) - S^c$; then we have $\dot{\delta}_i(t) = \dot{S}_i(t)$. The convergence of (13.28) is analyzed in the following theorem:

Lemma 13.1 *The consensus dynamics* (13.28) *converges to* S^c *in a finite-time, when the underlying topology is connected.*

Proof For the proof, let us consider a Lyapunov candidate as $V(t) = \frac{1}{2}\sum_{i=1}^n \|\delta_i\|_F^2$, which is radially unbounded, continuously differentiable, and $V = 0$ if and only if $S_i = S^c$, $\forall i \in \mathcal{V}$. With the relationship $S_j - S_i = \delta_j - \delta_i$, the derivative of $V(t)$ is obtained as

$$\begin{aligned} \dot{V}(t) &= \text{trace}\left(\sum_{i=1}^n \delta_i^T \dot{\delta}_i\right) \\ &= -\text{trace}\left(\sum_{i=1}^n \delta_i^T \sum_{j \in \mathcal{N}_i} \frac{1}{\|\delta_i - \delta_j\|_F^\alpha}(\delta_i - \delta_j)\right) \\ &= -\text{trace}\left(\sum_{\forall (i,j)^e \in \mathcal{E}} \frac{(\delta_i - \delta_j)^T(\delta_i - \delta_j)}{\|\delta_i - \delta_j\|_F^\alpha}\right) \\ &= -\sum_{\forall (i,j)^e \in \mathcal{E}} \frac{\|\delta_i - \delta_j\|_F^2}{\|\delta_i - \delta_j\|_F^\alpha} \\ &= -\sum_{\forall (i,j)^e \in \mathcal{E}} \|\delta_i - \delta_j\|_F^{2-\alpha} \end{aligned}$$

With the help of inequality $(\sum_{i=1}^{n} \xi_i)^p \le \sum_{i=1}^{n} \xi_i^p$, when $\xi_i \ge 0$ and $0 \le p \le 1$, the above equality can be further changed as

$$
\begin{aligned}
\dot{V}(t) &\le -\left(\sum_{\forall (i,j)^e \in \mathcal{E}} \|\delta_i - \delta_j\|_F^2 \right)^{1-\alpha/2} \\
&= -\left(\text{trace}\left(\sum_{\forall (i,j)^e \in \mathcal{E}} (\delta_i - \delta_j)^T (\delta_i - \delta_j) \right) \right)^{1-\alpha/2} \\
&= -\left(\text{trace}\left(\delta^T (\mathbb{L} \otimes \mathbb{I}_d) \delta \right) \right)^{1-\alpha/2} \\
&= -\left(\sum_{i=1}^{d} \delta(:, i)^T (\mathbb{L} \otimes \mathbb{I}_d) \delta(:, i) \right)^{1-\alpha/2}
\end{aligned}
$$

where $\delta(:, i)$ is the ith column vector of the matrix δ and $\mathbb{L}$ is the Laplacian corresponding to the underlying topology. Since null$(\mathbb{L} \otimes \mathbb{I}_d) = $ span(range$(\mathbf{1}_n \otimes \mathbb{I}_d)$), we have $\delta(:, i) \perp$ range$(\mathbf{1}_n \otimes \mathbb{I}_d)$ as far as $\delta^T (\mathbb{L} \otimes \mathbb{I}_d) \delta \neq 0$. Thus, denoting λ_{min} as the minimum nonzero eigenvalue of $(\mathbb{L} \otimes \mathbb{I}_d)$, we can have

$$
\begin{aligned}
\dot{V}(t) &\le -\left(\lambda_{min} \sum_{i=1}^{d} \|\delta(:, i)\|^2 \right)^{1-\alpha/2} \\
&\le -\lambda_{min}^{1-\alpha/2} \left(\sum_{i=1}^{n} \text{trace}(\delta_i^T \delta_i) \right)^{1-\alpha/2} \\
&= -(2\lambda_{min})^{1-\alpha/2} (2V(t))^{1-\alpha/2}
\end{aligned}
$$

Thus, we can see that $V(t) \to 0$ in a finite-time by Lemma 2.15, which implies that $S_i(t)$, $\forall i \in \mathcal{V}$ converges to S^c in a finite-time.

Now, with Lemma 13.1, we can make the following main result.

Theorem 13.8 *Suppose that the underlying graph is connected and the orientation localization is performed by* (13.26). *Also, suppose that we have computed* $\hat{R}_i$ *by the Gram–Schmidt process (see Chap. 8) of* Z_i. *Then,* $\hat{R}_i$ *converges to* $R_i R^c$ *in a finite-time where* $R^c \in SO(d)$ *is a constant matrix.*

Proof From Lemma 13.1, we see that Z_i and S_i converge to some constants in a finite-time. Let these constants be denoted as Z_i^∞ and S^c, respectively. By taking the Gram–Schmidt process of Z_i and S_i, we can have $\hat{R}_i$ and $\hat{R}_{S_i}$. Thus, due to the orthonormal property, we can have $\hat{R}_i = R_i \hat{R}_{S_i}$, from $S_i = R_i^{-1} Z_i$. Let us take the Gram–Schmidt process of S^c, which is denoted as R^c. Then, it is clear that $\hat{R}_i$ converges to $R_i R^c$ in a finite-time. Also since the Lebesgue measure of $C(\bar{\mathbb{L}} \otimes \mathbb{I}_d)$ is almost zero, it is almost globally asymptotically finite-time convergent.

The results so far can be extended to the directed graphs. The following result is the summary of the directed cases.

Lemma 13.2 *Let the underlying sensing topology have a single acyclic directed rooted-in tree with the root node* 1*. Then, the dynamics* (13.29) *would converge to the constant* $S_1(t_0)$ *for all* $i \in \mathcal{V}$ *in a finite-time.*

Proof Let us consider the first follower, which is the second agent. The dynamics of agent 2 is given as

$$\begin{aligned}\dot{S}_2(t) &= \sum_{j\in\mathcal{N}_2^O} \frac{1}{\|S_j - S_2\|_F^\alpha}(S_j - S_2)\\ &= \frac{1}{\|S_1 - S_2\|_F^\alpha}(S_1 - S_2)\end{aligned}$$

With the Lyapunov candidate $V_2(t) = 1/2\text{trace}((S_2 - S_1)^T(S_2 - S_1))$, we can see that

$$\begin{aligned}\dot{V}_2 &= \text{trace}((S_2 - S_1)^T(\dot{S}_2 - \dot{S}_1))\\ &= -\frac{\text{trace}((S_2 - S_1)^T(S_2 - S_1))}{\|S_2 - S_1\|_F^\alpha}\\ &= -\kappa_2 V_2^{1-\alpha/2}\end{aligned}$$

where $\kappa_2 = 2^{1-\alpha/2}$. Thus, $S_2(t)$ converges to S_1 in a finite-time. The dynamics of other follower agents are described by

$$\dot{S}_i(t) = \sum_{j\in\mathcal{N}_i^O} \frac{1}{\|S_j - S_i\|_F^\alpha}(S_j - S_i)$$

Taking the Lyapunov candidate $V_i(t) = 1/2\sum_{j\in\mathcal{N}_i^O}\text{trace}((S_j - S_i)^T(S_j - S_i))$, we can have $\dot{V}_i(t) = \sum_{j\in\mathcal{N}_i^O}\text{trace}((S_j - S_i)^T(\dot{S}_j - \dot{S}_i))$. Thus, since the rotation matrices are bounded, the states of orientation localization dynamics do not diverge. Moreover, it can be considered that the outgoing neighboring agents have been converged to S_1 in a finite-time. So, we can rewrite $\dot{V}_i$ such as

$$\begin{aligned}\dot{V}_i(t) &= |\mathcal{N}_i^O|\,\text{trace}((S_i - S_1)^T\dot{S}_i)\\ &= -|\mathcal{N}_i^O|\,\text{trace}\left((S_i - S_1)^T\sum_{j\in\mathcal{N}_i^O}\frac{1}{\|S_j - S_i\|_F^\alpha}(S_i - S_j)\right)\\ &= -|\mathcal{N}_i^O|\sum_{j\in\mathcal{N}_i^O}\text{trace}\left((S_i - S_1)^T\frac{1}{\|S_j - S_i\|_F^\alpha}(S_i - S_j)\right)\\ &= -\kappa_i V_i^{1-\alpha/2}\end{aligned}$$

where $\kappa_i = 2^{1-\alpha/2}|\mathcal{N}_i^O|$. Thus, the states $S_i(t),\ \forall i \geq 3,\ i \in \mathcal{V}$, also converge to S_1 in a finite-time.

Now, with the Lemma 13.2, we can make the following result:

Theorem 13.9 *Under the same assumption of Lemma 13.2, and with the update law* (13.26)*, when the rotation matrices $\hat{R}_i$ are computed by the Gram–Schmidt process of Z_i, it converges to $R_i R^c$ in a finite-time.*

Proof The proof is similar to the proof of Theorem 13.8.

13.5 Edge Localization

Thus far, in this monograph, the orientation localization has been achieved with bearing measurements. In this section, we would like to provide a discussion on orientation localization using subtended angles, which might be called *subtended angle-based orientation localization*. As already mentioned in the previous chapters, the orientations of agents have nothing to do with subtended angles. Let us consider a triangular group. Whatever the orientations of agents are, the three subtended angles α_1, α_2, and α_3 do not change. Thus, from the given subtended angles, we cannot decide the orientations of agents. Therefore, as commented in Chap. 9, without any additional sensing variables, it is not possible to decide the orientations; furthermore, it cannot conduct a formation control. So, we need to add some vector components to the problem formulation or into sensing or control variables. This argument can be clarified more technically, by observing an essential difference between the bearing vectors and subtended angles. A bearing vector is a function of positions and orientations of agents. From $g_{ji}^i = (p_j^i)/(\|p_j - p_i\|)$, it is clear that g_{ji}^i will change as p_j or p_i changes. Also since p_j^i is dependent upon the orientation of agent i, we can see that g_{ji}^i is a function of orientations of agents. However, for a subtended angle $\cos(\alpha_{jk}^i) = (\|p_{ij}\|^2 + \|p_{ik}\|^2 - \|p_{jk}\|^2)/(2\|p_{ij}\|\|p_{ik}\|)$, it is shown that it is only a function of positions of agents. Thus, the subtended angles can be controlled by way of changing the positions of agents. However, the change of positions of agents can be done only when the local coordinate frames of agents are aligned to a global coordinate frame. In such a case, without considering the orientations of agents, we can simply consider the agents as point masses. But, when the orientations of agents are not aligned, we need to consider relative orientations of agents. In this case, we need to have directional information such as relative displacements or bearing vectors, for the control of subtended angles. Actually, there is a trivial relationship between the subtended angles and bearing vectors. When an agent i measures two bearing vectors toward the neighboring agents j and k, denoting them as g_{ji}^i and g_{ki}^i, the subtended angle $\cos(\alpha_{jk}^i)$ is computed as $\cos(\alpha_{jk}^i) = (g_{ji}^i)^T g_{ki}^i$. Furthermore due to the fact that $(g_{ji}^i)^T g_{ki}^i = (g_{ji})^T g_{ki}$, we can have

$$\cos(\alpha_{jk}^i) = (g_{ji})^T g_{ki} = (g_{ji}^i)^T g_{ki}^i \tag{13.31}$$

Thus, if we define the orientation localization problem via subtended angles based on (13.31), it is to find g_{ji} and g_{ki} from α^i_{jk}. However, since α^i_{jk} is a scalar, and g_{ji} and g_{ki} are 2×1 vectors, it is not direct to compute g_{ji} and g_{ki} from $\cos(\alpha^i_{jk})$. To handle this problem, we may want to assume that an axis of agent's local coordinate frame points toward one of its neighboring agents. With respect to this axis (let us call this axis as *directional edge*), agent is supposed to be able to measure subtended angles to other neighboring agents. It is also assumed that agents can exchange the estimated orientations with neighboring agents. However, when an agent measures a subtended angle with respect to the directional edge, this angle is actually a bearing vector. Thus, under this setup, the overall localization procedure is exactly the same as the orientation localization via bearing measurements, which is a trivial problem.

As a nontrivial problem, let us assume that there are a set of subtended angle constraints $\mathcal{A}$, and each angle constraint is described as (13.31). Then, we define an edge localization problem as

Definition 13.1 (*Edge localization*) Given a set of angle constraints $\mathcal{A} = \{\ldots, \alpha^i_{jk}, \ldots\}$, find the bearing vectors g_{ji} for all edges $(i, j)^e \in \mathcal{E}$.

In the above edge localization problem, the constraints are subtended angles. So, if the graph constrained by the angles is rigid with angle constraints (see weak rigidity theory in Sect. 2.2.4 and refer to generalized weak rigidity concept [5]), then there will be a unique solution for the bearing vectors up to translations and dilations. The rigidity with a set of subtended angles is determined by Theorem 2.15. So, based on Theorem 2.15, let us assume that the realized graph is generalized infinitesimally weak rigid. The bearing vector for the edge $(i, j)^e$ is given as g_{ji}. If the edge $(i, k)^e$ is also an incident edge to the agent i, then the edges $(i, j)^e$ and $(i, k)^e$ are considered as adjacent edges. It can be also said that the edge $(i, j)^e$ has two incident vertices i and j. Let the bearing vectors g_{ji} be ordered as $g_1, \ldots, g_k, \ldots, g_m$. Then, each bearing vector g_k is mapped to the orientation angle θ_k, as

$$g_k \Longleftrightarrow e^{\mathrm{i}\theta_k} \tag{13.32}$$

Thus, the edge localization problem becomes the orientation estimation problem studied in Chap. 8. For a convenience, let the edge agents be denoted as i^e, and the vertex agents as i^v. Let the orientation of each edge agent i^e be characterized by the angle θ_{i^e}. Then, the neighboring edge agents of i^e are adjacent edges j^e of agent i^e when there are the subtended angle constraints such as $g^T_{i^e} g_{j^e}$.

When two edges i^e and j^e are neighboring edge agents, the difference of orientation angles θ_{i^e} and θ_{j^e}, i.e., $\theta_{ji^e} \triangleq \theta_{i^e} - \theta_{j^e}$, is equal to the subtended angle between two edges, in the counterclockwise direction. That is, $\theta_{ji^e} = \cos^{-1}(g^T_{i^e} g_{j^e})$, which is the measured subtended angle. Let the estimated orientation of θ_{i^e} be denoted as $\hat{\theta}_{i^e}$. Then, defining a virtual variable in Euclidean space as $\hat{z}_{i^e}$ matching to $\hat{\theta}_{i^e}$ as shown in Fig. 8.1, we can propose the following estimation law:

$$\dot{\hat{z}}_{i^e}(t) = \sum_{j^e \in \mathcal{N}_{i^e}} (e^{-\mathrm{i}\theta_{ji^e}} \hat{z}_{j^e}(t) - \hat{z}_{i^e}(t)) \tag{13.33}$$

Due to the same reason as Theorem 8.1 and (8.12), the orientations of bearing vectors can be localized up to a common offset. It is finally remarkable that when only the subtended angles are used for a localization, the orientation localization of agents is not doable; but the edge localization can be performed.

13.6 Summary

It can be said that formation control and network localization have a duality in the sense that they can be reformulated with each other under the same setups and the same property. The localized information can be immediately used for the formation control in local frames. This issue was mentioned in [1]. In Sect. 13.4, there were discussions on orientation localization and edge localization when there are only the subtended angle constraints. It was argued that the edge localization can be achieved, but the orientation localization is not possible. It will be interesting to examine the possibility of position localization only using subtended angle measurements. Since the subtended angles are functions of positions of agents, it appears that the localization could be done in a distributed way. There are two approaches for this. The first approach is to decompose the subtended angles into bearing vectors. Then, it is a direct bearing-based localization problem studied in Sect. 13.3.2; so, this approach looks trivial. Another approach is to find $\hat{p}_i$, $\forall i \in \mathcal{V}$, such that $\| \cos(\alpha^i_{jk}) - (\|\hat{p}_{ij}\|^2 + \|\hat{p}_{ik}\|^2 - \|\hat{p}_{jk}\|^2)/(2\|\hat{p}_{ij}\|\|\hat{p}_{ik}\|)\| = 0$ for all angle constraints when the underlying graph is generalized infinitesimally weak rigid. Then, the update of estimation may be implemented as

$$\begin{aligned} \dot{\hat{p}}_i(t) &= k_{\hat{p}} \sum_{j,\, k \in \mathcal{N}_i} \left(\cos(\alpha^i_{jk}) - \frac{\|\hat{p}_{ij}\|^2 + \|\hat{p}_{ik}\|^2 - \|\hat{p}_{jk}\|^2}{2\|\hat{p}_{ij}\|\|\hat{p}_{ik}\|} \right) \\ &= k_{\hat{p}} \sum_{j,\, k \in \mathcal{N}_i} \left(\frac{\|p_{ij}\|^2 + \|p_{ik}\|^2 - \|p_{jk}\|^2}{2\|p_{ij}\|\|p_{ik}\|} - \frac{\|\hat{p}_{ij}\|^2 + \|\hat{p}_{ik}\|^2 - \|\hat{p}_{jk}\|^2}{2\|\hat{p}_{ij}\|\|\hat{p}_{ik}\|} \right) \end{aligned} \tag{13.34}$$

But, defining the error of the estimation as $e_{\hat{p}_i} = \hat{p}_i - p_i$, it is quite difficult to analyze the convergence of $e_{\hat{p}_i}$ to a common value since the relationship in (13.34) is of highly nonlinear.

As another interesting problem, we may want to find the distances of edges from the given subtended angles. For this, we may define $\|\hat{p}_{ij}\|^2 - \|p_{ij}\|^2 = \bar{e}_{ij}$. Then, with the following relationship:

$$2\hat{p}_{ij}^T(\dot{\hat{p}}_i - \dot{\hat{p}}_j) = \dot{\bar{e}}_{ij} \tag{13.35}$$

we can replace (13.34) by the errors $\bar{e}_{ij}$. Although the relationship in error components becomes more obvious, it shows still a nonlinear relationship; so the analysis is still not that straightforward.

13.7 Notes

In this chapter, it was mentioned that the network localization is a dual problem of formation control, in its mathematical formulation. However, it may be necessary to note that there is a difference between these two setups. In the formation control problem, the agents evolve in physical domain, while in network localization problem, the agents are updated in a cyber domain mainly via communications. So, in formation control problem, a collision avoidance among agents may need to be further treated, while in network localization, for example, a cyber attack in communication networks may need to be elaborated. When the formation control and network localization are combined, the overall setup becomes network localization-based feedback formation control (see Chap. 7 and Sect. 10.2). This approach is analogous to the feedback control from estimated states (or observer-based dynamic feedback controller [2]). In the linear systems theory, when there are noises, the problem becomes the linear quadratic Gaussian (LQG). Similarly, we may be able to formulate a LQG problem in the network localization-based feedback formation control in the presence of noises.

In network localization, there could be mis-matchings between sensed physical information and communicated cyber-information. For example, in the leader–follower network localization, it is supposed that the leader agents provide their true position or orientation information to the follower agents. If the true information matches to the sensed physical information, the overall network would be localized correctly. However, some leader agents, which are supposed to deliver their true information, may deliver fake information to the follower agents. Then, the agents would not be localized correctly. Instead, the agents may attempt to localize their position or orientation continuously, without converging to fixed values. The agents may propagate the fake information with intention or the information may be biased due to cyber attack. Then, the overall system may not be stable. These problems would be meaningful for a further investigation in association with network systems theory.

The results of Sects. 13.2 and 13.3 have been reused from [6–8, 10], and the results of Sect. 13.4 are reproduced from [9]. The following copyright and permission notices are acknowledged.

- The text of Sect. 13.2 is © [2014] IEEE. Reprinted, with permission, from [8].
- The text of Sect. 13.3.2 is © [2018] IEEE. Reprinted, with permission, from [10].
- The text of Sect. 13.3.1 is © [2019] IEEE. Reprinted, with permission, from [7].
- The statements of Lemma 13.1, Theorems 13.8, and Lemma 13.2 with the proofs, and the text of Sect. 13.4 are © [2019] IEEE. Reprinted, with permission, from [9].

References

1. Ahn, H.-S.: Formation coordination for self-mobile localization: framework. In: Proceedings the IEEE International Symposium on Computational Intelligence in Robotics and Automation (CIRA), pp. 340–348 (2009)
2. Antsaklis, P.J., Michel, A.N.: A Linear Systems Primer. Birkhäuser, Boston (2007)
3. Aspnes, J., Eren, T., Goldenberg, D.K., Morse, A.S., Whiteley, W., Yang, Y.R., Anderson, B.D.O., Belhumeur, P.N.: A theory of network localization. IEEE Trans. Mob. Comput. **5**(12), 1663–1678 (2006)
4. Fang, J., Cao, M., Morse, A.S., Anderson, B.D.O.: Sequential localization of sensor networks. SIAM J. Control Optim. **48**(1), 321–350 (2009)
5. Kwon,S.-H., Ahn, H.-S.: Generalized weak rigidity: theory, and local and global convergence of formations, pp. 1–17. arXiv:1809.02562 [cs.SY]
6. Lee, B.-H., Ahn, H.-S.: Formation control and network localization via distributed global orientation estimation in 3-D, pp. 1–8 (2017). arXiv:1708.03591v1 [cs.SY]
7. Lee, B.-H., Kang, S.-M., Ahn, H.-S.: Distributed orientation estimation in SO(d) and applications to formation control and network localization. IEEE Trans. Control Netw. Syst. 1–11 (2019). https://doi.org/10.1109/TCNS.2018.2888999
8. Oh, K.-K., Ahn, H.-S.: Formation control and network localization via orientation alignment. IEEE Trans. Autom. Control **59**(2), 540–545 (2014)
9. Tran, Q.V., Trinh, M.H., Zelazo, D., Mukherjee, D., Ahn, H.-S.: Finite-time bearing-only formation control via distributed global orientation estimation. IEEE Trans. Control Netw. Syst. 1–11 (2019). https://doi.org/10.1109/TCNS.2018.2873155
10. Trinh, M.-H., Lee, B.-H., Ye, M., Ahn, H.-S.: Bearing-based formation control and network localization via global orientation estimation. In: Proceedings of the IEEE Conference on Control Technology and Applications, pp. 1084–1089 (2018)

Appendix
Proofs

Proof (Proof of Lemma 3.9) The agents 1, 2 and 3 are collinear. Let us consider the case when agent 1 is between agents 2 and 3, in the order of 2-1-3 as depicted in Fig. A.1a. Without loss of generality, we suppose that z_{12} is along the positive direction. Then, z_{13} and z_{23} are along the negative direction. From $\bar{e}_{12}z_{12} - \bar{e}_{23}z_{23} = 0$, assume that $\bar{e}_{12} < 0$. Then, $\bar{e}_{23} > 0$. But, from $\bar{e}_{12}z_{12} + \bar{e}_{13}z_{13} = 0$, it should be true that $\bar{e}_{13} < 0$. But, it is the contradiction that $\bar{e}_{12} < 0$, $\bar{e}_{13} < 0$, and $\bar{e}_{23} > 0$. Similarly, let us suppose that $\bar{e}_{12} > 0$ holds. In this case, we should have $\bar{e}_{32} < 0$ and $\bar{e}_{31} > 0$, which is also a contradiction. Thus, agent 1 cannot be between agents 2 and 3.

Next, let us suppose that agent 2 is between agents 1 and 3 in the order of 1-2-3 as depicted in Fig. A.1b. Suppose that z_{12} is along the positive direction. Then, z_{13} and z_{23} are also along the positive direction. From $\bar{e}_{12}z_{12} - \bar{e}_{23}z_{23} = 0$, assume that $\bar{e}_{12} < 0$. Then, $\bar{e}_{23} < 0$. But, from $\bar{e}_{12}z_{12} + \bar{e}_{13}z_{13} = 0$, it should be true that $\bar{e}_{13} > 0$. As another possible case, let us suppose that $\bar{e}_{12} > 0$. In this case, we should have $\bar{e}_{32} > 0$ and $\bar{e}_{31} < 0$; but this situation cannot happen due to the triangular inequality of the desired formation. Thus, overall, we can only have $\bar{e}_{12} = \|z_{12}\|^2 - \bar{d}_{12}^* < 0$, $\bar{e}_{23} = \|z_{23}\|^2 - \bar{d}_{23}^* < 0$, and $\bar{e}_{13} = \|z_{13}\|^2 - \bar{d}_{13}^* > 0$ under the order of 1-2-3. Let us now further scrutinize this situation. From the constraints $\bar{e}_{12}z_{12} + \bar{e}_{13}z_{13} = \bar{e}_{12}z_{12} - \bar{e}_{23}z_{23} = \bar{e}_{23}z_{23} + \bar{e}_{13}z_{13} = 0$, we have $|\bar{e}_{12}| > |\bar{e}_{13}|$ and $|\bar{e}_{23}| > |\bar{e}_{13}|$ since $\|z_{13}\| > \|z_{12}\|$ and $\|z_{13}\| > \|z_{23}\|$. Consequently, we have $\bar{e}_{23} + \bar{e}_{13} < 0$ and $\bar{e}_{12} + \bar{e}_{13} < 0$.

Proof (Proof of Lemma 5.4) First, for the analysis of convergence of $\|z_{13}\|$ and $\|z_{23}\|$, the potential function $\phi_3 = \frac{1}{4}(\bar{e}_{13}^2 + \bar{e}_{23}^2)$ is taken a derivative in the form

$$\begin{aligned}
\dot{\phi}_3 &= \frac{1}{2}(\bar{e}_{13}\dot{\bar{e}}_{13} + \bar{e}_{23}\dot{\bar{e}}_{23}) \\
&= \bar{e}_{13}z_{13}^T\dot{z}_{13} + \bar{e}_{23}z_{23}^T\dot{z}_{23} \\
&= -\bar{e}_{13}z_{13}^T Q(\bar{e}_{13}z_{13} + \bar{e}_{23}z_{23}) + \bar{e}_{23}z_{23}^T(\bar{e}_{12}z_{12} - Q(\bar{e}_{13}z_{13} + \bar{e}_{23}z_{23})) \\
&= -\bar{e}_{13}^2 z_{13}^T Q z_{13} - \bar{e}_{13}\bar{e}_{23}z_{13}^T(Q + Q^T)z_{23} - \bar{e}_{23}^2 z_{23}^T Q z_{23} + \bar{e}_{23}\bar{e}_{12}z_{23}^T z_{12}
\end{aligned}$$

H.-S. Ahn, *Formation Control*, Studies in Systems, Decision and Control 205, https://doi.org/10.1007/978-3-030-15187-4

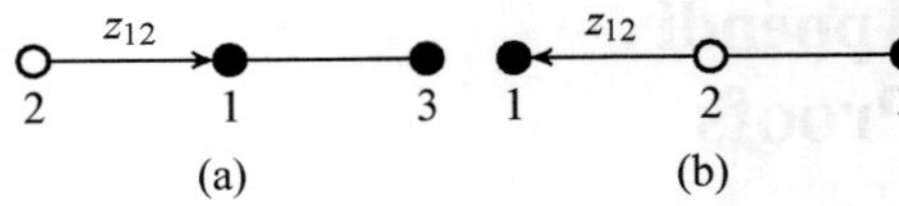

Fig. A.1 Formations in the order of 2-1-3 and in the order of 1-2-3

Since $z_i^T Qz_i = \|z_i\|^2 \cos\theta$ and $Q + Q^T = \text{diag}(2\cos\theta)$, the above equation can be further changed as

$$\dot{\phi}_3 = -\cos\theta\|\bar{e}_{13}z_{13} + \bar{e}_{23}z_{23}\|^2 + \bar{e}_{23}\bar{e}_{12}z_{23}^T z_{12} \tag{A.1}$$

From Lemma 5.3, it was shown that e_{12} converges to zero exponentially fast. Also as agent 3 stays away from agents 1 and 2, the term $\|\bar{e}_{13}z_{13} + \bar{e}_{23}z_{23}\|^2$ becomes dominated. It can be inferred that $-\cos\theta\|\bar{e}_{13}z_{13} + \bar{e}_{23}z_{23}\|^2 + \bar{e}_{23}\bar{e}_{12}z_{23}^T z_{12}$ is negative as agent 3 is away from other agents. Thus, it is clear that ϕ_3 is bounded, and all other states are bounded. From (A.1), the solution is obtained as

$$\begin{aligned}\phi_3(p(t)) = \phi_3(p(t_0)) &+ \int_{t_0}^{t} \bar{e}_{23}(\tau)\bar{e}_{12}(\tau)z_{23}^T(\tau)z_{12}(\tau)d\tau \\ &- \cos\theta \int_{t_0}^{t} \|\bar{e}_{13}(\tau)z_{13}(\tau) + \bar{e}_{23}(\tau)z_{23}(\tau)\|^2 d\tau\end{aligned} \tag{A.2}$$

Since $\bar{e}_{12}$ converges to zero exponentially fast, the second term of the right-hand side of (A.2) would converge to a constant. Also since $\phi_3(p(t))$ should be bounded below by zero, $\|\bar{e}_{13}(\tau)z_{13}(\tau) + \bar{e}_{23}(\tau)z_{23}(\tau)\|^2$ cannot be increased to infinity. Thus, it is inferred that $\|\bar{e}_{13}(\tau)z_{13}(\tau) + \bar{e}_{23}(\tau)z_{23}(\tau)\|^2$ would converge to zero as $t \to \infty$ by Barbalat's lemma (see Lemma 2.8). Now from Lemma 5.1, e_{13} and e_{23} exponentially converge to zero if agents are not collinear initially. Thus, $\dot{p}_3$ exponentially converges to zero. To further consider the collinearity case of agents (i.e., $\bar{e}_{13}z_{13} + \bar{e}_{23}z_{23} = 0$, $\bar{e}_{13} \neq 0$ or $\bar{e}_{23} \neq 0$), the linearization of (5.7–5.9) at $\bar{e}_{12} = 0$ is utilized as follows:

$$J_f = \left[\begin{array}{c|c|c} -Q(Z_{13} + \bar{e}_{13}\mathbb{I}_2) & -Q(Z_{23} + \bar{e}_{23}\mathbb{I}_2) & Z_{12} \\ \hline -Q(Z_{13} + \bar{e}_{13}\mathbb{I}_2) & -Q(Z_{23} + \bar{e}_{23}\mathbb{I}_2) & \mathbf{0}_{2\times 2} \\ \hline \mathbf{0}_{2\times 2} & \mathbf{0}_{2\times 2} & -Z_{01} \end{array}\right] \tag{A.3}$$

where $Z_{ij} = 2z_{ij}z_{ij}^T$. Note $-Z_{01}$ is negative semi-definite and it follows:

$$\det(\lambda\mathbb{I}_4 - K_f) = \lambda^2 \det(\lambda\mathbb{I}_2 + Q(Z_{13} + \bar{e}_{13}\mathbb{I}_2) + Q(Z_{23} + \bar{e}_{23}\mathbb{I}_2))$$

where

$$K_f = \left[\begin{array}{c|c} -Q(Z_{13} + \bar{e}_{13}\mathbb{I}_2) & -Q(Z_{23} + \bar{e}_{23}\mathbb{I}_2) \\ \hline -Q(Z_{13} + \bar{e}_{13}\mathbb{I}_2) & -Q(Z_{23} + \bar{e}_{23}\mathbb{I}_2) \end{array}\right] \tag{A.4}$$

Thus, K_f has two zero eigenvalues and other two eigenvalues that are eigenvalues of $L_f = -Q(Z_{13} + \bar{e}_{13}\mathbb{I}_2) - Q(Z_{23} + \bar{e}_{23}\mathbb{I}_2)$. Since $\bar{e}_{13}z_{13} + \bar{e}_{23}z_{23} = 0$, it is true that $Z_{23} = (\bar{e}_{13}/\bar{e}_{23})^2 Z_{13}$. Thus, L_f can be expressed as:

$$L_f = -Q\begin{bmatrix} 2x_{13}^2\left(\frac{\bar{e}_{13}^2+\bar{e}_{23}^2}{\bar{e}_{23}^2}\right) + \bar{e}_{13} + \bar{e}_{23} & 2x_{13}y_{13}\left(\frac{\bar{e}_{13}^2+\bar{e}_{23}^2}{\bar{e}_{23}^2}\right) \\ 2x_{13}y_{13}\left(\frac{\bar{e}_{13}^2+\bar{e}_{23}^2}{\bar{e}_{23}^2}\right) & 2y_{13}^2\left(\frac{\bar{e}_{13}^2+\bar{e}_{23}^2}{\bar{e}_{23}^2}\right) + \bar{e}_{13} + \bar{e}_{23} \end{bmatrix} \tag{A.5}$$

where $z_{ij} = (x_{ij}, y_{ij})^T$. Then using the fact that $\bar{e}_{13}^2\|z_{13}\|^2 = \bar{e}_{23}^2\|z_{23}\|^2$ and $\|z_{13} - z_{23}\| = d_{12}^*$ for $p \in \mathcal{U}_3$, the determinant of L_f can be obtained as

$$\begin{aligned} \det L_f &= (\bar{e}_{13} + \bar{e}_{23})(2\|z_{13}\|^2 + 2\|z_{23}\|^2 + \bar{e}_{13} + \bar{e}_{23}) \\ &= (\bar{e}_{13} + \bar{e}_{23})(-(d_{13}^*)^2 - (d_{23}^*)^2 + 3\|z_{13} - z_{23}\|^2 + 6z_{13}^T z_{23}) \\ &= (\bar{e}_{13} + \bar{e}_{23})(-\bar{d}_{13}^* - \bar{d}_{23}^* + 3\bar{d}_{12}^* + 6z_{13}^T z_{23}) \end{aligned} \tag{A.6}$$

Similarly, the trace of L_f can be obtained as

$$\text{trace} L_f = 2(\bar{d}_{13}^* + \bar{d}_{23}^* - 2\bar{d}_{12}^* - 4z_{13}^T z_{23})\cos\theta \tag{A.7}$$

From Lemma 5.2, it is shown that $\bar{e}_{13} + \bar{e}_{23} < 0$ for all $p \in \mathcal{U}_3$. Let us consider two cases. If $2\bar{d}_{12}^* + 4z_{13}^T z_{23} \leq \frac{2}{3}(\bar{d}_{13}^* + \bar{e}_{23}^2)$, then $\text{trace}L_f \geq \frac{2}{3}(\bar{d}_{13}^* + \bar{e}_{23}^2)\cos\theta$, which means that L_f has at least one positive real eigenvalue. On the other hand, if $2\bar{d}_{12}^* + 4z_{13}^T z_{23} > \frac{2}{3}(\bar{d}_{13}^* + \bar{e}_{23}^2)$, then $\det L_f < 0$, which means that the two eigenvalues have opposite signs. Therefore, $\mathcal{U}_3$ is unstable.

Proof (Proof of Lemma 5.7) Let $v_i v_j v_k v_l$ be a tetrahedron corresponding to the desired configuration of four vertices v_i, v_j, v_k and v_l. From now on, for a simplicity of presentation, in this proof, we write the vertices v_i, v_j, v_k and v_l just as i, j, k, and l. Then, $\|z_{ij}\| = d_{ij}^*$, $\|z_{ik}\| = d_{ik}^*$, and $\|z_{il}\| = d_{il}^*$ are the desired distances from agent i to the neighbor nodes. Also let $d_{jk}^* = \|z_{jk}\|$, $d_{kl}^* = \|z_{kl}\|$, and $d_{jl}^* = \|z_{jl}\|$. From Lemma 5.5, Lemma 5.6, and the triangular inequality, it follows that

$$\alpha_{ik}^j + \alpha_{il}^j > \alpha_{kl}^j \tag{A.8}$$

$$\alpha_{jk}^i + \alpha_{kl}^i + \alpha_{jl}^i < 2\pi \tag{A.9}$$

$$\alpha_{ik}^j + \alpha_{il}^j + \alpha_{kl}^j < 2\pi \tag{A.10}$$

$$d_{ij}^* + d_{il}^* > d_{jl}^*,\ d_{ij}^* + d_{jl}^* > d_{il}^*,\ d_{il}^* + d_{lj}^* > d_{ij}^* \tag{A.11}$$

When agent i converges to a position where i, j, k and l are coplanar, the configuration of the formation will be one of the configurations depicted in Fig. A.2. Denote the distances between vertices as $d_{i^u j} = \|z_{i^u j}\|$, $d_{i^u k} = \|z_{i^u k}\|$, and $d_{i^u l} = \|z_{i^u l}\|$. In what follows, we will investigate the geometrical properties of the cases of Fig. A.2. Let the position values of i, j, k and l be denoted as p_i, p_j, p_k and p_l.

- Figure A.2a: Let p_λ be the intersected point of the line containing i and j, and the line containing k and l. It can be written as $p_\lambda = \lambda p_k + (1-\lambda)p_l, \lambda \in (0,1)$ and the vector z_{ij} can be written as $z_{ij} = \alpha(p_i - p_\lambda)$ for some $\alpha > 1$. Then, $g_{ij}z_{ij} + g_{ik}z_{ik} + g_{il}z_{il} = 0$ can be expressed as

$$(\alpha\lambda g_{ij} + g_{ik})z_{ik} + (\alpha(1-\lambda)g_{ij} + g_{il})z_{il} = 0 \tag{A.12}$$

To satisfy the above equality, it should be true

$$\alpha\lambda g_{ij} + g_{ik} = 0 \tag{A.13}$$
$$\alpha(1-\lambda)g_{ij} + g_{il} = 0 \tag{A.14}$$

Further, to ensure the above equalities are satisfied, it is required that $\operatorname{sgn} g_{ij} = -\operatorname{sgn} g_{ik}$ and $\operatorname{sgn} g_{il} = -\operatorname{sgn} g_{ij}$. Thus, it is obtained as $\operatorname{sgn} g_{ij} = -\operatorname{sgn} g_{ik} = -\operatorname{sgn} g_{il}$. Assume that $g_{ij} < 0$. Then, $g_{ik} > 0$ and $g_{il} > 0$. Consequently, $d_{ij} < d_{ij}^*$, $d_{ik} > d_{ik}^*$, and $d_{il} > d_{il}^*$. Applying the results of Lemma 5.5 to the pair of triangles $v_iv_jv_k$ and $v_{i^u}v_jv_k$, and another pair of triangles $v_iv_jv_l$ and $v_{i^u}v_jv_l$, where v_i is the vertex corresponding to agent i when it is in a desired configuration, and v_{i^u} is the vertex corresponding to agent i when four agents are on the same plane, it follows:

$$\alpha_{ik}^j < \alpha_{i^uk}^j \text{ and } \alpha_{il}^j < \alpha_{i^ul}^j \tag{A.15}$$

where i is the case in a desired configuration and i^u is the case in coplanar configuration. Thus, it is obtained that $\alpha_{kl}^j = \alpha_{i^uk}^j + \alpha_{i^ul}^j > \alpha_{ik}^j + \alpha_{il}^j$. But, this is contradiction to (A.8). Next, suppose that $g_{ij} = 0$. Then, $g_{ik} = g_{il} = 0$. But, this is also contradiction to the assumption that the desired configuration is a tetrahedron. Lastly, suppose $g_{ij} > 0$. Then, it implies that $g_{ik} < 0$ and $g_{il} < 0$. Let k' and l' be the projections of k and l onto the edge connecting the vertices i^u and j. Since the tetrahedron $v_{i^u}v_kv_jv_l$ is a convex quadrilateral, it follows that $\|z_{i^uk'}\| < \|z_{i^uj}\|$ and $\|z_{i^ul'}\| < \|z_{i^uj}\|$. Now, the condition $g_{ij}z_{ij} + g_{ik}z_{ik} + g_{il}z_{il} = 0$ can be decomposed into two arbitrary orthogonal axes; one of them could be along the edge connecting i^u and j. Based on this argument, it is required to have

$$g_{ij}\|z_{i^uj}\| + g_{ik}\|z_{i^uk'}\| + g_{il}\|z_{i^ul'}\| = 0 \tag{A.16}$$

By inserting g_{ij} obtained from (A.16) to $g_{ij} + g_{ik} + g_{il}$, it is obtained as

$$g_{ij} + g_{ik} + g_{il} = \left(1 - \frac{\|z_{i^uk'}\|}{\|z_{i^uj}\|}\right) g_{ik} + \left(1 - \frac{\|z_{i^ul'}\|}{\|z_{i^uj}\|}\right) g_{il} < 0 \tag{A.17}$$

- Figure A.2b: Following the similar process as the case of Fig. A.2a, it is obtained as $\operatorname{sgn} g_{ij} = \operatorname{sgn} g_{ik} = \operatorname{sgn} g_{il}$. Assume that $g_{ij} \geq 0$, $g_{ik} \geq 0$, and $g_{il} \geq 0$. Consequently $d_{ij} \geq d_{ij}^*$, $d_{ik} \geq d_{ik}^*$, and $d_{il} \geq d_{il}^*$. Applying Lemma 5.5 to the three pairs of triangles $(v_jv_iv_k, v_jv_{i^u}v_k)$, $(v_kv_iv_l, v_kv_{i^u}v_l)$, and $(v_jv_iv_l, v_jv_{i^u}v_l)$, it follows

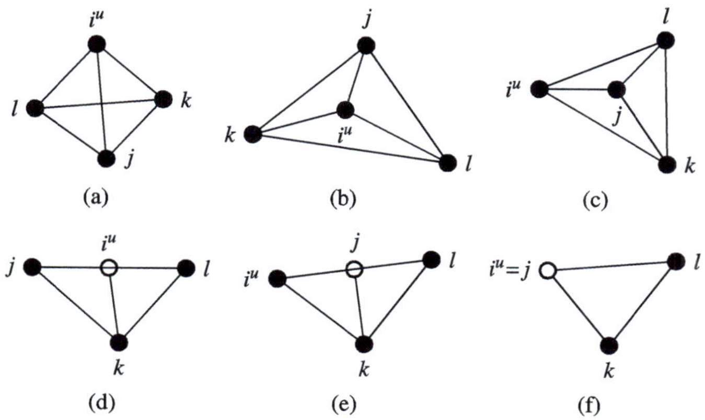

Fig. A.2 The possible coplanar cases when i is at the same plane with other vertices j, k, and l. © [2017] IEEE. Reprinted, with permission, from V. H. Pham, M. H. Trinh, and H.-S. Ahn. Distance-based directed formation control in 3-dimensional space. In *Proc. of the SICE Annual Conference*, pages 886–891, Sept. 2017

$$\alpha^i_{jk} \geq \alpha^{i^u}_{jk}, \ \alpha^i_{kl} \geq \alpha^{i^u}_{kl}, \ \text{and}\ \alpha^i_{jl} \geq \alpha^{i^u}_{jl} \tag{A.18}$$

which results in $\alpha^i_{jk} + \alpha^i_{kl} + \alpha^i_{jl} \geq \alpha^{i^u}_{jk} + \alpha^{i^u}_{kl} + \alpha^{i^u}_{jl} = 2\pi$. But this is a contradiction to (A.9). Thus, $g_{ij} < 0$, $g_{ik} < 0$, and $g_{il} < 0$.

- Figure A.2c: It is also similar to the case of Fig. A.2a. Following the similar process, it is obtained as $\operatorname{sgn} g_{ij} = -\operatorname{sgn} g_{ik} = -\operatorname{sgn} g_{il}$. Suppose that $g_{ij} \geq 0$. Then, $\alpha^j_{ik} \geq \alpha^j_{i^u k}$ and $\alpha^j_{il} \geq \alpha^j_{i^u l}$, which results in $\alpha^j_{ik} + \alpha^j_{il} + \alpha^j_{kl} \geq \alpha^j_{i^u k} + \alpha^j_{i^u l} + \alpha^j_{kl} = 2\pi$. This is a contradiction to (A.10). Thus, it should be $g_{ij} < 0$, $g_{ik} > 0$, and $g_{il} > 0$. Let k' and l' be the projections of k and l onto the edge connecting the vertices i^u and j. Then, $\|z_{i^u k'}\| \geq \|z_{i^u j}\|$ and $\|z_{i^u l'}\| \geq \|z_{i^u j}\|$. Thus, it is obtained as $g_{ij} + g_{ik} + g_{il} = \left(1 - \frac{\|z_{i^u k'}\|}{\|z_{i^u j}\|}\right) g_{ik} + \left(1 - \frac{\|z_{i^u l'}\|}{\|z_{i^u j}\|}\right) g_{il} < 0$.
- Figure A.2d: To satisfy the requirement $g_{ij}z_{ij} + g_{ik}z_{ik} + g_{il}z_{il} = 0$, it should be true that $g_{ik} = 0$, and $\operatorname{sgn} g_{ij} = \operatorname{sgn} g_{il}$. Suppose that $g_{ij} \geq 0$. Then, $g_{il} \geq 0$ and $d_{ij} \geq d^*_{ij}$ and $d_{il} \geq d^*_{il}$. It implies that $d^*_{ij} + d^*_{il} \leq d_{ij} + d_{il} = d_{jl} = d^*_{jl}$, which is a contradiction to (A.11). Thus, $g_{ij} < 0$ and $g_{il} < 0$.
- Figure A.2e: To satisfy the requirement $g_{ij}z_{ij} + g_{ik}z_{ik} + g_{il}z_{il} = 0$, it should be true that $g_{ik} = 0$, and $\operatorname{sgn} g_{ij} = -\operatorname{sgn} g_{il}$. Suppose that $g_{ij} \geq 0$. Then, $g_{il} \leq 0$, and $d_{ij} \geq d^*_{ij}$ and $d_{il} \leq d^*_{il}$. It implies that $d^*_{ij} + d^*_{jl} \leq d_{ij} + d^*_{jl} = d_{il} \leq d^*_{il}$, which is a contradiction to (A.11). Thus, $g_{ij} < 0$ and $g_{il} > 0$. Since $g_{ij}\|z_{i^u j}\| + g_{il}\|z_{i^u l}\| = 0$ and $\|z_{i^u j}\| < \|z_{i^u l}\|$, it should be true that $g_{ij} + g_{il} < 0$.
- Figure A.2f: From the requirement $g_{ij}z_{ij} + g_{ik}z_{ik} + g_{il}z_{il} = 0$, it is obtained as $g_{ik} = g_{il} = 0$. Thus, since $d_{ij} < d^*_{ij}$, it is true that $g_{ij} < 0$.

With the above arguments, the proof is completed.

Proof (Proof of Theorem 6.1) From the definition of $\text{dist}(x, \mathcal{E}_{2n})$ and since $e^{-k_p(\mathbb{L}\otimes\mathbb{I}_2)t}x$ converges to a point in $\mathcal{E}_{2n}$ exponentially fast, it is true that

$$\text{dist}(e^{-k_p(\mathbb{L}\otimes\mathbb{I}_2)t}x, \mathcal{E}_{2n}) \leq k_{\mathcal{E}} e^{-\lambda_{\mathcal{E}} t}\text{dist}(x, \mathcal{E}_{2n}) \tag{A.19}$$

It is clear that there exists an orthogonal subspace of $\mathcal{E}_{2n}$ in $\mathbb{R}^{2n}$, which is a Hilbert space, such as $\mathcal{E}_{2n}^{\perp} \in \mathbb{R}^{2n}$. Then, for any $x \in \mathbb{R}^{2n}$, x can be decomposed by $x^{\vdash} \in \mathcal{E}_{2n}$ and $x^{\perp} \in \mathcal{E}_{2n}^{\perp}$ in the sense $x = x^{\vdash} + x^{\perp}$. Thus, it is obvious that $\text{dist}(x, \mathcal{E}_{2n}) = \text{dist}(x^{\perp}, \mathcal{E}_{2n}) = \|x^{\perp}\|$. Hence for any $x \in \mathbb{R}^{2n}$ and $y \in \mathbb{R}^{2n}$, the following inequality holds:

$$\begin{aligned} \text{dist}(x + y, \mathcal{E}_{2n}) &= \|x^{\perp} + y^{\perp}\| \\ &\leq \|x^{\perp}\| + \|y^{\perp}\| \\ &= \text{dist}(x, \mathcal{E}_{2n}) + \text{dist}(y, \mathcal{E}_{2n}) \end{aligned} \tag{A.20}$$

Denoting $w(t) \triangleq k_p[\mathbb{I}_{2n} - D(R_e^i)^{-1}](\mathbb{L} \otimes \mathbb{I}_2)p^*$, the solution of (6.13) can be written as $e_{p^c}(t) = e^{-k_p(\mathbb{L}\otimes\mathbb{I}_2)t}e_{p^c}(t_0) + \int_{t_0}^{t} e^{-k_p(\mathbb{L}\otimes\mathbb{I}_2)(t-\tau)}w(\tau)d\tau$. Thus, using the property of the metric $\text{dist}(\cdot, \mathcal{E}_{2n})$, it can be shown that

$$\begin{aligned} \text{dist}(e_{p^c}(t), \mathcal{E}_{2n}) &\leq \text{dist}(e^{-k_p(\mathbb{L}\otimes\mathbb{I}_2)t}e_{p^c}(t_0), \mathcal{E}_{2n}) + \text{dist}(\int_{t_0}^{t} e^{-k_p(\mathbb{L}\otimes\mathbb{I}_2)(t-\tau)}w(\tau)d\tau, \mathcal{E}_{2n}) \\ &\leq k_{\mathcal{E}} e^{-\lambda_{\mathcal{E}} t}\text{dist}(e_{p^c}(t_0), \mathcal{E}_{2n}) + \int_{t_0}^{t} \text{dist}(e^{-k_p(\mathbb{L}\otimes\mathbb{I}_2)(t-\tau)}w(\tau), \mathcal{E}_{2n})d\tau \end{aligned} \tag{A.21}$$

Denoting $\eta(t) \triangleq \int_{t_0}^{t} \text{dist}(e^{-k_p(\mathbb{L}\otimes\mathbb{I}_2)(t-\tau)}w(\tau), \mathcal{E}_{2n})d\tau$, the following inequality can be obtained

$$\begin{aligned} \eta(t) &\leq k_{\mathcal{E}} e^{-\lambda_{\mathcal{E}} t} \int_{t_0}^{t} e^{\lambda_{\mathcal{E}}\tau}\text{dist}(w(\tau), \mathcal{E}_{2n})d\tau \\ &\leq k_{\mathcal{E}} e^{-\lambda_{\mathcal{E}} t} \left[\frac{e^{\lambda_{\mathcal{E}}\tau}}{\lambda_{\mathcal{E}}}\right]_{t_0}^{t} \sup_{t_0 \leq \tau \leq t} \{\text{dist}(w(\tau), \mathcal{E}_{2n})\} \\ &\leq \frac{k_{\mathcal{E}}}{\lambda_{\mathcal{E}}} \sup_{t_0 \leq \tau \leq t} \{\text{dist}(w(\tau), \mathcal{E}_{2n})\} \end{aligned} \tag{A.22}$$

Further, due to the fact that $\text{dist}(x, \mathcal{E}_{2n}) \leq \|x\|$, it follows that $\sup_{t_0 \leq \tau \leq t} \{\text{dist}(w(\tau), \mathcal{E}_{2n})\} \leq \sup_{t_0 \leq \tau \leq t}\{\|w(\tau)\|\}$. Also since $\|w(\tau)\|$ exponentially converges to zero, there exist constants k_w and γ_w such that

$$\sup_{t_0 \le \tau \le t} \{\text{dist}(w(\tau), \mathcal{E}_{2n})\} \le \sup_{t_0 \le \tau \le t} (k_w e^{-\lambda_w \tau} \|e_\theta(t_0)\|)$$
$$\le k_w \|e_\theta(t_0)\| \tag{A.23}$$

which implies $\eta(t) \le \frac{k_{\mathcal{E}} k_w}{\lambda_{\mathcal{E}}} \|e_\theta(t_0)\|$. Thus, (A.21) can be changed as

$$\text{dist}(e_{p^c}(t), \mathcal{E}_{2n}) \le k_{\mathcal{E}} e^{-\lambda_{\mathcal{E}}(t-t_0)} \text{dist}(e_{p^c}(t_0), \mathcal{E}_{2n}) + \frac{k_{\mathcal{E}} k_w}{\lambda_{\mathcal{E}}} \|e_\theta(t_0)\| \tag{A.24}$$

Now, by replacing t_0 in the above equation by $(t + t_0)/2$, it is shown that

$$\begin{aligned} \text{dist}(e_{p^c}(t), \mathcal{E}_{2n}) &\le k_{\mathcal{E}} e^{-\lambda_{\mathcal{E}}(t-t_0)/2} \text{dist}(e_{p^c}((t+t_0)/2), \mathcal{E}_{2n}) + \frac{k_{\mathcal{E}} k_w}{\lambda_{\mathcal{E}}} \|e_\theta((t+t_0)/2)\| \\ &\le k_{\mathcal{E}}^2 e^{-\lambda_{\mathcal{E}}(t-t_0)} \text{dist}(e_{p^c}(t_o), \mathcal{E}_{2n}) + k_{\mathcal{E}}^2 e^{-\lambda_{\mathcal{E}}(t-t_0)/2} \frac{k_w}{\lambda_{\mathcal{E}}} \|e_\theta(t_0)\| \\ &\quad + \frac{k_{\mathcal{E}} k_w}{\lambda_{\mathcal{E}}} k_{e_\theta} e^{-\frac{\lambda_{e_\theta}(t-t_0)}{2}} \|e_\theta(t_0)\| \end{aligned} \tag{A.25}$$

Thus, $e_{p^c}(t)$ exponentially converges to the set $\mathcal{E}_{2n}$. Also since $\dot{e}_{p^c} \to 0$, $e_{p^c}(t)$ converges to a point $e_{p^c}^\infty \in \mathcal{E}_{2n}$.

Proof (Proof of Lemma 7.3) For the simplicity of the presentation, let us define w_s such as

$$\begin{aligned} w_s \triangleq & -k_p[\mathbb{I}_{2n} - D(R_{e_\theta})^{-1}](e_{p^a} + e_{\hat{p}^a}) \\ & - k_{\hat{p}}[\mathbb{I}_{2n} - D(R_{e_\theta})](\mathbb{L} \otimes \mathbb{I}_2) e_{p^a} + k_{\hat{p}}[\mathbb{I}_{2n} - D(R_{e_\theta})](\mathbb{L} \otimes \mathbb{I}_2) p^* \end{aligned}$$

Then, from (7.7) and (7.8), it can be shown that

$$\dot{e}_{\hat{p}^a} = -k_{\hat{p}}(\mathbb{L} \otimes \mathbb{I}_2) e_{\hat{p}^a} + w_s \tag{A.26}$$

From Lemma 7.2, it is clear that e_{p^a} and $e_{\hat{p}^a}$ are bounded. Thus, there exists a constant $M_{w_s} > 0$ such that

$$\|w_s(t)\| \le k_\gamma e^{-\lambda_\gamma(t-t_0)} M_{w_s} \|e_\theta(t_0)\| \tag{A.27}$$

which shows that $\|w_s(t)\|$ exponentially converges to zero. Defining the state transition matrix of (A.26) as $\phi_{-\mathbb{L}}(t, t_0)$, from $e_{\hat{p}^a}(t) = \phi_{-\mathbb{L}}(t, t_0) e_{\hat{p}^a}(t_0) + \int_{t_0}^t \phi_{-\mathbb{L}}(t, \tau) w_s(\tau) d\tau$ and from the property of dist$(\cdot)$, it can be shown that

$$\begin{aligned} \text{dist}(e_{\hat{p}^a}(t), \mathcal{E}_{2n}) \le\ & \text{dist}(\phi_{-\mathbb{L}}(t, t_0) e_{\hat{p}^a}(t_0), \mathcal{E}_{2n}) \\ & + \text{dist}\Big(\int_{t_0}^{t} \phi_{-\mathbb{L}}(t, \tau) w_s(\tau) d\tau, \mathcal{E}_{2n}\Big) \end{aligned}$$

$$\leq k_{\mathcal{E}} e^{-\lambda_{\mathcal{E}}(t-t_0)} \text{dist}(e_{\hat{p}^a}(t_0), \mathcal{E}_{2n}) + k_{\mathcal{E}} \int_{t_0}^{t} e^{-\lambda_{\mathcal{E}}(t-\tau)} \text{dist}(w_s(\tau), \mathcal{E}_{2n}) d\tau$$

$$\leq k_{\mathcal{E}} e^{-\lambda_{\mathcal{E}}(t-t_0)} \text{dist}(e_{\hat{p}^a}(t_0), \mathcal{E}_{2n}) + \frac{k_{\mathcal{E}}}{\lambda_{\mathcal{E}}} \sup_{t_0 \leq \tau \leq t} \text{dist}(w_s(\tau), \mathcal{E}_{2n}) \tag{A.28}$$

Using (A.27), it can be further shown that

$$\sup_{t_0 \leq \tau \leq t} \text{dist}(w_s(\tau), \mathcal{E}_{2n}) \leq \sup_{t_0 \leq \tau \leq t} (\|w_s(\tau)\|) \leq k_\gamma M_{w_s} \|e_\theta(t_0)\| \tag{A.29}$$

Thus, the following inequality is obtained

$$\text{dist}(e_{\hat{p}^a}(t), \mathcal{E}_{2n}) \leq k_{\mathcal{E}} e^{-\lambda_{\mathcal{E}}(t-t_0)} \text{dist}(e_{\hat{p}^a}(t_0), \mathcal{E}_{2n}) + \frac{k_{\mathcal{E}} k_\gamma M_{w_s}}{\lambda_{\mathcal{E}}} \|e_\theta(t_0)\| \tag{A.30}$$

Replacing t_0 by $t/2$ in the above inequality yields

$$\text{dist}(e_{\hat{p}^a}(t), \mathcal{E}_{2n}) \leq k_{\mathcal{E}} e^{-\lambda_{\mathcal{E}}(t/2)} \text{dist}(e_{\hat{p}^a}(t/2), \mathcal{E}_{2n}) + \frac{k_{\mathcal{E}} k_\gamma M_{w_s}}{\lambda_{\mathcal{E}}} \|e_\theta(t/2)\| \tag{A.31}$$

Since there exist $k_{e_\theta}, \lambda_{e_\theta} > 0$ such that $\|e_\theta(t/2)\| \leq k_{e_\theta} e^{-\lambda_{e_\theta}(t/2-t_0)} \|e_\theta(t_0)\|$, and dist($e_{\hat{p}^a}(t/2), \mathcal{E}_{2n})$ is upper bounded as $\text{dist}(e_{\hat{p}^a}(t/2), \mathcal{E}_{2n}) \leq k_{\mathcal{E}} e^{-\lambda_{\mathcal{E}}(t/2-t_0)} \text{dist}(e_{\hat{p}^a}(t_0), \mathcal{E}_{2n}) + \frac{k_{\mathcal{E}} k_\gamma M_{w_s}}{\lambda_{\mathcal{E}}} \|e_\theta(t_0)\|$, the inequality of (A.31) can be further changed as

$$\text{dist}(e_{\hat{p}^a}(t), \mathcal{E}_{2n}) \leq k_{\mathcal{E}} e^{-\lambda_{\mathcal{E}}(t/2)} \left[k_{\mathcal{E}} e^{-\lambda_{\mathcal{E}}(t/2-t_0)} \text{dist}(e_{\hat{p}^a}(t_0), \mathcal{E}_{2n}) + \frac{k_{\mathcal{E}} k_\gamma M_{w_s}}{\lambda_{\mathcal{E}}} \|e_\theta(t_0)\| \right] + \frac{k_{\mathcal{E}} k_\gamma M_{w_s}}{\lambda_{\mathcal{E}}} k_{e_\theta} e^{-\lambda_{e_\theta}(t/2-t_0)} \|e_\theta(t_0)\| \tag{A.32}$$

Thus, it is certain that as $t \to \infty$, $e_{\hat{p}^a}(t)$ exponentially converges to $\mathcal{E}_{2n}$. In what follows, it will be shown that $e_{\hat{p}^a}(t)$ converges to a point. Since $e_{\hat{p}^a}$ converges to a subset in Euclidean space, as per Theorem 2.20, $e_{\hat{p}^a}(t)$ evolves in a compact set. Thus, for any subsequence of $\{e_{\hat{p}^a}(t_k)\}$, it is convergent as per Definition 2.27. Let the limit of the sequence $\{e_{\hat{p}^a}(t_k)\}$ be denoted by $\mathbf{1}_n \otimes e_{\hat{p}^a}^\infty$. Then, from the convergence in a compact set, it is true that $e_{\hat{p}^a}(t_k) \to \mathbf{1}_n \otimes e_{\hat{p}^a}^\infty$, which is a point in $\mathcal{E}_{2n}$. Moreover, given the state transition matrix $\phi_{-\mathbb{L}}(t, t_0)$, for any $\epsilon' > 0$, there exists a $\delta' > 0$ such that $\mathbb{I}_{2n} - \epsilon' \mathbf{1}_{2n \times 2n} \leq \phi_{-\mathbb{L}}(t_2, t_1) \leq \mathbb{I}_{2n} + \epsilon' \mathbf{1}_{2n \times 2n}$ for any $t_1 + \delta' \geq t_2 \geq t_1 \geq t_0$, where $\mathbb{I}_{2n}$ is the $2n \times 2n$ identity matrix and $\mathbf{1}_{2n \times 2n}$ is the $2n \times 2n$ matrix with all elements being one.

Also, since $w_s(t) \to 0$ as $t \to \infty$, for any $\epsilon'' > 0$, there exists an element $e_{\hat{p}^a}(t_k)$ of the sequence $\{e_{\hat{p}^a}(t_k)\}$ and $\tau > t_k$ such that

$$\|e_{\hat{p}^a}(t_k) - \mathbf{1}_n \otimes e_{\hat{p}^a}^{\infty}\| < \epsilon'' \tag{A.33}$$

and

$$(\tau - t_k)\|w_s(\tau)\| < \epsilon'' \tag{A.34}$$

Furthermore, since the sequence $\{e_{\hat{p}^a}(t_k)\}$ evolves in a dense set, for a given t, t_k can be chosen such that $t - t_k \leq \delta'$. Then, the state transition matrix $\phi_{-\mathbb{L}}(t, t_k)$ can be expressed as

$$\phi_{-\mathbb{L}}(t, t_k) = \mathbb{I}_{2n} + M_{2n\times 2n}(\epsilon') \tag{A.35}$$

where $M_{2n\times 2n}(\epsilon')$ is the $2n \times 2n$ matrix whose element magnitudes are less than or equal to ϵ'. With the above arguments, it can be shown that

$$\begin{aligned}
\|e_{\hat{p}^a}(t) - \mathbf{1}_n \otimes e_{\hat{p}^a}^{\infty}\| &= \|\phi_{-\mathbb{L}}(t, t_0)e_{\hat{p}^a}(t_0) + \int_{t_0}^{t} \phi_{-\mathbb{L}}(t, \tau)w_s(\tau)d\tau - \mathbf{1}_n \otimes e_{\hat{p}^a}^{\infty}\| \\
&= \|\phi_{-\mathbb{L}}(t, t_k)\phi_{-\mathbb{L}}(t_k, t_0)e_{\hat{p}^a}(t_0) + \int_{t_0}^{t_k} \phi_{-\mathbb{L}}(t_k, \tau)w_s(\tau)d\tau \\
&\quad + \int_{t_k}^{t} \phi_{-\mathbb{L}}(\tau, t_k)w_s(\tau)d\tau - \mathbf{1}_n \otimes e_{\hat{p}^a}^{\infty}\| \\
&= \|[\mathbb{I}_{2n} + M_{2n\times 2n}(\epsilon')]\phi_{-\mathbb{L}}(t_k, t_0)e_{\hat{p}^a}(t_0) + \int_{t_0}^{t_k} \phi_{-\mathbb{L}}(t_k, \tau)w_s(\tau)d\tau \\
&\quad + \int_{t_k}^{t} \phi_{-\mathbb{L}}(\tau, t_k)w_s(\tau)d\tau - \mathbf{1}_n \otimes e_{\hat{p}^a}^{\infty}\|
\end{aligned} \tag{A.36}$$

Because the sequence $\{e_{\hat{p}^a}(t_k)\}$ is convergent as $k \to \infty$, clearly there exists t_k that satisfies $\|\phi_{-\mathbb{L}}(t_k, t_0)e_{\hat{p}^a}(t_0) + \int_{t_0}^{t_k} \phi_{-\mathbb{L}}(t_k, \tau)w_s(\tau)d\tau - \mathbf{1}_n \otimes e_{\hat{p}^a}^{\infty}\| < \epsilon''$ for any ϵ''. Hence the equality (A.36) can be changed into an inequality such as:

$$\begin{aligned}
\|e_{\hat{p}^a}(t) - \mathbf{1}_n \otimes e_{\hat{p}^a}^{\infty}\| \leq& \|M_{2n\times 2n}(\epsilon')\phi_{-\mathbb{L}}(t_k, t_0)e_{\hat{p}^a}(t_0)\| \\
&+ \left\| \int_{t_k}^{t} \phi_{-\mathbb{L}}(\tau, t_k)w_s(\tau)d\tau \right\| + \epsilon''
\end{aligned} \tag{A.37}$$

Since all the signals are bounded, let the upper bound of the norm of the state transition matrix be denoted by $\bar{\phi}$ in the sense $\|\phi_{-\mathbb{L}}(t_2, t_1)\| \leq \bar{\phi}$, for all $t_2 \geq t_1 \geq t_0$. Then the right-hand side of (A.37) can be changed as

$$\begin{aligned}\|e_{\hat{p}^a}(t) - \mathbf{1}_n \otimes e_{\hat{p}^a}^{\infty}\| &\leq \|M_{2n\times 2n}(\epsilon')\|\bar{\phi}\|e_{\hat{p}^a}(t_0)\| \\ &\quad + \bar{\phi}(t - t_k) \sup_{t_k \leq \tau \leq t} \|w_s(\tau)\| + \epsilon'' \\ &\leq \|M_{2n\times 2n}(\epsilon')\|\bar{\phi}\|e_{\hat{p}^a}(t_0)\| + \bar{\phi}\epsilon'' + \epsilon'' \end{aligned} \tag{A.38}$$

Now, it is obvious that, for a given ϵ, there exist ϵ' and ϵ'' such that $\|M_{2n\times 2n}(\epsilon')\|\bar{\phi}$ $\|e_{\hat{p}^a}(t_0)\| + \bar{\phi}\epsilon'' + \epsilon'' \leq \epsilon$ does hold. Hence $e_{\hat{p}^a}(t) \to \mathbf{1}_n \otimes e_{\hat{p}^a}^{\infty}$ as $t \to \infty$.

Proof (Proof of Lemma 7.9) From (7.49), the solution $e(t)$ is given as

$$\begin{aligned} e(t) = e^{A(t-t_0)}e(t_0) + \int_{t_0}^{t} e^{A(t-\tau)}\Delta A(\Gamma(\tau))e(\tau)d\tau \\ + \int_{t_0}^{t} e^{A(t-\tau)}D(\Gamma(\tau))d\tau \end{aligned} \tag{A.39}$$

Based on Lemma 7.8, it is obvious that there exist $k_D, \lambda_D > 0$ such that $\|D(\Gamma(\tau))\| \leq k_D e^{-\lambda_D \tau}\|D(\Gamma(t_0))\|$. Thus, (A.39) can be changed as

$$\begin{aligned} \|e(t)\| &\leq \|e^{A(t-t_0)}\|\|e(t_0)\| + \int_{t_0}^{t} \|e^{A(t-\tau)}\|\|\Delta A(\Gamma(\tau))\|\|e(\tau)\|d\tau \\ &\quad + \int_{t_0}^{t} \|e^{A(t-\tau)}\|\|D(\Gamma(\tau))\|d\tau \\ &\leq M_A\|e(t_0)\| + \frac{k_D}{\lambda_D}M_A\|D(\Gamma(\tau))\| + M_A\int_{t_0}^{t} \|\Delta A(\Gamma(\tau))\|\|e(\tau)\|d\tau \end{aligned} \tag{A.40}$$

Then, using the Gronwall-Bellman lemma, it can be shown that

$$\|e(t)\| \leq M_A\|e(t_0)\|e^{\int_{t_0}^{t} M_A\|\Delta A(\Gamma(\tau))\|} + \frac{k_D}{\lambda_D}M_A\|D(\Gamma(t_0))\|e^{\int_{t_0}^{t} M_A\|\Delta A(\Gamma(\tau))\|d\tau} \tag{A.41}$$

From Lemma 7.8, it is clear that there exist $k_{\Delta A}, \lambda_{\Delta A} > 0$ such that $\|\Delta A(\Gamma(t))\| \leq k_{\Delta A} e^{-\lambda_{\Delta A}(t-t_0)} \|\Gamma(t_0)\|$. Thus, $\int_{t_0}^{t} \|\Delta A(\Gamma(\tau))\| d\tau$ is bounded for all $t \leq t_0$, which completes the proof.

Proof (Proof of Theorem 7.3) Let $w \triangleq -k_p[\mathbb{I}_{3n} - \Gamma](e_{p^c} + e_{\hat{p}^c}) - k_{\hat{p}}[\mathbb{I}_{3n} - \Gamma^{-1}](\mathbb{L} \otimes \mathbb{I}_3)e_{p^c} + k_{\hat{p}}[\mathbb{I}_{3n} - \Gamma^{-1}](\mathbb{L} \otimes \mathbb{I}_3)p^*$. Then, the dynamics of $e_{\hat{p}^c}$ can be written in a space state form:

$$\dot{e}_{\hat{p}^c} = -k_{\hat{p}}(\mathbb{L} \otimes \mathbb{I}_3)e_{\hat{p}^c} + w \tag{A.42}$$

From Lemma 7.8 and Lemma 7.9, there exist $k_{w_s}, \lambda_\gamma > 0$ such that

$$\|w(t)\| \leq k_{w_s} e^{-\lambda_\gamma t} \|w(t_0)\| \tag{A.43}$$

which shows $\|w(t)\|$ exponentially converges to 0. From the definition of dist$(\cdot)$, with the notation $\hat{\mathbb{L}} \triangleq k_{\hat{p}}(\mathbb{L} \otimes \mathbb{I}_3)$, there exist $k_{\mathcal{E}}, \lambda_{\mathcal{E}} > 0$ such that

$$\text{dist}(e^{-\hat{\mathbb{L}}t}x, \mathcal{E}_{3n}) \leq k_{\mathcal{E}} e^{-\lambda_{\mathcal{E}} t} \text{dist}(x, \mathcal{E}_{3n}) \tag{A.44}$$

The solution of (A.42) is given as

$$e_{\hat{p}^c}(t) = e^{-\hat{\mathbb{L}}(t-t_0)} e_{\hat{p}^c}(t_0) + \int_{t_0}^{t} e^{-\hat{\mathbb{L}}(t-\tau)} w(\tau) d\tau \tag{A.45}$$

Defining $\eta(t) \triangleq \int_{t_0}^{t} \text{dist}(e^{-\hat{\mathbb{L}}(t-\tau)} w(\tau) d\tau, \mathcal{E}_{3n}) d\tau$, the following inequality can be obtained:

$$\text{dist}(e_{\hat{p}^c}(t), \mathcal{E}_{3n}) \leq \text{dist}(e^{-\hat{\mathbb{L}}(t-t_0)} e_{\hat{p}^c}(t_0), \mathcal{E}_{3n}) + \eta(t) \tag{A.46}$$

Further using (A.44), $\eta(t)$ is upper bounded as

$$\begin{aligned} \eta(t) &\leq k_{\mathcal{E}} \int_{t_0}^{t} e^{-\lambda_{\mathcal{E}}(t-\tau)} \text{dist}(w(\tau), \mathcal{E}_{3n}) d\tau \\ &\leq \frac{k_{\mathcal{E}}}{\lambda_{\mathcal{E}}} \sup_{t_0 \leq \tau \leq t} (\text{dist}(w(\tau), \mathcal{E}_{3n})) \end{aligned} \tag{A.47}$$

Using the fact that $\text{dist}(x, \mathcal{E}_{3n}) \leq \|x\|$, it follows that $\sup_{t_0 \leq \tau \leq t}(\text{dist}(w(\tau), \mathcal{E}_{3n})) \leq \sup_{t_0 \leq \tau \leq t}(\|w(\tau)\|) \leq k_w \|w(t_0)\|$, which leads to

$$\text{dist}(e_{\hat{p}^c}(t), \mathcal{E}_{3n}) \leq k_{\mathcal{E}} e^{-\lambda_{\mathcal{E}}(t-t_0)} \text{dist}(e_{\hat{p}^c}(t_0), \mathcal{E}_{3n}) + \frac{k_{\mathcal{E}} k_w}{\lambda_{\mathcal{E}}} \|w(t_0)\| \tag{A.48}$$

Based on (A.48), by replacing t_0 by $(t + t_0)/2$, it can be obtained as

$$\begin{aligned} \text{dist}(e_{\hat{p}^c}(t), \mathcal{E}_{3n}) &\leq k_{\mathcal{E}} e^{-\lambda_{\mathcal{E}}(t-t_0)/2} \text{dist}(e_{\hat{p}^c}((t + t_0)/2), \mathcal{E}_{3n}) + \frac{k_{\mathcal{E}} k_w}{\lambda_{\mathcal{E}}} \|w((t + t_0)/2)\| \\ &\leq k_{\mathcal{E}} e^{-\lambda_{\mathcal{E}}(t-t_0)/2} \text{dist}(e_{\hat{p}^c}((t + t_0)/2), \mathcal{E}_{3n}) \\ &\quad + \frac{k_{\mathcal{E}} k_w^2}{\lambda_{\mathcal{E}}} e^{-\lambda_\gamma (t+t_0)/2} \|w(t_0)\| \end{aligned} \tag{A.49}$$

Also from

$$\text{dist}(e_{\hat{p}^c}((t + t_0)/2), \mathcal{E}_{3n}) \leq k_{\mathcal{E}} e^{-\lambda_{\mathcal{E}}((t-t_0)/2)} \text{dist}(e_{\hat{p}^c}(t_0), \mathcal{E}_{3n}) + \frac{k_{\mathcal{E}} k_w}{\lambda_{\mathcal{E}}} \|w(t_0)\| \tag{A.50}$$

it follows that

$$\begin{aligned} \text{dist}(e_{\hat{p}^c}(t), \mathcal{E}_{3n}) &\leq k_{\mathcal{E}}^2 e^{-\lambda_{\mathcal{E}}(t-t_0)} \text{dist}(e_{\hat{p}^c}(t_0), \mathcal{E}_{3n}) + \frac{k_{\mathcal{E}}^2 k_w}{\lambda_{\mathcal{E}}} e^{-\lambda_{\mathcal{E}}((t-t_0)/2)} \|w(t_0)\| \\ &\quad + \frac{k_{\mathcal{E}} k_w^2}{\lambda_{\mathcal{E}}} e^{-\lambda_\gamma (t+t_0)/2} \|w(t_0)\| \end{aligned} \tag{A.51}$$

Hence $e_{\hat{p}^c}(t)$ exponentially converges to a point of $\mathcal{E}_{3n}$ (denote it by $e_{\hat{p}^c_\infty}$) by the same argument of the proof of Theorem 7.1.

Let us show that $e_{p^c}(t) \to \mathbf{1}_n \otimes e_{\hat{p}^c_\infty}$ as $t \to \infty$. Let us denote $\xi_s \triangleq e_{p^c} - \mathbf{1}_n \otimes e_{\hat{p}^c_\infty}$ and $v \triangleq k_p[\mathbb{I}_{3n} - \Gamma^{-1}](e_{p^c} + \hat{e}_{p^c})$. Then, using the same procedure as the proof of Theorem 7.1, the following can be obtained

$$\xi_s(t) = e^{-k_p t} \xi_s(t_0) + k_p \int_{t_0}^{t} e^{-k_p(t-\tau)} (\hat{e}_{p^c}(\tau) - \mathbf{1}_n \otimes e_{\hat{p}^c_\infty}) d\tau + \int_{t_0}^{t} e^{-k_p(t-\tau)} v(\tau) d\tau \tag{A.52}$$

Thus, from the fact that $\hat{e}_{p^c}(t) \to \mathbf{1}_n \otimes e_{\hat{p}^c_\infty}$ and $v \to 0$, it is true that $e_{p^c}(t) \to \mathbf{1}_n \otimes e_{\hat{p}^c_\infty}$ as $t \to \infty$, which completes the proof.

Proof (Proof of Theorem 8.4) Define an equilibrium set as

$$\mathcal{E}_e \triangleq \{(x_1, \ldots, x_n) \in \mathbb{R}^{3n} : x_i = x_j, \forall i, j \in \mathcal{V}\} \tag{A.53}$$

Then, for a fixed $e^* \in \mathcal{E}_e$, there exist $k_l, \lambda_l > 0$ such that

$$\begin{aligned}\|e(t) - e^*\| &\leq \|e^{-(\mathbb{L}\otimes\mathbb{I}_3)(t-t_0)}e(t_0) - e^*\| + \|\int_{t_0}^{t} e^{-(\mathbb{L}\otimes\mathbb{I}_3)(t-\tau)}w(\tau)d\tau\| \\ &\leq k_l e^{-\lambda_l(t-t_0)}\|e(t_0) - e^*\| + \int_{t_0}^{t} \|e^{-(\mathbb{L}\otimes\mathbb{I}_3)(t-\tau)}\|\|w(\tau)\|d\tau\end{aligned} \tag{A.54}$$

Using the fact that $\|e^{-(\mathbb{L}\otimes\mathbb{I}_3)(t-\tau)}\| \leq \eta$, where η is a constant, and the inequality (8.38), it follows:

$$\begin{aligned}\int_{t_0}^{t} \|e^{-(\mathbb{L}\otimes\mathbb{I}_3)(t-\tau)}\|\|w(\tau)\|d\tau &\leq \int_{t_0}^{t} \eta k_w e^{-\lambda_w(\tau-t_0)}\|w(t_0)\|d\tau \\ &= \frac{-\eta k_w}{\lambda_w}\|w(t_0)\|(e^{-\lambda_w(t-t_0)} - 1) \\ &= \frac{\eta k_w}{\lambda_w}\|w(t_0)\|(1 - e^{-\lambda_w(t-t_0)})\end{aligned} \tag{A.55}$$

Thus, (A.54) can be changed as

$$\|e(t) - e^*\| \leq k_l e^{-\lambda_l(t-t_0)}\|e(t_0) - e^*\| + \frac{\eta k_w}{\lambda_w}\|w(t_0)\|(1 - e^{-\lambda_w(t-t_0)}) \tag{A.56}$$

Finally by replacing t_0 with $t_0 = (t + t_0)/2$, it is true that

$$\begin{aligned}\|e(t) - e^*\| &\leq k_l e^{-\lambda_l((t-t_0)/2)}\|e((t + t_0)/2) - e^*\| \\ &\quad + \frac{\eta k_w}{\lambda_w}\|w((t + t_0)/2)\|(1 - e^{-\lambda_w(t-t_0)/2}) \\ &\leq k_l^2 e^{-\lambda_l((t-t_0)/2)}\|e(t_0) - e^*\| \\ &\quad + \frac{\eta k_w^2}{\lambda_w}e^{-\lambda_w((t-t_0)/2)}\|w(t_0)\|(1 - e^{-\lambda_w(t-t_0)/2})\end{aligned} \tag{A.57}$$

Therefore from the above inequality, it is shown that $\|e(t) - e^*\| \to 0$ as $t \to \infty$, which completes the proof.

Proof (Proof of Lemma 9.4) First, from $z_{12} = d_{12}g_{12}$ and $g_{12}^T\mathbb{P}_{g_{12}} = (\mathbb{P}_{g_{12}}g_{12})^T = 0$, we can have

$$\begin{aligned}\frac{d}{dt}(d_{12}^2) &= \frac{d}{dt}(z_{12}^T z_{12})\\ &= 2z_{12}^T(\dot{p}_1 - \dot{p}_2)\\ &= 2z_{12}^T \mathbb{P}_{g_{12}} g_{12}^*\\ &= 0\end{aligned}$$

Next, for (9.8), the equilibrium points are obtained only when $g_{12} = g_{12}^*$ or $g_{12} = -g_{12}^*$. Let us denote the position of agent 2 when $g_{12} = g_{12}^*$ as p_{2a}^* and denote the position of agent 2 when $g_{12} = -g_{12}^*$ as p_{2b}^*. Consider the Lyapunov function $V = \frac{1}{2}\|p_2 - p_{2b}^*\|^2$, which is continuously differentiable for all $t \geq t_0$, as per Definition 2.21. The derivative of $V(t)$ along the trajectory of (9.8) is given as

$$\begin{aligned}\dot{V} &= (p_2 - p_{2b}^*)^T \dot{p}_2\\ &= -(p_2 - p_{2b}^*)^T \mathbb{P}_{g_{12}} g_{12}^*\\ &= (p_2 - p_{2b}^*)^T \mathbb{P}_{g_{12}} (p_1 - p_2 + p_2 - p_{2b}^*)/d_{21}\\ &= (p_2 - p_{2b}^*)^T \mathbb{P}_{g_{12}}/d_{12}(p_2 - p_{2b}^*) \geq 0\end{aligned} \tag{A.58}$$

since $\mathbb{P}_{g_{12}} z_{12} = 0$ and $\mathbb{P}_{g_{12}}$ is positive semidefinite. Therefore, we can see that $p_2 = p_{2b}^*$ is unstable by Chetaev theorem.[1] Similarly, let us consider the Lyapunov function $V(t) = \frac{1}{2}\|p_2 - p_{2a}^*\|^2$, which is continuously differentiable and radially unbounded. Taking the derivative yields

$$\begin{aligned}\dot{V} &= (p_2 - p_{2a}^*)^T \dot{p}_2\\ &= -(p_2 - p_{2a}^*)^T \mathbb{P}_{g_{12}} g_{12}^*\\ &= -(p_2 - p_{2a}^*)^T \mathbb{P}_{g_{12}}/d_{12}(p_2 - p_{2a}^*) \leq 0\end{aligned} \tag{A.59}$$

Since $\dot{V} = 0$ if and only if $p_2 = p_{2a}^*$ or $p_2 = p_{2b}^*$, and since p_{2b}^* is unstable, it is clear that p_{2a}^* is almost globally asymptotically stable due to the LaSalle's invariance principle (see Theorem 2.22). Moreover suppose that $p_2(t_0) \neq p_{2b}^*$. Then, based on the illustration in Fig. A.3, it is true that

$$\|(p_2 - p_{2a}^*)^T \mathbb{P}_{g_{12}}\| = \|p_2 - p_2^\dagger\| = \|p_2 - p_{2a}^*\| \sin\alpha \tag{A.60}$$

Since $\|p_2 - p_1\| = \|p_1 - p_{2a}^*\| = d_{21}$, we have $\angle(p_2, p_{2a}^*, p_1) = \angle(p_{2a}^*, p_2, p_1) = \alpha$. Then, (A.60) can be changed as

$$(p_2 - p_{2a}^*)^T \mathbb{P}_{g_{12}} (p_2 - p_{2a}^*) = \|p_2 - p_{2a}^*\|^2 (\sin\alpha)^2 = 2\sin^2\alpha V(t) \tag{A.61}$$

---

[1] Consider an autonomous system. If, for an equilibrium point x_0, there exists a continuously differentiable function $V(x)$ such that $V(x_0) = 0$, $V(x) > 0$ except x_0, and $\dot{V}(x) > 0$ holds in a neighborhood of x_0, then it is unstable at $x = x_0$.

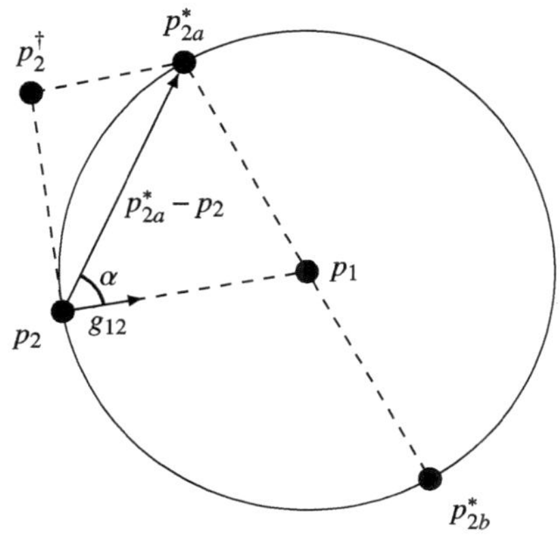

Fig. A.3 The geographical relationship of (A.60).

Moreover, from (A.58), the distance $\|p_2 - p_{2a}^*\|$ is decreasing continuously, which implies that $\alpha(t) \geq \alpha(t_0)$. Then, we have

$$\dot{V} = -2\sin^2\alpha/d_{21}V(t) \tag{A.62}$$

Thus, the convergence of $p_2(t)$ to p_{2a}^* is exponentially fast.

Index

H.-S. Ahn, *Formation Control*, Studies in Systems, Decision and Control 205, https://doi.org/10.1007/978-3-030-15187-4

MIX
Papier aus verantwortungsvollen Quellen
Paper from responsible sources
FSC® C105338

If you have any concerns about our products,
you can contact us on
ProductSafety@springernature.com

In case Publisher is established outside the EU,
the EU authorized representative is:
Springer Nature Customer Service Center GmbH
Europaplatz 3, 69115 Heidelberg, Germany

Printed by Libri Plureos GmbH
in Hamburg, Germany